CONFÉRENCIERS

G. FALKOVICH
M. FARGE
U. FRISCH
M.A. LESCHZINER
R. MANKBADI
O. MÉTAIS
H.K. MOFFATT
J. SOMMERIA
K.R. SREENIVASAN
A. YAGLOM

UJF NATO ASI

LES HOUCHES

Session LXXIV

2000

NEW TRENDS IN TURBULENCE

TURBULENCE : NOUVEAUX ASPECTS

ÉCOLE DE PHYSIQUE DES HOUCHES – UJF & INPG – GRENOBLE

a NATO Advanced Study Institute

LES HOUCHES

SESSION LXXIV

31 July – 1 September 2000

New trends in turbulence

Turbulence : nouveaux aspects

Edited by

M. LESIEUR, A. YAGLOM and F. DAVID

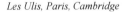

SCIENCES

Springer

Les Ulis, Paris, Cambridge

Berlin, Heidelberg, New York,
Barcelona, Hong Kong, London
Milan, Paris, Tokyo

Published in cooperation with the NATO Scientific Affair Division

ENGI

0 11680581

ISBN 3-540-42978-6   Springer-Verlag Berlin Heidelberg New York
ISBN 2-86883-547-3   EDP Sciences Les Ulis

LES HOUCHES - ÉCOLE DE PHYSIQUE
ÉCOLE D'ÉTÉ DE PHYSIQUE THÉORIQUE

SERVICE INTER-UNIVERSITAIRE COMMUN
À L'UNIVERSITÉ JOSEPH FOURIER DE GRENOBLE
ET À L'INSTITUT NATIONAL POLYTECHNIQUE
DE GRENOBLE, SUBVENTIONNÉ PAR LE MINISTÈRE
DE L'ÉDUCATION NATIONALE, DE LA RECHERCHE
ET DE LA TECHNOLOGIE, LE CENTRE NATIONAL
DE LA RECHERCHE SCIENTIFIQUE ET LE COMMISSARIAT
À L'ÉNERGIE ATOMIQUE

ÉCOLE D'ÉTÉ DE PHYSIQUE THÉORIQUE
SESSION LXXIV

INSTITUT D'ÉTUDES AVANCÉES DE L'OTAN
NATO ADVANCED STUDY INSTITUTE

EURO SUMMER SCHOOL

31 juillet — 1 septembre 2000

Directeurs Scientifiques de la session : M. LESIEUR, *LEGI/INPG, Grenoble, France ;* A. YAGLOM, *Institute of Atmospheric Physics, Russian Academy of Sciences, Moscow, Russia and Department of Aeronautics & Astronautics, MIT, Cambridge, U.S.A. ;* François DAVID, *SPhT, CEA Saclay, France*

SESSIONS PRÉCÉDENTES

1951 - 2000

---

[*] Sessions ayant reçu l'appui du Comité Scientifique de l'OTAN.

---

[*] Sessions ayant reçu l'appui du Comité Scientifique de l'OTAN.

[*] Sessions ayant reçu l'appui du Comité Scientifique de l'OTAN.
Publishers: Session VIII: Dunod, Wiley, Methuen; Sessions IX & X: Herman, Wiley – Session XI: Gordon and Breach, Presses Universitaires – Sessions XII-XXV: Gordon and Breach – Sessions XXVI-LXVIII: North-Holland – Session LXIX-LXXIII: EDP Sciences, Springer.

ORGANIZERS

M. LESIEUR, LEGI/INPG, Institut de Mécanique, UMR 101, BP. 53X, 38041 Grenoble Cedex, France

A. YAGLOM, Institute of Atmospheric Physics, Russian Academy of Sciences, Pyzhevsky 3, 109017 Moscow, Russia, and Department of Aeronautics & Astronautics, MIT 33-219, 77 Massachusetts Avenue, Cambridge, MA 02139-9307, U.S.A.

F. DAVID, SPhT, CEA Saclay, 91191 Gif-sur-Yvette Cedex, France

LECTURERS

P. COULLET, INLN, 1361 route des Lucioles, Sophia Antipolis, 06560 Valbonne, France

G. FALKOVITCH, Department of Nuclear Physics, Weizmann Institute of Science, P.O. Box 26, Rehovot 76100, Israel

M. FARGE, Laboratoire de Météorologie Dynamique, ENS, 24 rue Lhomond, 75231 Paris Cedex 05, France

U. FRISCH, Observatoire de la Côte d'Azur, BP. 42229, 06304 Nice Cedex 4, France

M. LESCHZINER, Department of Engineering, Queen Mary & Westfield College, University of London, Mile End Road, London E1 4NS, U.K.

R. MANKBADI, Technical Leader, CAA, ICOMP, NASA Gleen Research Center, Cleveland, Ohio 44135, U.S.A., and Aerospace Engineering Embry-Riddle Aeronautical University, Daytona Beach, FL 32174, U.S.A.

O. METAIS, LEGI, Domaine Universitaire, BP. 53, 38041 Grenoble Cedex 9, France

K. MOFFATT, Director, Isaac Newton Institute for Mathematical Sciences, University of Cambridge, 20 Clarkson Road, Cambridge CB3 OEH, U.K.

J. SOMMERIA, LEGI/CORIOLIS, 21 avenue des Martyrs, 38000 Grenoble, France

K. SREENIVASAN, Department of Mechanical Engineering, Mason Laboratory, Yale University, P.O. Box 2159, New Haven, CT 06520, U.S.A.

J. WESFREID, ESPCI/HMP, 10 rue Vauquelin, 75231 Paris Cedex 05, France

SEMINAR SPEAKERS

F. MAGAGNATO, Fachgebiet Stfömungsmaschinen, Universität Karlsruhe, Kaiserstrasse 12, Karlsruhe 76128, Germany

J.M. REDONDO, Department de Fisica Aplicada, GDF, U.P.C campus Nord B4-B5, Barcelona 08034, Spain

L. TUCKERMAN, LIMSI-CNRS, BP. 133, 91403 Orsay Cedex, France

PARTICIPANTS & FREE AUDITORS

M. BARTA, Astronomical Institute of the Academy of Sciences of the Czech Republic, 251 65 Ondrejov, Czech Republic

J. BEC, Laboratoire Cassini, Observatoire de la Côte d'Azur, BP. 4229, 06304 Nice cedex 4, France

A. BILSKY, Institute of Thermophysics   SB RAS, Lavrentyev Avenue 1, 630090 Novosibirsk, Russia

E. BOURNASKI, Bulgarian Academy of Sciences, Institute of Water Problems, Acad. G. Bontchev Street Block 1, BG 1113 Sofia, Bulgaria

M. CHAVES, Laboratoire Cassini, Observatoire de la Côte d'Azur, BP. 4229, 06304 Nice Cedex 4, France

J. DAVOUDI, Max-Planck-Institute for Complex Systems, Noethnitzer Str. 38, 01187 Dresden, Germany

C. DE LANGHE, Ghent University, Department of Flow, Heat and Combustion Mechanics, Sint–Pietersnieuwstraat 41, 9000 Gent, Belgium

A. DEMENKOV, Institute of Computational Technologies, Siberian Division, Russian Academy of Sciences, Lavrentjev Avenue 6, 630090 Novosibirsk, Russia

N. DENISSOVA, Moscow State University, Mechanics and Mathematics, Department of Mathematical Statistics and Random Process, Leninskie Gory, 119899 Moscow, Russia

B. DINTRANS, Nordic Institute for Theoretical Physics, Blegdamsvej 17, DK 2100 Copenhagen, Denmark

K. DOMELEVO, Laboratoire MIP, UMR 5674 du CNRS, Université Paul Sabatier, 118 route de Narbonne, 31062 Toulouse Cedex 4, France

A. DOSIO, Environment Institute, Joint Research Centre of the Commission of the European Communities, Via E. Fermi, 21020 Ispra, Italy

A. FOUXON, Physics of Complex Systems, Weizmann Institute of Sciences, Rehovot 76100, Israel

N. GARNIER, SPEC/CEA Saclay, L'Orme des Merisiers, 91191 Gif-sur-Yvette Cedex, France

P. GIULIANI, Dipartimento di Fisica, Università della Calabria, 87036 Rende, Italy, and The Niels Bohr Institute, Blegdamsvej 17, 2100 Copenhagen, Denmark

A. GOLOVIN, Centro Internacional de Fisica, Apartado Aero 4948, Bogota, Colombia

N. GOSPODINOVA, Institute of Electronics, Bulgarian Academy of Sciences, Blvd "Tzarigradsko Chausse" N72, 1784 Sofia, Bulgaria

B. GRITS, Institute of Earth Sciences, Hebrew University of Jerusalem, Department of Atmospheris Sciences, Jerusalem, Israel

P. GUALTIERI, Dipartimento di Meccanica e Aeronautica, Universita di Roma "La Sapienza", via Eudossiania 18, 00184 Roma, Italy

D. HEITZ, Cemagref, 17 avenue de Cucillé, CS 64427, 35044 Rennes, France

P. HORVAI, École Normale Supérieure, 45 rue d'Ulm, 75005 Paris, France

M. JASZCZUR, University of Mining and Metallurgy, Department Theoretical Metallurgy and Metallurgical Engineering, 30 Mickiewicza Avenue, 30-059 Krakow, Poland

P. JOHANSSON, Thermo and Fluid Dynamics, Chalmers University of Technology, SE 412 96 Göteborg, Sweden

M.-C. JULLIEN, LPS/ENS, 24 rue Lhomond, 75231 Paris Cedex 05, France

S. KATO, Graduate School of Mathematical Sciences, University of Tokyo, 3-8-1 Komaba, Meguro-ku, Tokyo 153-8914, Japan

V. KORABEL, Kharkov National University, Svoboda sq. 4, Kharkov 310077, Ukraine

L. KREMINSKAYA, Centro Internacional de Fisica, Apartado Aereo 4948, Bogota, Colombia

S. KUBACKI, Institute of Thermal Machinery, Technical University of Czestochowa, al. Armii Krajowej 21, 42-200 Czestochowa, Poland

D. LAVEDER, Laboratoire Cassini, Observatoire de la Côte d'Azur, BP. 4229, 06304 Nice Cedex 4, France

J. LOPEZ, Ocean & Coastal Research Group/ Etsi de Caminos CY.P, Universidad de Cantabria, av. Los Castros s/n, 39005 Santander, Spain

J. LUKOVICH, Department of Physics, University of Toronto, 60 St George Street, Toronto, Ontario, Canada M5S 1A7

S. MEDVEDEV, Institute of Computational Technologies, SB RAS, Novosibirsk 630090, Russia

D. MITRA, Department of Physics, Indian Institute of Science, Bangalore 560012, India

A. MOHAMMED, Oldenburg University, Department of physics, 26111 Oldenburg, Germany

F. MOISY, Laboratoire de Physique Statistique, ENS, 24 rue Lhomond, 75231 Paris Cedex 05, France

C. NEVES BETTENCOURT DA SILVA, Instituto Superior Tecnico, Pav. Mequines 1 audor, Av. Rovisco Pais, 1096 Lisboa Cedex, Portugal

S. OSSIA, LEGI, 38041 Grenoble Cedex 09, France

G. PASMANIK, Nizhny Novgorod State University, Radiophysical Department, Gagarin Ave. 23, Nizhny Novgorod 603000, Russia

F. PENEAU, Ceram EAI Tech, 157 rue Albert Einstein, BP. 085, 06902 Sophia Antipolis Cedex, France

J. RACHWALSKI, Institute of Fluid Machinery, Polish Academy of Sciences, Ul. Gen. J. Fiszera 14, PL-80 952 Gdansk, Poland

M.R. RAHIMI TABAR, IPM, P.O. Box 19325, 5531 Tehran, Iran

G. ROUSSEAUX, Laboratoire de Physique et Mécanique des Milieux Hétérogènes, ESPCI, 10 rue Vauquelin 75231 Paris Cedex 05, France

A. SCHEKOCHIHIN, Department of Astrophysical Sciences and Plasma Physics Laboratory, Princeton University, P.O. Box 451, Princeton, New Jersey 08543, U.S.A.

R. SEMANCIK, Joint Institute for Nuclear Research, Moscow Region, 141-980 Dubna, Russia

A. SOBOLEVSKI, Quantum Statistics and Field Theory Department, Physics Faculty of the M.V Lomonosov Moscow State University, Vorobjievy Gory, 119899 Moscow, Russia

A. STANISLAVSKY, Institute of Radio Astronomy of National Ukrainian Academy of Sciences, 4 Chervonopraporna St., 61002 Kharkov, Ukraine

M. STEPANOV, Physics of Complex Systems, Weizmann Institute of Sciences, Rehovot 76100, Israel

C. VAN DOORNE, Laboratory for Aero and Hydro-dynamics, Delft University of Technology, Leeghwaterstraat 21, 2628 CB Delft, The Netherlands

K. WIESE, FB7, Uni. Essen, 45117 Essen, Germany

B. WINGATE, Los Alamos National Laboratory, CIC3/MS B256, Los Alamos, NM 87545, U.S.A.

D. WIROSOETISNO, Department of Mathematics and Statistics, University of Edinburgh, Mayfield Road, King's Buildings, Edinburgh EH9 3JZ, U.K.

V. ZAKHAROV, Institute of Continuous Media Mechanics of Russian Academy of Science, 1 Acad. Korolyov Str., Perm 614013, Russia

D. BARKLEY, Mathematics Institute, University of Warwick, Coventry CV4 7AL, U.K.

J. CASTRO, CINVESTAV, Departamento de Fisica, Av. I.P.N 2508, Col. San Pedro Zacatenco, CP 07000, Mexico D.F.

Préface

Le phénomène de la turbulence en mécanique des fluides est connu depuis des siècles. Il est en effet déjà discuté par le poète latin Lucrèce dans le « de natura rerum », qui décrit comment une petite perturbation – « clinamen » – peut être à l'origine de l'apparition d'un ordre turbulent dans une rivière initialement laminaire composée d'atomes s'agitant de façon désordonnée. Plus récemment Léonard de Vinci a décrit et dessiné des tourbillons, et des croquis similaires se retrouvent dans l'école « Utagawa » d'artistes japonais du XIXe siècle, croquis qui influencèrent certainement V. van Gogh dans « La nuit étoilée ». Cependant, en dépit des contributions déterminantes de Bénard, Reynolds, Prandtl, van Karman, Richardson et Kolmogorov, ce problème est encore largement ouvert : il n'existe pas de dérivation rigoureuse de la fameuse cascade de Kolmogorov en $k^{-5/3}$ vers les petites échelles, ni de la valeur du nombre de Reynolds de transition à la turbulence dans un conduit. Par delà ces questions fondamentales, la turbulence est associée à des problèmes pratiques fondamentaux en hydraulique, aérodynamique (réduction de la traînée des voitures, des trains et des avions), combustion (amélioration du rendement des moteurs et réduction des pollutions), acoustique (la réduction du bruit induit par la turbulence est un problème essentiel pour les moteurs d'avions), études de l'environnement et du climat (rappelons nous les dommages causés en Europe par les tempêtes de la fin 1999), et astrophysique (la grande tache rouge de Jupiter et la granulation solaire sont des manifestations de la turbulence). Il est en conséquence urgent de développer des modèles permettant de mieux prédire et contrôler les effets de la turbulence.

Au cours des dix dernières années, des avancées spectaculaires ont conduit à une meilleure compréhension physique de la turbulence, en particulier en ce qui concerne l'auto-organisation cohérente de tourbillons résultant des interactions non-linéaires fortes. Ces avancées ont été dues à la fois aux progrès considérables des simulations numériques (en particulier dans les simulations des grands tourbillons) et à l'apparition de nouvelles techniques expérimentales comme les méthodes de vélocimétrie par visualisation numérique de particules. De nouveaux outils théoriques sont apparus pour étudier la turbulence fluide en trois et en deux dimensions, parfois empruntées à la physique statistique ou à la physique de la matière condensée : analyse multifractale, dynamique

Lagrangienne et mélange, états d'entropie maximale, ondelettes, équations non-linéaires d'amplitude... Tous ces progrès ont justifié l'organisation de cette École d'été des Houches « Turbulence : nouveaux aspects » et ce livre de cours qui en résulte. Les aspects théoriques ont été associés à des points de vue plus appliqués, où on considère l'influence sur la turbulence des conditions aux bords, de la compressibilité, des effets de courbure et de rotation, de l'hélicité et des champs magnétiques. Des applications variées de la modélisation et du contrôle de la turbulence pour certains problèmes industriels et de l'environnement ont également été considérées.

Le premier chapitre introductif de Akiva Yaglom traite de « Un siècle de théorie de la turbulence ». Le problème de la turbulence y est passé en revue, à la fois en ce qui concerne les points de vue linéaires et non-linéaires, la découverte par van Karman et Prandtl de profils logarithmiques de vitesse dans les couches limites, et la théorie de Kolmogorov-Obukhov du régime statistique universel pour la turbulence à petite échelle et grand nombre de Reynolds (théorie révisée en 1962 par les auteurs pour tenir compte de l'intermittence, et qui a conduit à prédire un comportement d'échelle pour les fonctions de structures). Akiva Yaglom fut un étudiant de Kolmogorov et a contribué au développement de ce dernier sujet ; sa présentation est donc basée sur une connaissance de première main de l'histoire de ces découvertes classiques et de leur statut actuel.

Ce traitement des bases de la théorie de la turbulence est complété par le cours de K. Sreenivasan et S. Kurien (Cours 2), qui couvre de façon très complète les résultats expérimentaux sur la turbulence développée à très grand nombre de Reynolds, et en particulier sur les déviations au comportement d'échelle de Kolmogorov.

Dans le Cours 3, Métais présente en détail les simulations numériques des grandes échelles (« Large-eddy simulations » – LES) de la turbulence, qui permettent des simulations déterministe de la formation et de l'évolution des tourbillons cohérents, ainsi que les méthodes d'identification des tourbillons et les simulations des écoulements sans cisaillement et des couches limites. Il s'intéresse aussi aux aspects particuliers liés à la rotation et à la stratification thermique, qui s'appliquent à la formation des eaux profondes dans les océans et aux tempêtes atmosphériques. Il conclut par les applications des LES à la turbulence dans les gaz compressibles : rentrée dans l'atmosphère des corps hypersoniques et refroidissement des moteurs fusées.

Dans le Cours 4, Michael Leschnizer décrit les méthodes de description statistique des écoulements industriels, où la géométrie est trop complexe pour permettre d'appliquer les LES. Il montre comment ces méthodes peuvent être validées, dans des situations géométriques simples, par comparaison avec les méthodes DNS et LED. Les résultats présentés sont très encourageants et conduisent à des applications intéressantes aux écoulements dans les turbomachines. Toujours en ce qui concerne les aspects industriels, Reda Mankbadi (NASA) donne dans le Cours 5 une revue très complète de l'aéroacoustique numérique (méthodes linéaires, nonlinéaires et LES), avec de nombreuses applications au problème du bruit engendré par les avions. Les cours restants sont de nature plus fondamentale. Keith Moffatt (Cours 6) présente les bases de la dynamique topologique des fluides et souligne l'importance de l'hélicité dans les écoulements neutres et magnéto-hydro-dynamiques. Dans ce dernier cas il montre comment l'hélicité aux petites échelles peut engendrer un champ magnétique aux grandes échelles (α-effect), mécanisme qui pourrait expliquer les champs magnétiques stellaires et planétaires. Moffatt discute également la possibilité d'apparition des singularités à temps fini dans les équations d'Euler et de Navier–Stokes. Dans un exposé oral, il présente l'analogie de ceci avec le disque d'Euler. Sa démonstration pratique fut un des sommets de cette école, surtout lorsque les participants réalisèrent qu'il était possible de faire l'expérience avec les assiettes du restaurant, ce qui fut moins apprécié par le cuisinier... Uriel Frisch et J. Bec (Cours 7) ont traité aussi des singularités à temps fini, surtout sur la base de l'équation de Burgers et de l'apparition de chocs multiples, des méthodes analytiques et numériques de résolution de ces équations, et du problème de la décroissance de l'énergie cinétique. Ils montrent comment l'équation de Burgers 3D s'applique en cosmologie au problème de la formation des grandes structures de l'Univers.

Le Cours 8 est un cours très complet de la turbulence bidimensionnelle par Joël Sommeria (Grenoble). De nombreux exemples de turbulence 2D sont d'abord présentés (expériences d'écoulements en rotation ou MHD, plasmas, exemples dans les océans, les atmosphères planétaires comme la grande tache rouge de Jupiter) en insistant sur l'absence de dissipation de l'énergie cinétique dans ces écoulements. La double cascade d'enstropie (cascade directe) et d'énergie (cascade inverse) proposée par Kraichnan et Batchelor est ensuite introduite et discutée, ainsi que l'analyse statistique des vortex ponctuels de Onsager, et la généralisation de ce modèle à des zones de vorticité (principe d'entropie maximale). Sommeria montre les applications de ce modèle aux couches de mélange et aux écoulements océaniques 2D avec rotation différentielle. Le Cours 9 est une présentation très utile par Marie Farge et K. Schneider (École Normale Supérieure) des méthodes d'analyse en ondelettes et de leurs applications. Dans le Cours 10, Gregory Falkovich, K. Gawedzki et M. Vergassola (Institut Weizmann) discutent des problèmes de mélange pour les scalaires passifs, et la dispersion relative de plusieurs particules. En particulier, pour deux particules la loi de Richardson est rediscutée à la lumière de l'analyse de renormalisation de Kraichnan.

Enfin, deux cours par Eduardo Weisfreid (ESPCI, Paris) et par Pierre Coullet (INLN, Nice) ont porté respectivement sur les effets centrifuges dans les écoulements en rotation et sur la turbulence et le chaos dans les systèmes à petits nombres de degrés de liberté. Une version écrite de ces cours n'a malheureusement pas pu être incluse dans ce volume.

Enfin terminons cette préface par quelques informations concernant le déroulement de cette session 2000 les Houches. Un centre de calcul a été mis en place par Patrick Bégou (LEGI/INPG, Grenoble) pendant toute la session. Les machines (Compaq DS20/2proc, Sun Ultra80/4proc, IBM 44p/2proc, HP J5600/2proc et Kayak Linux/2proc) nous ont été généreusement prêtées par les industriels. Avec cette puissance de calcul, 5 groupes de travail d'étudiants ont été organisés, sous la direction d'un professeur ou d'un scientifique confirmé, en utilisant les bases de données engendrées par plusieurs codes de calculs. Ces groupes consistaient en : (1) jets et sillages (Olivier Métais) ; (2) turbulence isotrope (Marcel Lesieur) ; (3) applications industrielles (Franco Magagnato) ; (4) transition à la turbulence (Laurette Tukerman) ; (5) applications à l'environnement (Elisabeth Wingate). Parmi les sujets traités dans ces groupes de travail citons : analyse de l'hélicité dans un sillage (groupe 1), tourbillons dans des LES de turbulence 3D isotrope (groupe 2), simulation d'écoulements dans des situations géométriques complexes (automobiles) à l'aide du code SPARC de F. Magagnato (groupe 3), écoulements de Couette planaires transitionnels (groupe 4), dispersion de marqueurs dans la stratosphère (groupe 5). Le centre de calcul étant toujours rempli en dehors des heures de cours. Il a aussi fourni aux étudiants une très bonne formation au calcul scientifique et au traitement d'image.

Le niveau des étudiants était extrêmement bon (leurs questions avisées embarrassaient parfois les meilleurs professeurs), et leurs pays d'origine très variés. Les relations étudiants-professeurs étaient très bonnes, et le niveau des discussions (durant les pauses, les repas, les soirées et parfois sur les sentiers) bien plus élevé que Le Prarion. Une participation importante d'étudiants venant d'Europe de l'Est a donné à cette école une saveur particulière. Les contacts entre les étudiants ont été nombreux, amicaux et remplis de tolérance.

Remerciements

Nous remercions beaucoup les professeurs pour tout leur travail pour préparer et présenter leurs cours, et rédiger dans les temps leurs notes de cours. Nous remercions également Compaq, Sun, IBM et Hewlett-Packard, qui nous ont prêté et ont assuré la maintenance de machines très efficaces. Des remerciements tout particuliers sont dus à Patrick Bégou (LEGI-Grenoble) pour l'immense énergie qu'il a dépensée pour prendre contact avec les industriels, mettre en place le centre de calcul, installer les machines aux Houches, les faire fonctionner sans problème pendant toute la session, et enfin (et ce n'est pas le moindre) tout désinstaller à la fin de la session.

L'organisation pratique fournie par l'École de Physique des Houches (en particulier grâce au travail et à la patience de Ghyslaine d'Henry, Isabel Lelièvre et Brigitte Rousset) était excellente, et avec de très bonnes conditions de logement. Les repas étaient remarquables, et le restaurant nous a fourni une atmosphère très amicale, grâce à Claude Cauneau et à son équipe, qui nous ont préparé d'inoubliables fondues, raclettes et tartiflettes. Ce dernier plat, à base de Reblochon, un fromage excellent, et pas si connu, est excellent.

Cette école a été financée par l'Université de Grenoble, la Commission Européenne en tant que « High Level Scientific Conference », l'OTAN en tant que « Institut d'Études Avancées », le CEA, le CNRS en tant que « École Thématique » de la Formation Permanente du CNRS, et le Ministère de la Défense. Nous les remercions tous de leur soutien, sans lequel cette école « Turbulence : nouveaux aspects » n'aurait pas pu être organisée. Nous espérons que le lecteur appréciera ce livre qui offre une présentation actuelle de ce que Richard Feynman a appelé le dernier problème non résolu de la physique classique.

Marcel Lesieur
Akiva Yaglom
François David

Preface

The phenomenon of turbulence in fluid mechanics has been known for many centuries. Indeed, it was for instance discussed by the Latin poet Lucretius who described in "de natura rerum" how a small perturbation ("clinamen") could be at the origin of the development of a turbulent order in an initially laminar river made of randomly agitated atoms. More recently, Leonardo da Vinci drew vortices. Analogous vortices were sketched by the Japonese school of artists called Utagawa in the 19th century, which certainly influenced van Gogh in "The Starry Night". However, and notwithstanding decisive contributions made by Bénard, Reynolds, Prandtl, von Karman, Richardson and Kolmogorov, the problem is still wide open: there is no exact derivation of the famous so-called Kolmogorov $k^{-5/3}$ cascade towards small scales, nor of the value of the transitional Reynolds number for turbulence in a pipe. Besides these fundamental aspects, turbulence is associated with essential practical questions in hydraulics, aerodynamics (drag reduction for cars, trains and planes), combustion (improvement of engine efficiency and pollution reduction), acoustics (the reduction of turbulence-induced noise is an essential issue for plane reactors), environmental and climate studies (remember the huge damage caused by severe storms in Europe at the end of 1999), and astrophysics (Jupiter's Great Red Spot and solar granulation are manifestations of turbulence). Therefore, there is an urgent need to develop models that allow us to predict and control turbulence effects.

During the last 10 years, spectacular advances have been made towards a better physical understanding of turbulence, concerning in particular coherent-vortex self-organization resulting from strong nonlinear interactions. This is due both to huge progress in numerical simulations (with the development of large-eddy simulations in particular) and the appearance of new experimental techniques such as digitized particle-image velocimetry methods. New theoretical tools for the study of fluid turbulence in three and two dimensions have appeared, sometimes borrowed from statistical thermodynamics or condensed-matter physics: multifractal analysis, Lagrangian dynamics and mixing, maximum-entropy states, wavelet techniques, nonlinear amplitude equations... These advances motivated the organization of the Les Houches 2000 School on "New Trends in Turbulence", and this book. Theoretical aspects have been blended with more practical viewpoints, where the influence on turbulence of boundaries, compressibility, curvature and rotation, helicity, and magnetic fields is looked at. Various applications of turbulence modelling and control to certain industrial or environmental issues are also considered.

The first introductory course by Akiva Yaglom deals with a "century of turbulence theory". Here the problem of turbulence is reviewed from both the linear and nonlinear points of view. Then the two main achievements of 20th century turbulence theory are considered, namely the logarithmic velocity profiles (found by T. von Karman and L. Prandtl) for the intermediate layers of flows in circular pipes, plane channels and boundary layers without pressure gradients, and the Kolmogorov–Obukhov theory of the universal statistical regime of small-scale turbulence in any flow with a large enough Reynolds number (revised in 1962 by the authors to consider intermittency, which led to predictions of the scaling for structure functions of various orders). Since Yaglom was a student of Kolmogorov and participated in the developments of the second of the above-mentioned subjects, his presentation is based on first-hand knowledge of the history of these classic discoveries and of their present status.

The basis of turbulence theory is complemented by Katepalli Sreenivasan and S. Kurien (Course 2), who provide also a very complete account of fully developed turbulence experimental data (concerning structure functions and spectra of velocity and passive scalars) both in the laboratory and the atmosphere at very high Reynolds number (up to $R_\lambda = 20\,000$). Sreenivasan insists on departures from Kolmogorov scaling in the light of anisotropy effects, which are studied with the aid of a SO(3) decomposition. This allows us to extract an isotropic component from inhomogeneous turbulence.

Olivier Métais presents in detail "Large-eddy simulations of turbulence" (LES) in Course 3. These new techniques are powerfull tools allowing us to simulate deterministically the coherent vortices formation and evolution. He reviews also methods for vortex identification based upon vorticity, pressure and the second invariant of the velocity-gradient tensor. He applies direct-numerical simulations and LES to free-shear flows and boundary layers. Métais is also involved in particular aspects related to thermal stratification and rotation, with application to deep-water formation in the ocean and atmospheric severe storms. He concludes with applications of LES to compressible turbulence in gases, with reentry of hypersonic bodies into the atmosphere, and cooling of rocket engines.

In Course 4, Michael Leschziner describes methods used for predicting statistical industrial flows, where the geometry is right now too complex to allow the use of LES. He shows how these methods can be, in simple geometric cases, assessed by comparison with DNS and LES methods. He presents very encouraging results obtained with nonlinear eddy-viscosity models and second-moment closures. Interesting applications to turbomachinery flows are provided. Still on the industrial-application side, Reda Mankbadi (NASA) provides in Course 5 a very informative review of computational aeroacoustics, with many applications to aircraft noise (in particular jet noise in plane engines). He shows linear, nonlinear and full LES methods.

The remaining chapters are more fundamental. Keith Moffatt (Course 6) presents the basis of topological fluid dynamics and stresses the importance of helicity in neutral and magnetohydrodynamic (MHD) flows. He shows in the latter case how helicity at small scales can generate a large-scale magnetic field (α-effect), a mechanism which might explain the magnetic-field generation in planets and stars. Moffatt discusses also the possibility of a finite-time singularity within Euler and even Navier–Stokes equations. During his oral presentation, he made an analogy with Euler's disk. His practical demonstration of the latter was one of the School's highlights, especially when people realized that the experiment could be done as well with the restaurant plates, which was less appreciated by the cook. Uriel Frisch and J. Bec (Course 7) speak also of finite-time singularities, but mostly on the basis of Burgers equations in one or several dimensions, with the formation of multiple shocks. They describe methods that allow us to solve these equations analytically and numerically, and the kinetic-energy decay problem. He shows how Burgers equations (in three dimensions) can, in cosmology, apply approximately to the formation of large scales in the universe.

Course 8 is a very complete account of two-dimensional turbulence provided by Joël Sommeria (Grenoble). He first presents numerous examples of 2D turbulence in the laboratory (rotating or MHD flows, plasmas), in the ocean and in planetary atmospheres (Jupiter), insisting in particular on the absence of kinetic-energy dissipation in these flows. He reviews the double cascade of enstrophy (direct) and energy (inverse) proposed by Kraichnan and Batchelor. He presents also very clearly the point-vortex statistical-thermodynamics analysis of Onsager, and the generalization of this model to finite blobs of vorticity (maximum-entropy principle). Sommeria shows applications of this model to mixing layers and to 2D oceanic flows with differential rotation. Course 9 (Marie Farge and K. Schneider) is a useful presentation of wavelet techniques, a further interesting application of which (not detailed in the book) concerns data compression. In Course 10 (Gregory Falkovich, K. Gawedzki and M. Vergassola), the Lagrangian mixing of passive scalars and the relative dispersion of several particles is discussed. For two particles in particular, Richardson's pioneering law (predicting a dispersion rate proportional to the separation raised to the 4/3 power) is revisited in the light of Kraichnan's renormalized analysis.

Let us finish this preface with some information regarding the Les Houches 2000 School "New Trends in Turbulence". A computing centre was set up thanks to Patrick Begou during the duration of the programme. The machines (Compaq ds20/2proc, Sun ultra80/4proc, IBM 44p/2proc, HP j5600/2proc, HP kayak linux/2proc) were kindly lent by the respective companies. With this important computing power, we could organize for students five working groups under the direction of a professor or senior researcher, and using databases generated using several computational programs. These groups were: (1) Jets and wakes (Olivier Métais); (2) Isotropic turbulence (Marcel Lesieur); (3) Industrial applications (Franco Magagnato); (4) Transition (Laurette Tuckerman); and (5) Environment (Elisabeth Wingate). Among the topics treated in these groups were: analysis of helicity in a wake (group 1), vortices in LES of 3D isotropic turbulence (group 2), simulation of flows in complex geometries (around cars) using Magagnato's SPARC code (group 3), transitional plane Couette flow (group 4), and dispersion of tracers in the stratosphere (group 5). The computing centre was always crowded in the non-course periods. It also gave students a very good formation in scientific computing and image processing.

The level of the students (who came from a wide range of countries) was extremely good, and their sharp questions embarrassed sometimes even the best professors. Relations between the students and the professors were excellent, and discussions continued between the courses and during the meals, in the evenings, and even on the mountain trails above Le Prarion. We had a wide participation of students coming from countries in eastern Europe, and it gave the School a distinct flavour. Contacts between the students were numerous, friendly and extremely tolerant.

Acknowledgments

We thank very much the lecturers for their efforts in preparing and delivering their courses, and then writing the above lecture notes. We thank also Compaq, Sun, IBM and Hewlett-Packard, who lent and maintained the (extremely efficient) computers. Particular thanks go to Patrick Begou (LEGI-Grenoble) for the huge energy he spent contacting the computer suppliers, setting up the computing centre, installing the machines, making them work perfectly during the programme, and (last but not least) de-installing everything at the end.

The practical organization provided by the Les Houches Physics School was very good (especially thanks to the efforts and the patience of Ghyslaine d'Henry, Isabel Lelièvre and Brigitte Rousset), with excellent housing conditions. The food was great, and the restaurant offered a very friendly atmosphere, thanks to Claude Cauneau and his staff, who prepared unforgettable fondues, raclettes and tartiflettes. The latter dish, made of Reblochon, an excellent and not very well known cheese, is excellent.

The School was sponsored by Grenoble University, the European Commission as a High-Level Scientific Conference, NATO as an Advanced Study Institute, the CEA, CNRS (*formation permanente*), and the French Ministry for Defense. We thank them very much for their support, without which "New Trends in Turbulence" could never have been organized. We hope that the reader will enjoy this book which provides an up-to-date account on what Feynman called the last unsolved problem of classical physics.

Marcel Lesieur
Akiva Yaglom
François David

CONTENTS

Course 3. Large-Eddy Simulations of Turbulence
by O. Métais *113*

Course 4. Statistical Turbulence Modelling for the Computation of Physically Complex Flows

by M.A. Leschziner   *187*

Course 5. Computational Aeroacoustics
by R. Mankbadi   *259*

Course 6. The Topology of Turbulence
by H.K. Moffatt   *319*

Course 7. "Burgulence"
by U. Frisch and J. Bec   *341*

Course 8. Two-Dimensional Turbulence
by J. Sommeria

385

Course 9. Analysing and Computing Turbulent Flows Using Wavelets
by M. Farge and K. Schneider

449

III Computation   478

IV Conclusion   500

Course 10. Lagrangian Description of Turbulence
by G. Falkovich, K. Gawedzki and M. Vergassola   *505*

COURSE 1

THE CENTURY OF TURBULENCE THEORY: THE MAIN ACHIEVEMENTS AND UNSOLVED PROBLEMS

A. YAGLOM

*Institute of Atmospheric Physics,
Russian Academy of Sciences,
Pyzhevsky 3, 109017 Moscow, Russia,
and Department of Aeronautics
& Astronautics, MIT 33-219,
77 Massachusetts Avenue, Cambridge,
MA 02139-9307, U.S.A.*

Contents

THE CENTURY OF TURBULENCE THEORY: THE MAIN ACHIEVEMENTS AND UNSOLVED PROBLEMS

A. Yaglom

1 Introduction

The flows of fluids actually met both in nature and engineering practice are turbulent in the overwhelming majority of cases. Therefore, in fact the humanity began to observe the turbulence phenomena at the very beginning of their existence. However only much later some naturalists began to think about specific features of these phenomena. And not less than 500 years ago the first attempts of qualitative analysis of turbulence appeared – about 1500 Leonardo da Vinci again and again observed, described and sketched diverse vortical formations ("coherent structures" according to the terminology of the second half of the 20th century) in various natural water streams. In his descriptions this remarkable man apparently for the first time used the word "turbulence" (in Italian "la turbolenza", originating from Latin "turba" meaning turmoil) in its modern sense and also outlined the earliest version of the procedure similar to that now called the "Reynolds decomposition" of the flow fields into regular and random parts (see, *e.g.* [1, 2]). However, original Leonardo's studies did not form a "theory" in the modern meaning of this word. Moreover, he published nothing during all his life and even used in most of his writings a special type which could be read only in a mirror. Therefore his ideas became known only in the second half of the 20th century and had no influence on the subsequent investigations of fluid flows.

During the first half of the 19th century a number of interesting and important observation of turbulence phenomena were carried out (such as, *e.g.*, the early pipe-flow observations by Hagen [3]) but all of them were only the precursors of the future theory of turbulence. Apparently, the first theoretical works having relation to turbulence were the brilliant papers on hydrodynamic stability published by Kelvin and Rayleigh at the end of the 19th century (apparently just Kelvin who know nothing about Leonardo's secret writings, independently introduced the term "turbulence" into fluid mechanics). However, these papers only "had relation to turbulence", but did not concern the developed turbulence at all. First scientific description of turbulence was in fact given by Reynolds [4]. In his paper of 1883 he described the results of his careful observations of water flows in pipes, divided all pipe flows into the classes of "direct" and "sinuous" (laminar and turbulent in the modern terminology) flows and introduced the most important dimensionless flow characteristic (now called "the Reynolds number") $Re = UL/\nu$ where U and L are characteristic velocity and length scales, and ν is the kinematic viscosity of the fluid. Then Reynolds proposed the famous "Reynolds- number criterion", according to which the turbulence can exist only if $Re > Re_{cr}$ where the critical value Re_{cr} takes different values for different flows and different levels of initial disturbances. And the first serious purely theoretical investigation of the developed turbulence was due again to Reynolds [4]. In his classical paper of 1894 he strictly determined the procedure of "Reynolds decomposition", derived the "Reynolds equations" for the mean velocities of turbulent flows and made the first attempt to estimate theoretically the value of Re_{cr} with the help of Navier–Stokes (briefly NS) equations of fluid dynamics. (These equations assume that the fluid is incompressible and have constant density and kinematic viscosity; below only such fluids will be considered.) Therefore the year 1894 may with good reason be considered as the birth year of the modern turbulence theory. After this year turbulence theory was developing energetically during the whole 20th century but up to now it is very far from the completion. Thus, we have quite weighty reasons to call the 20th century *the century of the turbulence theory*.

Of course the 20th century deserves also to be called by more high-grade title of the century of science. In fact during this century the enormous advances were achieved in all sciences and many highly important new scientific domains emerged; Theory of Relativity, Quantum Physics, Nuclear Physics, Physical Cosmology, Molecular Biology are only a few examples. However, in spite of this the modern status of the turbulence theory is quite exceptional and differing from that of all other new sciences.

The reason of such exceptional status is that the other new sciences deal with some very special and complicated objects and processes relating to some extreme conditions which are very far from realities of the ordinary life. These objects and processes are connected, for example, with the movements having enormously high velocities, or manifestations of unprecedentedly high (or, on the contrary, low) energy changes, with extremally small (or large) sizes of involving objects, enormously large or imperceptibly small length- or/and time-scales, and so on. However turbulence theory deals with the most ordinary and simple realities of the everyday life such as, *e.g.*, the jet of water spurting from the kitchen tap. Therefore, the turbulence is well-deservedly often called "the last great unsolved problem of the classical physics".

Such statement was, in particular, often repeated by the famous physicist Feynman who even include it, in a slightly different wording, in his textbook [5] intended for high-school and undergraduate university students (the names of three other great physicist to whom this remark is sometimes attributed were indicated on page 3 of the book by Holmes *et al.* [6] and repeated by Gad-el-Hak [7]). One of these physicists, Sommerfeld, in the late 1940s once noted that he understood long ago the enormous difficulty of the turbulence problem and therefore proposed it in the 1920s to his most talented student Werner Heisenberg; however Heisenberg did not solve this problem which remains unsolved up to now. Finally, the extraordinary status of the problem of turbulence is reflected in the popular funny story about a famous scientist; several versions of this story are met in the available literature. According to Goldstein [8] the story reflects the statement made by Lamb in 1932 at some meeting in London where Goldstein was present. Goldstein's memory was that Lamb remarked then: "I am an old man now, and when I die and go to Heaven there are two matters on which I hope for enlightenment. One is quantum electrodynamics, and the other the turbulent motion of fluids. And about the former I am really optimistic". (In other versions of the story Lamb was replaced by Prandtl, Heisenberg, or Einstein, and the time and place of the event and sometimes also the first of the mentioned matters were changed; see, *e.g.* [7].) Let us consider, however, just the above version where turbulence was compared with quantum electrodynamics. It seems that 1932 was too early date for considering the quantum electrodynamics as the most important unsolved physical problem, however somewhat later, say in the late 1940s and early 1950s, it was exactly so – at that time all experts in theoretical physics were tormented with this problem. However, the solution of it was found not much later. The solution made three physicists (Feynman, Schwinger and Tomonaga) the recipients of the 1965 Nobel Prize in physics; then this problem was closed for ever. When recently a group of prominent physicists formulated 10 most important unsolved problems of modern physics (so-called "Physics Problems for the Next Millennium"; see http://feynman.physics.lsa.umich.edu), these problems showed very clearly how far away went the physics of today from the primitive science of 1930-50s when noncontradictory development of quantum electrodynamics seemed to be an unsolvable problem. However, up to now no cardinal changes occurred in the studies of turbulence. Of course, a lot of new particular interesting results relating to turbulence were found in the 20th century and many technical problems of high practical importance were solved, but there were no Nobel Prizes for turbulence studies and most of the riddles of turbulent motion remain mysterious.

In fact, even the precise content of the "problem of turbulence" is still far from being clear at present (a few remarks about this topic will be made at the end of this text).

Let us now return back to Reynolds' classical papers [4]. In the first of them it was stated that the exceeding by the Reynolds number Re of a laminar flow of the critical value Re_{cr} leads to flow turbulization but the mechanism of this transition to a new flow regime was not considered in any detail. In the second paper of 1894 the turbulization was connected with the growth of flow disturbances but only a very crude method of estimation of Re_{cr} was proposed there. In fact, the accurate determination of turbulization conditions was then, and is up to now, complicated by the obvious incompleteness of the mathematical theory of NS equations. Even the conditions guaranteeing the existence and uniqueness of the solutions of the most natural initial-value problems for NS equations were completely unknown at the end of the 19th century (and are far from being perfectly clear even today). Note in this respect, that the famous French mathematician Leray, who in the paper [9] and some other works of the 1930s and 1940s made very important contributions to the mathematical theory of NS equations, sometimes was inclined to assume that the transition to turbulence may be produced by the termination of the existence of the solution of NS equations corresponding to the laminar regime of fluid flow. However, this assumption was not confirmed afterwards and therefore the dominating position again became occupied by the old idea of Kelvin, Reynolds and Rayleigh who assumed that flow tubulization is caused not by the nonexistence of the laminar-flow solution of NS equations but by the instability of this solution to small exterior disturbances.

2 Flow instability and transition to turbulence

The early studies of flow instabilities to small disturbances used the simplest approach based on the linearization of the NS equations with respect to the disturbance velocities and pressure. Studies of the solutions of linearized dynamic equations for the disturbance variables which grow in time (or, in the case of a spatially formulated parallel-flow stability problem, in streamwise direction) form the so-called *linear theory of hydrodynamic stability*.

The initial approach to the study of linear stability of steady parallel laminar flows, which was proposed by Stokes, Kelvin and Rayleigh in the second half of the 19th century, is the *normal-mode method*. Here the eigensolutions of the system of linearized NS equations are studied. These solutions are proportional to $\exp(-i\omega t)$, where ω is an eigenvalue which may be real or complex. The considered laminar flow is called unstable with respect to small disturbances if the eigenvalue ω with $\Im m\,\omega \geq 0$ (where $\Im m$

denotes the imaginary part) does exist, while otherwise the flow is stable. In the spatial approach to the same problem the eigenfunctions proportional to $\exp(ikx)$ are studied where x is the streamwise coordinate of a parallel flow and k is a complex eigenvalue. Here the flow is called unstable if there exists an eigenvalue k with $\Im m\, k \leq 0$. Spatial approach was first sketched by Orr [10] but in 1907 the spatial eigenvalue problem seemed to be unsolvable and therefore such approach became popular only in the late 1970s. Note, however, that this approach generates some new mathematical problems (relating, *e.g.*, to the validity of the spatial version of the Squire theorem and to the completeness of the corresponding systems of eigenfunctions) and apparently not all of these problems are already solved.

Let us now revert to the classical temporal approach. According to Reynolds' conjecture at values of Re smaller than Re_{cr} all eigenvalues ω_j, $j = 1, 2, 3, ...$, have negative imaginary parts. Orr [10] and Sommerfeld [11] independently proposed in 1907-1908 to determine Re_{cr} as the smallest value of Re at which there exist at least one real eigenvalue ω_j. The linear equation determining in the case of a plane-parallel flow the eigenvalues ω_j is called therefore the Orr–Sommerfeld (OS) equation[1]. The papers by Orr and Sommerfeld led to numerous computations of the OS-eigenvalues, the values of Re_{cr}, and the "neutral curves" in (Re, k) and (Re, ω)-planes for various parallel and nearly parallel flows. These computations played central role in the development of the theory of hydrodynamic stability during the main part of the 20th century. However, the values of Re_{cr} given by the OS equation often exceeded very much values of Re at which real flow instability and transition to turbulence were observed. Moreover, in the cases of plane Couette and circular Poiseuille flows the OS-method led to conclusion that $Re_{cr} = \infty$ which contradict to experimental data showing that both these flows become turbulent at moderate values of Re. (By the way, although the validity of the relation $Re_{cr} = \infty$ for the Poiseuille pipe flow was confirmed by numerous computations with the 100% reliability, the rigorous mathematical proof of this result was not found yet and still represents an unsolved problem.)

---

[1]Note that both these authors considered only the simplest case of two-dimensional wave disturbances assuming that the eigenvalue ω with the smallest imaginary part must always correspond to a plane-wave disturbance (this assumption was rigorously proved only by Squire [12] in 1933). Moreover, they both in fact did not use the OS equation since only the case of a plane Couette flow was considered by them and in this case OS differential equation of the fourth order is reducing to a system of two second-order equations. General form of the OS equation (at once for general three-dimensional wave disturbances) was given by Kelvin [13] in 1887, who however made from this equation an incorrect conclusion.

The often observed disagreement between the OS estimates of Re_{cr} and the observed values of Re corresponding to transition to turbulence may have several reasons. It is clear, in particular, that the consideration of only the eigenfunctions of linearized NS equations in fact represents some oversimplification. Proportional to $e^{-i\omega t}$ eigenfunctions are only special solutions of the linarized NS equations which have amplitudes monotonically (more precisely, exponentially) growing or decaying with t. Moreover, already in 1887 the future Lord Kelvin [14] (at that time he was still called William Thomson) found a solution of the linearized NS equations for a plane Couette flow which "at first rises gradually from initial small value and only asymptotically tends to zero". The paper [14] contained some errors indicated by Rayleigh and Orr; apparently therefore all its results (including the correct ones) were long neglected. As to the nonmonotone Kelvin's solution, Orr [10] generalized it finding a whole family of such nonmonotone solutions (again for the case of plane Couette flow) some of which grew up (proportional to some positive powers of t) to quite large values before they began to decay. Orr even stated the assumption that such transient growth of small disturbances may explain the real instability of plane Couette flow. However this important remark also did not attract then any attention. As a results, the interesting results by Kelvin and Orr were long forgotten and some of them were independently rediscovered by other authors only in 1960s and 1970s.

Strong revival of interest to transient (algebraic in t) growth of disturbances arose at the end of the 20th century. During the last twenty years many dozens of papers about such growth were published (papers [15–21] represent only a few examples of them), while much attention to this topic was also given in the books [22, 23] and survey [24]. It was shown, in particular, that transient growth of nonmodal disturbances may exceed very much the growth of the linearly unstable wave modes. This circumstance gave rise to keen interest to "optimal disturbances" undergoing most intensive transient growth in a given laminar flow; see, e.g., papers [16, 25, 26] devoted to this subject. Note also that in the case of "subcritical fluid flow" with $Re < Re_{cr}$ all solutions of linearized disturbance equations tend to zero as $t \to \infty$. Therefore here transient growth of any disturbance determined by the linearized NS equations must be replaced by decay at some value t_0 of t. However, even before t_0 an initially small disturbance may grow so much that the linearized NS equations will be inapplicable to it and its further development will be governed by the nonlinear NS equations. Then it is possible that the nonlinear theory will show that the considered disturbance will continue to grow also at some times exceeding t_0. Moreover, it may also happen that growing nonlinearly disturbance will produce by nonlinear interactions some new small transiently growing formations maintaining the process of disturbance-energy growth (at the expense of the mean-flow energy) which finally will lead to transition to turbulence. This reason may sometimes explain the transition of a subcritical flow to turbulence. Some specific nonlinear models of such "subscritical transitions" (dealing usually not with NS partial differential equations but with more simple finite-dimensional nonlinear systems of ordinary differential equations) were considered, in particular, in papers [27–29] (however in [29] where the onset of turbulence in subcritical plane Poiseuille flow was discussed at length, results found for model equations were confirmed also by references to the results found by DNS of a disturbance development in a channel flow, *i.e.*, by solution of the corresponding initial-value problem for nonlinear NS equations). A similar scheme of transition to turbulence of the Poiseuille flow in a pipe, where only subcritical disturbances exist, was earlier outlined in [30] and compared with the results of simplified numerical analysis of disturbance development described by nonlinear NS equations. There were also some other numerical simulations of temporal or spatial development of flows in plane channels containing initial disturbances of various forms. These simulations showed, in particular, that the flow development may be rather different in the cases where initial disturbances had different forms; see, *e.g.*, typical papers [31–33] and discussion of this topic in the books [22, 23].

Let us now say a few words about the present state of the studies of the final stage of the flow transition to turbulence. Recent computations of transient disturbance growths followed by flow turbulization confirm the conclusion obtained earlier from the experimental data which showed that for any laminar flow there are several ways to turbulent regime which realizations depend on a number of often hardly controlled external factors. In the first half of the 20th century almost all performed theoretical studies of flow instability dealt only with linear and (rarely) weakly nonlinear development of disturbances and therefore the real mechanisms of transition to turbulence were then not considered at all. The first physical model of laminar-flow-transition was developed by Landau [34, 35] in the early 1940s when he began working on the volume of his fundamental *Course of Theoretical Physics* devoted to continuum mechanics. According to Landau's model transition is produced by a series of subsequent bifurcations of flow regime, where each bifurcation increases by one the number of periodic components of the quasi-periodic fluid motion arising at the preceding bifurcation. This simple model (which was in 1948 supplemented by Hopf [36] by a mathematical example of such instability development) was then almost unanimously accepted by turbulence community as the universal mechanism of flow turbulization. However, the further development of the mathematical theory of dynamic systems showed that Landau's model of flow development not only is nonuniversal but is exceptional in some important respects and thus is rarely observed.

Basing on the available in the early 1970s new results of the dynamic-system theory, Ruelle and Takens [37] proposed a new model of transition to turbulence cardinally differing from Landau's model. According to these authors, transition to turbulence is realized by a succession of a few (usually three) "normal" flow bifurcations of Landau–Hopf type, followed by a sudden appearance of a very intricate attracting set (called a "strange attractor") in the phase space of a flow. The flow states corresponding to phase points within the attractor are very irregular and can be characterized as being "chaotic" or "turbulent". Ruelle and Takens' model at first caused some doubts but later it was found that this model agrees quite satisfactorily with some (but not all) experimental data relating to transitions to turbulence and can also explain seemingly paradoxical data of the old numerical experiment by Lorenz [38] who considered a low-dimensional numerical model of a convective fluid flow. After this discovery the Ruelle-Takens model gained high popularity and stimulated a great number of further studies of temporal and spatial developments of nonlinear dynamic systems. As a result there appeared enormous (and rather sophisticated) literature on both the general theory of dynamic systems and its applications to flow developments; in this literature the words "chaos", "strange attractor" and some other new terms play the main part.

Results of this very extensive and diverse literature relating to transition to turbulence are, nevertheless, not fully satisfactory up to now. It was, in particular, discovered that there are several different "scenarios" for transition of a dynamic system to chaotic behavior as "parameter of nonlinearity" (*e.g.*, the Reynolds number of a fluid flow) increases. In addition to the scenario by Ruelle and Takens, the Feigenbaum scenario of a cascade of period-doubling bifurcations ([39, 40], *cf.* also the related model described in [41]), and the so-called intermittent-transition scenario by Pomeau and Manneville [42, 43] may be mentioned as examples. Note also subcritical-flow transition scenarios studied in [29, 30] which do not include any cascade of successive instabilities. During the last 15 years many hundreds of papers and many dozens of books and lengthy surveys appeared were these and some other scenarios of transition of dynamic systems to chaotic regimes are discussed (the books [44–48] and the papers [49–51] discussing the applicability of the concept of chaos to turbulence are only a few examples). Let us mention in this respect also a few laboratory and numerical studies [52–55] where some flow-transition phenomena were described which have features close to those of some of the proposed transition scenarios. However, all the results obtained up to now do not form a complete physical theory of the transition of fluid flows to turbulence. Note that up to

now there are no strict conditions for realization of various transition scenarios although it is known that sometimes different scenarios may take
place in the same flow depending on some poorly known circumstances.
And all the proposed scenarios were mostly compared with computations
relating to some finite-dimensional models much simpler than the very intricate infinite-dimensional dynamic system evolving in the space of vector-
functions of four variables in accordance with the NS equations (*cf.*, *e.g.*,
paper [56] where a scenario for the onset of space-time chaos in a flow was
studied on the model example of relatively simple nonlinear partial differential equation and it was shown that even in this case the transition to chaos
proves to be quite complicated). Up to now even the question about the existence and properties of strange attractors in the infinite-dimensional phase
spaces of real fluid flows is not answered satisfactorily enough (rich in content book [57] in fact covers only the attractor problem of two-dimensional
fluid dynamics; see in this respect also the books listed in [136]). Thus,
the completely new approaches to the transition-to-turbulence problem developed at the end of the 20th century generated, together with a number
of interesting new results, also a great number of new unsolved problems
which only confirm the popular assumption about the "insolvability of the
problem of turbulence".

3 Development of the theory of turbulence in the 20th century: Exemplary achievements

Calling the 20th century "the century of the turbulence theory" we stressed
that during this century very great progress was achieved in the studies of
turbulence phenomena. And there are two main trends (often overlapping
each other) of the turbulence-theory development in the 20th century –
elaboration of the methods allowing to determine the practical effects of
turbulence, and the investigation of fundamental physical laws controlling
the turbulent flows. Below only a few results relating to the second group
will be discussed; these results were long assumed to be the most important
achievements of the theory of turbulence but at the end of the century it
became clear that there are some quite reasonable doubts concerning the
classical results discussed below.

3.1 Similarity laws of near-wall turbulent flows

The class of near-wall parallel (or nearly parallel) laminar flows includes
such important examples as flows in broad plane channels (which may be
modeled with a good accuracy by plane Poiseuille flows produced by a pressure gradient in a layer between two infinite parallel walls), flows bounded

by parallel walls one of which is stationary and the other is moving with constant velocity (plane Couette flows), pressure gradient flows in long circular pipes (circular Poiseuille flows), and boundary-layer flows over flat plates in the absence of the longitudinal pressure gradient (Blasius boundary-layer flows). Plane Poiseuille, plane Couette, and Blasius boundary-layer flows are bounded by flat walls which for simplicity will be assumed to be smooth. Pipe flows are bounded by a cylindrical wall (also assumed to be smooth) but if pipe radius R is much larger than the "wall length-scale" $l_w = \nu/u_*$, where $u_* = (\tau_w/\rho)^{1/2}$ is the friction velocity, τ_w – the wall shear stress, and ρ – fluid density (only this case will be considered below), then it is usually possible to neglect, in a reasonable first approximation, the influence of wall curvature, $i.e.$ to consider again the wall as flat one. For fully turbulent near-wall flows the mean-velocity profiles $U(z)$ (where z is the normal-to-wall coordinate) and the skin-friction laws (giving the value of the friction, or drag, coefficient) were carefully studied in the late 1920s and early 1930s by Prandtl and von Kármán who combined a few simple semi-empirical hypotheses with the methods of dimensional analysis (based on definite assumptions about the list of physical parameters which are essential here). Apparently the most important discovery of the mentioned authors was the discovery of the $logarithmic$ $mean$-$velocity$ law for the values of z large with respect to l_w and small with respect to the vertical length scale L (equal to the half-distance between parallel walls, pipe radius, or the boundary-layer thickness). According to this law

$$U(z) = u_* \left[A \ln \left(\frac{z u_*}{\nu} \right) + B \right] \quad \text{for} \quad l_w \ll z \ll L, \tag{1}$$

where A and B are universal constants (and $\kappa = 1/A$ is called von $Kármán's$ $constant$).

Logarithmic law (1) was first announced by von Kármán in 1930 at the International Congress of Applied Mechanics at Stockholm. He derived it from a seemingly natural "similarity principle" while Prandtl in 1933 gave another more simple derivation of this law (see, $e.g.$ [8]). A still simpler purely dimensional derivation of this law was proposed in 1944 by Landau [35] This derivation was based on "rational arguments" stating that at $z \ll L$ the "external length scale" L cannot affect the flow structure, while at $z \gg l_w$ the velocity shear (but not the velocity itself) of a developed turbulent flow cannot be affected by ν (since at such z the velocity gradients are quite small and also the "eddy viscosity" is much greater than the molecular viscosity). Therefore, at $l_w \ll z \ll L$ the shear dU/dz can depend only on u_* (determining the vertical flux of momentum) and z. Thus $dU/dz = Au_*/z$ there and this implies equation (1). Similar arguments were applied by Landau [35] to the first derivation of the logarithmic law of the form

$$T(z) - T(0) = T_* \left[A_\mathrm{T} \ln \left(\frac{z u_*}{\nu} \right) + B_\mathrm{T}(\mathrm{Pr}) \right] \qquad (1a)$$

for the profile of mean temperature (or mean concentration of some passive admixture) $T(z)$ in a wall flow with a heat (or mass) transfer from the wall. Here $T_* = j_w/u_*$ is the heat-flux scale of temperature (for definiteness only the case of heat transfer will be mentioned in this paper), j_w is the temperature flux at the wall, while A_T is a new constant, and $B_\mathrm{T}(\mathrm{Pr})$ is a function of the Prandtl number $\mathrm{Pr} = \nu/\chi$, where χ is the coefficient of thermal diffusivity.

One more elegant derivation of the law (1) was proposed in 1937 by Izakson [58] who recalled that the rational arguments of dimensional analysis led Prandtl to the formulation of the general *wall law* of the form

$$U(z) = u_* f^{(1)}(u_* z/\nu) \qquad (2)$$

(where $f^{(1)}$ is an universal function) for velocity $U(z)$ at $z \ll L$. Similar dimensional arguments imply also the validity at $z \gg l_w$ of the *velocity defect law*

$$U_0 - U(z) = u_* f^{(2)}(z/L), \quad \text{where} \quad U_0 = U(L) \qquad (3)$$

(for a pipe flow the law (3) was first empirically detected in 1911 by Stanton and in 1930 it was justified by dimensional arguments by von Kármán). Then Izakson noted that if an *overlap layer* of not too small and not too large values of z exists where both laws (2) and (3) are simultaneously valid, then it is easy to show that in this layer the wall law (2) must have logarithmic form (1) while the velocity defect law (3) must be also logarithmic and have the form

$$U_0 - U(z) = u_* \left[-A \ln \left(\frac{z}{L} \right) + B^{(1)} \right]. \qquad (4)$$

Here again $A = 1/\kappa$, and $B^{(1)}$ is a new constant taking different values for flows in channels, pipes, boundary layers, and for plane Couette flow.

Izakson's derivation of two logarithmic laws quickly gained popularity. In particular, in 1938 Millikan [59] noted that Izakson's arguments may be applied to flows along both smooth and rough walls (in the latter case the coefficient B will depend on characteristics of wall roughness) and that adding together equations (1) and (4) one may easily derive the famous Prandtl–Nikuradse logarithmic skin-friction law for turbulent flows in smooth-wall and rough-wall pipes and plane channels. (Millikan also remarked that the same method can be applied to turbulent boundary layers. However he did not consider boundary-layer flows and the first derivation of Kármán's skin-friction law for boundary layers by the method sketched here was apparently due to Clauser [60].) Such derivation allows to determine the dependencies of the coefficients of skin-friction laws on logarithmic-law coefficients A, B and $B^{(1)}$; obtained results were found to be in agreement with the available data of velocity and skin-friction measurements. Some further developments of Izakson's method will be indicated slightly later.

The logarithmic velocity-profile and skin-friction laws for wall turbulent flows were conventionally considered as some of the most fundamental (and most valuable for the practice) achievements of the 20th-century turbulence theory. These theoretical results were many times compared with data of direct measurements of turbulent-flow characteristics in pipes, boundary layers and plane channels. As a rule, obtained results agreed more or less satisfactorily with logarithmic laws (see, *e.g.*, the recent survey [61]) but measured values of "universal coefficients" A, B, and $B^{(1)}$ of these laws proved (and prove up to now) to be rather scattered. During long time the most popular estimates of $\kappa = A^{-1}$ and B were these ones: $\kappa = 0.40$ (or 0.41, but the values in the range from 0.36 to 0.46 were also sometimes obtained), $B = 5.2$ (but many other values in the range from 4.8 to 5.7, and also some values outside of this range, were met in the literature). The values of $B^{(1)}$ were measured not so often; according to majority of estimates $B^{(1)} \approx 0.6$ for circular pipes and plane channels, and $B^{(1)} \approx 2.4$ for flat-plate boundary layers (see, *e.g.*, the surveys [62,63]). The range of z-values belonging to the so-called *logarithmic layer* of a wall flow, where equation (1) is valid, was also subjected to great scatter; most often it was suggested that this layer is extended from the lower limit at $z \approx 50 l_w$ (coefficient 50 was sometimes replaced by 30 or by 70) up to the upper limit at $z \approx 0.15L$ (instead of 0.15 the coefficients 0.2 and 0.3 were sometimes used). And at the present time the uncertainty relating to the coefficients $\kappa = A^{-1}$ and B and limits of the "logarithmic layer" did not become smaller; see below about this matter.

There were also many works extending and generalizing the theory of the logarithmic layer and Izakson's method of derivation of results relating to this layer. Von Mises [64] considered the cases of non-circular pipes while the applications of the same method to near-wall turbulent flows with heat (or mass) transfer were considered in [62, 63, 65]. Numerous applications to flows along rough walls and wall-flows with non-zero pressure gradients were discussed in surveys [62, 66]. Comprehensive generalization of the "logarithmic-layer theory" to the case of the near-wall layers of turbulent flows in stratified fluids with mean density $\rho(z)$ depending on the vertical coordinate z (first of all to atmospheric and oceanic surface layers) was developed by Monin and Obukhov; see, *e.g.*, Chapter 4 of the book [67]. Townsend in the book [68] published in 1956 formulated the general "Reynolds-number similarity principle" used then for the derivation of the similarity laws (2) and (3) and logarithmic law (1). Simultaneously he also sketched applications of the general similarity arguments to the second moments of velocity fluctuations and, in particular, investigated wall laws (5) indicated below for the second-order moments where $k+l+m = 2$. More detailed exposition of the applications of Izakson's arguments to moments of velocity-component fluctuations $(u_1, u_2, u_3) = (u, v, w)$ (and temperature fluctuations θ) was presented in the paper [62]. In this paper it was postulated that in the near-wall flow region, where $z \ll L$, the wall similarity law of the form

$$M_{klm}(z) \equiv \langle u^k v^l w^m \rangle = (u_*)^{k+l+m} f_{klm}(z u_*/\nu), \qquad (5)$$

is valid, while in the outer flow region, where $z \gg l_w = \nu/u_*$, the outer similarity law of the form

$$M_{klm}(z) = (u_*)^{k+l+m} g_{klm}(z/L) \qquad (6)$$

takes place. Here angular brackets denote the probabilistic (ensemble) averaging, while f_{klm} and g_{klm} are two families of universal functions of one variable. If an overlap layer, where $l_w \ll z \ll L$, exists in the considered flow, then both equations (5) and (6) must be valid there and this implied that in this layer the moments of velocity fluctuations take constant values, i.e.

$$M_{klm}(z) \equiv \langle u^k v^l w^m \rangle = a_{klm}(u_*)^{k+l+m} \qquad (7)$$

where a_{klm} are universal constants. Related similarity laws can be formulated for many other statistical characteristics of the velocity-component and temperature fluctuations in fully turbulent wall flows (e.g., for correlation functions, spectra and multipoint higher moments of these fluctuations).

Beginning from the 1930s logarithmic velocity and skin-friction laws were used in the engineering practice much more widely than any other scientific results related to turbulence, and very long they were universally treated as indisputable certainty. Such opinion was strongly supported by the unquestionable authority of famous scientists who independently proposed different derivations of these laws and then actively popularized them; the list of such scientists includes the names of Prandtl, von Kármán, Taylor, Landau, and Millikan. (By the way, Kolmogorov also highly estimated logarithmic velocity laws and their derivation from the overlap-layer arguments. He even elucidated these results and their application to the determination of skin-friction laws in two short notes of 1946 and 1952 published in

"Doklady of USSR Acad. Sci." and intended for engineers; see the list of his works on turbulence in [69].) However, at present the study of turbulence advanced very much in comparison to its state in the middle of the 20th century and this development produced some doubts in the universal validity of these classical results.

Prandtl's wall law (2) follows from the assumption that at $z \ll L$ the length L cannot affect the flow characteristics. This assumption seemed to be obvious not only in 1925, when the wall law was proposed by Prandtl, but also long after this year, but now it causes doubt by reasons which will be explained below. However the inclusion in the velocity-defect law (3) of the friction velocity u_* determined by flow condition at the wall did not always seem fully motivated and some scientists were long ago inclined to think that the law (3) is in fact of empirical origin. (For this reason some authors even proposed to replace the near-wall velocity scale u_* in (3) by a scale more appropriate to outer-flow conditions; one such example will be mentioned below.)   Reverting now to the independence of flow characteristics near the wall (at $z \ll L$) of the length L, let us note that such independence became to be non-obvious after the discovery of the important part playing in turbulent flows by the large-scale organized vortical structures (so called "coherent structures") which affect all regions of the flow. The study of these structures and of their role in turbulence was developed rapidly after the end of the World War II (at great degree under the influence of clear presentation of this topic in Townsend's important book [68] of 1956); for modern state of the studies of coherent structures see the recent book [6].

Slightly later Townsend's experimental studies of turbulent boundary layers [70] showed that the intensities $\langle u^2 \rangle$ and $\langle v^2 \rangle$ of the horizontal velocity fluctuations in the "logarithmic layer" (where $l_w \ll z \ll L$) sometimes take different values in two boundary layers with the same value of u_*. This result clearly contradicted the wall laws (7) corresponding to mean squares $\langle u^2 \rangle$ and $\langle v^2 \rangle$. Townsend explained this disagreement with the wall laws assuming that turbulent motion in the wall regions of turbulent boundary layers consist of the "active" component (which produced the shear stress $\tau = -\rho\langle uw \rangle$ and satisfies the usual wall laws (1), (2) and (5)), and "inactive" practically irrotational component which is produced by large-scale fluctuations in the outer region of boundary layer and depends on L (*i.e.* on the boundary-layer thickness, since in [70] only boundary-layer characteristics were discussed). Later Bradshaw [71] (see also [72]) confirmed Townsend's hypothesis by new experimental data and showed that it explains also some other experimental results inexplicable by the traditional theory. More-over, Bradshaw also repeated Townsend's statement that "inactive motions" contribute nothing to the mean-velocity profiles (and hence do not violate the logarithmic velocity laws) and to vertical (normal-to-wall) velocity fluctuations w. And in the second edition of 1976 of the book [68] Townsend [73] connected the inactive motions with the contributions to the fluid motions made by a definite family of similar to each other vortical structures differing by their length scales. Basing on this idea he derived new equations for quantities $\langle u^2 \rangle$ and $\langle v^2 \rangle$ within the logarithmic layer; these equations included in addition to constant right-hand sides $a_{200}(21_*)^2$ and $a_{020}(21_*)^2$ of equation (7), also terms proportional to $\ln(z/L)$ (together with small terms which depended on z/l_w and became negligible at very high Re). According to data by Perry and Li [74], Townsend's equations agree more or less satisfactorily with the results of measurements of mean squares of horizontal velocity fluctuations. (Note, however, that all experimental data relating to higher moments of velocity fluctuations are much more scattered and controversial than results of mean-velocity-profile measurements; $cf.$, $e.g.$ [61].) Arguments similar to those of Townsend [73] were later applied by the present author [75] to theoretical evaluation of intensities of the horizontal wind fluctuations in the unstably stratified atmospheric surface layer. This approach allowed to explain seemingly paradoxical dependence of the intensity of wind fluctuations at few-meter heights above the Earth's surface on the thickness of the planetary boundary layer having the order of 1–2 km.

Townsend's results show that the influence of large-scale coherent structures made incorrect at least some of the classical similarity laws postulating the negligible effect of the external length scale L on the flow characteristics within the flow region where $z \ll L$. Of course, in [73–75] only some particular violations of traditional wall laws were noted. Since, however, at present it is known that large-scale structures of many different types and length scales exist in developed turbulent wall flows, it may be expected that all the similarity laws which neglect the possible influence of the length L are of limited accuracy. And the other fundamental assumption used in the formulations of classical similarity laws of near-wall turbulent flows, according to which the molecular-viscosity effects must be negligibly small at $z \gg l_w = \nu/u_*$, also becomes questionable in the light of recent experimental findings.

Experiments (and numerical simulations) of 1990s definitely show that the developed turbulent flows at large values of Re always include a tangle of intense and very thin vortex filaments which diameters sometimes are of the order of the Kolmogorov length scale η. (This length scale characterizes the spatial extent of viscous influences; for its definition see Eq. (11) below, while more detailed discussion of the role of the filaments may be found, $e.g.$, in [2], Sect. 8.9, and [117], Sect. 5.) In other words, according to modern views the range of scales of organized vortical structures existing in fully-developed turbulent flows extends from the external length scale L up to Kolmogorov's internal length scale η. Since the topology and general structure of the tangle of filaments must depend on Re and the filaments are found in all regions of turbulent flows, the characteristics of the high-Reynolds-number turbulence also may everywhere depend on ν and $Re = UL/\nu$. (Moreover, Barenblatt [84] noted in his paper of 1999 that Prandtl in the remark made at the Intern. Congr. of Appl. Mech. of 1930 after the talk by von Kármán where the logarithmic form of the velocity profile first appeared, indicated that at moderate values of Re the influence of near-wall streaks on the flow at greater heights may generate mechanism of possible influence of viscosity ν on turbulence characteristics at $z \gg l_w$. However, later Prandtl apparently never mentioned this effect.)

The arguments presented above imply that the classical logarithmic mean-velocity (and mean-temperature) laws of wall turbulent flows possibly represent only some reasonable approximations the accuracy of which must be thoroughly checked. Barenblatt, Chorin and Prostokishin, who are apparently the most energetic modern opponents of logarithmic laws, reasonably noted (in [76] and a number of other publications) that the description of the mean velocity $U(z)$ of wall turbulent flows by power laws $U(z) \propto z^k$ was widely used long enough by scientists and engineers and, if the power k was properly chosen for all values of Re of interest, usually led to satisfactory agreement with the data over a wide range of z-values (in this respect usually Schlichting's book [77] is referred). Barenblatt et al. indicated also a great number of more recent publications containing the data illustrating the dependence on Re of the mean flow characteristics of turbulent near-wall flows. (A number of appropriate references may be found in the survey paper [78]; in a short subsequent remark [7] Gad-el-Hak also noted quite reasonably that since any doubt concerning logarithmic laws where long considered heresy, most of the papers containing such heresies were apparently rejected by editors of scientific journals.)

In [76] and the other related papers Barenblatt et al. suggested that logarithmic law should be replaced by laws of quite different form. This proposition was directly connected with some general ides introduced in 1972 by Barenblatt and Zeldovich [79]. It was noted in this paper that self-similar solutions of the form $V(x, t) = A(t)F[x/l(t)]$ (where x and t are some independent variables) are very often encountered in fluid dynamics and other branches of physics as "intermediate asymptotics" describing the behavior of the dependent variable V in regions where direct influence on it of peculiar features of the initial or/and boundary conditions is already lost but the system is still far from being in a state of equilibrium. It was then remarked that only a small part of such self-similar solutions may be determined by simple arguments of dimensional analysis. For this part of self-similar solutions the term "self-similar solutions of the first type" was proposed in [79], while all the other self-similar solutions were called "self-similar solutions of the second type". (Later also the terms "complete similarity" and "incomplete similarity" were sometimes used by Barenblatt for these two types of self-similar solutions.) In [79] and the subsequent publications of the same authors (in particular, in Barenblatt's book [80]) a great number of self-similar solutions of both types was indicated. The general form of a solution of the first type may be uniquely determined with the help of dimensional arguments; hence it can be easily found and usually includes some factors raised to definite integer (or simple fractional) powers. For a solution of the second type the situation is much more complicated; here only some supplementary physical arguments and experimental data may suggest the general form of the sought for solution which usually includes some factors raised to powers which may take arbitrary values. The corresponding exponents may be determined in some cases from solutions of some supplementary eigenvalue problems of physical origin (see examples in [80]) but very often they must be determined from results of data processing. And the conditions guaranteeing the existence of a self-similar solution of the second type and allowing to determine its form most often are unknown; here the physical intuition and the good luck of the explorer may play the decisive part.

The problem concerning self-similar solutions of the second type in turbulence theory is especially complicated. Recall that the evolution of a fluid flow is governed by the system of Navier–Stokes equations. These partial differential equations are very complicated, they cannot be easily analyzed and are insufficiently investigated up to now while their solutions corresponding to turbulent flow regimes are enormously intricate and completely nonexplored. Therefore, it seems that the dynamic equations could not help here in search of needed self-similar solutions. On the other hand, the abundance of self-similar solutions of the second type reliably established in other branches of continuum mechanics gives some reasons to expect that such solutions may play definite part in the turbulence theory too. To verify this expectation, it was only possible to perform the careful examination of the available experimental data of high enough quality.

Such examination of the pipe-flow turbulent data by Nikuradse [81] (which were indisputably the best ones available in the 1930s and are sometimes referred to even now) was carried out by Barenblatt in the early 1990s (see, *e.g.* [82]) and then presented at greater length in a number of papers (in particular, in the joint paper with Chorin and Prostokishin [76]). According to these data the velocity profile $U(z)$ of a turbulent flow in a pipe satisfied the simple equation of the form

$$U(z)/u_* = C(u_* z/\nu)^\alpha \tag{8}$$

over almost the whole pipe cross-section (except the thin "viscous sublayer" where $u_* z/\nu$ does not exceed some threshold value of the order of a few tens). In equation (8) parameters C and α do not depend on z but vary (rather slowly) with the flow Reynolds number $Re = U_\mathrm{m} D/\nu$ (where U_m is the flow velocity averaged over the pipe cross-section and $D = 2R$ is the pipe diameter). Careful examination of the Nikuradse data led Barenblatt to proposition of the following expressions for the functions $C(Re)$ and $\alpha(Re)$

$$C(Re) = \frac{\ln Re}{\sqrt{3}} + \frac{5}{2}, \quad \alpha(Re) = \frac{3}{2\ln Re} \cdot \tag{9}$$

Equations (8) and (9) were first obtained by treatment of the old data by Nikuradse. However in [76] these equations were compared with a number of more recent pipe-flow and boundary-layer turbulence data and according to results of this paper (which were not unanimously supported) all the considered data agree well with equations (8) and (9). Later results of more detailed comparison by Barenblatt *et al.* of equations (8) and (9) with velocity profiles $U(z)$ measured in various fully turbulent zero-pressure-gradient boundary layers on flat plates were presented in [83]. [In the case of boundary layers the pipe-flow equations (8) and (9) were used without any modification but now the value of Re was determined as that leading to the best fit of these equations with the available velocity data. This means that here a new "boundary-layer thickness" Λ was introduced by the condition that the substitution of $Re = \Lambda U_0/\nu$, where U_0 is the free-stream velocity, into equation (9) leads to good agreement of equation (8) with the measured mean-velocity profile $U(z)$.] In [83] it was found that the velocity profiles of turbulent boundary layers agree well with the power law (8) and (9) in the range of z-values extending from the upper edge of the viscous sublayer (located at $u_* z/\nu = 70$) to the upper edge of the whole boundary layer above which the homogeneous "free stream" begins. However, if the "free stream" is nonturbulent, then in the "upper sublayer" adjoining the "free stream" another power law is valid which differs from the law (8), (9) valid in the "intermediate layer" located between the viscous and upper sublayers. According to data analyzed in [83], the upper-sublayer velocity profiles have the form:

$$U(z)/u_* = B(u_* z/\nu)^\beta \tag{10}$$

where β is an universal constant which is close to $1/5$, while B takes different values in different experiments.

Let us now consider at greater length the power law of equations (8) and (9) proposed for the intermediate layers (where z takes not too high and not to low values) of flows in round tubes, plane channels and flat-plate boundary layers. The important questions about the declared universality of these equations and the ranges of z and Re values where they are applicable were widely discussed and up to now produce hot-spirited controversies.

Barenblatt *et al.* repeatedly stated that they regard the power law (8) as having the theoretical foundation of the same rigor as the foundation of the logarithmic law (1). I think that this statement is both correct and incorrect (even if the possibility to measure "the degree of rigor" will be accepted). It is true that both laws have no fully rigorous proofs. The results given by dimensional analysis which provided the humanity with so many physical laws of the first-rate importance, are always not completely rigorous for a captious mathematician, since they are based on unproved assumptions about the list of physical parameters really affecting the studied process. It is also true that very often the development of science leads to discovery of new factors which were fully neglected in the past and violate the correctness of laws which earlier seemed to be established forever. Nevertheless, the hypotheses used in the dimensional analysis are of physical character and as a rule are based on clear physical intuition without which a physicist cannot be a good scientist. Just physical base of dimensional arguments implying the logarithmic law (1) made this law long undisputed for great scientists listed above.

Of course, physical intuition may sometimes deceive great scientists too and may be questioned by new discoveries. In particular, the discovery in the second half of the 20th century of great part played in turbulence phenomena by organized vortical structures of various kinds and sizes has changed noticeably the situation. It is clear that such structures may depend on some dimensional physical parameters neglected in the traditional derivation of the logarithmic law and this circumstance can restrict the validity of the laws (1), (1a) and (4) or even make them incorrect. Unfortunately, at present there is not too much information about the organized structures which may affect the mean velocity profiles of steady near-wall turbulent flows. (Recall that Townsend [70] and Bradshaw [71] stressed that studied by them "inactive motions" do not affect the mean velocity; the same statement was also repeated in [72–74] in regard to "attached eddies" of various sizes.) Nevertheless, since not all coherent structures are known well enough and in principle some of them may affect $U(z)$, it is impossible to exclude the possibility that "classical similarity laws" represent only a reasonable first approximations valid only when the influences on the mean velocity $U(z)$ of the length L at $z \ll L$ and of the viscosity ν at $z \gg l_w$ may be neglected. Therefore, found in experiments precise validity of the logarithmic law (1) with universal values of coefficients A and B may be considered as a proof of the negligibility of these influences, while discovered violations of this law or nonuniversality of its coefficients show that there exist some nonnegligible such influences.

However, the power law (8) has only much more general grounds: it is supported by the very wide prevalence of "power laws" and "incomplite similarities" not only in physics but also in many other scientific fields. Many examples of such "incomplete similarities" which include power laws with "anomalous exponents" (which can take arbitrary values), are impressively demonstrated in [79, 80]. It was also correctly stressed there that in many cases where incomplete similarities were reliably detected, they could not be derived rigorously from some mathematical equations since such equations were lacking. Nevertheless, this circumstance does not mean that "incomplite similarities" represent an universal form of the laws of nature which take place everywhere and everywhen. Moreover, while the forms of "self-similar solutions of the first type" are usually determined by the dimensionality arguments with rather high degree of definiteness, "self-similar solutions of the second type" as a rule may have many different forms. Therefore, if even one is sure that such solution exists, this did not determine automatically its precise form which choice requires the use of some supplementary assumptions. At the same time, in many important cases the existence of a "self-similar solution of the second type" is not enough for determination of definite verifiable physical conditions and limits of its validity.

Reverting to the logarithmic velocity-profile law, one must say that at present it seems quite possible that the influence of organized structures of various types, which was neglected in conventional derivations of this law, will require to replace it by some more general incomplete-similarity law. On the other hand, such a possibility do not prove that the laws (1)–(4) in all cases must be considered as being incorrect and inappropriate for any practical use. Of course, the law (8) which contains two unknown functions which may be arbitrarily chosen, allows to get often rather good agreement with the experimental data. Therefore this law is an appropriate candidate for a new version of the velocity-profile equation which will describe the observed profiles $U(z)$ more accurately than the logarithmic law where only two constants may be varied. In [84] and some other papers by Barenblatt and Chorin the overlap-layer approach to derivation of equation (8) was considered; however here it proved to be necessary to add the argument Re to the arguments of functions $f^{(1)}$ and $f^{(2)}$ on the right-hand sides of equations (2) and (3). This addition implies that now the overlap layer equation (1) will be valid where however A and B will be not constants but arbitrary functions of Re. The presence of two arbitrary functions gives too many possibilities to fit these equations to the available experimental data. In [84] it was shown that equations (8) and (9) may be make consistent with the logarithmic law with coefficients A and B dependent on Re, if some consequences of the vanishing-velocity approach of Chorin [84, 117] will be additionally taken into account and some very small (and asymptotically negligible) terms will be omitted. Therefore, the derivation of equations (8) and (9) with the help of the overlap-layer method is possible but such derivation uses some additional assumptions and therefore is less convincing than very elementary Izakson's derivation of the law (1) which is based on the use of more simple (but maybe incorrect) assumptions. Note in this respect, that George and Castillo [98] (this paper will be considered below) also tried to apply the overlap-layer method to a situation where both functions $f^{(1)}$ and $f^{(2)}$ depend additionally on Re, but using some other supplementary assumptions they got quite different form of the overlap-layer velocity profile.

Of course, the derivation of equations (1) and (4) uses some empirical facts too; moreover, the value of coefficients A and B also must be determined here from experimental data. Therefore, it may be said that the logarithmic laws are to some degree of empirical origin. However, it is clear that equations (8) and (9) are empirical to greater degree than the logarithmic law (1). (The possibility to measure "the degree of empiricism" is somewhat vague but the general sense of this expression is rather clear.) In spite of the connection of equations (8) and (9) with Chorin's vanishing-velocity method of physical origin, the empirical part of the arguments leading to these equations remains to be quite considerable. Of course, the empirical laws are often of a great importance and there is always a hope that a purely physical base for such a law will be determined later. In the case of equation (8) subsequent physical arguments maybe will help to determine the strict conditions of its validity. Moreover, if some necessary, or sufficient, conditions of validity will be found for the law (8), they probably will also help to estimate quantitatively its accuracy.

The accuracy estimate is important for equations (8) and (9) since the degree of their agreement with the available experimental data is up to now a point of controversy. Experimental studies of near-wall turbulent flows continue to be popular and recently several such investigations claiming to be quite accurate were carried out but this did not clarify the situation. Here we will only mention often cited recent papers by Zagarola and Smits [85] and Österlund et al. [86] which both stated that their data confirm the validity of the logarithmic law (1) and both gave rise to a controversy.

Zagarola and Smits' measurements were made in the "superpipe" at Princeton University where strongly compressed air was used as a working fluid. The compression decreases the kinematic viscosity ν of air and thus make possible the study of pipe flows in a wide range of very high Reynolds numbers (data used in [85] covered the range $31 \times 10^3 \leq Re \leq 35 \times 10^6$ where Re is based on the average flow velocity U_m and pipe diameter $2R$). The authors found that at $Re > 4 \times 10^5$ logarithmic law (1) (with coefficients

$\kappa = 1/A = 0.436$, $B = 6.15$) was valid for values of z in the range $600l_w < z < 0.07R$. Note that found by Zagarola and Smits limits of the logarithmic layer and the values of "universal coefficients" A and B differ considerably from "traditional estimates" of previous investigators (who usually observed log-law at smaller values of Re). And for the range $60l_w < z < 500l_w$ (or $60l_w < z < 0.15R$ if Re is not great enough), which was earlier always considered as a part of (or even the whole) logarithmic layer, it was found that there at all values of Re the velocity profile $U(z)$ has the power form $U(z)/u_* = 8.7(z/l_w)^{0.137}$ – this result clearly disagrees with all previous pipe-flow data. As to the velocity defect law (3), the authors recommended to replace in it the near-wall velocity scale u_* by the outer velocity scale $U_0 = U_{\max} - U_m$ where $U_{\max}$ is the mean velocity at the pipe axis. According to [85], this replacement makes the function $f^{(2)}$ really independent on Re while at large values of Re it changes nothing since then the ratio U_0/u_* takes constant value.

Österlund et al. [86] summarized results of independent experimental studies of flat- plate boundary layers in two wind tunnels: one at the Royal Institute of Technology in Stockholm and the other at the Illinois Institute of Technology in Chicago. These studies covered the range $2500 < Re < 27\,000$ of Reynolds numbers $Re = U_0\delta^{**}/\nu$ (where U_0 is the free-stream velocity and δ^{**} is the momentum thickness of boundary layer). According to [86], results of both experiments excellently agree with each other and show that in the studied range of Re-values there exists an "overlap layer" where logarithmic laws (1) and (4) are both valid with independent of Re constant coefficients: $\kappa = A^{-1} = 0.38$, $B = 4.1$, $B^{(1)} = 3.6$ (as the external length-scale L now the thickness $\delta = \delta_{95}$ of the layer where $U(z) \leq 0.95U_0$ was used). In the experiments by Österlund et al. the overlap layer corresponded to the conditions: $200l_w < z < 0.15\delta$. Note that values of coefficients of logarithmic laws and of the overlap-layer limits coincide here neither with values found by Zagarola and Smits nor with conventional values of previous authors.

Barenblatt et al. in [76] and some other papers asserted that Princeton data for $Re > 10^6$ contain a systematic error due to neglect of the wall-roughness influence which becomes important at high Re-values, while all the other data of Princeton group agree very well with equations (8) and (9). However Smits and Zagarola rejected in [87] the accusation that the wall roughness affected substantially their data relating to high values of Re and in [88] they disagreed with the assertion that the low-Re Princeton data confirm the validity of equation (8). (According to [88] their data agree with the logarithmic law (1) better than with the power law (8) even in the case where optimal values of functions $C(Re)$ and $\alpha(Re)$ were determined anew by processing of the Princeton, and not Nikuradse's, data.) Answering to [88], Barenblatt and Chorin published comments [89] repudiating the arguments in this paper, and just then Smits and Zagaropa declared in [87] their disagreement with statements presented in [89]. As to the paper [86], Barenblatt *et al.* [83] presented some diagrams obtained by processing of the original data used in [86] and showing that these data agree very well with equations (8) and (9). Later, in the note [90] they tried to show that data processing used in [86] had serious defects while correct processing leads to results supporting conclusions formulated in [83] and [76]. However, the note [90] again did not close the polemic: it caused the comments [91] rejecting the made accusations and presenting a diagram showing that the data used in [86] agree with the logarithmic law (1) not worse (maybe even slightly better) than with the power law (8).

The prolonged controversy on the true form of the turbulent-wall-flow velocity profiles was continued at the 53d Annual Meeting of the APS Division of Fluid Dynamic in Washington, DC (November 19-21, 2000). The Invited Lecture by Chorin there was devoted again to his and Barenblatt's theory of the mean-velocity profiles in turbulent boundary layers. The critical estimation of this theory was reflected in three short talks by Buschmann and Gad-el-Hak, Panton, and Nagib *et al.* [92]. Buschmann and Gad-el-Hak analyzed the experimental and DNS data of mean-velocity measurements or calculations in fully turbulent zero-pressure-gradient boundary layers (with $300 \leq Re \leq 6200$, where again $Re = U_0 \delta^{**}/\nu$) obtained by six independent research groups. These data were compared with the results following from both traditional logarithmic laws and recently proposed power laws. The authors found that the log law and power law both agree well with the data within considerable but somewhat different ranges of z values. The log law becomes to be applicable at lower distances from the wall while the power law continues to have a good accuracy in some part of the boundary layer placed above the "logarithmic layer" where the log law is not valid. However, there is a quite considerable flow zone where both laws agree well with all the available data and have there practically the same accuracy.

Panton's talk [92] (at greater length its contents is described in the informal document [93]) was devoted mainly to studies of the velocity profile of a turbulent pipe flow. Here the traditional overlap-layer arguments were supplemented by corrections taking into account the influence of finite (but high) value of Re. To compute such corrections Panton used the method of matched asymptotic expansion which has many applications to fluid mechanics (see, *e.g.* [94,95] and short discussions of its applications to high-Reynolds-number turbulent flows in the books [96,97] and surveys [62, 63]). Panton considered only the first approximation of this method which he presented in a special form (corresponding to the uniformly valid so-called Poincaré expansion), while the initial profile equation included in his analysis both the log low in the overlap layer and the wake law in a zone adjacent to this layer. Then he showed that the approximation considered by him leads to results describing with a good accuracy numerous experimental and DNA data (including, in particular, the data of papers [85, 86] on the mean velocity and Reynolds-stress profiles $U(z)$ and $\tau(z) = -\langle uw \rangle(z)$). Composite velocity profiles $U(z)$ obtained in a wide range of Re values agreed rather well with the available data and also with the logarithmic law within the traditional "logarithmic layer" of z values (where the use of the conventional coefficients $\kappa = 0.41$ and $B = 5.25$ did not lead in most of the cases to disagreement with the data). Moreover, in the case of pipe flows this profile $U(z)$ agrees also well enough with Barenblatt's equations (8) and (9) but in another range of z values which includes the outer part of the "logarithmic layer" and the inner part of the "wake layer". Since Panton found that these two laws are valid in different regions, he concluded that it is not appropriate to ask which of these two laws is correct. As to the boundary-layer flows, Panton came to conclusion that the method used by Barenblatt *et al.* for the determination of the most appropriate value of $Re = U_0 \Lambda / \nu$ do not lead to values of $C(Re)$ and $\alpha(Re)$ which make equation (8) to agree well with Österlund's experimental data (at greater length this conclusion is considered in the second document [93]).

Finally, in the talk by Nagib *et al.* [92] it was stated that the experiments described in [86] were continued by the present authors in the range of very high values of $Re = U_0 \delta^{**} / \nu$ exceeding 50 000. The new measurements showed that the mean velocity distribution in the overlap layer of the flat-plate boundary layer for these Reynolds numbers continues to be accurately enough described by the Reynolds-number-independent log law with the same unconventional values of the coefficients $\kappa = 0.38$ and $B = 4.1$ as in [86].

What may be said in conclusion of this lengthy many-sided discussion? It shows clearly that advocates of two different similarity models cannot convince each other in the correctness of their point of view. Both side refer to (often the same) experimental data trying to prove to opponents that these data confirm their model. This makes an impression that at present the reached accuracy of the available data on near-wall turbulent velocity profiles is simply insufficient for obtaining of a convincing unique conclusion about the real form of the mean-velocity profile in the intermediate layer of not-too-small and not-too-large values of z. However it seems also that great (and continued to grow) scatter of experimental values for the coefficients A, B and $B^{(1)}$ and for the limits of the logarithmic layer (*cf.*, *e.g.*, the strongly differing results of [85] and [86] which both asserted that their data have high precision), contradicts the idea of an universal overlap layer with logarithmic velocity profile having always the same constant coefficients.

Barenblatt *et al.* [90] remarked in this respect that too low value $\kappa = 0.38$ of von Kármán constant found in [86] contradicts the logarithmic-law universality. Österlund *et al.* [91] in their answer noted that used by them inner (*i.e.* lower) limit of the logarithmic layer corresponded best to their data but was much greater than its "traditional" value; moreover, their data also covered a wider range of high Re values than those used in earlier studies. According to [91], using only the part of their data which corresponded to "traditional" low range of Re values and "traditional" overlap-layer limits, the authors got the usual estimate $\kappa = 0.41$. Does this mean that just the further increase of the used values of Re and of the lower limit of the overlap layer implies still greater value $\kappa \approx 0.44$ found in [85]? In fact, the dependence of the value of κ (and of other coefficients of laws (1) and (4)) on the range of Re-values and limits of the considered "overlap layer" means that either these laws are not universal or the corresponding experimental data are inexact. If the first explanation is true, then the velocity shear dU/dz in the "intermediate layer" of a wall flow depends not only on u_* and z but also on some other physical quantities which must be directly indicated. Note also that the conventional "overlap-layer arguments" do not imply conclusions agreeing satisfactorily with the available data when these arguments are applied not to mean-velocity profiles but to more complicated statistical characteristics of wall turbulence (for more details see the concluding paragraph of this subsection). This remark also decreases the confidence in the universal validity of the uniquely determined logarithmic law for the overlap-layer velocity.

Let us now mention one more group of researchers who independently studied the mean-velocity profiles $U(z)$ in near-wall turbulent flows. This group, headed by George, also modified the traditional "overlap-layer similarity assumptions" and used a more complicated method for analysis of the nonclosed Reynolds equation for the mean velocity $U(z)$ of a turbulent wall flow. The results obtained by them relating to zero-pressure-gradient boundary layers and to pipe (or channel) flows were summarized in papers [98] and [99], respectively. According to the new theory indicated here, Reynolds number strongly affects all flow regions; therefore the argument Re must be again included in the list of arguments of functions $f^{(1)}$ and $f^{(2)}$ on the right-hand sides of the wall and defect laws (2) and (3). This makes impossible the direct determination of the form of functions $f^{(1)}$ and $f^{(2)}$ in the "overlap layer" and requires to use here some supplementary hypotheses. Proposed in [98,99] hypotheses implied that the velocity profile $U(z)$ takes in the intermediate "overlap layer" quite different forms in the cases of boundary-layer flows and flows in pipes and channels: in the first case $U(z)$ satisfies the power-law with respect to the variable $z + a$, and in the second case – the logarithmic law again with respect to $z + a$. (Here a is

an auxiliary parameter describing the vertical shift of the coordinate origin
and taking different values in different wall flows.) We have no possibility
to consider here these rather unexpected results at greater length; note only
that physical intuition (which may be incorrect) makes one to be surprised
by cardinal difference between the near-wall flow structures in boundary-
layer and pipe (or channel) high-Reynolds-number flows. It was also stated
in [98, 99] that found there results agree satisfactorily with the available
experimental data. (This statement was confirmed also by Had-el-Hak [7]
who found that results of [98] "are elegant".) The found agreement of quite
different velocity-profile equations with the same data shows once more that
at present the accuracy of the existing data does not permit to determine
reliably the true forms of wall-flow velocity profiles.

Completing the discussion of the present situation concerning the choice
of the most appropriate theoretical equation for the velocity profiles $U(z)$
of steady turbulent wall-bounded flows one must say that at present there
is no equation which will satisfy everybody and will be unanimously recognized as the best one. From this point of view, the situation now is
even worse than it was up to the 1980s when the discovery of the logarithmic velocity-profile equation was unanimously considered as one of the most
fundamental scientific achievements of the 20th century which solved forever
the problem about the form of velocity profile in turbulent wall flows. Now
it seems clear that the accuracy of the available experimental and numerical
data is insufficient for the determination of the unique "correct solution"
of the problem. At the same time, the great scatter of the found values of
logarithmic-law parameters and limits of its validity makes one to suppose
that this law represents only a reasonable first approximation which may
be useful for engineering practice but cannot be considered as a rigorously
established physical law. Therefore, the old velocity-profile problem which
tortured Prandtl, Taylor and von Kármán in the first quarter of the 20th
century, now again became actual and apparently requires supplementary
studies of physical mechanisms leading to possible violations of the logarithmic law and to reliably detected violations of related similarity laws for
higher-order statistical characteristics of wall-bounded turbulent flows.

Before the appearance of much more accurate experimental (and/or
DNS) data (and even after it too), better understanding of the main factors determining the form the velocity profiles in various turbulent flows
undoubtedly requires more direct use of the physical arguments concerning the mechanisms of turbulent mixing. This is an arduous task: physics
of turbulence phenomena is very complicated and even mysterious up to
now, dynamic equations are nonclosed and requiring additional hypotheses.
Therefore it is not surprising that all approaches discussed above did not
use the Navier–Stokes equations of fluid dynamics at all. For this reason the

attempt by Nazarenko with coworkers [100] to consider some simplified physical mechanisms producing the near-wall turbulence with logarithmic (or power-law) mean-velocity profiles is worth to be mentioned. These authors studied near-wall turbulence produced by a weak small-scale external forcing. They found that the mean velocity profile of such forced turbulence is very sensitive to the properties of the initial near-wall vorticity penetrating into the outer flow regions. For the case of a simplified dynamic model derived from NS equations the authors found specific conditions guaranteeing the existence of an exact analytic solution of model equations corresponding to the logarithmic (or to power-law) velocity profile. Thus here for the first time it was shown that sometimes these two types of velocity profiles may be obtained under definite conditions from dynamic equations derived from the NS equations. Results of this work in fact stressed again that classical derivations of logarithmic law by Prandtl, Kármán, Izakson, and Millikan in no way can be considered as the conclusive solution of the problem of the velocity profile of near-wall turbulent flows. Such derivations must be also supported by careful physical analysis based on dynamic equations which maybe will explain the interrelation between the power-law and logarithmic velocity profiles.

Above only the mean-velocity profiles $U(z)$ of the near-wall turbulent flows were considered. However any turbulent flow in addition to mean-velocity profile has also a lot of "statistical characteristics of higher orders" such as higher moments, correlation and structure functions, spectra of fluid-dynamic fields, probability density functions (pdf) of turbulent fluctuations and so on. All these characteristics are peculiar just to given flow and knowledge of many of them may be necessary for solution of some important practical problems. However up to now the higher-order statistical characteristics of wall turbulent flows are poorly known since relating to them experimental data are either missing or are very scattered and unreliable. Moreover, the applications of the "standard dimensional arguments" of wall-turbulence theory to the higher-order flow characteristics usually lead to results which agree with the available data much worse than results relating to mean-velocity profiles. Recall that the first violations of the "classical similarity laws" for the "overlap layer" of near-wall turbulence which were detected by Townsend [70] and Bradshaw [71] (and confirmed by Perry and Li [74]) concerned not the mean-velocity profile but profiles of the second-order moments $\langle u^2 \rangle$ and $\langle v^2 \rangle$. Since the mentioned here similarity laws were based on the same seemingly obvious dimensional arguments which imply the logarithmic velocity-profile law, the discovery of their violations is very important for future studies of real properties of near-wall turbulence. It has been already mentioned above that Fernholz and Finley noted in the review [61] that the available mean-velocity data for zero-pressure-gradient boundary layers agree quite satisfactorily with the logarithmic laws (1), (4) (and more general laws (2) and (3)) but the data relating to higher moments of velocity fluctuations are very scattered and disorderly. Note that, nevertheless, in early reviews [62, 63] an attempt was made to collect some preliminary (not too reliable) data relating to functions $f_{klm}(zu_*/\nu)$, $g_{klm}(z/L)$ and constants a_{klm} for the cases where $k + l + m = 2$. In particular, it was stated there that apparently $a_{200} \approx 5.5$, $a_{020} \approx 3$, $a_{200} \approx 1$, while $a_{101} = -1$, $a_{110} = a_{011} = 0$. However, later it was stressed in [101] that in fact much data disagree with these estimates (and with general Eqs. (5–7) too). As an example the atmospheric data by Högström [102] and Smedman [103] were presented in [101] which show that in the near-earth logarithmic layer of the atmosphere $\langle u^2 \rangle^{1/2}/u_*$ often decreases and $\langle w^2 \rangle^{1/2}/u_*$ increases with height in direct contradiction to equation (7). Many more recent data relating to various higher-order statistical characteristics of near-wall laboratory or atmospheric turbulence may be found, *e.g.*, in the papers [104–108]. These data show that similarity laws (5–7) (and similarity laws of the same type corresponding to other characteristics of near-wall turbulence) often disagree with the experimental data or, in the best case, may be considered only as some rough approximations. (In particular, the dependence of statistical characteristics of turbulence on the value of Re was often observed in both the inner, near-wall, and the outer flow regions.) Therefore the search for similarity laws adequately describing higher-order statistical characteristics of wall turbulent flows represents a very difficult problem requiring much further work.

3.2   Kolmogorov's theory of locally isotropic turbulence

Kolmogorov's theory of 1941 (so-called K41 theory, or briefly K41) was first stated in two short notes (of 4 and 3 pages) in "Doklady Akad. Nauk SSSR" ("Reports of USSR Acad. Sci."). These notes undoubtedly represented one of the highest achievements of the theory of turbulence which, luckily, became very early known in the West. (Up to 1946 Russian "Doklady" were simultaneously published under the title "C. R. Acad. Sci. URSS" in translations to one of three main Western languages. One day in the early 1940s young Cambridge student Batchelor by chance found these "C. R." in the London library, read Kolmogorov's notes, at once understood their enormous importance and became an urgent popularizer of this work.) So, seven printed pages glorified Kolmogorov as the brilliant physicists and mechanicians, while earlier he was known only as a famous mathematician. (In fact K41 was the unique achievement in the field of turbulence which was seriously discussed as a work worth the Nobel prize in physics, and probably Kolmogorov would get the Nobel prize if he did not die too early.)

Kolmogorov's theory was based on very clear and convincing physical ideas represented in the form of two hypotheses concerning the mechanisms producing the small-scale turbulent fluctuations. When this theory was developed by Kolmogorov, there were no experimental data to compare with conclusions following from his theory; all of them have the character of pure predictions. Only later numerous experiments confirmed the perfect validity (with the then attainable accuracy) of the main results of Kolmogorov's theory (see, *e.g.*, the books [2, 67, 96]). Let us stress, however, that the K41 theory did not use at all the dynamic NS equations. In fact, here only intuitive physical reasons were used where the principal part was again played by dimensional arguments. Physical intuition prompted Kolmogorov the idea that the small-scale turbulence fluctuations are produced by a cascade process of energy transfer from the mean flow and the large flow structures to more and more smaller such structures. If so, then it was natural to assume that in the case of very high Reynolds numbers, where cascade process includes many steps, this process must make the small-scale turbulence (corresponding to distances r much smaller than the typical length L of the large-scale flow nonhomogeneities) locally homogeneous, isotropic and depending, in the case of incompressible fluid, only on two dimensional physical parameters. These two parameters are the mean rate ε of the energy transfer over the cascade of eddies (which must be equal to the mean rate of viscous dissipation of the kinetic energy of velocity fluctuations) and the kinematic viscosity of fluid ν. And dependence of only two parameters allows to use dimensional analysis very effectively. In particular, dimensional considerations imply the following result

$$E_{11}(k) = A\varepsilon^{2/3}k^{-5/3}\phi(k\eta), \quad \text{where} \quad \eta = (\nu^3/\varepsilon)^{1/4}, \ \phi(0) = 1, \quad (11)$$

$E_{11}(k)$ is the one-dimensional spatial *spectrum* of the streamwise velocity fluctuations, k – the streamwise wave number, η – Kolmogorov's length scale (which has been already met above when the range of length scales of vortical structures was discussed), and A and ϕ are some universal constant and function. Equation (11) is valid in flows with large values of Re for $k \gg 1/L$ (since only such values of k correspond to small-scale turbulence) and it follows from this equation that in the *inertial range* $1/L \ll k \ll 1/\eta$ of wave numbers k spectrum $E_{11}(k)$ has the following simple form:

$$E_{11}(k) = A\varepsilon^{2/3}k^{-5/3} = Bk^{-5/3}, \quad \text{where} \quad B = A\varepsilon^{-5/3}. \quad (11a)$$

Equation (11a) represents the famous *five-thirds law* determining the form of the velocity spectrum in the inertial range of wave numbers; this law is one of the most important conclusions following from K41 theory.

First attempts of experimental checking of K41 theory led to confirmation of theoretical predictions; in particular, it was found that velocity spectra of atmospheric turbulence (where Re always takes very high value) are almost always proportional to $k^{-5/3}$ in a wide range of wave numbers. However in the late 1950s the researchers working at Moscow Institute of Atmospheric Physics noted that nevertheless some results of their measurements disagree with original Kolmogorov predictions. The first found disagreement concerned the coefficient B of equation (11a). According to K41, at a fixed point of a steady turbulent flow coefficient B must have a constant value. However, real measurements at fixed points of the Earth's atmosphere showed that B fluctuates very strongly – a new spectral measurement made slightly later (say, after 15–20 min) gave again a spectrum of the form (11a) but coefficient B often took then quite different value.

This observation led to formulation by Obukhov and Kolmogorov in 1962 of a new, modified, theory of small-scale turbulence, which is now often called the K62 theory (for more details see [2] or [67], Sect. 25). The main idea of it consists in the replacement of the mean dissipation rate ε by the spatially averaged local dissipation rate ε_r. Here $r = 2\pi/k$ is the wave length corresponding to wave number k, and ε_r is obtained by averaging of the local energy dissipation rate $\varepsilon(\mathbf{x}, t)$ over a spherical volume of points $\mathbf{x}$ having the radius $r/2$ and the center at the point to which the considered spectrum $E_{11}(k)$ corresponds.

Let us consider not the one-dimensional spectrum $E_{11}(k)$ but more simple velocity *structure function* of the second order:

$$D_2(r) \equiv \langle [u_1(\mathbf{x} + \mathbf{r}) - u_1(\mathbf{x})]^2 \rangle, \qquad r = |\mathbf{r}| \qquad (12)$$

(here u_1 is velocity component in the direction of vector $\mathbf{r}$ and, as usual, angular brackets denote ensemble averaging). Then, according to K41 for $r \ll L$

$$D_2(r) = C\varepsilon^{2/3} r^{2/3} f_2(r/\eta), \qquad (13)$$

where f_2 is an universal function, $f_2(\infty) = 1$, and $C \approx 4A$ is an universal constant. From (13) it follows that in the inertial range $L \gg r \gg \eta$ of distances r Kolmogorov's *two thirds law* of the form

$$D_2(r) = C\varepsilon^{2/3} r^{2/3} \qquad (13a)$$

is valid. On the other hand, according to K62 theory for $r \ll L$

$$D_2(r) = C\langle (\varepsilon_r)^{2/3} \rangle r^{2/3} f_2(r/\eta_r), \quad \text{where} \quad \eta_r = \nu^{3/4} (\varepsilon_r)^{-1/4}. \qquad (14)$$

In the inertial range $L \gg r \gg \eta_r$ (the length η_r fluctuates but usually it is of the same order as η) $f_2(r/\eta_r) = 1$, and hence

$$D_2(r) = C\langle (\varepsilon_r)^{2/3} \rangle r^{2/3}, \qquad \text{where} \quad \langle (\varepsilon_r)^{2/3} \rangle \neq \langle \varepsilon_r \rangle^{2/3} = \varepsilon^{2/3}. \qquad (15)$$

According to equation (15) dimensional coefficient $D = C\langle(\varepsilon_r)^{2/3}\rangle$ of the two-thirds law may fluctuate producing variations of the value of the dimensionless coefficient $C_0 = D/\varepsilon^{2/3}$ (where ε is strictly constant "mean dissipation rate"). The same arguments may explain the observed variability of the coefficient B of the law (11a).

In his work of 1962 Obukhov assumed that ε_r has lognormal probability distribution with variance depending on r and used this model for a crude estimation of $\langle(\varepsilon_r)^{2/3}\rangle$. Kolmogorov in his version of K62 theory, sketched some general similarity hypotheses which generalized the hypotheses used in K41 (namely, instead of the local isotropy of the velocity field $\mathbf{v}(\mathbf{x}, t)$ assumed in K41 he suggested the assumption that the probability distributions of the ratios of velocity differences in two pairs of points are invariant with respect to all motions and mirror reflections of this group of points). However, this last hypothesis was never developed to a state of a completed theory. Moreover, Kolmogorov also proposed to use Obukhov's lognormal assumption not only in equation (15) but also in the more general equation for the structure function $D_n(r)$ of the arbitrary order n (defined by presented below Eq. (16)). This proposition implied the following approximate estimate of the form of the velocity structure functions of arbitrary orders in the inertial range:

$$D_n(r) \equiv \langle[u_1(\mathbf{x} + \mathbf{r}) - u_1(\mathbf{x})]^n\rangle = C_n(\mathbf{x})(\varepsilon r)^{n/3}(L/r)^{\mu n(n-3)/18}. \qquad (16)$$

Here $\varepsilon = \langle\varepsilon_r\rangle$ is the mean rate of the energy dissipation, μ is an universal constant, and $C_n(\mathbf{x})$ depends on the flow macrostructure (and is practically constant in regions of a size much smaller than L). Old K41 theory corresponds to the case where C_n are universal constants and $\mu = 0$; note also for $n = 3$ both theories imply the same result.

At present it is clear that the lognormal assumption accepted in 1962 by both Kolmogorov and Obukhov was only a crude approximation. (In fact both authors also considered it as only an example allowing to illustrate the possible influence of the dissipation-rate intermittency on the inertial-range spectra and structure functions.) After 1962 a number of attempts were made by different authors to replace this assumption by some more general model of the self-similar cascade process of sequential breakdown of smaller and smaller eddies (the early stage of this development was summarized in Sect. 25 of the book [67]; see also [2]). From all this material only the result due to Novikov [109] will be presented here. Novikov considered three spatial volumes similar to each other (for definiteness let us assume that they are spherical) of radii $\rho < r < R$ contained within each other and corresponding to them three averaged dissipation rates ε_ρ, ε_r and ε_R (which are fluctuating random variables). He postulated that self-similarity of the cascade breakdown process is represented by the fact that if all three radii ρ,

r, and R belong to the inertial range of lengths, then the random ratios $\varepsilon_\rho/\varepsilon_r$ and $\varepsilon_r/\varepsilon_R$ are statistically independent from each other and have probability distributions depending only on ratios ρ/r and r/R, respectively. Then he showed that from such self-similarity it follows that in the inertial range of distances r

$$D_n(r) = C_n(\varepsilon r)^{n/3}\left(\frac{r}{L}\right)^{\xi_n} \propto r^{\zeta_n}, \qquad \zeta_n = n/3 + \xi_n. \qquad (17)$$

A number of measured in various turbulent flows or determined from numerical simulations values of *scaling exponents* ζ_n corresponding to different values of n was found during the 1980s and 1990s, in particular, by F. Anselmet *et al.* (*J. Fluid Mech.* **140** (1984) 60-89), R. Benzi *et al.* (*Phys. Rev.* **E 48** (1993) R29-R32), G. Stolovitzky *et al.* (*Phys. Rev.* **E 48** (1993) R3217-R3220), and J.A. Herweijer and W. van de Water (*Phys. Rev. Lett.* **14** (1995) 4651-4654). The first analytical models of the scaling-exponent function $\zeta_n = \zeta(n)$ was proposed by Kolmogorov in 1962 (see Eq. (16)); its agreement with the subsequently found values of the exponents ζ_n proved to be quite poor. Note that according to equation (16) ξ_2 is positive and apparently small (μ is positive by definition but hardly large), $\xi_3 = 0$, and ξ_n are negative for $n > 3$ and $|\xi_n|$ grow very quickly with n. The available data shows that the signs of corrections ξ_n were predicted by equation (16) correctly (but ξ_2 is so small, that it is sometimes assumed to be zero), but for higher-order corrections with $n > 3$ values of $|\xi_n|$ are always much smaller than they must be according to equation (16).

Later many other "theoretical models" of scaling exponents corresponding to various particular self-similar models of the cascade process of eddy breakdowns were given by a number of authors; the papers by U. Frisch *et al.* (*J. Fluid Mech.* **87** (1978) 719-736), R. Benzi *et al.* (see G. Paladin and A. Vulpiani, *Phys. Rev. A* **35** (1987) 1971-1973), S. Kida (*J. Phys. Soc. Jpn* **60** (1990) 5-8), Z.-S. She and E. Lévêque (*Phys. Rev. Lett.* **72** (1994) 336-339), B. Dubrulle (*Phys. Rev. Lett.* **73** (1994) 959-962), Z.-S. She (*Progr. Theor. Phys. Suppl.* **130** (1998) 87-102), J. Jiménez (*J. Fluid Mech.* **409** (2000) 99-120), the book [2] and short survey by O.N. Bortav (*Phys. Fluids* **9** (1997) 1206-1208) represent only a small part of the material relating to this topic. Many of the proposed quite different analytic models led to results which agreed more or less satisfactorily with available experimental and numerical estimates of the exponents ζ_n, if the model parameters were appropriately chosen. This agreement shows again that up to now available data on high-Reynolds-number turbulence very often do not allow to select uniquely the best of the various proposed theoretical models.

Let us now made some general comments. The K41 theory was based on definite hypotheses which were not (and apparently cannot be) proved rigorously (*i.e.*, derived directly from equations of fluid mechanics). However, these hypotheses seemed, at least, to be quite natural and consistent with physical intuition. In contrast, the reformulation by Obukhov and Kolmogorov of K41 theory as a new K62 theory is far less evident and physically convincing. Of course, the Kolmogorov-Obukhov's attempt of crude estimation of the intermittency effect with the help of replacement of the constant dissipation rate ε by depending on the length r fluctuating characteristic ε_r was a brilliant piece of work, but it was based on a plausible guess only and could not be considered as an adequate physical theory. Therefore it was only natural that at the end of his paper of 1962 Kolmogorov set up a problem of elimination of the quantity ε_r from K62 theory and proposed to use for this purpose two new similarity hypotheses remarking simultaneously that apparently they must be also supplemented by something else. However, the realization of this program is clearly a difficult task and this was not done yet. A partial progress was connected with the appearance of the multifractal formalism of Parisi and Frisch (see [2] about it) where ε_r was not mentioned explicitly. However this formalism represents some idealization of the real situation and it requires the introduction of some supplementary hypotheses.

A modification of the K41 theory differing from K62 theory was proposed by Barenblatt and his co-authors (see, *e.g.* [110–112]). In the paper [110] with Goldenfeld based on some general arguments and the analogy with the problems concerning the near-wall velocity profile and some physical problems of quite different origin the authors assumed that maybe more appropriate correction of the classical two-thirds law (15) of K41 than that of equation (17) with $n = 2$, will be given by an equation of the form

$$D_2(r) = C(\ln Re)(\varepsilon r)^{2/3} \left(\frac{r}{L}\right)^{\alpha(\ln Re)} \tag{18}$$

where L has the same meaning as in equation (17) but now coefficient $C_2 = C$ and exponent $\xi_2 = \alpha$ are not constants but functions of $\ln Re$ (*i.e.*, slowly changing functions of Re). Expanding these functions in powers of a small parameter $(\ln Re)^{-1}$, the authors assumed that $\alpha(Re) = \alpha_1/\ln Re + O[(\ln Re)^{-2}]$, $C(\ln Re) = C_0 + C_1/\ln Re + O[(\ln Re)^{-2}]$ (constant term was omitted in the series for $\alpha(Re)$ to guarantee the validity of the K41 scaling when $Re \to \infty$). For crude estimate of the function $C(\ln Re)$ the data by Praskovsky and Onsley [113] were used. These authors combined results of spectral measurements of velocity fluctuations in the atmospheric surface layer and in two high-Reynolds-number wind-tunnel flows to verify the possibility of dependence of the Kolmogorov constant $C = C_2$ on the value of the Reynolds number $Re_\lambda = u'\lambda/\nu$ (where $u' = \langle u^2\rangle^{1/2}$ is the root-mean-square value of the streamwise, corresponding to Ox direction, velocity fluctuation and $\lambda = [\langle u^2\rangle/\langle(\partial u/\partial x)^2\rangle]^{1/2}$ is the so-called Taylor length microscale). According to [113] values of the coefficient C in eight flows with $2 \times 10^3 \le Re_\lambda \le 12.7 \times 10^3$ are weakly decreasing with Re_λ (approximately as $(Re_\lambda)^{-0.1}$). This dependence on Re differs from that assumed by Barenblatt and Goldenfeld. However, since the results

of [113] had low precision (note that the summary tables of the measured
C-values collected in [114, 115] showed that these values are very scattered
but gave no indications of their dependence on Re), it was concluded in [110]
that these results may be also crudely approximated by some equation for
$C(\ln Re)$ proposed in this paper. As such approximations even two version
of proposed in [110] equation were considered: one with $C_0 = 0$ and the
other with $C_0 \neq 0$. Note that if $C_0 \neq 0$, then equation (18) implies that at
$Re \to \infty$ limiting regime of "fully developed turbulence" is realized where
Kolmogorov's "two-thirds law" is valid, while if $C_0 = 0$, then such regime
do not exist.

Later Barenblatt and Chorin [84, 111, 112] generalized equation (18) and
given above approximate models of the functions $\alpha(Re)$ and $C(Re)$ to the
case of the velocity structure functions $D_n(r)$ of orders $n \geq 4$, suggesting the
following approximate equation for values of these functions in the inertial
range $\eta \ll r \ll L$:

$$D_n(r) = \left(C_n + C_n^1 / \ln Re\right) (\varepsilon r)^{n/3} \left(\frac{r}{L}\right)^{\alpha_n / \ln Re}, \qquad n = 4, 5, ..., \qquad (19)$$

where C_n, C_n^1 and α_n are some constants. (For $n = 2$ proposed in [110]
equation of the same form as (19) was used; as to the case where $n = 3$,
here the known Kolmogorov's equation $D_3(r) = -(4/5)\varepsilon r$ was used in the
inertial range of lengths r.)

Equations (18) and (19) correspond to definite concept of the passage to
the zero-viscosity limit in fluid mechanics (see, *e.g.* [111, 112]). Recall that
according to the K62 small-scale spatial intermittency of the field $\varepsilon(\mathbf{x}, t)$
leads to the appearance of small (but finite) changes of "classical" spectral
and structure-function exponents $-5/3$ and $2/3$. (These changes have the
same absolute value but opposite signs: they diminish the spectral exponent
but increase the structure-function exponent.) At the same time intermittency also produces changes of the form (17) of exponents describing the
forms of structure functions of higher orders in the inertial range of lengths
r. This prediction of K62 was widely discussed during the last two decades
(see, *e.g.*, papers cited after Eq. (17)). However it was also sometimes
contested (*e.g.*, in [116, 117]), and equations (19) (and similar equation for
$n = 2$) also corresponds to the assumption that "intermittency corrections"
of the inertial-range exponents tend to zero as $Re \to \infty$. (Just the accep-
tance of this assumption forced the authors to require that in equation (18)
$\alpha(Re) \to 0$ as $Re \to \infty$.) Since the available experimental and numerical estimates of "intermittency corrections" are scattered and small, the reliable
verification of equations (18) and (19) is apparently impossible at present.
Let us consider, for example, the situation relating to the "intermittency
correction" ξ_2 corresponding to the second-order structure function $D_2(r)$.
The first experimental estimate of ξ_2 given in [118] was close to 0.04, while

at present the available non-zero estimates cover the range from 0.05 to 0.02, but zero value is also sometimes accepted. (In particular, Praskovsky and Onsley [113] found that ξ_2 is close to zero at all inspected by them values of Re_λ, and there are also other authors who supposed that the available data are insufficient for proving that $\xi_2 \neq 0$.) Barenblatt *et al.* [119] tried to use for the verification of their assumption about the dependence of $\alpha = \xi_2$ on Re the data by Benzi *et al.* [120] who measured the values of functions $D_2(r)$ and $D_3(r)$ in four different flows with $Re = 5000, 6000, 18\,000$, and $300\,000$ (where different definitions of Re were used for different flows). In [120] it was found that to the summary collection of all obtained data corresponded the practically constant correction $\xi_2 \approx 0.03$. Barenblatt *et al.* separated data points corresponding to individual experiments and their processing of four separate (rather small) groups of points led to conclusion that the corrections ξ_2 differ in the cases of different experiments decreasing with the growth of Re and possibly tending to zero as $Re \to \infty$. However, Benzi *et al.* in their reply [121] to the note [119] disagreed with such interpretation of their data. At the beginning they rightly noted that from a theoretical point of view, the dependence of the exponent ξ_2 on Re and its convergence to zero as $Re \to \infty$ does not seem impossible. However then they stated that their experimental data, and also analyzed by them additional data of some other authors covering a larger range of high Re values, show that $\xi_2 \approx 0.03$ in all studied flows and it does not change with the increase of Re. Moreover, it was also noted in [121] that according to data presented in [122] the higher-order scaling exponents ξ_n with $n \leq 7$ also do not depend on Re. (In [122] an attempt was made to collect results of approximate evaluations of values of ξ_n, $n \leq 7$, based on data of seven experiments corresponding to quite different turbulent flows and values of Re_λ between 300 and 5000.)

Of course, the experimental results presented in [120, 122] cannot be considered as a strict proof of the independence of scaling exponents ξ_n and $\zeta_n = \xi_n + n/3$ on Re. All the measurements of these exponents are rather crude and their results may depend on the choice of the "inertial range" where the structure functions satisfy the power laws. Note also that in [120–122] the scaling exponents were determined indirectly basing on the "extended self-similarity" (ESS) hypothesis by Benzi *et al.* [120] generalizing the concept of the inertial range where structure functions $D_n(r)$ satisfy power laws (17). Equation (17) implies that in the inertial range any function $D_n(r)$ is proportional to the function $D_m(r)$ raised to the power ζ_n/ζ_m. ESS stated that the proportionality of $D_n(r)$ to $[D_m(r)]^{\zeta_n/\zeta_m}$ is often valid over an unexpectedly wide range of scales r extending far beyond the small-scale limit of the inertial range. (In practical applications it is usually assumed that $m = 3$; then $\zeta_m = 1$ and within the inertial range the ESS representation is equivalent to that of Eq. (17).) The use of the ESS method allows to simplify and make more easy the determination of exponents ζ_n from the experimental data, but in principle values of ζ_n found by this method may be somewhat affected by the extension of the range of r values considered. However, even more important is the absence of any explanation of the ESS phenomenon. ESS clearly represents a surprising similarity property which must be somehow connected to similarity of organized structures determining the shapes of structure functions in the covered by ESS range of lengths r. This generalization of the following from K62 equation (17) may be compared with proposed by Barenblatt, Chorin and Goldenfeld equations (18) and (19) which validity also must reflect some unknown symmetry features of flow structures determining the velocity differences. Moreover, equation (17) by itself is also a similarity relation the derivation of which in the framework of K62 is based on the use of some unproved and physically somewhat vague hypotheses. Therefore it is not surprising that Sreenivasan and Dhruva [123] even tried to investigate whether the scaling (17) really exists in high-Reynolds-number turbulence or not. Their measurements in the atmospheric surface layer at $10^4 \leq Re_\lambda \leq 2 \times 10^4$ led them to the conclusion that apparently in atmospheric turbulence there exists an inertial range where equation (17) is valid but its validity is often disturbed by velocity shear and finiteness of Re (see also the discussion of the results of the paper [128] below). However, the paper [123] did not clarify the origin of the similarity law (17).

One more generalization of the K62 scaling (17) for the case of $n = 2$ was proposed by Gamard and George [124]. According to their theory the scaling exponent ξ_2 and Kolmogorov's coefficient $C = C_2$ depend on the Reynolds number Re and ξ_2 tends to zero while C tends to a non-zero constant C_0 as $Re \to \infty$. Thus, this theory stated that the "classical" turbulent regime of K41 theory is valid in the limiting case of very high Reynolds numbers. The authors applied to the problem considered by them hypotheses of the same type as those used in the papers [98,99] for the evaluation of velocity profiles in turbulent pipe, channel and boundary-layer flows. Results obtained in [124] results proved to be in good agreement with the experimental results by Mydlarski and Warhaft [125] relating to spectral measurements in the isotropic turbulent flow produced in a relatively small wind tunnel by an "active grid" generating intensive turbulent fluctuations. The data by Mydlarski and Warhaft corresponded to a limited range of not too large Reynolds numbers; therefore even the existence here of the intermediate range of wave numbers k where $E_{11}(k) \propto k^{-\alpha}$, $\alpha > 0$, was somewhat unexpected. Note also that in this case the found corrections which must be added to the "Kolmogorov exponent" $-5/3$ prove to be positive while according to K62 the intermittency corrections of the spectral exponent are always negative (equal to $-\xi_2$). For this reason the results of this work cannot be compared with the results discussed above which were relating to flows with much higher values of Re.

The present state of the considered above investigations of the K41 theory and of the similarity laws for near-wall turbulent flows, produces an impression that at the end of the 20th century the fundamental achievements of Prandtl, Kármán, Kolmogorov and other giants laying, seemingly for ever, the foundations of the modern theory of turbulence, began to stagger producing doubts and the feeling of uncertainty. Thus, at present the theory of turbulence seems to be more neglected than it was in the middle of the 20th century when the great discoveries of the 1930s, 1940s and 1950s produced universal enthusiasm. Let us nevertheless hope that arising difficulties will be get over and will lead to great progress in understanding of turbulence phenomena in the initial part of the 21st century.

4   Concluding remarks; possible role of Navier–Stokes equations

It has been already stressed above that both the theory of logarithmic layer of wall-bounded fully turbulent flows developed by Kármán, Prandtl, Izakson, and Millikan in the 1930s and Kolmogorov's K41 theory of locally-isotropic turbulence were based on some seemingly plausible physical hypotheses and dimensionality consideration, while the exact NS equations of fluid dynamics were not used there at all. Both these theories were shortly after their appearance confirmed by seemingly faultless experimental data, became very popular and were unanimously accepted as a final truth by scientific community. It is worth noting that physical basis of the K41 theory at first stimulated enthusiasm only within the community of physicists, while many fluid mecanicians were much in doubt. The closeness of this theory to physical manner of thinking was reflected in a remarkable fact that some results of this theory were later independently obtain also by two famous physicists, both the Nobel-prize winners, namely by Onsager (in 1945) and Heisenberg (in 1947). Moreover, Kolmogorov's theory was first included in textbooks also by famous physicists – in courses of the continuum mechanics written by Landau in Russia (then USSR) and by Sommerfeld in Germany as parts of the general courses of theoretical physics in many volumes. However later the K41 theory was accepted by everybody and became an important part of modern fluid mechanics.

When it was found in the late 1950s that some of the results of K41 disagree with the data of spectral measurements in the lower atmosphere, Obukhov and Kolmogorov developed a modified K62 theory. As it was told above, this new theory included some description of the influence of the external length scale L (equal to the typical length of large-scale flow non-homogeneities) on the small-scale turbulence but preserved the assumption about the spatial homogeneity and isotropy of turbulence within small spatial regions of diameters $l \ll L$. This assumption was also left inviolable

in the subsequent modifications of K62 by Barenblatt *et al.* and by some
other authors and in numerous studies of cascade models of small-scale intermittency and of scaling exponents (the careful studies [126] of anisotropic
contributions to structure functions of various orders and to their scaling
laws were rare exceptions in this respect). However, now there is a lot of data
showing that the fundamental Kolmogorov's assumption about the isotropy
of turbulent fluctuations of scales $l \ll L$ in any high-Reynolds-number flow
is quite often violated.

Let us note, for example, that the local isotropy implies that the cospec-
tra $E_{ij}(k)$ of velocity components u_i and u_j, where $i \neq j$, must vanish in the
inertial range of wave numbers, *i.e.*, at $|k| \gg 2\pi/L$. However in the lower
atmosphere, where Re takes very high value, the cospectrum $E_{13}(k)$ of the
horizontal (in the mean-wind direction) and vertical wind components always takes non-zero values in the range of values of k where spectra $E_{11}(k)$
and $E_{33}(k)$ are proportional to $k^{-5/3}$. (Cospectrum $E_{13}(k)$ decreases in this
range approximately as $k^{-7/3}$, *i.e.* faster than spectra $E_{11}(k)$ and $E_{33}(k)$
but not fast enough to become negligibly small; see, *e.g.* [127].) The simultaneous validity of K41 theory for $E_{11}(k)$ and $E_{33}(k)$ and non-validity for
$E_{13}(k)$ requires special explanation which is lacking up to now.

In addition to this, Shen and Warhaft [128] measured recently a number
of small-scale characteristics of velocity fluctuations in a homogeneous shear
flow (with constant shear dU/dz where U is the mean velocity) behind an
active grid. These measurements covered the range $100 \leq Re_\lambda \leq 1100$
of high enough Reynolds numbers Re_λ. For the normalized moments of
streamwise-velocity derivative $\partial u/\partial z$

$$S_{2m+1} = \langle (\partial u/\partial z)^{2m+1} \rangle \left[\langle (\partial u/\partial z)^2 \rangle^{(2m+1)/2} \right]^{-1} \tag{20}$$

they found that S_3 is decreasing with Re_λ (and possibly tends to zero as
$Re_\lambda \to \infty$), while S_5 does not decrease with Re_λ (and is close to 10 at $Re_\lambda \approx$
1000), while S_7 increases with Re_λ. These results clearly show that studied
turbulence is not locally isotropic in the dissipation range of lengths (since
at local isotropy all moments S_n of odd orders n must vanish). At the same
time it was found that lateral structure functions $D_n(r) = \langle [u(x, y, z+r) -
u(x, y, z)]^n \rangle$ of odd orders $n = 3, 5$, and 7 take non-zero values in the inertial
range of lengths r (*i.e.*, for $\eta \ll r \ll L$); hence the studied homogeneous-
shear-flow turbulence is anisotropic also in the inertial range of lengths.
Thus, results of [128] show that the shear-flow turbulence is locally non-
isotropic, at least up to $Re_\lambda \approx 1000$, and demonstrates no tendency to
become isotropic at higher values of Re_λ. Here again the question arises
how the discovered local anisotropy can be combined with the validity of the
ordinary laws of two and five thirds which was confirmed by data relating
to very different high-Reynolds-number shear flows.

Strong deviations from the predictions of K41 theory were in fact first detected in studies of small-scale fluctuations of temperature (or other passive scalars) in high-Reynolds- number turbulent flows[2]. In particular, at the end of the 1960s it was discovered that the skewness of temperature derivative $S_T = \langle (dT/dx)^3 \rangle / [\langle (dT/dx)^2 \rangle]^{3/2}$ is different from zero (being of the order of 1) in the atmospheric flows with very high values of Re although for locally- isotropic temperature fluctuations $S_T = 0$; see, $e.g.$ [129]. (This excellent survey of the modern studies of passive-scalar fluctuations in turbulent flows contains a long list of references. This fact allows us to omit here all references to papers on this subject, with the exception of very recent papers [130] appearing after the publication of [129].) Later it was found that S_T practically does not depend on Re, $i.e.$ it takes rather high values in all flows. Moreover, also the structure functions of temperature

$$D_{T,n}(r) = \langle [T(\mathbf{x} + \mathbf{r}) - T(\mathbf{x})]^n \rangle, \quad r = |\mathbf{r}|, \tag{21}$$

of odd orders $n = 2m+1$ were found to be different from zero, though the local isotropy implies that all these functions must vanish. There were many attempts to explain these violations of the local isotropy of temperature fluctuations by the influence of "temperature ramps" (where slow temperature growth is suddenly replaced by very rapid decrease or $vice\ versa$) and some other strongly asymmetric large-scale temperature structures. However, these attempts were not fully successful and also the origin of the asymmetric temperature structures in scalar turbulence remains enigmatic up to now. Let us note in this respect described in [129] results of the numerical simulation by Holzer and Siggia of the development of temperature fluctuations in a homogeneous Gaussian velocity field without any appreciable structures accompanied with a constant gradient of the mean temperature. It was found that in this case the temperature "ramp structures" of unknown origin also appeared regularly. In any case, the available at present data relating to small-scale temperature fluctuations show that Kolmogorov's assumption about the isotropy of small-scale turbulent fluctuations in all flows with high enough Reynolds (and Peclét) numbers is usually invalid in the real flow turbulence.

Detected at the end of the 20th century strong deviations of the results of careful measurements of turbulent-flow characteristics from the previous predictions of great scientists are very disturbing for all modern fluid

---

[2]Generalization of the K41 theory to temperature and other scalar fields (for simplicity, only temperature field will be mentioned here) was carried out independently by Obukhov and Corrsin in 1949-51; see, $e.g.$ [67], Chapter 8. It was found, in particular, that the temperature structure functions and one-dimensional spectra in the inertial ranges of lengths and wave numbers satisfy the same two-thirds and five-thirds laws as structure functions and spectra of velocity.

mechanicians. These deviations make highly desirable the comparison of the old theoretical results, based on physically convincing but unproved hypotheses, with conclusions following directly from rigorous dynamic equations of fluid motions. Unfortunately, this natural desire cannot be satisfied easily since the derivation of the specific results relating to high-Reynolds-number fluid flows from the dynamic equations met with unexpected resistance. Below, as everywhere above, only the incompressible fluid flows satisfying the Navier–Stokes equations will be considered. Very complicated properties of these equations have been already noted earlier, and now this complexity becomes especially evident in view of some recently appearing new curious developments relating to this subject.

In the Introduction to these lectures the so-called "Physics Problems for the Next Millennium" have been already mentioned. Let us now explain that the appearance of these problems was stimulated by publication slightly earlier by the Clay Mathematics Institute of a list of seven "Mathematics Millennium Prize Problems" (first announced during the "Millennium Meeting" of mathematicians at the Collège de France in Paris in May 2000). It was announced there that the solution of any of these problem will be rewarded by a prize of $1 million (see [131] and http://www.claymath.org/prize_problems). Clay Institute Problems were considered by their authors as the continuation of the famous "Hilbert's Problems" – a list of 23 then unsolved problems set up by the famous German mathematician Hilbert at the International Mathematical Congress of 1900 in Paris for solution in the 20th century. For the subject discussed here it is only of importance that seven Clay Institute Prize Problems include a problem called "Navier–Stokes Equations". A short explanation accompanying the problem title at the internet announcement states that "Our understanding of the Navier–Stokes equations remains minimal. The challenge is to make substantial progress toward a mathematical theory which will unlock the secrets hidden in these equations". This is somewhat vague formulation for a problem whose solution is estimated in one million dollars, but it is clear that it is supposed here that the solution must explain the inexplicable features of fluid flows, both laminar and, especially, much more mysterious turbulent. Brief summary of the same prize problem in [131] was expressed as follows: "Prove or disprove the existence and smoothness of solutions to the three-dimensional Navier–Stokes equations (under reasonable boundary and initial conditions)". A little more detailed discussion of this problem by Prof. Fefferman accompanying the internet notice paid again most attention to unsolved problems relating to the existence, smoothness, and possible singularities of the solutions of three-dimensional NS equations. Moreover, in even more detailed discussions of this problem by Constantin and by Yudovich [132] much attention was again paid to existence problems for smooth solutions of the NS (and Euler's, where $\nu = 0$) equations, but at the same time some problems on the asymptotic behavior of solutions at large times (closely connected with the secrets of flow instability) and on mysteries of turbulence were also briefly described there. All this is told here to pay attention of the readers to the remarkable fact that mathematical problems of fluid motions were included in a short list of major unsolved mathematical problems which the science of the 20th century left for solution to the 21st century.

Let us now revert to possible applications of the NS equations to studies of turbulence phenomena. A number of difficulties met on this way was discussed by L'vov and Procaccia in 1997 (see [133]). These two scientists have long been trying to develop the hydrodynamic theory of turbulence and, in particular, to apply the NS equations to the proof of the existence of a range of the power-law behavior of the velocity structure functions and to the estimation of the corresponding scaling exponents ζ_n (see, *e.g.*, the second paper in [133] and the cited there papers on this subject). Their work showed clearly how complicated this problem is and how difficult it is to obtain here even a modest success. Another very interesting discussion of the problems arising in the hydrodynamic theory of turbulence was published by Foias [134] also in 1997. In the title of the paper [134] it was asked: "What do the Navier–Stokes equations tell us about turbulence?", and in the first sentence of it the following answer was proposed: "Until the early eighties, very little; since then, quite a lot". It seems, however, that this answer is a little too optimistic, though it is impossible to neglect serious successes in this field reached during the last twenty years.

The main purpose of Foias in [134] was to make an attempt to find rigorous proofs based on the NS equations of some remarkable results of the turbulence theory which were earlier derived from some combination of the physical intuition with the purely empirical evidence. As the appropriate examples of such theoretical results Kolmogorov's (relating to K41) and Kraichnan's [135] inertial-range laws for three-dimensional (3D) and two-dimensional (2D) turbulence were chosen. (Kraichnan's 2D results were also included since the 2D NS equations are much simpler than the 3D ones.) Some elementary model of the cascade process of energy transfer from larger to smaller eddies was included in Foias' analysis but all intermittency effects were fully neglected. Under this condition the author was able to give almost fully rigorous proofs of the K41 and Kraichnan's $k^{-5/3}$ and k^{-3} laws for the energy spectrum $E(k)$ in the inertial ranges of wave numbers and of the equations determining the dissipation length scales in three and two dimensions. However, these proofs proved to be rather complicated and they nevertheless included some purely heuristic arguments.

Quite impressive successes were achieved in the studies of the asymptotic behavior of the solutions of the NS equations and of the structure of the corresponding "attractors" in the infinite-dimensional phase spaces of fluid flows; see, *e.g.*, the books [57, 136] where some of the results relating to this topic were considered. (Here again advances were most impressive in the case of 2D turbulence.) However, the development of the rigorous mathematical theory of the high-Reynolds-number turbulence is apparently up to now only in its initial stage.

In the case of developed turbulence most interesting are not individual solutions describing the time evolution of separated flow fields but "statistical solutions" corresponding to time evolution of the probability measure in the space of all possible fluid-dynamics fields when the initial measure at the time $t = 0$ is given. Instead of the difficult for mathematical treatment probability measure in the infinite-dimensional space of turbulent fields, it is much more convenient to consider corresponding to this measure *characteristic functional* (first introduced, for the case of a random function of one variable, long ago by Kolmogorov [137]). Spatial characteristic functional of the velocity field $\mathbf{u}(\mathbf{x}, t) = \{u_1(\mathbf{x}, t), u_2(\mathbf{x}, t), u_3(\mathbf{x}, t)\}$ of a turbulent flow is given by the equation

$$\Phi[\Theta(\mathbf{x}, t] \equiv \Phi\left[\theta_1(\mathbf{x}), \theta_2(\mathbf{x}), \theta_3(\mathbf{x}), t\right]$$

$$= \left\langle \exp\left\{ i \iiint \sum_{k=1}^{3} \theta_k(\mathbf{x}) u_k(\mathbf{x}, t) \mathrm{d}x_1 \, \mathrm{d}x_2 \, \mathrm{d}x_3 \right\} \right\rangle \quad (22)$$

(here $\mathbf{x} = (x_1, x_2, x_3)$ and integration is extended over the whole space of points $\mathbf{x}$ while the functions $\theta_k(\mathbf{x})$, $k = 1, 2, 3$, are chosen to provide convergence of the integral on the right in (22)). Angular brackets, as usual, denote in (22) the ensemble averaging, *i.e.*, the integration with respect to probability measure. Note that the moments of all orders (both one-point and multipoint) of the velocity field $\mathbf{u}(\mathbf{x}, t)$ (where t is fixed) may be easily expressed in terms of the partial *functional derivatives* of various orders of the functional $\Phi[\Theta(\mathbf{x}), t]$ (see, *e.g.* [67], Sect. 4.4, or any of cited below other books where Hopf equation is considered). For determination of the multitime velocity moments relating to velocity values at various space and time points, the spatial-temporal characteristic functional $\Phi[\Theta(\mathbf{x}, t)]$ may be used. This functional is given by similar to (22) equation where the functions $\theta_k(\mathbf{x})$ are replaced by functions $\theta_k(\mathbf{x}, t)$ and integration is taken over the four-dimensional space of points $(\mathbf{x}, t)$. However, such functionals (introduced in the paper [138]) will be not considered below.

Characteristic functional determines uniquely the probability measure of the turbulent velocity field and its time evolution is governed by linear functional derivative equation derived in 1952 by Hopf [139]. *Hopf equation* may be written in the form

$$\frac{\partial \Phi[\Theta(\mathbf{x}), t]}{\partial t} = i \left(\hat{\theta}_k \frac{\partial D_k D_m \Phi}{\partial x_m} \right) + \nu \left(\hat{\theta}_k \Delta D_k \Phi \right) \tag{23}$$

where $D_k = D_k(\mathbf{x}) = \delta/\delta\theta_k(\mathbf{x})d\mathbf{x}$ is the functional derivative with respect to the component $\theta_k(\mathbf{x})$ of the vector $\Theta(\mathbf{x})$, Δ is the Laplace operator, the summation is performed over the three values of the twice appearing indices k and m, and $\hat{\theta}_k(\mathbf{x})$ are the components of the vectorial function $\hat{\Theta}(\mathbf{x})$ which may be obtained from the vectorial function $\Theta(\mathbf{x})$ by means of some simple linear operation. Equation (23) seems to be very attractive, since it is linear, not very clumsy, and determined the whole probability distribution of the velocity field. Unfortunately, the mathematical theory of functional derivative equations was quite undeveloped in the fifties (*e.g.*, nothing was known then about the solvability of such equations and the conditions for the uniqueness of their solutions, and there were no methods for solution computation). Therefore, at first the practical usefulness of the Hopf equation seemed rather questionable. However, during almost a half century separating our time from the early fifties the mathematical theory of the linear functional derivative equations advanced considerably (to a considerable degree just in the connection with induced by Hopf's paper active development of mathematically-oriented studies of statistical fluid mechanics) and this made the situation much less hopeless. A number of results of these studies may be found, in particular, in the papers [140], in the fundamental monograph by Vishik and Fursikov [141] and in the recent books [142] on mathematical fluid mechanics and turbulence theory which include analysis of the Hopf equation.

It is worth to say here a few words about the paper by Foias, Manley and Temam of 1987 (see [140]), which did not used Hopf's equation directly, but referred to it repeatedly and was ideologically connected with the functional approach to statistical fluid mechanics. In this paper for the case of isotropic turbulence an attempt was made to connect the derivation of Kolmogorov's "five-thirds law" for the energy spectrum with the study of statistical solutions of Navier–Stokes equations and even to use the found connection for the determination of lower bound of the range of Reynolds numbers at which the inertial range of wave numbers exists. However, apparently there were no attempts to explain with the help of Navier–Stokes dynamic equations the observed anomalous scaling of the velocity structure functions (*i.e.*, the appearance of the non-zero scaling corrections ξ_n to Kolmogorov's exponents $n/3$).

Let us now made a small remark of general character at the end of this long text. It is clear that characteristic functional of a random function is a natural generalization of the *characteristic function* of a random variable (or random vector). Method of characteristic functions was used very fruitfully in probability theory by the famous Russian scientist Lyapunov almost exactly one hundred years ago (about 1900) when he applied this method to the first rigorous proof of the Central Limit Theorem of this theory under very general conditions. Later it was found that this method represents an universal tool (of very high efficiency) for the study of the asymptotic behavior of the families of random variables and random functions depending on a parameter tending to infinity. During the 20th century many hundreds of papers (and several dozens of books) were published where characteristic functions were widely used for this purpose. Of course, characteristic functionals are analytically much more complicated than characteristic functions, but the power of analytic methods today also exceeds very much their possibilities in the Lyapunov's time. Let us therefore hope that the method of characteristic functionals will have in the new century a development comparable to that of the method of characteristic functions in the previous century. (Note that in the turbulence theory the investigation of the asymptotic behavior of fluid-dynamical fields as $t \to \infty$ or/and $Re \to \infty$ always plays a very important part.) Since NS equations are very complicated, it is reasonable to elaborate at first the new analytical methods in application to simpler models; from this point of view the numerous recent studies of "nonphysical" Burgers turbulence (to this subject, in particular, the lectures by Uriel Frisch at this summer school were devoted) may be very useful.

It seems natural to expect now that the 21st century will be a century of an astonishingly large growth of turbulent investigations. However, crude dimensional arguments, playing such important part in most fundamental achievements of the previous century, apparently will be of secondary importance for the future development of our science but much more important part will play the deep physical insight and very artful analytical technique.

Many colleagues helped me in preparation of this text discussing with me some topics touched in here and/or sending me written materials related to the contents of my talks. I wish to thank here Peter Bradshaw, Friedrich Busse, Peter Constantin, Siegfried Grossmann, John Lumley, Ron Panton, Peter Schmid, Mark Vishik, Zellman Warhaft and Victor Yudovich. I am very grateful also to Grisha Barenblatt who regularly sent me his interesting papers on turbulence, but I understand that he will disagree with a number of my opinions. I wish also to express my gratitude to the Department of Aeronautics and Astronautics of MIT and the Department Head Prof. E.F. Crawley, who do everything to help my systematic work at MIT. The last but not the least, I want to express my deep gratitude to Dr. John William Poduska, Sr., and the Poduska Family Foundation who presented a grant to MIT for financial support of my work; without this support the preparation of this text would be impossible.

References

[1] J.L. Lumley, *Phys. Fluids A* **4** (1992) 203-211.

[2] U. Frisch, *Turbulence. The Legacy of A.N. Kolmogorov* (Cambridge University Press, 1995).

[3] G. Hagen, *Pogg. Ann.* **46** (1839) 423-442.

[4] O. Reynolds, *Phil. Trans. Roy. Soc. London* **174** (1883) 935-982; O. Reynolds, *Phil. Trans. Roy.Soc. London* **186** (1884) 123-161.

[5] R.P. Feynman, *The Feynman Lectures on Physics* (by Feynman, Leighton and Sands, Addison-Wesley, Redwood City, 1964).

[6] P. Holmes, J.L. Lumley and G. Berkooz, *Turbulence, Coherent Structures, Dynamical Systems and Symmetry* (Cambridge University Press, 1996).

[7] M. Gad-el-Hak, *Appl. Mech. Rev.* **50** (1997) i, ii.

[8] S. Goldstein, *Ann. Rev. Fluid Mech.* **1** (1969) 1-28.

[9] J. Leray, *Acta Math.* **63** (1934) 193-248.

[10] W.M. Orr, *Proc. Roy. Irish Acad. A* **27** (1971) 9-68, 69-138.

[11] A. Sommerfeld, *Proc. 4th Int. Congr. Math. Rome* **III** (1908) 116-124.

[12] H. Squire, *Proc. Roy. Soc. London A* **142** (1933) 621-628.

[13] Kelvin Lord (W. Thomson), *Phil. Mag.* **24** (1887) 272-278.

[14] Kelvin Lord (W. Thomson), *Phil. Mag.* **24** (1887) 188-196.

[15] M.T. Landahl, *J. Fluid Mech.* **98** (1980) 243-251.

[16] K.M. Butler and B.F. Farrell, *Phys. Fluids* **4** (1992) 1637-1650.

[17] S.C. Reddy and D.S. Henningson, *J. Fluid Mech.* **252** (1993) 209-238.

[18] L.N. Trefethen, A.E. Trefethen, S.C. Reddy and T.A. Driscol, *Science* **261** (1993) 578-584.

[19] S. Grossmann, Instability Without Instability?, in: *Nonlinear Physics of Complex Systems. Current Status and Future Trends*, edited by J. Parisi, S.C Müller and W. Zimmermann (Springer, Berlin, 1996) pp. 10-22.

[20] W.O. Criminale, T.L. Jackson, D.G. Lasseigne and R.D. Joslin, *J. Fluid Mech.* **339** (1997) 55-75.

[21] D.G. Lasseigne, R.D. Joslin, T.L. Jackson and W.O. Criminale, *J. Fluid Mech.* **381** (1999) 89-119.

[22] P. Schmid and D.S. Hennigson, *Stability and Transition in Shear Flows* (Springer, New York, 2001).

[23] W.O. Criminale, T.L. Jackson and R.D. Joslin, *Hydrodynamic Stability: Theory and Computations* (Cambridge University Press, in preparation).

[24] A.M. Yaglom, *More About Instability Theory; Studies of the Initial-Value Problem*, Chapter 3 of the fully revised new edition of *Statistical Fluid Mechanics*, edited by A.S. Monin and A.M. Yaglom (CTR Monograph, Center for Turb. Res., Stanford, 1998).

[25] P. Andersson, M. Berggren and D.S. Henningson, *Phys. Fluids* **11** (1999) 134-150.

[26] P. Luchini, *J. Fluid Mech.* **404** (2000) 289-309.

[27] J.S. Baggett, T.A. Driscoll and L.N. Trefethen, *Phys. Fluids* **7** (1995) 833-838.

[28] T. Gebhardt and S. Grossmann, *Phys. Rev. E* **50** (1994) 3705-3711.

[29] S. Grossmann, *Rev. Mod. Phys.* **72** (2000) 603-618.

[30] I. Boberg and U. Brosa, *Z. Naturforsch. A* **43** (1988) 697-726.

[31] T.A. Zang and S.E. Krist, *Theoret. Comput. Fluid Dynamics* **1** (1989) 41-64.

[32] N.D. Sandham and L. Kleiser, *J. Fluid Mech.* **245** (1992) 319-348.

[33] L. Bergström, *Phys. Fluids* **11** (1999) 590-601.

[34] L.D. Landau, *Dokl. Akad. Nauk SSSR* **44** (1944) 339-342. English Translation in *C.R. Acad. Sci. URSS* **44** (1944), and in *Collected Papers by L.D. Landau* (Pergamon, Oxford, 1965).

[35] L.D. Landau and E.M. Lifshitz, *Continuum Mechanics* (Gostekhizdat, Moscow, 1944, in Russian); see also any of the subsequent editions of *Fluid Mechanics* by Landau and Lifshitz.

[36] E. Hopf, *Comm. Pure Appl. Math.* **1** (1948) 303-322.

[37] D. Ruelle and F. Takens, *Comm. Math. Phys.* **20** (1971) 167-192; D. Ruelle, *Turbulence, Strange Atractors and Chaos* (World Scientific, Singapore, 1995).

[38] E.N. Lorenz, *J. Atmos. Sci.* **20** (1963) 130-141.

[39] M.J. Feigenbaum, *J. Stat. Phys.* **21** (1979) 669-706.

[40] M.Y. Feigenbaum, *Comm. Math. Phys.* **77** (1980) 65-86.

[41] S. Grossmann and S. Thomae, *Z. Naturforsch. A* **32** (1977) 1353-1363.

[42] Y. Pomeau and P. Manneville, *Comm. Math. Phys.* **74** (1980) 189-197.

[43] P. Manneville and Y. Pomeau, *Physica D* **1** (1980) 219-226.

[44] G.I. Barenblatt, G. Iooss and D.D. Joseph (eds.), *Nonlinear Dynamics and Turbulence* (Pitman, Boston, 1983).

[45] P. Bergé, Y. Pomeau and C. Vidal, *L'Ordre dans la Chaos. Vers une Approche Déterministe de la Turbulence* (Hermann, Paris, 1988).

[46] A.J. Lichtenberg and M.A. Lieberman, *Regular and Chaotic Dynamics*, 2nd Edition (Springer, New York, 1992).

[47] L.P. Kadanoff, *From Order to Chaos. Essays: Critical, Chaotic and Otherwise* (World Scientific, Singapore, 1993).

[48] R.C. Hilborn, *Chaos and Nonlinear Dynamics. An Introduction for Scientists and Engineers* (Oxford University Press, New York, 1994).

[49] A.C. Newell, Chaos and Turbulence: Is There a Connection?, in *Mathematics Applied to Fluid Mechanics and Stability* (Soc. Ind. Appl. Math., Philadelphia, 1986).

[50] Y. Tsuji, K. Hondu, I. Nakamura and S. Sato, *Phys. Fluids A* **3** (1991) 1941-1946.

[51] C. Menevau, *Phys. Fluids A* **4** (1992) 1587-1588.

[52] P. Blodeaux and G. Vittori, *Phys. Fluids A* **3** (1991) 2492-2495.

[53] D. Rockwell, F. Nuzzi and C. Magness, *Phys. Fluids A* **3** (1992) 1477-1478.

[54] A.G. Tomboulides, G.S. Triantafyllow and G.E. Karniadakis, *Phys. Fluids A* **4** (1992) 1333-1335.

[55] A.M. Guzmán and C.H. Amon, *Phys. Fluids* **6** (1994) 1994-2002.

[56] G. Goren, J.-P. Eckmann and I. Procaccia, *Phys. Rev. E* **57** (1998) 4106-4134.

[57] A.V. Babin and M.I. Vishik, *Attractors of Evolution Equations* (North-Holland, Amsterdam, 1992).

[58] A.A. Izakson, *Zh. Eksper. Teor. Fiz.* **7** (1937) 919-924 (in Russian, English translation in *Tech. Phys. USSR* **4** (1937) 155-159).

[59] C.B. Millikan, A Critical Discussion of Turbulent Flows in Channels and Circular Tubes, in *Proc. 5th Intern. Congr. Appl. Mech.*, edited by J.P. Den Hartog and H. Peters (Wiley, New York, 1939) pp. 386-392.

[60] F.H. Clauser, *Adv. Appl. Mech.* **4** (1954) 1-51.

[61] H.-H. Fernholz and P.J. Finley, *Prog. Aerospace Sci.* **32** (1986) 245-311.

[62] A.M. Yaglom, *Ann. Rev. Fluid Mech.* **11** (1979) 505-540.

[63] B.A. Kader and A.M. Yaglom, Similarity Laws for Turbulent Wall Flows, in *Developments in Science and Technology, Ser. Mech. Liquid and Gas*, Vol. 15 (1980) 81-153 (Soviet Inst. Sci. and Engn. Inform., Moscow, in Russian).

[64] R. von Mises, *Some Remarks on the Laws of Turbulent Motion in Tubes*, in: T. von Kármán Anniversary Volume (Calif. Inst. Techn. Press, Pasadena, CA) pp. 317-327.

[65] B.A. Kader and A.M. Yaglom, *Int. J. Heat and Mass Transfer* **15** (1972) 2329-2351.

[66] B.A. Kader and A.M. Yaglom, Roughness and Pressure-Gradient Effects on Turbulent Boundary Layers, in *Developments in Science and Technology, Ser. Mech. Liquid and Gas*, Vol. 18 (1984) 3-111 (Soviet Inst. Sci. and Engn. Inform., Moscow; in Russian).

[67] A.S. Monin and A.M. Yaglom, *Statistical Fluid Mechanics*, Vols. 1 and 2 (MIT Press, Cambridge, MA, 1971 and 1975).

[68] A.A. Townsend A.A., *The Structure of Turbulent Shear Flow* (Cambridge University Press, 1956).

[69] A.M. Yaglom, *Ann. Rev. Fluid Mech.* **26** (1994) 1-22.

[70] A.A. Townsend, *J. Fluid Mech.* **11** (1961) 97-120.

[71] P. Bradshaw, *J. Fluid Mech.* **30** (1967) 241-258.

[72] J.F. Morrison, C.S. Subramanian and P. Bradshaw, *J. Fluid Mech.* **241** (1992) 75-108.

[73] A.A. Townsend, *The Structure of Turbulent Shear Flow*, 2nd Edition (Cambridge University Press, 1976).

[74] A.E. Perry and J.D. Li, *J. Fluid Mech.* **218** (1990) 405-438.

[75] A.M. Yaglom, *Phys. Fluids* **6** (1994) 962-972.

[76] G.I. Barenblatt, A.J. Chorin and V.M. Prostokishin, *Appl. Mech. Rev.* **50** (1997) 413-429.

[77] H. Schlichting, *Boundary Layer Theory* (McGraw-Hill, New York, 1968).

[78] M. Gad-el-Hak and P.R. Bandyopadhyay, *Appl. Mech. Rev.* **47** (1994) 307-366.

[79] G.I. Barenblatt and Ya.B. Zeldovich, *Ann. Rev. Fluid Mech.* **4** (1972) 285-312.

[80] G.I. Barenblatt, *Scaling, Self-Similarity and Intermediate Asymptotics* (Cambridge University Press, 1996).

[81] J. Nikuradze, *Gesetzmässigkeiten der turbulente Strömung in glatten Rohren*, VDI Forschugheft No. 356 (1932).

[82] G.I. Barenblatt, *C. R. Acad. Sci. Paris Ser. II* **313** (1991) 307-312; G.I. Barenblatt, *J. Fluid Mech.* **248** (1993) 513-520; G.I. Barenblatt and V.M. Prostokishin, *J. Fluid Mech.* **248** (1993) 521-529.

[83] G.I. Barenblatt, A.J. Chorin and V.M. Prostokishin, *J. Fluid Mech.* **410** (2000) 263-283.

[84] G.I. Barenblatt and A.J. Chorin, *Comm. Pure Appl. Math.* **50** (1997) 381-398; G.I. Barenblatt, *J. Eng. Math.* **36** (1999) 361-384.

[85] M. Zagarola and A.J. Smits, *J. Fluid Mech.* **373** (1998) 33-79.

[86] J.M. Österlund, A.V. Johansson, H.M. Nagib and M.H. Hites, *Phys. Fluids* **12** (2000) 1-4.

[87] A.J. Smits and M.V. Zagarola, *Phys. Fluids* **10** (1998) 1045-1046.

[88] M. Zagarola, A.E. Perry and A.J. Smits, *Phys. Fluids* **9** (1997) 2094-2100.

[89] G.I. Barenblatt and A.J. Chorin, *Phys. Fluids* **10** (1998) 1043-1044.

[90] G.I. Barenblatt, A.J. Chorin and V.M. Prostokishin, *Phys. Fluids* **12** (2000) 2159-2161.

[91] J.M. Österlund, A.V. Johansson and H.M. Nagib, *Phys. Fluids* **12** (2000) 2360-2363.

[92] M. Buschmann and M. Gad-el-Hak, *Bull. Amer. Phys. Soc.* **45** (2000) 160; R.L. Panton, *ibid.* **45** (2000) 160-161; H. Nagib, C. Christophorou, J. Österlund and P. Monkewitz, *ibid.* **45** (2000) 161.

[93] R.L. Panton, *Some Issues Concerning Wall Turbulence*, Informal document distributed by the author; R.L. Panton, Comments on "A Note on the Intermediate Region in a Turbulent Boundary Layers" (submitted to *Phys. Fluids*).

[94] J.D. Cole, *Perturbation Methods in Applied Mathematics* (Blaisdell, Waltham, MA, 1968).

[95] M. Van Dyke, *Perturbation Methods in Fluid Mechanics* (Parabolic Press, Stanford, CA, 1975).

[96] H. Tennekes and J.L. Lumley, *A First Course in Turbulence* (MIT Press, Cambridge, MA, 1972).

[97] R.L. Panton, *Incompressible Flow*, 2nd Editon (Wiley, New York, 1996).

[98] W.K. George and L. Castillo, *Appl. Mech. Rev.* **50** (1997) 689-729.

[99] M. Wosnik, L. Castillo and W.K. George, *J. Fluid Mech.* **421** (2000) 115-145.

[100] S. Nazarenko, N.K.-R. Kevlahan and B. Dubrulle, *Physica D* **139** (2000) 158-176.

[101] A.M. Yaglom, Similarity Laws for Wall Turbulent Flows: Their Limitations and Generalizations, in: *New Approaches and Concepts in Turbulence*, edited by Th. Dracos and A. Tsinober (Birkhäufer, Basel, 1993) pp. 7-27.

[102] U. Högström, *J. Atmos. Sci.* **47** (1990) 1949-1972.

[103] A.-S. Smedman, *J. Atmos. Sci.* **48** (1991) 856-868.

[104] R.L. Panton, *J. Fluid Eng.* **119** (1997) 325-330.

[105] M. Fisher, J. Jovanvic and F. Durst, *Int. J. Heat and Fluid Flow* **21** (2000) 471-479.

[106] D.B. DeGraaff and J.K. Eaton, *J. Fluid Mech.* **422** (2000) 319-346.

[107] J.C.R. Hunt and J.F. Morrison, *Eur. J. Mech. B./ Fluids* **19** (2000) 673-694.

[108] J.C.R. Hunt and P. Carlotti, *Flow, Turb. and Comb.* (to be published in 2001).

[109] E.A. Novikov, *Appl. Math. Mech.* **35** (1971) 266-277 (in Russian).

[110] G.I. Barenblatt and N. Goldenfeld, *Phys. Fluids* **7** (1995) 3078-3082.

[111] G.I. Barenblatt and A.J. Chorin, *SIAM Rev.* **40** (1998) 265-291.

[112] G.I. Barenblatt and A.J. Chorin, *Proc. Symp. Appl. Math.* **54** (1998) 1-25.

[113] A. Praskovsky and S. Onsley, *Phys. Fluids A* **6** (1994) 2886-2888.

[114] A.M. Yaglom, Izv. Akad. Nauk SSSR, *Fiz. Atmos. i Okeana* **17** (1981) 1235-1257 (in Russian, Engl. transl. in *Izvestia, Atmos. Oceanic Phys.* **17** (1981) 919 935).

[115] K.R. Sreenivasan, *Phys. Fluids* **7** (1995) 2778-2784.

[116] R.H. Kraichnan, *J. Fluid Mech.* **62** (1974) 305-330.

[117] A.J. Chorin, *Vorticity and Turbulence* (Springer, New York, 1994).

[118] A.M. Yaglom, *Dokl. Akad. Nauk SSSR* **166** (1966) 49-52 (Engl. transl. in *Sov. Phys. - Doklady* **11**, 26-29).

[119] G.I. Barenblatt, A.J. Chorin and V.M. Prostokishin, *Physica D* **127** (1999) 105-110.

[120] R. Benzi, C. Ciliberto, C. Baudet and G. Ruiz Chavarria, *Physica D* **80** (1995) 385-398.

[121] R. Benzi, C. Ciliberto, C. Baudet and G. Ruiz Chavarria, *Physica D* **127** (1999) 111-112.

[122] A. Arneodo *et al.*, *Europhys. Lett.* **34** (1996) 411-416.

[123] K.R. Sreenivasan and B. Dhruva, *Prog. Theor. Phys. Suppl.* **130** (1998) 103-120.

[124] S. Gamard and W.K. George, *Flow, Turb. and Combus.* **63** (1999) 443-477.

[125] L. Mydlarski and Z. Warhft, *J. Fluid Mech.* **320** (1996) 331-368.

[126] I. Arad, B. Dhruva, S. Kurien, V.S. L'vov, I. Procaccia and K.R. Sreenivasan, *Phys. Rev. Lett.* **81** (1998) 5330-5333; S. Kurien, V.S. L'vov, I. Procaccia and K.R. Sreenivasan, *Phys. Rev. E* **61** (2000) 407-421; S. Kurien and K.R. Sreenivasan, *Phys. Rev. E* **62** (2000) 2206-2212.

[127] J.C. Kaimal, J.C. Wyngaard, Y. Izumi and O.R. Coté, *Quart. J. Roy. Meteor. Soc.* **98** (1972) 563-589.

[128] X. Shen and Z. Warhaft, *Phys. Fluids* **12** (2000) 2976-2989.

[129] Z. Warhaft, *Ann. Rev. Fluid Mech.* **32** (2000) 203-240.

[130] M. Gonzalez, *Phys. Fluids* **12** (2000) 2302-2310; B.I. Shraiman and E.D. Siggia, *Nature* **405** (2000) 639-646.

[131] E. Jackson, *Notices Amer. Math. Soc.* **47** (2000) 871-879.

[132] P. Constantin, Some Open Problems and Research Directions in the Mathematical Study of Fluid Dynamics, to be published in *Mathematics Unlimited – 2001 and Beyond* (Springer, New York, 2001); V.I. Yudovich, Global Regularity *vs.* Collapse in Dynamics of Incompressible Fluid, to be published, in Russian and in English, in *The Main Mathematical Events of the 20th Century* (Fazis, Moscow, 2001).

[133] V. L'vov and I. Procaccia, Hydrodynamic Turbulence: A 19th Century Problem with a Challenge for the 21st Century, in *Turbulence Modeling and Vortex Dynamics*, edited by O. Bortav *et al.* (Springer, Berlin, 1997) pp. 1-16; V. L'vov and I. Procaccia, *Phys. Rev. E* **62** (2000) 8037-8051.

[134] C. Foias, What Do the Navier–Stokes Equations Tell Us about Turbulence?, in *Harmonic Analysis and Nonlinear Differential Equations*, edited by M.L. Lapidus *et al.*, Contemporary Mathematics, Vol. 208 (Amer. Math. Soc., Providence, RI, 1997) pp. 151-180.

[135] R.H. Kraichnan, *Phys. Fluids* **10** (1967) 1417-1423.

[136] P. Constantin, C. Foias and R. Temam, *Attractors Representing Turbulent Flows*, Memoirs Amer. Math. Soc. **53**, No. 314 (Amer. Math. Soc., Providence, R.I., 1985); P. Constantin and C. Foias, *Navier–Stokes Equations* (Chicago Univ. Press, 1988); R. Temam, *Infinite Dimensional Dynamical Systems in Mechanics and Physics* (Springer, New York, 1998); R. Temam, *Navier–Stokes Equations and Nonlinear Functional Analysis* (Soc. Ind. Appl. Math., Philadelphia, 1994); A. Eden, C. Foias, B. Nicolaenko and R. Temam R., *Exponential Attractors for Dissipative Evolution Equations* (Wiley, New York, 1994).

[137] A.N. Kolmogorov, *C. R. Acad. Sci. Paris* **200** (1935) 1717-1718.

[138] R.M. Lewis and R.H. Kraichnan, *Comm. Pure Appl. Math.* **15** (1962) 397-411.

[139] E. Hopf, *J. Rat. Mech. Anal.* **1** (1952), 87-123.

[140] C. Foias, *Rend. Sem. Mat. Univ. Padova* **48** (1973) 219-349; **49** (1973) 9-123; O.A. Ladyzhenskaya and A.M. Vershik, *Ann. Scuola Norm. Sup. Pisa Ser. IV* **4** (1977) 209-230; C. Foias, O.P. Manley and R. Temam, *Phys. Fluids* **30** (1987) 2007-2020; A. Inoue, A Tiny Step Towards a Theory of Functional Derivative Equations - A Strong Solution of the Space-Time Hopf Equation, in *The Navier–Stokes Equations II - Theory and Numerical Methods*, edited by J.G. Heywood *et al.* (Springer, Berlin, 1991) pp. 246-261; A.V. Fursikov, Time Periodic Statistical Solutions of the Navier–Stokes Equations, in *Turbulence Modeling and Vortex Dynamics*, edited by O. Bortav *et al.* (Springer, Berlin, 1997) pp. 123-147.

[141] M.I. Vishik and A.V. Fursikov, *Mathematical Problems of Statistical Hydrodynamics* (Kluwer, Dordrecht, 1988).

[142] M. Capinski and N.J. Cutland, *Nonstandard Methods for Stochastic Fluid Mechanics* (Word Scientific, Singapore, 1995); C.R. Doering and J.D. Gibbon, *Applied Analysis of the Navier–Stokes Equations* (Cambridge Univ. Press, 1995); A.S. Monin and A.M. Yaglom, *Statisticheskaya Gidromekhanika* (Statistical Fluid Mechanics), Vol. 2, 2nd Russian Ed. (Gidrometizdat, St.-Peterburg, 1996, in Russian).

COURSE 2

MEASURES OF ANISOTROPY AND THE UNIVERSAL PROPERTIES OF TURBULENCE

K.R. SREENIVASAN

*Department of Mechanical
Engineering, Mason Laboratory,
Yale University, P.O. Box 2159,
New Haven, CT 06520, U.S.A.*

Contents

MEASURES OF ANISOTROPY AND THE UNIVERSAL PROPERTIES OF TURBULENCE

S. Kurien[1] and K.R. Sreenivasan[2]

Abstract

Local isotropy, or the statistical isotropy of small scales, is one of the basic assumptions underlying Kolmogorov's theory of universality of small-scale turbulent motion. The literature is replete with studies purporting to examine its validity and limitations. While, until the mid-seventies or so, local isotropy was accepted as a plausible approximation at high enough Reynolds numbers, various empirical observations that have accumulated since then suggest that local isotropy may not obtain at any Reynolds number. This throws doubt on the existence of universal aspects of turbulence. Part of the problem in refining this loose statement is the absence until now of serious efforts to separate the isotropic component of any statistical object from its anisotropic components. These notes examine in some detail the isotropic and anisotropic contributions to structure functions by considering their SO(3) decomposition. After an initial discussion of the status of local isotropy (Sect. 1) and the theoretical background for the SO(3) decomposition (Sect. 2), we provide an account of the experimental data (Sect. 3) and their analysis (Sects. 4–6). Viewed in terms of the relative importance of the isotropic part to the anisotropic parts of structure functions, the basic conclusion is that the isotropic part dominates the small scales at least up to order 6. This follows from the fact that, at least up to that order, there exists a hierarchy of increasingly larger power-law exponents, corresponding to increasingly higher-order anisotropic sectors of the SO(3) decomposition. The numerical values of the exponents deduced from experiment suggest that the anisotropic parts in each order roll off less sharply than previously thought by dimensional considerations, but they do so nevertheless.

---

[1]Physics Department, Yale University, New Haven, CT 06520, U.S.A.
[2]Mason Laboratory, Yale University, New Haven, CT 06520, U.S.A.

1   Introduction

Local isotropy, or the isotropy of small scales of turbulent motion, is one of the assumptions at the core of the belief that small scales attain some semblance of universality [11]. This is a statistical concept, and is not necessarily opposed to the idea of structured geometry of small scales [9]. The important question about local isotropy is not whether the small scales are strictly isotropic, but the degree to which the notion becomes a better approximation as the scales become smaller [20]. Aside from the generic requirement that the flow Reynolds number be high (so that small scales as a distinct range may exist independent of the large scale), the question of how or if local isotropy becomes a good working approximation depends on the nature of large-scale anisotropy, on whether or not there are other body forces, on the nearness to physical boundaries, etc. All of this was well understood by the time of publication of Monin and Yaglom [19]. The general consensus at the time seems to have been that small scales indeed attain isotropy far from the boundary at high enough Reynolds numbers, at least when one considered second-order quantities.

The situation changed perceptibly when the small scales of scalar fluctuations were experimentally found to be anisotropic in many types of shear flows even at the highest Reynolds numbers of measurement. For a summary, see [23]. Since that time, various other pieces of evidence are slowly accumulating to suggest that small-scale velocity is also anisotropic; see, for example [22]. The claim is that the previously held belief – which, to some degree, was comforting – loomed large only because we had not explored the right statistical quantities. The notion that anisotropy persists at all scales at all Reynolds numbers (though manifested only in certain statistical parameters) puts a strong damper on any theory that purports to consider small-scale turbulence as a universal object. This is somewhat of an impasse.

Since the problems began with passive scalars, we might as well consider the evidence in that case a little more closely. The evidence, collected over many years by many people, is reproduced in Figure 1 from [24]. The argument is that if the temperature fluctuation θ is locally isotropic, its derivative $\partial\theta/\partial x$, being a small-scale quantity by construction, should be isotropic. Taking reflection symmetry as part of isotropy, we should have $\langle(\partial\theta/\partial x)^n\rangle = -\langle(\partial\theta/\partial x)^n\rangle$ for all odd values of n. This means that all the odd moments must be zero. In particular, $\langle(\partial\theta/\partial x)^3\rangle = 0$, or "small" in practice. But small compared to what? The standard thing to do is to normalize $\langle(\partial\theta/\partial x)^3\rangle$ by $\langle(\partial\theta/\partial x)^2\rangle^{3/2}$, or to examine the behavior of the skewness S of $\partial\theta/\partial x$. These are the data shown in Figure 1. The data suggest, despite some large scatter, that the skewness remains to be of

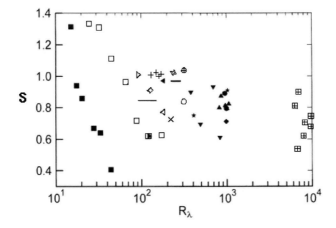

Fig. 1. The magnitude of the skewness of the temperature derivative in turbulent shear flows. The figure is taken from [24]. The Taylor microscale Reynolds number R_λ is proportional to the square root of the large scale Reynolds number. It is defined more precisely in Section 2.

the order unity even at the largest Reynolds numbers for which measurements are available.

There are two points to be made. The first is that the Taylor microscale Reynolds number R_λ may not be the right parameter against which to plot S. As articulated by Hill [8], the reasons are that the velocity that appears in R_λ is a large-scale quantity and the length scale λ cannot be defined independent of the large scale velocity. (In practice, the definition of λ may not require the large scale, see [26], though this is not a result that can be shown formally; but the comment on the velocity remains valid in any case.) We believe, however, that the situation will probably not change qualitatively even if we adopted a different abscissae for Figure 1. The second point is that, if one used $\langle(\partial\theta/\partial x)^4\rangle^{3/4}$ instead of $\langle(\partial\theta/\partial x)^2\rangle^{3/2}$ for normalizing $\langle(\partial\theta/\partial x)^3\rangle$, the normalized quantity will vanish with Reynolds number, roughly according to some power of the Reynolds number. Why is it not legitimate to compare the third moment to the next high-order even moment, instead of the neighboring low-order moment, or, perhaps to the geometric mean of the lower and higher order even moments? In either case, we will have a quantity that diminishes with the Reynolds number.

These are not elegant arguments, and seem like desperate efforts made to save the situation at any cost. There is a further argument to be made, however. Leaving aside the technicalities for a moment, let us suppose that, for a given statistical quantity to be measured, there are at all scales an isotropic part and an anisotropic part. We might now ask whether the ratio of the anisotropic part to the isotropic part vanishes as the scale size becomes smaller. We are no longer asking if the third moment vanishes with respect to (a suitable power of) some even order moment, but if, within a moment of a given order, the isotropic part eventually dominates the anisotropic part as the scale size vanishes. If this is indeed so, it allows us to say that, while non-universal anisotropic parts may always be present, the isotropic part dominates at small enough scales. If, of course, the isotropic part contributes exactly zero to the statistical quantity being considered, the anisotropic part will no doubt prevail at any finite Reynolds number, but this is not in contradiction to the statement just made. This is a richer point of view to take, potentially less inhibiting for the development of a sensible theory of turbulence. In such a picture, the universal theory that may emerge holds for the isotropic part alone, but, for it to be applicable to other types of turbulence, the anisotropic 'correction' has to be added in some suitable way.

The questions of interest, then, are obvious: what is a good way to decompose any statistical object of choice into isotropic and anisotropic parts? How do these two parts vary with scale size, relative to each other? A plausible method for answering these questions was proposed in [5] by using the SO(3) decomposition of tensorial objects usually considered in turbulence. One can argue as to whether this is the best perspective, but there is no doubt that it provides one framework within which our questions can be posed and answered. An experimental assessment of this issue is the broad topic of these notes. This is preceded by a detailed description of the theoretical issues involved.

The notes were part of the lectures presented by KRS at Les Houches, and form part of the Ph.D. work of SK. We note in advance that the literature cited is limited to a few key articles. We thank Itamar Procaccia and Victor L'vov for introducing us to the subject discussed here, and Christopher White and Brindesh Dhruva for their help in acquiring some of the data. KRS thanks the organizers of the School for the invitation to deliver the lectures.

2 Theoretical tools

2.1 The method of SO(3) decomposition

A familiar example of decomposition into the irreducible representations of the SO(3) symmetry group is the solution of the Laplace equation in spherical coordinates

$$\nabla^2 \psi(\mathbf{r}) = 0, \qquad (2.1)$$

where ψ is a scalar function defined over the sphere of radius r. This equation is linear, homogeneous and isotropic. The solutions are separable, and may be written as products of functions of r, θ and ϕ. The solutions form a linear space, a possible basis for which are derived from the spherical harmonics $Y_{lm}(\theta, \phi)$

$$\psi_{lm}(\mathbf{r}) = r^l Y_{lm}(\theta, \phi) = r^l Y_{lm}(\hat{\mathbf{r}}) \tag{2.2}$$

where the index $l = 0, 1, 2, ..., \infty$ denotes the degree of the harmonic polynomial. The angular dependence (θ, ϕ) is equivalent to the unit vector $\hat{\mathbf{r}}$, or the unit sphere. In the angular dependence, the index m denotes the elements of the orthonormal basis space that span each harmonic polynomial of degree l. For each l there are $2l+1$ elements indexed by $m = -j, -j+1, \ldots, j$ which are the Y_{lm}. Equation (2.2) is a useful basis space of solutions for the Laplace equation because the $\psi_{lm}(\hat{\mathbf{r}})$ are each solutions of the Laplace equation and are orthonormal for different l,m. Each basis element has a definite behavior under rotations, that is, the action on it by an element of SO(3) preserves the indices. In other words, the rotated basis element will also be indexed by the same l,m. The general solution of (2.1) is given by

$$\psi(\mathbf{r}) = \sum_{l,m} a_{lm} \psi_{lm}(\mathbf{r}). \tag{2.3}$$

The coefficients a_{lm} are obtained from the boundary conditions on $\psi(\mathbf{r})$ and, for a particular solution, any of them may be zero. In the theory of group representations, $\psi_{lm}(\mathbf{r})$ is said to be the $2l+1$ dimensional *irreducible representation* of the group of all rotations, the SO(3) symmetry group, in the space of scalar functions over the sphere. The space of scalar functions is called a "carrier" or "target" space for the SO(3) group representations. For each irreducible representation indexed by $l = 0...\infty$ there are $2l + 1$ components indexed by $m = -l... + l$. A representation of a symmetry group is said to be irreducible if it does not contain any subspaces that are invariant under the transformation associated with that symmetry.

Here we make a distinction between the behaviors under rotation of the *equation* (2.1) and of its *solutions* $\psi(\mathbf{r})$. The statement that the equation is isotropic means that it will hold true in a rotated frame with the Laplacian operator $\overline{\nabla}$ and the function $\overline{\psi}(\overline{\mathbf{r}})$ properly defined in the new coordinate system, $\overline{\nabla}^2 \overline{\psi}(\overline{\mathbf{r}}) = 0$. However, the function $\psi(\mathbf{r})$ may contain both isotropic and anisotropic components. All symmetry groups possess a one-dimensional representation in the carrier space, which is invariant under the transformations of that group. For the representations of the SO(3) symmetry group in the space of scalar functions over the unit sphere, the one-dimensional representation, indexed by $l = 0$, is $Y_{00} = constant$; it will clearly remain unchanged under proper rotations. This is the *isotropic* representation. The higher-dimensional representations Y_{lm}

(dimension $3, 5, 7, \ldots$ corresponding to $l = 1, 2, 3, \ldots$) have functional forms in θ and ϕ which are *altered* under rotation even though the degree (l), and hence the dimension, of the representation is preserved. These are the *anisotropic* representations. We examine this point in more detail with further discussion of the simple example of the scalar functions ψ.

For a particular rotation Λ in Euclidean space which tells us how to rotate a vector $\mathbf{x}$ into a new coordinate system where it is denoted by $\overline{\mathbf{x}}$,

$$\overline{x}^\alpha = \Lambda^\alpha_\beta x^\beta, \tag{2.4}$$

we can define an operator O_Λ which tells us how to rotate the function $\psi(\mathbf{r})$ to $\overline{\psi}(\overline{\mathbf{r}})$. If the function $\psi(\mathbf{r})$ is written in terms of a linear combination of its basis elements as in equation (2.3), then the rotation operation is written as

$$\overline{\psi}(\overline{\mathbf{r}}) = O_\Lambda \psi(\mathbf{r}) = O_\Lambda \sum_{l,m} a_{lm} \psi_{lm}(\mathbf{r}) = \sum_{l,m} O_\Lambda a_{lm} \psi_{lm}(\mathbf{r}). \tag{2.5}$$

The O_Λ for each l representation is a $(2l+1) \times (2l+1)$ matrix denoted by $D^{(l)}_{m',m}(\Lambda)$. The transformation is written as

$$O_\Lambda \psi_{lm}(\mathbf{r}) = \sum_{m'=-l}^{l} D^{(l)}_{m',m}(\Lambda) \psi_{lm'}(\mathbf{r}). \tag{2.6}$$

Thus, when a rotation of the function $\psi(\mathbf{r})$ into a new coordinate system is to be performed, the rotation matrices $D^{(j)}_{m,m'}(\Lambda)$ are all that is needed in order to transform each of the irreducible representations of the SO(3) group that form the basis functions of $\psi(\mathbf{r})$. As can be seen from equation (2.6), the $j = 0$ (one-dimensional) irreducible representation is the one which is *invariant* to all rotations in that its functional form is always preserved. The $D^{(0)}(\Lambda)$ matrix is merely a number, independent of Λ, multiplying the $j = 0$ representation. The $j = 0$ irreducible representation is the *isotropic* component of the SO(3) decomposition. For all higher-order j's the rotation preserves the dimension of the transformed component, *i.e.* the indices j and m are retained, but the functional form is altered. The $D^{(j)}(\Lambda)$ for $j > 0$ are true $(2j+1) \times (2j+1)$ matrices that mix the various m contributions of the original, unrotated basis.

The above simple example involved the case of a scalar function ψ that depended on a single unit vector $\hat{\mathbf{r}}$. The method of SO(3) decomposition may be carried over to more complicated objects. The theory is given in detail in [5]. In general we can imagine an nth order tensor function which depends on p unit vectors $T^{\alpha_1 \alpha_2 \cdots \alpha_n}(\hat{\mathbf{r}}_1, \hat{\mathbf{r}}_2, \ldots \hat{\mathbf{r}}_p)$ and is the solution to an isotropic equation. Then, the rules of SO(3) decomposition carry over in the following manner. The rotation operator O_Λ is now defined through the relation

$$\overline{T}^{\alpha_1\alpha_2\ldots\alpha_n}\left(\widehat{\mathbf{r}_1},\widehat{\mathbf{r}_2},\ldots\widehat{\mathbf{r}_\mathbf{p}}\right) = O_\Lambda T^{\alpha_1\alpha_2\ldots\alpha_n}(\hat{\mathbf{r}}_1,\hat{\mathbf{r}}_2,\ldots,\hat{\mathbf{r}}_\mathbf{p})$$
$$= \Lambda^{\alpha_1}_{\beta_1}\Lambda^{\alpha_2}_{\beta_2}\ldots\Lambda^{\alpha_n}_{\beta_n}T^{\beta_1\beta_2\ldots\beta_n}\left(\Lambda^{-1}\widehat{\mathbf{r}_1},\Lambda^{-1}\widehat{\mathbf{r}_2},\ldots,\Lambda^{-1}\widehat{\mathbf{r}_\mathbf{p}}\right). \quad (2.7)$$

The tensor function T may be written in terms of the irreducible representations of the SO(3) symmetry group. The basis elements are denoted by $B^{\alpha_1\alpha_2\ldots\alpha_n}_{jm}(\hat{\mathbf{r}})$ where the j index plays the same role as the l index in the example of the scalar function. The basis elements are more complicated than the ψ_{lm} because the space of functions T is the direct product of n Euclidean three-dimensional vector spaces (manifest in the indices $\alpha_1\ldots\alpha_n$) with p infinite dimensional spaces of continuous single-variable functions over the unit sphere (the $\hat{\mathbf{r}}_1\ldots\hat{\mathbf{r}}_\mathbf{p}$). If the constituent spaces are also written in the SO(3) decomposition, the rules of angular momentum addition, familiar from quantum mechanics, may be used in taking the direct product. The three-dimensional Euclidean vector space is a $j = 1$ space while each of the infinite-dimensional spaces functions over the unit sphere is the sum of the $j = 0, j = 1,\ldots,j \to \infty$ irreducible representations of SO(3) (recall the above example of $\psi(\hat{\mathbf{r}})$) with each j representation appearing once.

In tensor product notation, the product space of two vector spaces V_1 and V_2 is denoted by $V_1 \otimes V_2$. In the case of vector spaces in the SO(3) notation the spaces are named uniquely by their index j. Using tensor product notation, the n three-dimensional Euclidean spaces form a space $1\otimes1\otimes\ldots\otimes1$ (n times). The tensor notation indicating a linear sum of tensors with SO(3) representation indices j_1 and j_2 is $j_1 \oplus j_2$. Using this notation, each of the infinite dimensional spaces is $0\oplus1\oplus2\oplus\ldots$ The direct product of p of these is $(0\oplus1\oplus2\oplus\ldots)\otimes(0\oplus1\oplus2\oplus\ldots)\otimes\ldots\otimes(0\oplus1\oplus2\oplus\ldots)$ (p times). Thus, the final product space for $B^{\alpha_1,\ldots,\alpha_n}_{jm}(\hat{\mathbf{r}})$ is written in tensor-product notation as $1\otimes1\otimes\ldots\otimes1$ (n times) $\otimes(0\oplus1\oplus2\oplus\ldots)\otimes(0\oplus1\oplus2\oplus\ldots)\ldots(0\oplus1\oplus2\oplus\ldots)$ (p times). Now, the tensor product of spaces j_1 and j_2 will contain new spaces whose SO(3) indices are given in the following manner. We recall the rules of angular momentum addition familiar from quantum mechanics. The total angular momentum of an SO(3) representation space is given by its index j. The rules of angular momentum addition dictate how the product of two spaces of angular momentum (SO(3) index) j_1 and j_2 may be added and the possible j indices of the resulting spaces,

$$j_1 \otimes j_2 = |j_1 - j_2| \oplus \ldots \oplus (j_1 + j_2). \quad (2.8)$$

Equation (2.8) says that the direct product of two spaces each belonging to a particular j will generate a sum of new spaces with only those j indices allowed by the rule. The operations $\otimes$ and $\oplus$ are distributive like the corresponding arithmetic operators. For example

$$1 \otimes 1 = 0 \oplus 1 \oplus 2$$
$$1 \otimes 1 \otimes 1 = (0 \oplus 1 \oplus 2) \otimes 1$$
$$= (0 \otimes 1) \oplus (1 \otimes 1) \oplus (2 \otimes 1). \quad (2.9)$$

If we now apply the angular momentum addition rules (2.8) to each of the product terms on the right hand side of equation (2.9), we get

$$1 \otimes 1 \otimes 1 = 1 \oplus 0 \oplus 1 \oplus 2 \oplus 1 \oplus 2 \oplus 3$$
$$= 0 \oplus (1 \times 3) \oplus (2 \times 2) \oplus 3. \quad (2.10)$$

When taking the product of more than two spaces, there will be several ways to arrive at a particular j in the final product space. We see this in the final rearrangement of the right hand side of equation (2.10), where there is only one representation each of $j = 0$ and $j = 3$, but *three* of $j = 1$ and *two* of $j = 2$. Going back to our more complicated target space for the basis elements $B_{jm}^{\alpha_1 \alpha_2 \dots \alpha_n}(\hat{\mathbf{r}})$, which is $1 \otimes 1 \otimes \dots \otimes 1$ (n times) $\otimes (0 \oplus 1 \oplus 2 \oplus \dots) \otimes (0 \oplus 1 \oplus 2 \oplus \dots) \dots (0 \oplus 1 \oplus 2 \oplus \dots)$ (p times), we expect that there may be many ways to obtain a particular j in the final product space. To indicate this, for each basis element we associate a further index q giving $B_{qjm}(\hat{\mathbf{r}})$. These are important when we start to actually calculate the basis elements. The machinery used is the Clebsch–Gordan method well-known in quantum mechanics and the reader is referred to [5] for the details. The tensor T may also depend on *vectors* $\mathbf{r}_i$ with magnitudes r_i but for each rotation operation O_Λ these may be treated as parameters included in the weight associated with each basis element. The T are represented as

$$T^{\alpha_1 \alpha_2 \dots \alpha_n}(\mathbf{r_1}, \mathbf{r_2}, \dots, \mathbf{r_p}) = \sum_{q,j,m} a_{qjm}(r_1, r_2, \dots, r_p)$$
$$\times B_{qjm}^{\alpha_1 \alpha_2 \dots \alpha_n}(\hat{\mathbf{r}}_1, \hat{\mathbf{r}}_2, \dots, \hat{\mathbf{r}}_p). \quad (2.11)$$

In the next two sections, we move away from the very general formalism of SO(3) decomposition discussed above, and apply it to the specific case of the statistical tensor quantities in fluid turbulence and their dynamical equations obtained from the Navier–Stokes equations. We are motivated by the observation that the SO(3) decomposition when applied in the correct manner would allow (a) separation of the scaling variable r from the angular dependence, and (b) the separation of the isotropic from the anisotropic parts of the tensor.

2.2 Foliation of the structure function into j-sectors

We would like to use the formalism of the SO(3) representation, as presented in general terms in the previous section, to study the structure function tensor in turbulence theory. The n-th order structure function is defined by an ensemble average of the n-th moment of the difference of velocity components across scales r_p. It is in fact a tensor function over p unit spheres. We define the velocity difference as

$$w^{\alpha_i}(\mathbf{r_k}) = u^{\alpha_i}(\mathbf{x} + \mathbf{r_k}) - u^{\alpha_i}(\mathbf{x}), \qquad (2.12)$$

where the α_i denotes the component of the velocity vector in the direction i in a defined coordinate system, and the subscript k on $\mathbf{r_k}$ denotes a particular choice of the vector. The n-th order structure function for p such choices of vector scale $\mathbf{r}$ is then

$$S^{\alpha_1 \alpha_2 \ldots \alpha_n}(\mathbf{r_1}, \mathbf{r_2}, \ldots, \mathbf{r_p}) = \langle w^{\alpha_1}(\mathbf{r_i}) w^{\alpha_2}(\mathbf{r_j}) \ldots \times w^{\alpha_n}(\mathbf{r_k}) \rangle, \qquad (2.13)$$

where the subscripts i, j, ... denote any of the vectors from $\mathbf{r_1}$ to $\mathbf{r_p}$.

In this section we review briefly the reason why the structure functions may be written as a linear combination of the irreducible representations of the SO(3) symmetry group. For details the reader is again referred to [5]. The dynamical equation for the structure function may be derived from the Navier-Stokes equations. Similar to the solutions of the Laplace equation, the solution to the n-th order structure function equation forms a linear space with the basis chosen to be the irreducible representations $B_{qjm}(\mathbf{r})$ of the SO(3) group as shown in previous sections (see Eq. (2.11)). Since the basis is orthonormal for different j, m, and the equation is isotropic, we obtain a *hierarchy* of dynamical equations each with terms of a *given j* and m. This demonstrates that the dynamical equations themselves *do not* mix the various j and m contributions, and that the dynamical equation for the *isotropic* part of the structure function is different from the dynamical equations for any of the higher j contributions. Further, it may be shown from solvability conditions on this set of equations that the scaling part of the function changes between j-sectors while remaining the same within a j-sector, for different m. This motivates the postulate that the different j sectors scale with exponents different from the *isotropic* ζ_n. We denote these by $\zeta_n^{(j)}$ where subscript n indicates the rank of the tensor and superscript (j) indicates the number of the irreducible representation. We will leave out the superscript in the case of $j = 0$ since then we recover the known isotropic scaling exponent ζ_n.

2.3   The velocity structure functions

In the previous section we provided a heuristic justification for the use of the SO(3) decomposition for this tensor using the fact that it is the solution of an isotropic equation. Now, we use the rules of SO(3) (see Eq. (2.11)) decomposition to write it as

$$S^{\alpha_1\alpha_2\ldots\alpha_n}(\mathbf{r_1},\mathbf{r_2},\ldots,\mathbf{r_p}) \;=\; \sum_{q,j,m} a_{qjm}(r_1,r_2,\ldots,r_p)$$
$$\times B^{\alpha_1\alpha_2\ldots\alpha_n}_{qjm}(\hat{\mathbf{r}}_1,\hat{\mathbf{r}}_2,\ldots,\hat{\mathbf{r}}_p). \quad (2.14)$$

In this form, as was demonstrated for the scalar function ψ, the $j = 0$ representation is the isotropic one. Its prefactor $a_{q00}(r_1,r_2,\ldots,r_p)$ contains the scale dependence. For $p > 1$ there will be infinitely many ways to obtain a particular j for the product space as may be seen from the angular momentum addition rules in equation (2.8) and the direct-product representation. Therefore, q ranges from 1 to ∞. This is a difficult hurdle computationally but, fortunately, most experimental and theoretical work deals with the dependence on a single vector, and $p = 1$. In that case, and for rank $n = 2$, we obtain the second order structure function $S^{\alpha\beta}(\mathbf{r})$ quite easily.

2.3.1　The second-order structure function

We consider the structure function of equation (2.14) for $n = 2$ and $p = 1$. The tensor product space is

$$
\begin{aligned}
1 \otimes 1 \otimes (0 \oplus 1 \oplus 2 \oplus \ldots) \;&=\; (0 \oplus 1 \oplus 2) \otimes (0 \oplus 1 \oplus 2 \oplus \ldots) \\
&=\; (0 \otimes 0) \oplus (0 \otimes 1) \oplus (0 \otimes 2) \oplus \ldots \\
&\oplus\; (1 \otimes 0) \oplus (1 \otimes 1) \oplus (1 \otimes 2) \oplus \ldots \\
&\oplus\; (2 \otimes 0) \oplus (2 \otimes 1) \oplus (2 \otimes 2) \oplus \ldots \;(2.15)
\end{aligned}
$$

As demonstrated in (2.10) using the addition rule (2.8), there could be more than one way of obtaining a particular j representation is obtained in the product space. To count these for a given j, we had to add there the index q to the basis tensors (see for example Eq. (2.11)). We find for the second-order tensor over a single sphere that

- $j = 0$ has a total of 3 representations;

- $j = 1$ has a total of 7 representations;

- $j > 1$ has a total of 9 representations.

The Clebsch–Gordon machinery tells us in addition about the symmetry (in the indices) and parity (in r) of each q contribution. Using this information, the terms of the basis may be constructed. As in the case of the scalar function, the $B^{\alpha\beta}_{qjm}$ are orthonormal for different j, m and q. In practice, the Clebsch–Gordon method of constructing these objects is rather tedious and so we follow the alternative method offered in [5]. We make use of the

Clebsch–Gordon methods only to obtain the number of representations in each j, and its parity and symmetry properties. Armed with this information, the $B_{qjm}^{\alpha\beta}$ may be constructed by a more convenient means. The idea is to use the fact that we already know an orthonormal basis in the SO(3) representation for the scalar function over the sphere (Eq. (2.2)). We can now "add indices" to this basis in a way to be described, while retaining the properties under rotation, in other words its j and m values. To add indices in a way that does not change the j, m values is to perform contraction with the objects $\delta^{\alpha\beta}$, r^{α}, $\epsilon^{\alpha\beta\gamma}$ and the partial derivative operator ∂_{α}. This method automatically takes care of the j and m properties; and all that needs to be done now is to apply the above operators to obtain the different q terms with the right symmetry and parity properties. This gives us a complete set of basis elements which are different than the Clebsch–Gordan method. The orthonormality among different q for the same j and m is lost, but the orthonormality among different j and m is maintained because we start out with a basis equation (2.2) which already possesses these properties. However, different q elements are linearly independent and span a given j, m-sector. The details on how this is done using the rules from [5] is presented in Appendix A. In what follows we simply write down the components calculated using that method.

The second order structure function tensor is

$$S^{\alpha\beta}(\mathbf{r}) = \langle (u^{\alpha}(\mathbf{x}+\mathbf{r}) - u^{\alpha}(\mathbf{x}))(u^{\beta}(\mathbf{x}+\mathbf{r}) - u^{\beta}(\mathbf{x})) \rangle \qquad (2.16)$$

which we decompose using the SO(3) irreducible representations $B_{qjm}^{\alpha\beta}(\hat{\mathbf{r}})$ as

$$\begin{aligned}
S^{\alpha\beta}(\mathbf{r}) &= S_{j=0}^{\alpha\beta}(\mathbf{r}) + S_{j=1}^{\alpha\beta}(\mathbf{r}) + S_{j=2}^{\alpha\beta}(\mathbf{r}) + \dots \\
&= \sum_{q,j,m} a_{qjm}(r) B_{qjm}^{\alpha\beta}(\alpha\beta)(\hat{\mathbf{r}}).
\end{aligned} \qquad (2.17)$$

A further constraint is provided by the incompressibility condition

$$\partial_{\alpha} S^{\alpha\beta}(\mathbf{r}) = 0. \qquad (2.18)$$

Each j, m-sector must separately satisfy the incompressibility condition since taking the partial derivative preserves the rotation properties and does not mix the j, m sectors. Therefore, the incompressibility condition provides a constraint among the different q contributions within a given j, m-sector. For example, the $j = 0$ contribution has three different representations ($q = \{1, 2, 3\}$), two of which are symmetric and one antisymmetric in the indices α, β. By definition, the structure function is symmetric in the indices, therefore the antisymmetric contribution will not appear giving only two q contributions. The incompressibility constraint gives

$$\partial_\alpha \sum_q a_{q00} r^{\zeta_2} B_{q00}^{\alpha\beta}(\hat{\mathbf{r}}) = 0,$$

$$(2.19)$$

and

$$\partial_\alpha \left(a_{100} r^{\zeta_2} \delta^{\alpha\beta} + a_{200} r^{\zeta_2} \frac{r^\alpha r^\beta}{r^2} \right) = 0. \qquad (2.20)$$

Motivated by the expectation that the structure functions scale as powers of r in the inertial range, we assumed in equation (2.20) that the scale-dependent prefactor is of the form $a_{q00} r^{\zeta_n}$ where a_{q00} is a flow-dependent constant. Equation (2.20) results in a relationship between a_{100} and a_{200} giving the final form of the isotropic $j = 0$ sector with just one unknown coefficient c_0 to be determined by the flow boundary conditions. That is,

$$S_{j=0}^{\alpha\beta}(\mathbf{r}) = c_0 r^{\zeta_2} \left[(2 + \zeta_2)\delta^{\alpha\beta} - \zeta_2 \frac{r^\alpha r^\beta}{r^2} \right]. \qquad (2.21)$$

Here, $\zeta_2 \approx 0.69$ is the known empirically known anomalous second-order scaling exponent. We would like to assume a similar scaling form for the prefactor $a_{qjm}(r)$ for $j > 1$. In such a formulation, there is a hierarchy of scaling exponents which we denote by $\zeta_2^{(j)} \neq \zeta_2$ corresponding to the higher-order j sectors. Successive j's indicate increasing degrees of anisotropy. The following section provides a justification for such a classification of the scaling of the various sectors. It is the larger goal of this article to examine these aspects of the theory with the help of high-Reynolds-number data.

2.4 Dimensional estimates for the lowest-order anisotropic scaling exponents

In this section we present dimensional considerations to determine the "classical" values expected for $\zeta_2^{(1)}$ and $\zeta_2^{(2)}$ the spirit of Kolmogorov's 1941 theory (henceforth called K41). We work on the level of K41 to produce the value $\zeta_2^{(0)} = 2/3$. Ignoring intermittency corrections is justified to the lowest order because the differences between any two values $\zeta_2^{(j)}$ and $\zeta_2^{(j')}$ for $j \neq j'$ are considerably larger than the intermittency corrections to either of them.

It is easiest to produce a dimensional estimate for $\zeta_2^{(2)}$. One simply asserts [15] that the $j = 2$ contribution is the first one appearing in $S^{\alpha\beta}(\mathbf{r})$ due to the existence of a shear. Since the shear is a second rank tensor, it can appear linearly in the $j = 2$ contribution to $S^{\alpha\beta}(\mathbf{r})$. We thus have for any m, $-j \leq m \leq j$,

$$S_{j=2}^{\alpha\beta}(\mathbf{r}) \sim T^{\alpha\beta\gamma\delta} \frac{\partial \bar{U}^\gamma}{\partial r^\delta} f(r, \bar{\epsilon}). \tag{2.22}$$

Here $T^{\alpha\beta\gamma\delta}$ is a dimensionless tensor made of $\delta^{\alpha\beta}$, $\frac{r^\alpha}{r}$, and *bilinear* contributions made of the three unit vectors $\hat{\mathbf{p}}$, $\hat{\mathbf{m}}$, $\hat{\mathbf{n}}$, which form our coordinate system as defined in Appendix A. The form of equation (2.22) means that the dimensional function $f(r, \bar{\epsilon})$ stands for the response of the second-order structure function to a small external shear. Within the K41 dimensional reasoning, this function in the inertial interval can be made only of the mean energy flux per unit time and mass, $\bar{\epsilon}$ and r itself. The only combination of $\bar{\epsilon}$ and r that yields the right dimensions of the function f is $\bar{\epsilon}^{1/3} r^{4/3}$. Therefore

$$S_{j=2}^{\alpha\beta}(\mathbf{r}) \sim T^{\alpha\beta\gamma\delta} \frac{\partial \bar{U}^\gamma}{\partial r^\delta} \bar{\epsilon}^{1/3} r^{4/3}. \tag{2.23}$$

We thus find a "classical K41" value of $\zeta_2^{(2)}$ to be 4/3.

To obtain the value of $\zeta_2^{(1)}$ we cannot proceed in the same way. We need a contribution that is linear (rather than bilinear) in the unit vectors $\hat{\mathbf{p}}$, $\hat{\mathbf{m}}$, $\hat{\mathbf{n}}$. We cannot construct a contribution that is linear in the shear and yet does not vanish due to the incompressibility constraint. Thus there is a fundamental difference between the $j = 2$ term and the $j = 1$ term. While the former can be understood as an inhomogeneous term linear in the forced shear, the $j = 1$ term, being more subtle, may perhaps be connected to a solution of some homogeneous equation well within the inertial interval.

We therefore need to consider some quantity other than the shear which could contribute to the anisotropy. One invariance in the inviscid limit is as given by Kelvin's circulation theorem. In the so-called Clebsch representation, one writes the Euler equation in terms of one complex field $a(\mathbf{r}, t)$, see for example [16]. In the $\mathbf{k}$-representation, the Fourier component of the velocity field $\mathbf{u}(\mathbf{k}, t)$ is determined from a bilinear combination of the complex field

$$\mathbf{u}(\mathbf{k}, t) = \frac{1}{8\pi^3} \int d^3k_1 d^3k_2 \, \mathbf{\Psi}(\mathbf{k_1}, \mathbf{k_2}) a^*(\mathbf{k_1}, t) a(\mathbf{k_2}, t) \tag{2.24}$$

$$\mathbf{\Psi}(\mathbf{k_1}, \mathbf{k_2}) = \frac{1}{2} \left(\mathbf{k_1} + \mathbf{k_2} - (\mathbf{k_1} - \mathbf{k_2}) \frac{k_1^2 - k_2^2}{|\mathbf{k_1} - \mathbf{k_2}|^2} \right). \tag{2.25}$$

It was argued in [16] that this representation reveals a local conserved integral of motion given by

$$\mathbf{\Pi} = \frac{1}{8\pi^3} \int d^3k \, \mathbf{k} a^*(\mathbf{k}, t) a(\mathbf{k}, t). \tag{2.26}$$

Note that this conserved quantity is a vector, and so it cannot have a finite mean in an isotropic system. Consider a correction to the second order structure function due to a flux of the integral of motion $\bar{\pi}$. The dimensionality of $\bar{\pi}$ is $[\bar{\pi}] = [\bar{\epsilon}^{2/3}/r^{1/3}]$, and therefore now the dimensionless factor is $\bar{\pi} \, r^{1/3}/\bar{\epsilon}^{2/3}$. As we did with the shear in the case of the $j = 2$, we assume here the expandability of $\delta_\pi S$ at small values of the flux $\bar{\pi}$ and find

$$S_{j=1}^{\alpha\beta}(\mathbf{r}) \sim T^{\alpha\beta\gamma} \, \bar{\pi}^\gamma \, r \qquad (2.27)$$

where $T^{\alpha\beta\gamma}$ is a constant dimensionless tensor that is *linear* in the unit vectors $\hat{\mathbf{p}}$, $\hat{\mathbf{m}}$, $\hat{\mathbf{n}}$. We thus find the "classical K41" value $\zeta_2^{(1)}$ is 1. The derivation of the exponents $\zeta_2^{(1)}$ and $\zeta_2^{(2)}$ is given in [12]. It is concluded that dimensional analysis predicts values of 2/3, 1 and 4/3 for $\zeta_2^{(j)}$ with $j = 0$, 1 and 2 respectively. It is not known at present how to continue this line of argument for $j > 2$.

For the higher-order structure functions $S^{\alpha_1\alpha_2\cdots\alpha_n}(\mathbf{r})$ where $n > 2$, similar dimensional analysis may be performed in order to find the K41 contribution to scaling due to shear (corresponding to the $j = 2$ component). For each n, the lowest order correction to scaling that is linear in the shear is $(n + 2)/3$.

2.5 Summary

The technique of SO(3) decomposition may be used in order to write the structure function tensor in terms of its isotropic part, indexed by $j = 0$, and higher-order anisotropic parts, indexed by $j > 0$. The dynamical equation for the structure function of order n foliates into a set of equations each of different j, m. This motivates the postulate that the different sectors scale differently. The theory also indicates that any scaling behavior would depend only the j index and be independent of the m within that sector. There is allowance for dependence on boundary conditions, specific kinds of forcing and so on, in the unknown coefficients a_{qjm} in the SO(3) expansion (Eq. (2.11)). We first consider the second-order structure function in a detailed manner in the light of this group representation. The analysis may be implemented for structure functions of any order but the task becomes computationally and conceptually more difficult for $n > 2$. In Section 6, we present a means of circumventing these problems.

We have reviewed the theoretical estimates for the low-order anisotropic scaling exponents for the second-order structure function. The method may be carried over to the higher-order objects as well, and we have derived the lowest order shear-dependent scaling contributions to the nth order structure function to be $r^{(n+2)/3}$. For the second-order structure function, we have shown that the $j = 2$ contribution corresponds to a low-order shear-dependency. Again for the second-order structure function, we have presented a conserved quantity in the Clebsch-representation which provides the correct dimensional properties for the $j = 1$ scaling contribution to be r. The arguments used are purely dimensional in the K41 sense, and do not take account of anomalous scaling. We expect that real turbulent flows will exhibit anomalous scaling in the isotropic sector, *and* in every anisotropic sector in the SO(3) hierarchy. The issue of anomalous exponents in turbulence has now multiplied several-fold, to all the j sectors, in light of the apparent universality that this work suggests. Thus, the theoretically predicted anisotropic exponents are to be treated merely as estimates, especially for higher-order structure functions $(n > 2)$, since the scaling anomalies are expected to increase with the order n, similar to the isotropic case.

In subsequent sections we demonstrate the use of experimental data in order to test the predictions of the theory. In particular, we provide explicit calculations of measurable tensor quantities and extract scaling corrections due to anisotropy.

3   Some experimental considerations

3.1   Background

Experimental studies of turbulent flows at very high Reynolds numbers are usually limited in the sense that one measures the velocity field at a single spatial point, or a few spatial points, as a function of time [19], and uses Taylor's hypothesis to identify velocity increments at different times with those across spatial length scales, r. The standard outputs of such single-point measurements are the longitudinal two-point differences of the Eulerian velocity field and their moments. In homogeneous and isotropic turbulence, these structure functions are observed to vary as power-laws in r, with scaling exponents ζ_n [7].

Recent progress in measurements and in simulations has begun to offer information about the tensorial nature of structure functions. Ideally, one would like to measure the tensorial n-th order structure functions defined in equations (2.12) and (2.13). Such information should be useful in studying the anisotropic effects induced by all practical means of forcing.

3.2   Relevance of the anisotropic contributions

In analyzing experimental data the model of "homogeneous and isotropic small-scale" is universally adopted, but it is important to examine the relevance of this model for realistic flows. As we will demonstrate in the next section, our data we use exhibit anisotropy down to fairly small scales [25]. We have shown mathematically that keeping the tensorial information helps significantly in disentangling different scaling contributions to structure functions. In the light of the SO(3) representation of Section 2, where it is shown that anisotropy might lead to different scaling exponents for different tensorial components, a careful study of the various contributions is needed. This is our goal in the rest of this article.

3.3   The measurements

In order to extract a particular j contribution and the associated scaling exponent, one would ideally like to possess the statistics of the velocity at all points in three-dimensional space. One could then extract the j contribution of particular interest by multiplying the full structure function by the appropriate B_{qjm} and integrating over a sphere of radius r. Orthogonality of the basis functions ensures that only the j contribution survives the integration. One could then perform this procedure for various r and extract the scaling behaviors.

The method just described was adopted successfully in [2] using data from direct numerical simulations of channel flows. The Taylor microscale Reynolds number R_λ for the simulations was about 70. This is not large enough for a clear inertial scaling range to exist. The authors of [2] resort to extended self-similarity (ESS) in both the isotropic and the anisotropic sectors. Nonetheless, the results indicated a scaling exponent of about 4/3 in the $j = 2$ sector. While the experimental data are limited to a few points in space, and the integration over the sphere is not possible, we are able to attain very high Reynolds numbers especially in the atmosphere under steady conditions ($R_\lambda \approx 10\,000 - 20\,000$). Despite the advantage of extended scaling range, we are however faced with a true superposition of contributions from various j sectors with no simple way of disentangling them as was done with the numerical data. However, as we will show in Section 6, we can make a judicious choice of the tensor components studied, and obtain access to the anisotropic contributions.

The tensor structure of the velocity structure functions is lost in the computation of the usual single-point single-component measurement of longitudinal and transverse objects. This is because the part of the expansion that is dependent on the angle θ is hidden: the longitudinal and transverse components set the value of θ to a constant. The boundary-dependent prefactors now collapse to a single number as

$$S_{j=2}^{\alpha\beta}(r) = ar^{\zeta_2^{(j)}}. \tag{3.1}$$

On the one hand, this is a simple expression. On the other, the angular dependence is completely lost and there is no formal difference from the

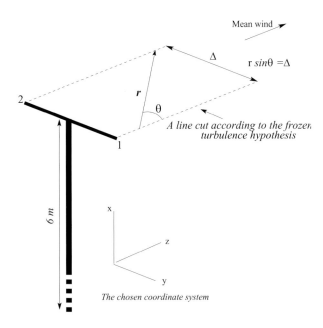

Fig. 2. Schematic of the experimental set-up. Shown is the positioning of the single wire probes 1 and 2 with respect to the mean wind, and a visual explanation of how Taylor's hypothesis is employed. The coordinate system chosen is different from what is conventional in turbulence – where the mean-wind is taken to lie along the x-axis. The pictured choice of coordinates simplifies the calculations involving the spherical harmonics.

isotropic object

$$S_{j=0}^{\alpha\beta}(r) = c_0 r^{\zeta_2}. \tag{3.2}$$

In order to see the true tensor character of the structure function we need an angular variation of the scale separation r. In the atmospheric boundary layer which offers the highest Reynolds numbers available, the simplest configuration that would allow us to do this is a two hot-wire combination separated by distance Δ in the spanwise direction (y), orthogonal to the mean-wind. By Taylor's frozen flow hypothesis, such a set-up will provide two simultaneous one-dimensional cuts through the flow. Therefore, one can measure the correlation between the two probes, across a scale r that makes an angle θ with the mean-wind direction. As r varies, the angle with respect to the mean-wind will also vary, giving some functional dependence on θ for the coefficients in the SO(3) decomposition. A schematic of the experimental configuration is presented in Figure 2.

A final consideration in measurements aimed at measuring anisotropic contributions is the homogeneity of the flow. The incompressibility

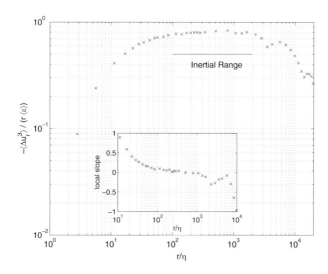

Fig. 3. A typical compensated longitudinal third-order structure function. The inset indicates the region of isotropic inertial range scaling. (From Ref. [6].)

condition may only be applied as a constraint on the structure function coefficients if the flow is homogeneous in the r direction. If we consider, for instance, probes separated in the shear (vertical) direction, r will have a component in the inhomogeneous direction and incompressibility may not be used to constrain the terms in the SO(3) expansion, leaving too many free parameters. However, in some instances, it is important to consider quantities that are not constrained by the incompressibility condition. For example, the $j = 1$ contribution to the second order structure function will not appear in the experimental configuration described above. This is so because only one symmetric, even-parity, term exists in $j = 1$ by the Clebsch–Gordon rules, and this must vanish by the incompressibility constraint. Probe-separation in the shear direction would be needed to produce a non-zero contribution of the $j = 1$ term. In short, one can understand the nature of the optimal experimental configuration from carefully studying the tensor decomposition of the structure function.

We analyze measurements in atmospheric turbulence at various heights above the ground (data sets I, II and III). Sets I and III were acquired from flow over a long fetch in the salt flats in Utah. The site of measurements was chosen to provide steady wind conditions. The surface of the desert was smooth and even: the water that floods the land during spring recedes uniformly and leaves the ground hard and essentially smooth during early

Table 1. Data sets I (first line), II (second line) and III (third-fifth lines). The various symbols have the following meanings: $\overline{U}$ = local mean velocity, u' = root-mean-square velocity, $\langle \varepsilon \rangle$ = energy dissipation rate obtained by the assumption of local isotropy and Taylor's hypothesis, η and λ are the Kolmogorov and Taylor length scales, respectively, the microscale Reynolds number $R_\lambda \equiv u'\lambda/\nu$, and f_s is the sampling frequency.

Height meters	$\overline{U}$ m s^{-1}	u' m s^{-1}	$10^2 \langle \varepsilon \rangle$, m^2 s^{-3}	η mm	λ cm	R_λ	f_s, per channel, Hz	# of samples
6	4.1	1.08	1.1	0.75	15	10 500	10 000	4×10^7
35	8.3	2.30	7.8	0.45	13	19 500	5000	4×10^7
0.11	2.7	0.47	6.6	0.47	2.8	900	5000	8×10^6
0.27	3.1	0.48	2.8	0.6	4.4	1400	5000	8×10^6
0.54	3.5	0.5	1.5	0.7	6.2	2100	5000	8×10^6

summer. The boundary layer on the desert floor in early summer is thus quite similar to that on a smooth flat plate [10]. The measurements were made in the early summer season roughly between 6 PM and 9 PM during which nearly neutral stability conditions prevailed. Data set II was acquired over a rough terrain with ill-defined fetch at the meteorological tower at the Brookhaven National Laboratory. In sets I and II, data were acquired at heights 6 m and 35 m respectively. They were recorded simultaneously from two single hot-wire probes separated in the spanwise direction y by 55 cm and 40 cm respectively. In both cases, the separation distance was within the inertial range, and was set nominally orthogonal to the mean wind direction (see below). Set III was acquired from an array of three cross-wires, arranged *above* each other at heights 11 cm, 27 cm and 54 cm respectively. These measurements are thus much closer to the desert floor. The hot-wires, about 0.7 mm in length and 6 μm in diameter, were calibrated just prior to mounting them on the meteorology towers and checked immediately after dismounting. The hot-wires were operated on DISA 55M01 constant-temperature anemometers. The frequency response of the hot-wires was typically good up to 20 kHz. The voltages from the anemometers were suitably low-pass filtered and digitized. The voltages were constantly monitored on an oscilloscope to ensure that they did not exceed the digitizer limits. Also monitored on-line were power spectra from an HP 3561A Dynamic Signal Analyzer. The wind speed and direction were independently monitored by a direction indicator mounted on the tower (sets I and III), or a vane anemometer a few meters away (set II). The real-time durations of data records were limited only by the degree of constancy demanded of the wind speed and its direction.

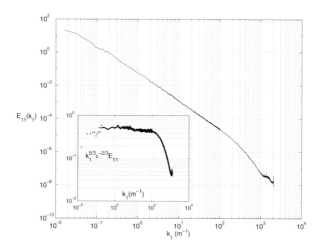

Fig. 4. The one-dimensional energy spectrum computed from data set II. (From Ref. [6].)

The Taylor microscale Reynolds number was 10 000 for set I and about 20 000 for set II [6]. For set III, the corresponding numbers were 900, 1400 and 2100, respectively, at the three heights. Table 1 lists a few relevant facts about the data records analyzed here. Figure 3 shows a compensated third order longitudinal structure function and indicates the region over which the inertial range is expected to hold. As another example of the nature of the data, we show in Figure 4 the longitudinal energy spectrum which displays extensive range of scaling. The slope is slightly larger than 5/3, as the compensated spectrum in the inset is supposed to clarify.

4   Anisotropic contribution in the case of homogeneity

4.1   General remarks on the data

We analyze data sets I and II described in Section 3 (see Tab. 1) to extract the lowest order anisotropic contribution to scaling [4, 12]. First, some preliminary tests and corrections need to be made. To test whether the separation between the two probes is indeed orthogonal to the mean wind, we computed the cross-correlation function $\langle u_1(t + \tau)u_2(t)\rangle$. Here, u_1 and u_2 refer to velocity fluctuations in the direction of the mean wind, for the arbitrarily numbered probes 1 and 2 respectively (see Fig. 2). If the separation were precisely orthogonal to the mean wind, the cross-correlation function should be maximum for $\tau = 0$. Instead, for set I, we found the maximum shifted slightly to $\tau = 0.022$ s, implying that the separation was not precisely orthogonal to the mean wind. To correct for this effect, the data from the second probe were time-shifted by 0.022 s. This amounts to a change in the actual value of the orthogonal distance. We computed this effective distance to be $\Delta \approx 54$ cm (instead of the 55 cm that was set physically). For set II, the effective separation distance was estimated to be 31 cm (instead of the physically set 40 cm). Next, we tested the isotropy of the flow for separations of the order of Δ. Define the "transverse" structure function across Δ as $S_T(\Delta) \equiv \langle [u_1(\bar{U}t) - u_2(\bar{U}t)]^2 \rangle$ and the "longitudinal" structure function as $S_L(\Delta) \equiv \langle [u_1(\bar{U}t + \bar{U}t_\Delta) - u_1(\bar{U}t)]^2 \rangle$ where $t_\Delta = \Delta/\bar{U}$. If the flow were isotropic we would expect [19]

$$S_T(\Delta) = S_L(\Delta) + \frac{\Delta}{2}\frac{\partial S_L(\Delta)}{\partial \Delta}. \tag{4.1}$$

In the isotropic state both longitudinal and transverse components scale with the same exponent, $S_{T,L}(\Delta) \propto \Delta^{\zeta_2}$, and the ratio S_T/S_L is computed from (4.1) to be $1 + \zeta_2/2 \approx 1.35$, because $\zeta_2 \approx 0.69$ (see below). The experimental ratio was found to be 1.32 for set I, indicating that the anisotropy at the scale Δ is small. This same ratio was about 1.8 for set III, indicating higher degree of anisotropy in that scale range. The differences between the two data sets seem attributable partly to differences in the terrain and other atmospheric conditions, and partly to the different distances from the ground.

Lastly, we needed to assess the effects of high turbulence level on Taylor's hypothesis. A comparison was made of the structure functions of two signals with turbulence levels differing by a factor of 2 and no difference was found. The correction scheme proposed in [27] also showed no changes. For a few separation distances, the statistics of velocity increments from two probes separated along $\mathbf{n}$, the direction of the mean wind, agreed with Taylor's hypothesis. More details can be found in [6].

The use of the isotropy equation (4.1) presents some evidence regarding the prevalence of modest amounts of anisotropy in the second-order statistics. This persistence is more explicit if one considers another familiar object, namely the cross-spectral density between horizontal and vertical velocity components. We consider the one-dimensional cross-spectral density (or shear-stress cospectrum) $E_{13}(k_3)$, which is zero in the case of isotropy. From dimensional considerations, the scaling exponent for this quantity is $-7/3$ (see [15] for theory and [21] for experimental tests). Figure 5 shows the cospectrum computed for the height of 0.54 m. The inset shows that the cospectrum compensated with a scaling exponent of -2.1 is flat. To the extent that this is numerically smaller than $7/3$, the decay of anisotropy is slower than expected for second-order quantities. Even allowing for the fact that the dimensional analysis assumes K41 scaling and therefore does

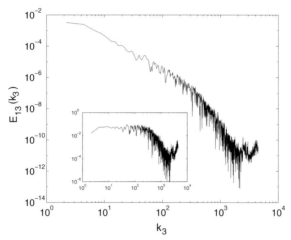

Fig. 5. Log-log plot of the shear-stress cospectrum $E_{13}(k_3)$ computed at 0.54 m. The inset shows a log-log plot of $k_3^{2.1} E_{13}(k_3)$ *vs.* k_3. The flat region indicates a region of scaling with exponent -2.1.

not account for possible intermittency corrections in the anisotropic sectors, it is still clear that the small scales do not attain isotropy as fast as dimensional considerations suggest. We would like to use the properties of the SO(3) decomposition in order to disentangle the isotropic from the anisotropic contributions and better quantify the anisotropy. We first consider the second-order structure function in detail.

4.2   The tensor form for the second-order structure function

4.2.1   The anisotropic tensor component derived under the assumption of axisymmetry

To obtain a theoretical form of the structure function tensor we first select a natural coordinate system. Our choice is to have the mean-wind direction **n** along the z-axis which we label as 3. The angle with respect to the 3-axis is the polar angle θ. The second axis is given by the separation vector **Δ** between the two probes. We make the simplifying assumption that the main symmetry broken in the flow is the cylindrical symmetry about the mean-wind direction. In other words, the main anisotropic contribution is cylindrically symmetric about the mean-wind direction. It is shown *a posteriori* that this assumption probably accounts for most of the anisotropy *for this particular geometrical set-up*. In the next section, we provide a complete analysis with no assumptions about the symmetries of the flow. We conclude from that analysis that the simplification of the present section is not inconsistent with the properties of the full tensor.

Next, we write down the tensor form for the general second-order structure function (defined by Eq. (2.13) for $n = 2$) in terms of irreducible representations of the SO(3) rotation group. Since we are far from the wall, our interest is in relatively modest anisotropies, and so we focus on the lowest order correction to the isotropic ($j = 0$) contribution. We therefore write

$$S^{\alpha\beta}(\mathbf{r}) = S^{\alpha\beta}_{j=0}(\mathbf{r}) + S^{\alpha\beta}_{j=2}(\mathbf{r}) + \ldots \qquad (4.2)$$

We do not have a $j = 1$ term because it vanishes from parity considerations (the structure function itself is even in r in our experimental configuration) or by the incompressibility constraint. Assumed cylindrical symmetry of the anisotropic $j = 2$ contribution implies that we only consider the $m = 0$ subspace in this contribution. Now, the most general form of the tensor can be written down by inspection. The case $j = 0$ is well known (see Eq. (2.21)) to be

$$S^{\alpha\beta}_{j=0}(\mathbf{r}) = c_0(r)\left[(2 + \zeta_2)\delta^{\alpha\beta} - \zeta_2\frac{r^\alpha r^\beta}{r^2}\right] \qquad (4.3)$$

where $c_0(r) = c_0 r^{\zeta_2}$, and c_0 is a non-universal numerical coefficient that needs to be obtained from fits to the data. The $j = 2$ component has six independent tensor forms and corresponding coefficients. These correspond to the different a_{q00} described in Section 1. As with $j = 0$, the $j = 2$ component can be simplified by imposing conditions of incompressibility and orthogonality with the $j = 0$ part of the tensor. This leaves us (in the case of cylindrical symmetry) with two independent coefficients which we call a and b, giving,

$$
\begin{aligned}
S^{\alpha\beta}_{j=2}(\mathbf{r}) = {} & ar^{\zeta_2^{(2)}}\left[(\zeta_2^{(2)} - 2)\delta^{\alpha\beta} - \zeta_2^{(2)}(\zeta_2^{(2)} + 6)\right. \\
& \times \delta^{\alpha\beta}\frac{(\mathbf{n}\cdot\mathbf{r})^2}{r^2} + 2\zeta_2^{(2)}(\zeta_2^{(2)} - 2)\frac{r^\alpha r^\beta(\mathbf{n}\cdot\mathbf{r})^2}{r^4} \\
& + ([\zeta_2^{(2)}]^2 + 3\zeta_2^{(2)} + 6)n^\alpha n^\beta \\
& \left. - \frac{\zeta_2^{(2)}(\zeta_2^{(2)} - 2)}{r^2}(r^\alpha n^\beta + r^\beta n^\alpha)(\mathbf{n}\cdot\mathbf{r})\right] \\
& + br^{\zeta_2^{(2)}}\left[-(\zeta_2^{(2)} + 3)(\zeta_2^{(2)} + 2)\delta^{\alpha\beta}(\mathbf{n}\cdot\mathbf{r})^2\right. \\
& + (\zeta_2^{(2)} - 2)\frac{r^\alpha r^\beta}{r^2} + (\zeta_2^{(2)} + 3)(\zeta_2^{(2)} + 2)n^\alpha n^\beta \\
& + (2\zeta_2^{(2)} + 1)(\zeta_2^{(2)} - 2)\frac{r^\alpha r^\beta(\mathbf{n}\cdot\mathbf{r})^2}{r^4} \\
& \left. - ([\zeta_2^{(2)}]^2 - 4)(r^\alpha n^\beta + r^\beta n^\alpha)(\mathbf{n}\cdot\mathbf{r})\right].
\end{aligned}
\qquad (4.4)
$$

We note that the forms for the tensor were derived on the basis of the assumption of cylindrical symmetry. We used neither the Clebsch–Gordon method nor the "adding indices" method of Section 1. This is why they were not automatically orthogonal to the $j = 0$ contribution; that condition had to be explicitly imposed. However, in the end, this method also yields the same number of independent coefficients as the formal Clebsch–Gordon methods. This merely shows that there are different ways of writing the tensor forms for the basis elements. However, the Clebsch–Gordon rules tell us the number, parity and symmetry of the terms we must in the end have for a given j, m sector.

Finally, we note that in the present experimental set-up only the component of the velocity in the direction of $\mathbf{n}$ (the 3-axis) is measured. In the coordinate system chosen above we can read from equations (2.21) and (4.4) the relevant component to be

$$
\begin{aligned}
S^{33}(r,\theta) &= S^{33}_{j=0}(r,\theta) + S^{33}_{j=2}(r,\theta) \\
&= c_0 \left(\frac{r}{\Delta}\right)^{\zeta_2} \left[2 + \zeta_2 - \zeta_2 \cos^2\theta\right] \\
&\quad + a \left(\frac{r}{\Delta}\right)^{\zeta_2^{(2)}} \left[(\zeta_2^{(2)} + 2)^2 - \zeta_2^{(2)}(3\zeta_2^{(2)} + 2)\cos^2\theta \right. \\
&\qquad \left. + 2\zeta_2^{(2)}(\zeta_2^{(2)} - 2)\cos^4\theta\right] \\
&\quad + b \left(\frac{r}{\Delta}\right)^{\zeta_2^{(2)}} \left[(\zeta_2^{(2)} + 2)(\zeta_2^{(2)} + 3) - \zeta_2^{(2)}(3\zeta_2^{(2)} + 4)\right. \\
&\qquad \left. \times \cos^2\theta + (2\zeta_2^{(2)} + 1)(\zeta_2^{(2)} - 2)\cos^4\theta\right].
\end{aligned}
\tag{4.5}
$$

Here θ is the angle between $\mathbf{r}$ and $\mathbf{n}$, and r has been normalized by Δ, making all the coefficients dimensional, with units of $(\text{m/sec})^2$. Taylor's hypothesis allows us to obtain components of $S^{\alpha\beta}$ from (4.5), with $\theta = 0$ and variable, respectively. In other words,

$$
S^{33}(r, \theta = 0) = \langle [u_1(\bar{U}t + \bar{U}t_r) - u_1(\bar{U}t)]^2 \rangle,
\tag{4.6}
$$

where $t_r \equiv r/\bar{U}$, and

$$
S^{33}(r, \theta) = \langle [u_1(\bar{U}t + \bar{U}t_{\tilde{r}}) - u_2(\bar{U}t)]^2 \rangle.
\tag{4.7}
$$

Here $\theta = \arctan(\Delta/\bar{U}t_{\tilde{r}})$, $t_{\tilde{r}} = \tilde{r}/\bar{U}$, and $r = \sqrt{\Delta^2 + (\bar{U}t_{\tilde{r}})^2}$.

The quantities on the left hand side of equations (4.6) and (4.7) were computed from the experimental data and fitted to the theoretical expression (4.5) using the appropriate values of θ. The fits were performed in the range $1 < r/\Delta < 10$ (0.54 m $< r <$ 5.4 m) for set I and $1 < r/\Delta < 25$ (0.31 m $< r <$ 8 m) for set II. The ranges were based on the constancy

Table 2. The scaling exponents and the three coefficients in units of $(\text{m/s})^2$ as determined from the nonlinear fit of equation (7) to data sets I (first line) and II (second line).

ζ_2	$\zeta_2^{(2)}$	c_0	a	b
0.69	1.38 ± 0.15	0.023 ± 0.001	−0.0051 ± 0.0006	0.0033 ± 0.0005
0.69	1.36 ± 0.10	0.112 ± 0.001	−0.052 ± 0.004	0.050 ± 0.004

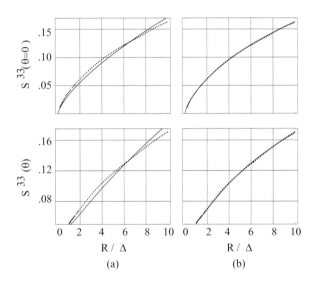

Fig. 6. The structure functions S^{33} for $\theta = 0$ and for non-zero θ computed for set I. The dots are for experimental data and the line is the analytic fit. Panel (a) presents fits to the $j = 0$ component only, and panel (b) to components $j = 0$ and $j = 2$ together.

of the third order structure function (see Fig. 3 for an example). Panels (a) of Figures 6 and 7 show, for data sets I and II respectively, a comparison between the measured $S^{33}(r, \theta = 0)$ and the $j = 0$ form of the equation. The comparison shows that the agreement is modest, and that the best-fit yields the exponent ζ_2 to be 0.69. A careful analysis of the data elsewhere [25] shows that, if the effect of the shear is removed in a plausible way, the power-law fit is excellent over a range of scale separations. To include the $j = 2$ contribution, we fixed ζ_2 to be 0.69 and performed the following analysis. For given values of the variables r and θ, we guessed the second exponent $\zeta_2^{(2)}$ and estimated the unknown coefficients c_0, a and b by

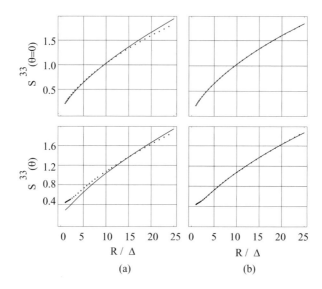

Fig. 7. The structure functions S^{33} for $\theta = 0$ and for non-zero θ computed for set III. The dots are for experimental data and the line is the analytic fit. Panel (a) presents fits to the $j = 0$ component only, and panel (b) to the sum of the components $j = 0$ and $j = 2$.

using a linear regression algorithm. We followed this procedure repeatedly for different values of $\zeta_2^{(2)}$ ranging from 0 to 2. We then chose the value of $\zeta_2^{(2)}$ that minimized χ^2 (the sum of the squares of the differences between the experimental data and the fitted values).

In Figure 8 we present χ^2 values as a function of $\zeta_2^{(2)}$. The optimal value of this exponent and the uncertainty determined from these plots is $\zeta_2^{(2)} \approx 1.38 \pm 0.15$ from set I, and $\zeta_2^{(2)} \approx 1.36 \pm 0.1$ from set II. The best numerical values for the coefficients are presented in Table 2. Panels (b) in Figures 6 and 7 show fits to the sum of the $j = 0$ and $j = 2$ contributions to the experimental data. Even though the $j = 2$ contributions are small, they improve the fits tremendously. This situation lends support to the essential correctness of the present analysis.

The figures show that the purely longitudinal structure function corresponding to $\theta = 0$ is somewhat less affected by the anisotropy than is the finite θ structure function (see especially, Fig. 7). The reason is the closeness of the numerical absolute values of the coefficients a and b (see Tab. 2). For $\theta = 0$ the two tensor forms multiplied by a and b coincide, and the $j = 2$ contribution becomes very small. The value of $\zeta_2 = 0.69$ quoted above can be obtained from such a fit to the $\zeta_2 = 0$ part alone; as

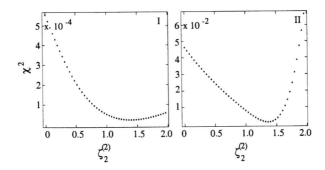

Fig. 8. The determination of the exponent $\zeta_2^{(2)}$ from a least-square fit of $S^{33}(r,\theta)$ to its analytic form. From set I we obtain a numerical value of the best fit, $\zeta_2^{(2)} = 1.38 \pm 0.15$ while from set III the best fit value is 1.36 ± 0.1 both of which are in close agreement with the theoretical expectation of 4/3 of Section 2.4 (without intermittency corrections). The differences in the nature of the minima are not understood.

long as one measures only this component, it seems reasonable to proceed with just that one exponent. However, the inclusion of the second exponent $\zeta_2^{(2)}$ improves the fit even for the longitudinal case; for the finite θ case, this inclusion appears essential for a good fit. In fact, the fit using $j = 0$ and $j = 2$ for data set II extends all the way to 40 m (beyond the range shown in Fig. 7). This indicates that the "inertial-range" where scaling theory applies is much longer than anticipated by traditional log-log plots. The close agreement with the theoretical expectation of 4/3 (*e.g.* [12,15]), and the apparent reproducibility of the result for two different experiments is a strong indication that this exponent may be universal.

It should be understood that, for the objects considered in this section, the exponent $\zeta_2^{(2)}$ is just the smallest exponents in a possible hierarchy $\zeta_2^{(j)}$ that characterizes higher order irreducible representations indexed by j. (The case of $\zeta_2^{(1)}$ will be discussed in Sect. 5.) As discussed earlier, we expect the exponents to be a non-decreasing function of j, and that the highest values of j are being peeled off quickly when r decreases. Nevertheless, the lower order values of $\zeta_2^{(j)}$ can be measured and computed. The results of this section were first reported in [4]. Results are also in agreement with the subsequent analysis of numerical simulations [2].

4.2.2 The complete $j = 2$ anisotropic contribution

In the previous section, preliminary results on the scaling exponent $\zeta_2^{(2)}$, were obtained under the assumption of cylindrical symmetry of the dominating anisotropic contribution. The analysis here is more complete, and takes into account the full tensorial structure. We show that this is feasible, and the final results are in agreement with those presented in the previous section.

The method used to extract the unknown anisotropic scaling exponent is essentially the same as in the previous section. Since we look for the lowest order anisotropic contributions in our analyses, we perform a two-stage procedure to separate the various sectors. First we look at the small scale region of the inertial range to determine the extent of the fit with a single (isotropic) exponent. We then seek to extend this range by including appropriate anisotropic tensor contributions, and obtain the additional scaling exponents using least-squares fitting procedure. This procedure is self-consistent.

Reference [18] presents a detailed analysis of the consequences of Taylor's hypothesis on the basis of an exactly soluble model. It also proposes ways for minimizing the systematic errors introduced by the use of Taylor's hypothesis. In light of that analysis we will use an "effective" wind, U_{eff}, for surrogating the time data. This velocity is a combination of the mean wind $\overline{U}$ and the root-mean-square u',

$$U_{\text{eff}} \equiv \sqrt{\overline{U}^2 + (bu')^2} \qquad (4.8)$$

where b is a dimensionless parameter.

In the second order structure function defined already, *viz.*,

$$S^{\alpha\beta}(\mathbf{r}) = \langle (u^\alpha(\mathbf{x}+\mathbf{r}) - u^\alpha(\mathbf{x}))(u^\beta(\mathbf{x}+\mathbf{r}) - u^\beta(\mathbf{x}))\rangle, \qquad (4.9)$$

the $j = 2$ component of the SO(3) symmetry group corresponds to the lowest order anisotropic contribution that is symmetric in the indices, and has even parity in $\mathbf{r}$ (due to homogeneity). Although the assumption of axisymmetry used in [4] and in the previous section seemed to be justified from the excellent qualities of fits obtained, we attempt to fit the same data (set I) with the *full* tensor form for the $j = 2$ contribution. The derivation of the full $j = 2$ contribution to the symmetric, even parity, structure function appears in Appendix A. We then find the range of scales over which the structure function

$$S^{33}(r, \theta = 0) = \langle (u_1^{(3)}(x+r) - u_1^{(3)}(x))^2 \rangle, \qquad (4.10)$$

with the subscript 1 denoting one of the two probes, can be fitted with a single exponent. To find the $j = 2$ anisotropic exponent we need to use data taken from both probes. To clarify the procedure, for the geometry shown in Figure 2, what is computed is actually

$$S^{33}(r, \theta) = \langle [u_1^{(3)}(U_{\text{eff}}t + U_{\text{eff}}t_{\tilde{r}}) - u_2^{(3)}(U_{\text{eff}}t)]^2 \rangle. \qquad (4.11)$$

Here $\theta = \arctan(\Delta/U_{\text{eff}}t_{\tilde{r}})$, $t_{\tilde{r}} = \tilde{r}/U_{\text{eff}}$, and $r = \sqrt{\Delta^2 + (\bar{U}_{\text{eff}}t_{\tilde{r}})^2}$. U_{eff} is defined by equation (4.8) with the optimal value of b taken from model studies to be 3. For simplicity we shall refer from now on to such quantities as

$$S^{33}(r,\theta) = \langle (u_1^{(3)}(x+r) - u_2^{(3)}(x))^2 \rangle. \tag{4.12}$$

Next, we fix the scaling exponent of the isotropic sector to be 0.69 and find the $j = 2$ anisotropic exponent that results from fitting to the full $j = 2$ tensor contribution. We fit the objects in equations (4.10) and (4.12) to the sum of the $j = 0$ (with scaling exponent $\zeta_2 = 0.69$) and the $j = 2$ contributions. The sum is given by (see Appendix A)

$$
\begin{aligned}
S^{33}(r,\theta) = {} & S^{33}_{j=0}(r,\theta) + S^{33}_{j=2}(r,\theta) = c_0 \left(\frac{r}{\Delta}\right)^{\zeta_2} \left[2 + \zeta_2 - \zeta_2 \cos^2\theta\right] \\
& + a \left(\frac{r}{\Delta}\right)^{\zeta_2^{(2)}} \left[(\zeta_2^{(2)} + 2)^2 - \zeta_2^{(2)}(3\zeta_2^{(2)} + 2)\cos^2\theta \right. \\
& \left. \qquad + 2\zeta_2^{(2)}(\zeta_2^{(2)} - 2)\cos^4\theta\right] \\
& + b \left(\frac{r}{\Delta}\right)^{\zeta_2^{(2)}} \left[(\zeta_2^{(2)} + 2)(\zeta_2^{(2)} + 3) - \zeta_2^{(2)}(3\zeta_2^{(2)} + 4)\cos^2\theta \right. \\
& \left. \qquad + (2\zeta_2^{(2)} + 1)(\zeta_2^{(2)} - 2)\cos^4\theta\right] \\
& + a_{9,2,1} \left(\frac{r}{\Delta}\right)^{\zeta_2^{(2)}} \left[-2\zeta_2^{(2)}(\zeta_2^{(2)} + 2)\sin\theta\cos\theta \right. \\
& \left. \qquad + 2\zeta_2^{(2)}(\zeta_2^{(2)} - 2)\cos^3\theta\sin\theta\right] \\
& + a_{9,2,2} \left(\frac{r}{\Delta}\right)^{\zeta_2^{(2)}} \left[-2\zeta_2^{(2)}(\zeta_2^{(2)} - 2)\cos^2\theta\sin^2\theta\right] \\
& + a_{1,2,2} \left(\frac{r}{\Delta}\right)^{\zeta_2^{(2)}} \left[-2\zeta_2^{(2)}(\zeta_2^{(2)} - 2)\sin^2\theta\right].
\end{aligned}
\tag{4.13}
$$

We fit the experimentally generated functions to the above form using values of $\zeta_2^{(2)}$ ranging from 0.5 to 3. Each iteration of the fitting procedure involves solving for the six unknown, non-universal coefficients. The best value of $\zeta_2^{(2)}$ is the one that minimizes the χ^2 value for these fits; we obtain that to be 1.38 ± 0.15. The fits with this choice of exponent are displayed in Figure 9.

The corresponding values of the six fitted coefficients is given in Table 2. The range of scales that are fitted to this expression is $0.2 < r/\Delta < 25$ for the $\theta = 0$ (single-probe) structure function and $1 < r/\Delta < 25$ for the $\theta \neq 0$ (two-probe) structure function. We are unable to fit equation (4.14) to scales larger than about 12 m without losing the quality of the fit in the small scales. This limit is roughly twice the height of the

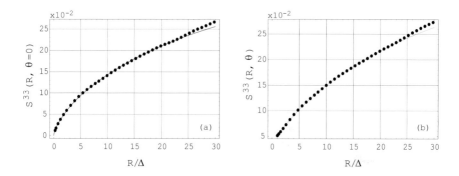

Fig. 9. The structure functions computed from data set I and fit with the $j = 0$ and full $j = 2$ tensor contributions using the best fit values of exponents $\zeta_2 = 0.68$ and $\zeta_2^{(2)} = 1.38$. Panel (a) shows the fit to the single-probe ($\theta = 0$) structure function in the range $0.2 < r/\Delta < 25$ and panel (b) shows the fit to the θ-dependent structure function in the range $1 < r/\Delta < 25$.

probe from the ground. Based on the earlier discussion, we should be in the regime of the largest scales up to which the three-dimensional theory would hold. (Beyond this scale the boundary layer is dominated by the sloshing motions which are quasi two-dimensional in nature.) Therefore this limit to the fitting range is consistent with our expectations for the maximum scale of three-dimensional turbulence. We conclude that the structure functions exhibit scaling behavior over the whole scaling range, but this important fact is missed if one does not consider a superposition of the $j = 0$ and $j = 2$ contributions. We thus conclude that the estimate for the $j = 2$ scaling exponent $\zeta_2^{(2)} \approx 1.38$. This same estimate was obtained in [4] and in the previous section using only the axisymmetric terms. The value of the coefficients a and b are again close in magnitude but opposite in sign – just as in [4], giving a small contribution to $S^{33}(r, \theta = 0)$. The non-axisymmetric contributions vanish in the case of $\theta = 0$. The contribution of these terms to the finite θ function is relatively small because the angular dependence appears as $\sin\theta$ and $\sin^2\theta$, both of which are small for small θ (large r); it was for this reason that we were able to obtain in the previous section a good fit to just the axisymmetric contribution. Lastly, we note that the total number of free parameters in this fit is 7 (6 coefficients and 1 exponent). The relative "flatness" of the χ^2 function near its minimum (see, especially the left panel of Fig. 8) may be indicative of the large number of free parameters in the fit. However, the value of the exponent is perfectly in agreement with the analysis of numerical simulations [2] in which one can properly integrate the structure function against the basis functions, eliminating all contributions except that of the $j = 2$ sector. Furthermore, fits to the data in the vicinity of $\zeta_2^{(2)} = 1.38$ show enough divergence from experiment that we are satisfied about the genuineness of the χ^2 result. The results of this analysis taking into account the full $j = 2$ tensor were first presented in [12].

4.3 Summary

We consider the lowest order anisotropic contribution to the second order tensor functions of velocity in the atmospheric boundary layer. The $j = 1$ contribution is absent in our experimental configuration because of the incompressibility constraint. The $j = 2$ sector is expected to be the dominant contribution to anisotropy. First we make the *a priori* assumption that the cylindrical symmetry is broken in the first deviation from anisotropy. The use of the Clebsch–Gordon rules tell us the number of terms that must be included. We derive on this basis the tensor form for the $j = 2$ contribution that is axisymmetric. The conditions of orthogonality with $j = 0$ and incompressibility are used to constrain the unknown coefficients giving two unknown coefficients and one unknown exponent in the anisotropic sector. Using this form along with the known isotropic contribution, we can extract the anisotropic scaling exponent from the experimental data. We have also used the full form of the $j = 2$ tensor including all its $2j + 1$ components in order to examine if the results from the initial analysis were justified. The excellent agreement between the two strengthens our confidence in the value of the $j = 2$ exponent $\zeta_2^{(2)} \approx 1.36$, and in our initial assumption of the breaking of axisymmetry of the flow.

5   Anisotropic contribution in the case of inhomogeneity

5.1   Extracting the $j = 1$ component

The homogeneous structure function defined in equation (4.9) is known from properties of symmetry and parity to possess no contribution from the $j = 1$ sector (see Sect. B.2), the $j = 2$ sector being its lowest order anisotropic contributor. In order to isolate the scaling behavior of the $j = 1$ contribution in atmospheric shear flows we must either (a) explicitly construct a new tensor object which will allow for such a contribution, or (b) extract it from the structure function itself computed in the presence of *inhomogeneity*. Adopting the former approach, we construct the tensor

$$T^{\alpha\beta}(\mathbf{r}) = \langle (u^\alpha(\mathbf{x} + \mathbf{r}) - u^\alpha(\mathbf{x}))(u^\beta(\mathbf{x} + \mathbf{r}) + u^\beta(\mathbf{x})) \rangle. \qquad (5.1)$$

This object vanishes both when $\alpha = \beta$ and when $\mathbf{r}$ is in the direction of homogeneity, viz., the streamwise direction. From data set III we can calculate this function for non-homogeneous scale-separations (in the shear direction). In general, this will exhibit mixed parity and symmetry; the incompressibility condition does not reduce our parameter space. Therefore, to minimize the final number of fitting parameters, we examine only the antisymmetric contribution. In Section B.2, we have derived the antisymmetric tensor contributions in the $j = 1$ sector, and used this to fit for the unknown $j = 1$ exponent. We describe the results of this effort below. This can be used to find $j = 1$ exponent for the inhomogeneous structure function which is symmetric but has mixed parity. We do not present the result of that analysis here essentially because it is consistent with those from the antisymmetric case.

Returning to consideration of the antisymmetric part of the tensor object defined in equation (5.1), viz.,

$$\widetilde{T}^{\alpha\beta}(\mathbf{r}) = \frac{T^{\alpha\beta}(\mathbf{r}) - T^{\beta\alpha}(\mathbf{r})}{2}$$
$$= \langle u^{\alpha}(\mathbf{x})u^{\beta}(\mathbf{x}+\mathbf{r})\rangle - \langle u^{\beta}(\mathbf{x})u^{\alpha}(\mathbf{x}+\mathbf{r})\rangle, \qquad (5.2)$$

it is easy to see that it will only have contributions from the antisymmetric $j = 1$ basis tensors. An additional useful property of this object is that, for the configuration of data set II, it does not have any contribution from the isotropic $j = 0$ sector spanned by $\delta^{\alpha\beta}$ and $r^{\alpha}r^{\beta}$ since these objects are symmetric in the indices. This allows us to isolate the $j = 1$ contribution and determine its scaling exponent $\zeta_2^{(1)}$ starting from the smallest scales available. Using data (set III) from the probes at 0.27 m (probe 1) and at 0.11 m (probe 2) we calculate

$$\widetilde{T}^{31}(\mathbf{r}) = \langle u_2^{(3)}(\mathbf{x})u_1^{(1)}(\mathbf{x}+\mathbf{r})\rangle - \langle u_1^{(3)}(\mathbf{x}+\mathbf{r})u_2^{(1)}(\mathbf{x})\rangle \qquad (5.3)$$

where again superscripts denote the velocity component and subscripts denote the probe by which this component is measured. The goal is to fit this experimental object to the tensor form derived in Appendix B, equation (B.8), namely,

$$\widetilde{T}^{31}(r, \theta, \phi = 0) = -a_{3,1,0}r^{\zeta_2^{(1)}}\sin\theta + a_{2,1,1}r^{\zeta_2^{(1)}} + a_{3,1,-1}r^{\zeta_2^{(1)}}\cos\theta. \qquad (5.4)$$

Figure 10 gives the χ^2 minimization of the fit as a function of $\zeta_2^{(1)}$. We obtain the best value to be 1 ± 0.15 for the final fit. This is shown in Figure 11. The fit in Figure 11 peels off at around $r/\Delta = 2$. The values of the coefficients corresponding to the exponent $\zeta_2^{(1)} = 1$ are given in Table 3. The maximum range of scales over which the fit works is of the order of the

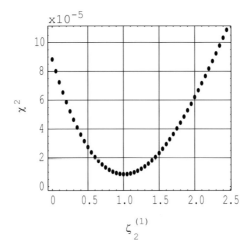

Fig. 10. The χ^2 minimization by the best-fit value of the exponent $\zeta_2^{(1)}$ of the $j = 1$ anisotropic sector from the fit to θ-dependent $\widetilde{T}^{31}(r, \theta)$ function in the range $1 < r/\Delta < 2.2$.

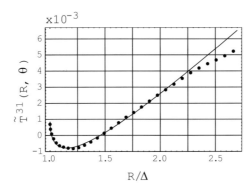

Fig. 11. The fitted $\widetilde{T}^{31}(r, \theta)$ function. The dots indicate the data and the line is the fit.

height of the probes from the ground, consistent with the considerations presented earlier. This value of the scaling exponent of the $j = 1$ sector is in agreement with the theoretically expected value of unity (Sect. 2.4). Again, we have satisfied ourselves that a different value of the exponent yields a substantially poorer fit to the data. These findings significantly strengthen the proposition [5] that the scaling exponents in the various sectors (at least up to $j = 2$) are likely to be universal.

Table 3. The values of the exponents and coefficients (in units of $(m/s)^2$) obtained from the fit to the function $\widetilde{T}^{31}(r, \theta)$.

$\zeta_2^{(1)}$	$a_{3,1,0}$	$a_{2,1,1}$	$a_{3,1,-1}$
1 ± 0.15	0.0116 ± 0.001	0.0124 ± 0.001	-0.0062 ± 0.001

6   The higher-order structure functions

6.1   Introduction

The method of SO(3) decomposition is applicable to tensor functions of any rank. However, computing the anisotropic terms in the higher order structure functions ($n > 2$) becomes an increasingly difficult task. The number of contributions to a given j (*i.e.*, the q index) quickly becomes too large. If we were to use the methods adopted in the case of the second order structure function, the large number of free parameters available for fitting the data guarantees over-fitting. This would not allow us to unambiguously extract the anisotropic scaling behavior. In order to circumvent this problem we recall two properties of the tensor decomposition. First, for the single-point measurements, with $\theta = 0$ (no angular dependence), although the tensorial information is lost, the free coefficients collapse to a single number, see equations (3.2) and (3.1). Second, the *isotropic* part of the structure function tensor is very easily obtained, as we will show below. Examination of this part of the SO(3) expansion will show us which tensor components do not contribute to the isotropic sector. We then extract anisotropic exponents by considering only those tensor components that are explicitly zero in the isotropic sector, so that whatever is measured derives its contribution *entirely* from the anisotropic sector. We use an interpolation formula to compensate for the large-scale encroachment of inertial-range scales. This allows us to examine the lowest order anisotropic scaling behavior. The resulting anisotropic exponents for a given tensorial order are larger than those known for the corresponding isotropic part. One conclusion that emerges is that the anisotropy effects diminish with decreasing scale, although more slowly than previously thought.

We can use the present method in principle to examine the anisotropic contribution of tensors of *any* order without requiring the knowledge of the particular mathematical form of the anisotropic sectors of these tensors. This is a considerable advantage theoretically because the high-order tensors are non-trivial to compute; it is an advantage experimentally because, unlike in numerical simulations, one can measure only some components for simple geometric arrangements of probes. The results of this analysis were first presented in [13].

6.2   Method and results

In this part of the analysis we use data set II described in Section 2. We first consider the second-order tensor $S^{\alpha\beta}(\mathbf{r})$. Isotropy implies that this tensor can be expressed as a linear combination of two terms, $\delta^{\alpha\beta}$ and $r^\alpha r^\beta$. As is well known, both terms give non-zero contributions to longitudinal as well as transverse components, corresponding to $\alpha = \beta$. For $\alpha \neq \beta$ these two terms are identically zero if $\mathbf{r}$ is taken to be in the streamwise direction 3. Therefore, we compute the so-called mixed structure function

$$S^{31}(r) = \langle (u^3(x+r) - u^3(x))(u^1(x+r) - u^1(x)) \rangle, \qquad (6.1)$$

where, as already noted, the superscripts 1 and 3 denote the vertical and streamwise components respectively. This object is identically zero in the isotropic sector, and so, any non-zero value comes from anisotropy. In particular, any scaling behavior that it obeys should relate solely to anisotropy. By computing equation (6.1) and examining its scaling, we intend to extract the purely anisotropic scaling behavior in the $j = 2$ sector, uncontaminated by any isotropic scaling, in contrast to the case of either longitudinal or transverse structure functions.

6.2.1   The second-order structure function

The previous paragraph provides the motivation for examining the measured structure functions $S^{31}(r)$. However, as we shall see shortly, apart from the expected r^2 behavior in the dissipative range and saturation at some large scale, there appears to be no distinct inertial range scaling. We suspect that this happens because there is poor scale separation, since the probes are fairly close to the ground; in fact, the large scales (which we expect to be of the order of the height of the probe from the ground and larger (see previous section)) may be encroaching significantly into the inertial range. We would be aided materially in our search for scaling if, somehow, the large-scale effects can be separated. One way of doing this is to write down an interpolation function that models the entire structure function in its three different scaling regions – a dissipative range that scales like r^2 when r is of the order of the Kolmogorov scale η, a large-scale behavior that tends to saturate (indicating decorrelation) as r gets to be larger than L, and the intermediate inertial range for $\eta \ll r \ll L$ which may exhibit scaling. Through the use of the interpolation formula, one can extract the scaling part in a natural way. This is described below.

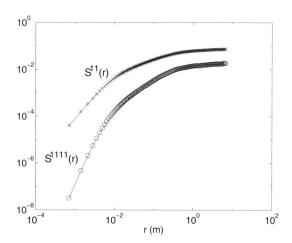

Fig. 12. Log–log plots of transverse structure functions at 0.54 m. ($\times$) denote the second order, ($\circ$) the fourth order and the solid lines represent the interpolation fit.

A suitable form of the interpolation function is given in [6] for structure functions of arbitrary order. It has the form

$$S^{\alpha_1\alpha_2\ldots\alpha_n}(r) = \frac{A_n\eta^n(r/\eta)^n}{(1 + B_n(r/\eta)^2)^{C_n}}(1 + D_n(r/L))^{2C_n-n}, \qquad (6.2)$$

where A_n, B_n, C_n and D_n are variable parameters. This formula is an extension of that given in [28] and includes a large-scale term. Such extensions have been attempted earlier (*e.g.* [14]). Antonia *et al.* [1] have successfully tested the function (6.2). Dhruva [6] has shown that the present interpolation formula works extremely well for longitudinal structure functions of order 2, 4 and 6. To reinforce this point, we test its performance by comparing it to the measured transverse structure function, $\alpha = \beta = 1$, r in the direction 3. For each data set, the height of the probe is assumed to be the large scale L. The fit is shown for the transverse structure function of orders 2 and 4 at the 0.54 m probe in Figure 12. The agreement between the formula and the data is excellent. Taken together with similar conclusions in [6] for longitudinal structure functions, we conclude that the interpolation formula describes the familiar structure functions very well. For this pragmatic reason, we shall adopt it for our purposes here, and test the robustness of the results obtained in the Appendix C.

In the formula (6.2), the large scale behavior is given by the factor $(1 + D_2(r/L))^{2C_2-2}$. If the measured structure function is divided by this factor,

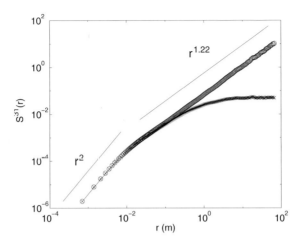

Fig. 13. Log–log plot of second order mixed structure function at 0.54 m. ($\times$) denote data, the solid line is the interpolation fit (not visible beyond an r of 10^{-1} m because of the closely packed symbols), and ($\circ$) correspond to the large-scale compensated function.

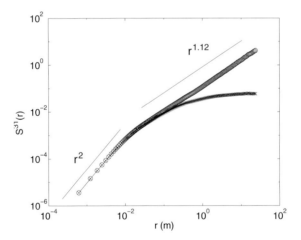

Fig. 14. Log–log plot of second order mixed structure function at 0.27 m. The legend is the same as for Figure 13.

we should recover the contribution of the remaining parts – in particular the inertial range part, with the leading order scaling exponent given by $2 - 2C_2$. Figures 13 and 14 display a second-order anisotropic structure function for two heights above the ground. Presumably because of the finiteness of the Reynolds number and the relatively large shear effect, the scaling in the intermediate range $\eta \ll r \ll L$ is not apparent. However, by dividing out the large scale contribution as described above, we see two distinct regions of scaling; the dissipative range of $\sim r^2$ and the extended mid-range which scales with exponent between 1.22 and 1.12. The advantage of the scaling function is thus evident: it has allowed us to extract a scaling exponent that is most likely to be due to anisotropy. The values of the fitted parameters and the corresponding $\zeta_2^{(2)}$ are given in Tables 4 and 5 for the probes at 0.54 m and at 0.27 m respectively. The error on the measurement of the exponent C_2 at 0.54 m is about 0.05 while at 0.27 m it is about 0.08. This gives an error on the estimates of $\zeta_2^{(2)}$ of 0.07 and 0.11 respectively, and places the theoretically expected value of $\approx 4/3$ within 1.5 to 2 standard deviations of the present value. This result is consistent with general expectations [15] and the findings of the previous section.

6.2.2   Higher-order structure functions

In general, the tensor forms contributing to the $j = 0$ sector for tensors of *any* rank n are composed of linear combinations of the Kronecker-δ and the components of **r** along the tensor indices. The following is a list of isotropic tensor contributions for rank 3 through 6:

- $n = 3$: $\delta^{\alpha\beta}r^\gamma$ + permutations, and $r^\alpha r^\beta r^\gamma$;

- $n = 4$: $\delta^{\alpha\beta}\delta^{\gamma\delta}$ + permutations, $\delta^{\alpha\beta}r^\gamma r^\delta$ + permutations, and $r^\alpha r^\beta r^\gamma r^\delta$;

- $n = 5$: $\delta^{\alpha\beta}\delta^{\gamma\delta}r^\mu$ + permutations, $\delta^{\alpha\beta}r^\gamma r^\delta r^\mu$ + permutations, and $r^\alpha r^\beta r^\gamma r^\delta r^\mu$;

- $n = 6$: $\delta^{\alpha\beta}\delta^{\gamma\delta}\delta^{\mu\nu}$ + permutations, $\delta^{\alpha\beta}\delta^{\gamma\delta}r^\mu r^\nu$ + permutations, $\delta^{\alpha\beta}r^\gamma r^\delta r^\mu r^\nu$ + permutations, and $r^\alpha r^\beta r^\gamma r^\delta r^\mu r^\nu$.

Based on the above considerations, it can be expected that the structure function components that are zero in the $j = 0$ sector are:

- $n = 3$: S^{111} (transverse), S^{331};

- $n = 4$: S^{3331}, S^{3111};

- $n = 5$: S^{11111} (transverse), S^{33111}, S^{33331};

- $n = 6$: $S^{333111}, S^{311111}, S^{333331}$.

Table 4. Structure function calculated and the anisotropic scaling exponents for the data at 0.54 m.

Order n	Tensor	A_n	B_n	C_n	D_n	$\zeta_n^{(2)} = n - 2C_n$	ζ_n
2	S^{31}	3.9	0.014	0.39	0.67	1.22	0.7
3	S^{111}	2400	0.010	0.93	2.28	1.14	1
4	S^{3331}	5200	0.014	1.21	0.27	1.58	1.26
5	S^{11111}	1.22×10^7	0.029	1.59	3.09	1.82	1.56
6	S^{333111}	3.75×10^7	0.041	1.93	0.50	2.14	1.71

Table 5. Structure function calculated and the anisotropic scaling exponents for the data at 0.27 m.

Order n	Tensor	A_n	B_n	C_n	D_n	$\zeta_n^{(2)} = n - 2C_n$	ζ_n
2	S^{31}	9.4	0.005	0.44	0.52	1.12	0.7
3	S^{111}	6940	0.015	0.89	2.78	1.21	1
4	S^{3331}	2.1×10^4	0.014	1.23	0.23	1.54	1.26
5	S^{11111}	5.9×10^7	0.028	1.58	3.52	1.84	1.56
6	S^{333111}	2.7×10^8	0.038	2.00	0.34	2.00	1.71

Note that the odd-order transverse structure function is *always* zero in the isotropic sector. The functions we shall now consider are given in the second column of Tables 4 and 5. For the case of the third and fifth order transverse structure functions we use the moments of the *absolute value* of the velocity differences in order to obtain better convergence. In using the interpolation function we assume that the inertial range scaling of these anisotropic components is given by a single exponent $\zeta_n^{(j)}$ where the superscript denotes an isotropic exponent without reference to the precise j sector. The compensated functions (with large-scale effects removed) are shown in Figures 15–22. The errors on the value of $\zeta_n^{(2)}$ obtained are about 7% at 0.54 m and about 9% at 0.27 m. For comparison, the last column in Tables 4 and 5 gives the empirical values of the isotropic scaling exponent of the same order as given in [6]. The entries in this column are measurably smaller than the corresponding non-isotropic exponents. This suggests that the isotropic component alone survives at very small scales.

6.3 Summary

We have presented a new method of extracting anisotropic exponents that avoids mixing with contributions from the isotropic sector. We do this by

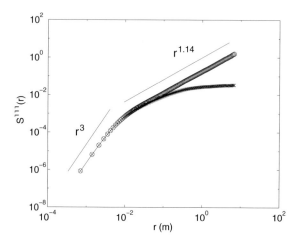

Fig. 15. Log–log plot of third-order transverse structure function at 0.54 m. The legend is the same as for Figure 13.

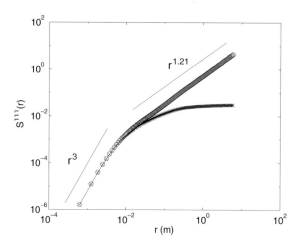

Fig. 16. Log–log plot of third-order transverse structure function at 0.27 m. The legend is the same as for Figure 13.

explicitly constructing those tensors that are zero in the isotropic sector. An operational step in the extraction of the scaling exponents is the use of an interpolation formula in the spirit of a "scaling function". This method has allowed us to examine anisotropic effects in structure function tensors of order greater than 2. The resulting anisotropic exponents are consistently

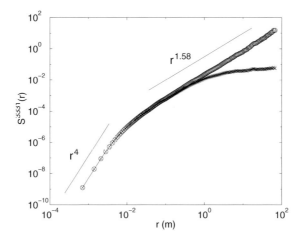

Fig. 17. Log–log plot of fourth-order mixed structure function at 0.54 m. The legend is the same as for Figure 13.

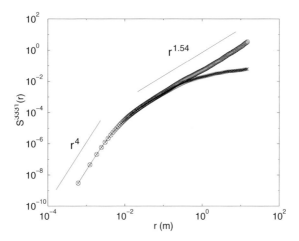

Fig. 18. Log–log plot of fourth-order mixed structure function at 0.27 m. The legend is the same as for Figure 13.

larger than those known for isotropic parts at all orders. This suggests that anisotropy effects decrease with decreasing scale. However, the rate of decrease is much slower than expected from dimensional arguments (which yield 4/3, 5/3, 2, 7/3 and 8/3 for orders 2 through 6, to be compared with the values obtained at 0.54 m of 1.22, 1.14, 1.58, 1.82, 2.14).

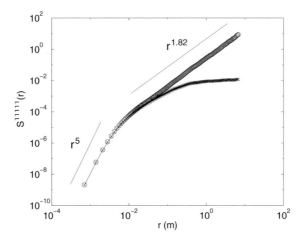

Fig. 19. Log–log plot of fifth-order transverse structure function at 0.54 m. The legend is the same as for Figure 13.

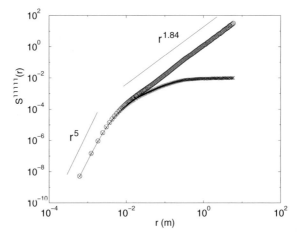

Fig. 20. Log–log plot of fifth-order transverse structure function at 0.27 m. The legend is the same as for Figure 13.

Our conclusions are based on the use of the interpolation formula, given by equation (6.2). While this formula is not based on a solid theoretical framework, we have shown that it works very well in describing the measured structure functions. We have also performed tests of its robustness by fitting it to smaller sections of the data in order to detect changes in

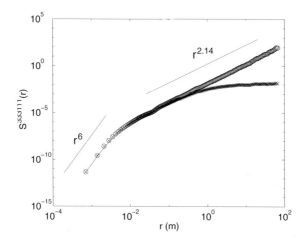

Fig. 21. Log–log plot of sixth-order mixed structure function at 0.54 m. The legend is the same as for Figure 13.

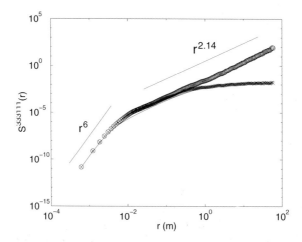

Fig. 22. Log–log plot of sixth-order mixed structure function at 0.27 m. The legend is the same as for Figure 13.

the exponent. A discussion of these checks and their results is presented in Appendix C. To the lowest order, the results are independent of the r-segment to which the formula is fitted (except, of course, when the fit is entirely for the dissipation range or in the large-scale range). Any other formula that works equally well will yield similar results. Even so, the formula is empirical, which is why we have not paid much attention to the fact that the scaling exponents obtained for the two probe positions are slightly different, and that the second-order exponent for 0.54 m is slightly larger than that obtained for the third-order. On the whole, the trend is that the anisotropic exponents become larger for larger orders of the structure function.

An interesting conclusion of the present work is that the effects of anisotropy vanish with decreasing scale more slowly than expected. The expectation in the light of the SO(3) formalism is that a hierarchy of increasingly larger exponents, corresponding to increasingly higher-order anisotropic sectors [17], would exist. This expectation appears to be true in the case of the passively advected vector field [3] where a discrete spectrum of anisotropic scaling exponents is obtained theoretically for all anisotropic sectors. In the present experiments, the fact that anisotropic effects can be fitted reasonably well by power laws suggests that the high-order effects may be small. It is perhaps true, however, that the power-laws described here may contain high-order corrections, and that the exponents deduced for the behavior of anisotropy may indeed undergo some revision when contributions from other sectors of the SO(3) decomposition are also considered. In spite of this possibility, we wish to emphasize that the anisotropy effects *for each order* of the structure function appear to be well described by something close to a power law with a single exponent. The magnitudes of the anisotropic exponents in each order indicate that the roll-off from isotropy happens less sharply than previously thought, but the roll-off occurs nevertheless. The higher-order objects considered here have not been studied extensively in the light of anisotropy.

7   Conclusions

We have used the SO(3) decomposition of tensorial objects such as structure functions measured in fluid turbulence experiments. This enables us to separate, for any given order structure function, the isotropic part from the anisotropic part. We have used experimental data at very high Reynolds numbers (Taylor microscale Reynolds numbers of up to 20 000) to carry through this decomposition.

We considered the lowest order anisotropic contribution to the second order tensor functions of velocity in the atmospheric boundary layer. For second order structure functions, we have extracted the anisotropic scaling exponent from the experimental data. We also used the full form of the $j = 2$ tensor, as well as the axisymmetric version of it. In both instances, we have shown that the $j = 2$ exponent $\zeta_2^{(2)} \approx 1.36$. This compares favorably with the estimate based on purely dimensional grounds. It should be noted that the dimensional estimates do not take into account possible anomalous scaling. We expect that turbulence flows will indeed exhibit anomalous scaling in the isotropic sector, and in every anisotropic sector in the SO(3) hierarchy. By suitably arranging the measurement configuration, we have also extracted the $j = 1$ contribution, and find the appropriate scaling exponent to be close to the dimensional estimate of unity.

For high orders, we have presented a new method for the extraction of anisotropic exponents that avoids mixing with contributions from the isotropic sector. We do this by explicitly constructing those tensors that are zero in the isotropic sector. The resulting anisotropic exponents are consistently larger than those known for isotropic parts at all orders. This suggests that anisotropy effects decrease with decreasing scale. However, the rate of decrease is slower than expected from dimensional arguments.

Armed with these details, it is now possible to say something of value about local isotropy. These statements are interspersed through out the text, but the most important conclusion is that the effects of anisotropy within a structure function of a given order do vanish with decreasing scale, though more slowly than expected. This provides a rich perspective on the notion of local isotropy and removes a hurdle towards the search for a universal theory of small-scale turbulence.

Appendix

A   Full form for the $j = 2$ contribution for the homogeneous case

Each index j in the SO(3) decomposition of an n-rank tensor labels a $2j + 1$ dimensional SO(3) representation. Each dimension is labeled by $m = -j, -j + 1, \ldots j$. The $j = 0$ sector is the isotropic contribution while higher order j's should describe any anisotropy. The $j = 0$ terms are well known to be

$$S_{j=0}^{\alpha\beta}(\mathbf{r}) = c_0 r^{\zeta_2} \left[(2 + \zeta_2)\delta^{\alpha\beta} - \zeta_2 \frac{r^\alpha r^\beta}{r^2} \right] \tag{A.1}$$

where $\zeta_2 \approx 0.69$ is the known universal scaling exponent for the isotropic contribution, and c_0 is an unknown coefficient that depends on the boundary conditions of the flow. For the $j = 2$ sector, which is the lowest contribution to anisotropy to the homogeneous structure function, the $m = 0$ (axisymmetric) terms were derived from constraints of symmetry, even parity (because of homogeneity) and incompressibility on the second order structure function to be [4]

$$S_{j=2,m=0}^{\alpha\beta}(\mathbf{r}) = ar^{\zeta_2^{(2)}} \left[(\zeta_2^{(2)} - 2)\delta^{\alpha\beta} - \zeta_2^{(2)}(\zeta_2^{(2)} + 6) \right.$$

$$\times \delta^{\alpha\beta} \frac{(\mathbf{n} \cdot \mathbf{r})^2}{r^2} + 2\zeta_2^{(2)}(\zeta_2^{(2)} - 2)\frac{r^\alpha r^\beta (\mathbf{n} \cdot \mathbf{r})^2}{r^4}$$

$$+([\zeta_2^{(2)}]^2 + 3\zeta_2^{(2)} + 6)n^\alpha n^\beta$$

$$-\frac{\zeta_2^{(2)}(\zeta_2^{(2)} - 2)}{r^2}(r^\alpha n^\beta + r^\beta n^\alpha)(\mathbf{n} \cdot \mathbf{r})\Big] \tag{A.2}$$

$$+br^{\zeta_2^{(2)}}\Big[-(\zeta_2^{(2)} + 3)(\zeta_2^{(2)} + 2)\delta^{\alpha\beta}(\mathbf{n} \cdot \mathbf{r})^2 + \frac{r^\alpha r^\beta}{r^2}$$

$$+(\zeta_2^{(2)} + 3)(\zeta_2^{(2)} + 2)n^\alpha n^\beta + (2\zeta_2^{(2)} + 1)(\zeta_2^{(2)} - 2)$$

$$\times \frac{r^\alpha r^\beta (\mathbf{n} \cdot \mathbf{r})^2}{r^4} - ([\zeta_2^{(2)}]^2 - 4)(r^\alpha n^\beta + r^\beta n^\alpha)(\mathbf{n} \cdot \mathbf{r})\Big].$$

Here $\zeta_2^{(2)}$ is the universal scaling exponent for the $j = 2$ anisotropic sector and a and b are independent unknown coefficients to be determined by the boundary conditions. We would now like to derive the remaining $m = \pm 1$, and $m = \pm 2$ components

$$S_{j=2,m}^{\alpha\beta} = \sum_q a_{q,2,m} r^{\zeta_2^{(2)}} B_{q,2,m}^{\alpha\beta}(\hat{\mathbf{r}}), \tag{A.3}$$

where ζ_2^2 is the scaling exponent of the $j = 2$ SO(3) representation of the $n = 2$ rank correlation function. The $B_{q,j,m}^{\alpha\beta}(\hat{\mathbf{r}})$ are the basis functions in the SO(3) representation of the structure function. The q label denotes the different possible ways of arriving at the the same j and runs over all such terms with the same parity and symmetry (a consequence of homogeneity and hence the constraint of incompressibility); see [5]. In our case, even parity and symmetric in the two indices. In all that follows, we work closely with the procedure outlined in [5]. Following the convention in [5] the q's to sum over are $q = \{1, 7, 9, 5\}$. The incompressibility condition $\partial_\alpha u^\alpha = 0$ coupled with homogeneity can be used to give relations between the $a_{q,j,m}$ for a given (j, m). That is, for $j = 2$, $m = -2 \ldots 2$, we have

$$(\zeta_2^{(2)} - 2)a_{1,2,m} + 2(\zeta_2^{(2)} - 2)a_{7,2,m} + (\zeta_2^{(2)} + 2)a_{9,2,m} = 0$$

$$a_{1,2,m} + (\zeta_2^{(2)} + 3)a_{7,2,m} + \zeta_2^{(2)}a_{5,2,m} = 0. \tag{A.4}$$

We solve equations (A.4) in order to obtain $a_{5,2,m}$ and $a_{7,2,m}$ in terms of linear combinations of $a_{1,2,m}$ and $a_{9,2,m}$:

$$a_{5,2,m} = \frac{a_{1,2,m}([\zeta_2^{(2)}]^2 - \zeta_2^{(2)} - 2) + a_{9,2,m}([\zeta_2^{(2)}]^2 + 5\zeta_2^{(2)} + 6)}{2\zeta_2^{(2)}(\zeta_2^{(2)} - 2)}$$

$$a_{7,2,m} = \frac{a_{1,2,m}(2 - \zeta_2^{(2)}) - a_{9,2,m}(2 + \zeta_2^{(2)})}{2(\zeta_2^{(2)} - 2)}. \tag{A.5}$$

We note here that the choice of constants eliminated using the above so-lutions is arbitrary. We could have solved for *any* two of the a_{qjm} and obtained equivalent results. Using the above constraints on the coefficients, we are now left with a linear combination of just two linearly independent tensor forms *for each m*:

$$
\begin{aligned}
S^{\alpha\beta}_{j=2,m} =\ & a_{9,2,m} r^{\zeta^{(2)}_2} [-\zeta^{(2)}_2 (2 + \zeta^{(2)}_2) B^{\alpha\beta}_{7,2,m}(\hat{\mathbf{r}}) \\
& + 2\zeta^{(2)}_2 (\zeta^{(2)}_2 - 2) B^{\alpha\beta}_{9,2,m}(\hat{\mathbf{r}}) \\
& + ([\zeta^{(2)}_2]^2 + 5\zeta^{(2)}_2 + 6) B^{\alpha\beta}_{5,2,m}(\hat{\mathbf{r}})] \\
& + a_{1,2,m} r^{\zeta^{(2)}_2} [2\zeta^{(2)}_2 (\zeta^{(2)}_2 - 2) B^{\alpha\beta}_{1,2,m}(\hat{\mathbf{r}}) \\
& - \zeta^{(2)}_2 (\zeta^{(2)}_2 - 2) B^{\alpha\beta}_{7,2,m}(\hat{\mathbf{r}}) \\
& + ([\zeta^{(2)}_2]^2 - \zeta^{(2)}_2 - 2) B^{\alpha\beta}_{5,2,m}(\hat{\mathbf{r}})].
\end{aligned} \tag{A.6}
$$

The task remains to find the explicit form of the basis tensor functions $B^{\alpha\beta}_{q,2,m}(\hat{\mathbf{r}})$, $q \in \{1,7,9,5\}$, $m \in \{\pm 1, \pm 2\}$

- $B^{\alpha\beta}_{1,2,m}(\hat{\mathbf{r}}) \equiv r^{-2} \delta^{\alpha\beta} r^2 Y_{2m}(\hat{\mathbf{r}})$

- $B^{\alpha\beta}_{7,2,m}(\hat{\mathbf{r}}) \equiv r^{-2}[r^\alpha \partial^\beta + r^\beta \partial^\alpha] r^2 Y_{2m}(\hat{\mathbf{r}})$

- $B^{\alpha\beta}_{9,2,m}(\hat{\mathbf{r}}) \equiv r^{-4} r^\alpha r^\beta r^2 Y_{2m}(\hat{\mathbf{r}})$

- $B^{\alpha\beta}_{5,2,m}(\hat{\mathbf{r}}) \equiv \partial^\alpha \partial^\beta r^2 Y_{jm}(\hat{\mathbf{r}})$.

We obtain the $m = \{\pm 1, \pm 2\}$ basis functions in the following derivation. We first note that it is more convenient to form a real basis from the $r^2 Y_{2m}(\hat{\mathbf{r}})$ since we ultimately wish to fit to real quantities and extract real best-fit parameters. We therefore form the $r^2 \tilde{Y}_{2k}(\hat{\mathbf{r}})$ ($k = -2, -1, 0, 1, 2$) as follows:

$$
\begin{aligned}
r^2 \tilde{Y}_{2\,0}(\hat{\mathbf{r}}) =\ & r^2 Y_{2\,0}(\hat{\mathbf{r}}) = r^2 \cos^2 \theta = r_3{}^2 \\
r^2 \tilde{Y}_{2\,-1}(\hat{\mathbf{r}}) =\ & r^2 \frac{Y_{2\,-1}(\hat{\mathbf{r}}) - Y_{2\,+1}(\hat{\mathbf{r}})}{2} \\
=\ & \frac{r^2}{2} \Big[(\cos\phi - i\sin\phi)\cos\theta\sin\theta \\
& + (\cos\phi + i\sin\phi)\cos\theta\sin\theta \Big] \\
=\ & r^2 \cos\theta\sin\theta\cos\phi = r_3 r_1 \\
r^2 \tilde{Y}_{2\,+1}(\hat{\mathbf{r}}) =\ & r^2 \frac{Y_{2\,-1}(\hat{\mathbf{r}}) + Y_{2\,+1}(\hat{\mathbf{r}})}{-2i} \\
=\ & \frac{r^2}{-2i} \Big[(\cos\phi - i\sin\phi)\cos\theta\sin\theta
\end{aligned}
$$

$$-(\cos\phi + i\sin\phi)\cos\theta\sin\theta\Big]$$

$$= r^2\cos\theta\sin\theta\sin\phi = r_3 r_2$$

$$r^2\widetilde{Y}_{2\,-2}(\hat{\mathbf{r}}) = r^2\frac{Y_{2\,2}(\hat{\mathbf{r}}) - Y_{2\,-2}(\hat{\mathbf{r}})}{2i}$$

$$= \frac{r^2}{2i}\Big[(\cos 2\phi + i\sin 2\phi)\sin^2\theta - (\cos 2\phi - i\sin 2\phi)\sin^2\theta\Big]$$

$$= r^2\sin 2\phi\sin^2\theta = 2r_1 r_2$$

$$r^2\widetilde{Y}_{2\,+2}(\hat{\mathbf{r}}) = r^2\frac{Y_{2\,2}(\hat{\mathbf{r}}) + Y_{2\,-2}(\hat{\mathbf{r}})}{2}$$

$$= \frac{r^2}{2}\Big[(\cos 2\phi + i\sin 2\phi)\sin^2\theta + (\cos 2\phi - i\sin 2\phi)\sin^2\theta\Big]$$

$$= r^2\cos 2\phi\sin^2\theta = r_1^2 - r_2^2. \tag{A.7}$$

This new basis of $r^2\widetilde{Y}_{2k}(\mathbf{r})$ is equivalent to the $r^2 Y_{jm}(\mathbf{r})$ themselves as they form a complete, orthogonal set (in the new k's). We omit the normalization constants for the spherical harmonics for notational convenience. The subscripts on r denote its components along the 1 (m), 2 (p) and 3 (n) directions; $\mathbf{m}$ denotes the shear direction, $\mathbf{p}$ the horizontal direction parallel to the boundary and orthogonal to the mean wind direction and $\mathbf{n}$ the direction of the mean wind. This notation simplifies the derivatives when we form the different basis tensors; the only thing to remember is that

$$\partial^\alpha r_1 = \partial^\alpha(\mathbf{r}\cdot\mathbf{m}) = \mathbf{m}^\alpha$$
$$\partial^\alpha r_2 = \partial^\alpha(\mathbf{r}\cdot\mathbf{p}) = \mathbf{p}^\alpha$$
$$\partial^\alpha r_3 = \partial^\alpha(\mathbf{r}\cdot\mathbf{n}) = \mathbf{n}^\alpha. \tag{A.8}$$

We use the above identities to proceed to derive the basis tensor functions

$$B^{\alpha\beta}_{1,2,-1}(\hat{\mathbf{r}}) = r^{-2}\delta^{\alpha\beta}(\mathbf{r}\cdot\mathbf{n})(\mathbf{r}\cdot\mathbf{m})$$

$$B^{\alpha\beta}_{7,2,-1}(\hat{\mathbf{r}}) = r^{-2}[(r^\alpha m^\beta + r^\beta m^\alpha)(\mathbf{r}\cdot\mathbf{n}) + (r^\alpha n^\beta + r^\beta n^\alpha)(\mathbf{r}\cdot\mathbf{m})]$$

$$B^{\alpha\beta}_{9,2,-1}(\hat{\mathbf{r}}) = r^{-2}r^\alpha r^\beta(\mathbf{r}\cdot\mathbf{n})(\mathbf{r}\cdot\mathbf{m})$$

$$B^{\alpha\beta}_{5,2,-1}(\hat{\mathbf{r}}) = n^\alpha m^\beta + n^\beta m^\alpha$$

$$B^{\alpha\beta}_{1,2,1}(\hat{\mathbf{r}}) = r^{-2}\delta^{\alpha\beta}(\mathbf{r}\cdot\mathbf{n})(\mathbf{r}\cdot\mathbf{p})$$

$$B^{\alpha\beta}_{7,2,1}(\hat{\mathbf{r}}) = r^{-2}[(r^\alpha p^\beta + r^\beta p^\alpha)(\mathbf{r}\cdot\mathbf{n}) + (r^\alpha n^\beta + r^\beta n^\alpha)(\mathbf{r}\cdot\mathbf{p})]$$

$$B^{\alpha\beta}_{9,2,1}(\hat{\mathbf{r}}) = r^{-2}r^\alpha r^\beta(\mathbf{r}\cdot\mathbf{n})(\mathbf{r}\cdot\mathbf{p})$$

$$B^{\alpha\beta}_{5,2,1}(\hat{\mathbf{r}}) = n^\alpha p^\beta + n^\beta p^\alpha$$

$$B^{\alpha\beta}_{1,2,-2}(\hat{\mathbf{r}}) = 2r^{-2}\delta^{\alpha\beta}(\mathbf{r}\cdot\mathbf{m})(\mathbf{r}\cdot\mathbf{p})$$

$$B^{\alpha\beta}_{7,2,-2}(\hat{\mathbf{r}}) = 2r^{-2}[(r^\alpha p^\beta + r^\beta p^\alpha)(\mathbf{r}\cdot\mathbf{m}) + (r^\alpha m^\beta + r^\beta m^\alpha)(\mathbf{r}\cdot\mathbf{p})]$$

$$B_{9,2,-2}^{\alpha\beta}(\hat{\mathbf{r}}) = 2r^{-2}r^{\alpha}r^{\beta}(\mathbf{r}\cdot\mathbf{m})(\mathbf{r}\cdot\mathbf{p})$$

$$B_{5,2,-2}^{\alpha\beta}(\hat{\mathbf{r}}) = 2(m^{\alpha}p^{\beta}+m^{\beta}p^{\alpha})$$

$$B_{1,2,2}^{\alpha\beta}(\hat{\mathbf{r}}) = r^{-2}\delta^{\alpha\beta}[(\mathbf{r}\cdot\mathbf{m})^{2}-(\mathbf{r}\cdot\mathbf{p})^{2}]$$

$$B_{7,2,2}^{\alpha\beta}(\hat{\mathbf{r}}) = 2r^{-2}[(r^{\alpha}m^{\beta}+r^{\beta}m^{\alpha})(\mathbf{r}\cdot\mathbf{m})-(r^{\alpha}p^{\beta}+r^{\beta}p^{\alpha})(\mathbf{r}\cdot\mathbf{p})]$$

$$B_{9,2,2}^{\alpha\beta}(\hat{\mathbf{r}}) = r^{-2}r^{\alpha}r^{\beta}[(\mathbf{r}\cdot\mathbf{m})^{2}-(\mathbf{r}\cdot\mathbf{p})^{2}]$$

$$B_{5,2,2}^{\alpha\beta}(\hat{\mathbf{r}}) = 2(m^{\alpha}m^{\beta}-p^{\alpha}p^{\beta}). \tag{A.9}$$

Substituting these tensors forms into equation (A.6) we obtain the full tensor forms for the $j=2$ non-axisymmetric terms, with two independent coefficients for each k:

$$\begin{aligned}
S_{j=2,k=-1}^{\alpha\beta}(\mathbf{r}) = \ & a_{9,2,-1}r^{\zeta_{2}^{(2)}}\Big[([\zeta_{2}^{(2)}]^{2}+5\zeta_{2}^{(2)}+6)(n^{\alpha}m^{\beta}+n^{\beta}m^{\alpha}) \\
& +2\zeta_{2}^{(2)}(\zeta_{2}^{(2)}-2)r^{-4}r^{\alpha}r^{\beta}(\mathbf{r}\cdot\mathbf{n})(\mathbf{r}\cdot\mathbf{m}) \\
& -\zeta_{2}^{(2)}(2+\zeta_{2}^{(2)})r^{-2}[(r^{\alpha}m^{\beta}+r^{\beta}m^{\alpha})(\mathbf{r}\cdot\mathbf{n}) \\
& +(r^{\alpha}n^{\beta}+r^{\beta}n^{\alpha})(\mathbf{r}\cdot\mathbf{m})]\Big] \\
& +a_{1,2,-1}r^{\zeta_{2}^{(2)}}\Big[([\zeta_{2}^{(2)}]^{2}-\zeta_{2}^{(2)}-2)(n^{\alpha}m^{\beta}+n^{\beta}m^{\alpha}) \\
& +2\zeta_{2}^{(2)}(\zeta_{2}^{(2)}-2)r^{-2}\delta^{\alpha\beta}(\mathbf{r}\cdot\mathbf{n})(\mathbf{r}\cdot\mathbf{m}) \\
& -\zeta_{2}^{(2)}(\zeta_{2}^{(2)}-2)r^{-2}[(r^{\alpha}m^{\beta}+r^{\beta}m^{\alpha})(\mathbf{r}\cdot\mathbf{n}) \\
& +(r^{\alpha}n^{\beta}+r^{\beta}n^{\alpha})(\mathbf{r}\cdot\mathbf{m})]\Big] \\[2mm]
S_{j=2,k=1}^{\alpha\beta}(\mathbf{r}) = \ & a_{9,2,1}r^{\zeta_{2}^{(2)}}\Big[([\zeta_{2}^{(2)}]^{2}+5\zeta_{2}^{(2)}+6)(n^{\alpha}p^{\beta}+n^{\beta}p^{\alpha}) \\
& +2\zeta_{2}^{(2)}(\zeta_{2}^{(2)}-2)r^{-4}r^{\alpha}r^{\beta}(\mathbf{r}\cdot\mathbf{n})(\mathbf{r}\cdot\mathbf{p}) \\
& -\zeta_{2}^{(2)}(2+\zeta_{2}^{(2)})r^{-2}[(r^{\alpha}p^{\beta}+r^{\beta}p^{\alpha})(\mathbf{r}\cdot\mathbf{n}) \\
& +(r^{\alpha}n^{\beta}+r^{\beta}n^{\alpha})(\mathbf{r}\cdot\mathbf{p})]\Big] \\
& +a_{1,2,1}r^{\zeta_{2}^{(2)}}\Big[([\zeta_{2}^{(2)}]^{2}-\zeta_{2}^{(2)}-2)(n^{\alpha}p^{\beta}+n^{\beta}p^{\alpha}) \\
& +2\zeta_{2}^{(2)}(\zeta_{2}^{(2)}-2)r^{-2}\delta^{\alpha\beta}(\mathbf{r}\cdot\mathbf{n})(\mathbf{r}\cdot\mathbf{p}) \\
& -\zeta_{2}^{(2)}(\zeta_{2}^{(2)}-2)r^{-2}[(r^{\alpha}p^{\beta}+r^{\beta}p^{\alpha})(\mathbf{r}\cdot\mathbf{n}) \\
& +(r^{\alpha}n^{\beta}+r^{\beta}n^{\alpha})(\mathbf{r}\cdot\mathbf{p})]\Big] \\[2mm]
S_{j=2,k=-2}^{\alpha\beta}(\mathbf{r}) = \ & a_{9,2,-2}r^{\zeta_{2}^{(2)}}\Big[([\zeta_{2}^{(2)}]^{2}+5\zeta_{2}^{(2)}+6)(m^{\alpha}p^{\beta}+m^{\beta}p^{\alpha}) \\
& +2\zeta_{2}^{(2)}(\zeta_{2}^{(2)}-2)r^{-4}r^{\alpha}r^{\beta}(\mathbf{r}\cdot\mathbf{p})(\mathbf{r}\cdot\mathbf{m}) \\
& -2\zeta_{2}^{(2)}(2+\zeta_{2}^{(2)})r^{-2}[(r^{\alpha}p^{\beta}+r^{\beta}p^{\alpha})(\mathbf{r}\cdot\mathbf{m})
\end{aligned}$$

$$+(r^\alpha m^\beta + r^\beta m^\alpha)(\mathbf{r}\cdot\mathbf{p})]\Big]$$

$$+a_{1,2,-2}r^{\zeta_2^{(2)}}\Big[2([\zeta_2^{(2)}]^2 - \zeta_2^{(2)} - 2)(m^\alpha p^\beta + m^\beta p^\alpha)$$

$$+2\zeta_2^{(2)}(\zeta_2^{(2)} - 2)r^{-2}\delta^{\alpha\beta}(\mathbf{r}\cdot\mathbf{m})(\mathbf{r}\cdot\mathbf{p})$$

$$-2\zeta_2^{(2)}(\zeta_2^{(2)} - 2)r^{-2}[(r^\alpha p^\beta + r^\beta p^\alpha)(\mathbf{r}\cdot\mathbf{m})$$

$$+(r^\alpha m^\beta + r^\beta m^\alpha)(\mathbf{r}\cdot\mathbf{p})]\Big]$$

$$S^{\alpha\beta}_{j=2,k=2}(\mathbf{r}) = a_{9,2,2}r^{\zeta_2^{(2)}}\Big[2([\zeta_2^{(2)}]^2 + 5\zeta_2^{(2)} + 6)(m^\alpha m^\beta - p^\beta p^\alpha)$$

$$+2\zeta_2^{(2)}(\zeta_2^{(2)} - 2)r^{-4}r^\alpha r^\beta[(\mathbf{r}\cdot\mathbf{m})^2 - (\mathbf{r}\cdot\mathbf{p})^2]$$

$$-2\zeta_2^{(2)}(2 + \zeta_2^{(2)})r^{-2}[(r^\alpha m^\beta + r^\beta m^\alpha)(\mathbf{r}\cdot\mathbf{m})$$

$$-(r^\alpha p^\beta + r^\beta p^\alpha)(\mathbf{r}\cdot\mathbf{p})]\Big]$$

$$+a_{1,2,2}r^{\zeta_2^{(2)}}\Big[2([\zeta_2^{(2)}]^2 - \zeta_2^{(2)} - 2)(m^\alpha m^\beta - p^\beta p^\alpha)$$

$$+2\zeta_2^{(2)}(\zeta_2^{(2)} - 2)r^{-2}\delta^{\alpha\beta}[(\mathbf{r}\cdot\mathbf{m})^2 - (\mathbf{r}\cdot\mathbf{p})^2]$$

$$-2\zeta_2^{(2)}(\zeta_2^{(2)} - 2)r^{-2}[(r^\alpha m^\beta + r^\beta m^\alpha)(\mathbf{r}\cdot\mathbf{m})$$

$$-(r^\alpha p^\beta + r^\beta p^\alpha)(\mathbf{r}\cdot\mathbf{p})]\Big]. \tag{A.10}$$

Now we wish to use this form to fit for the scaling exponent $\zeta_2^{(2)}$ in the structure function $S^{33}(\mathbf{r})$ from data set I where $\alpha = \beta = 3$ and the azimuthal angle of $\mathbf{r}$ in the geometry is $\phi = \pi/2$.:

$$S^{33}_{j=2,k=-1}(r,\theta,\phi = \pi/2) = 0$$

$$S^{33}_{j=2,k=1}(r,\theta,\phi = \pi/2) = a_{9,2,1}r^{\zeta_2^{(2)}}[-2\zeta_2^{(2)}(\zeta_2^{(2)} + 2)\sin\theta\cos\theta$$
$$+2\zeta_2^{(2)}(\zeta_2^{(2)} - 2)\cos^3\theta\sin\theta]$$

$$S^{33}_{j=2,k=-2}(r,\theta,\phi = \pi/2) = 0$$

$$S^{33}_{j=2,k=2}(r,\theta,\phi = \pi/2) = a_{9,2,2}r^{\zeta_2^{(2)}}[-2\zeta_2^{(2)}(\zeta_2^{(2)} - 2)\cos^2\theta\sin^2\theta]$$
$$+a_{1,2,2}r^{\zeta_2^{(2)}}[-2\zeta_2^{(2)}(\zeta_2^{(2)} - 2)\sin^2\theta]. \tag{A.11}$$

We see that choosing a particular geometry eliminates certain tensor contributions. In the case of set I we are left with 3 independent coefficients for $m \neq 0$, the 2 coefficients from the $m = 0$ contribution (Eq. (A.2)), and the single coefficient from the isotropic sector (A.1), giving a total of 6 fit parameters. The general forms in (A.10) can be used along with the $k = 0$ (axisymmetric) contribution (A.1) to fit to any second order tensor object. For convenience, Table 6 shows the number of independent coefficients that a few different experimental geometries we have will allow in the $j = 2$

Table 6. The number of free coefficients in the $j = 2$ sector for homogeneous turbulence and for different geometries.

k	$\phi = \pi/2, \alpha = \beta = 3$		$\phi = 0, \alpha = \beta = 3$		$\phi = 0, \alpha = \beta = 1$		$\phi = 0, \alpha = 3, \beta = 1$	
	$\theta \neq 0$	$\theta = 0$	$\theta \neq 0$	$\theta = 0$	$\theta \neq 0$	$\theta = 0$	$\theta \neq 0$	$\theta = 0$
0	2	2	2	2	2	2	2	0
-1	0	0	1	0	1	0	2	2
1	1	0	0	0	0	0	0	0
-2	0	0	0	0	0	0	0	0
2	2	0	2	0	2	2	2	0
Total	5	2	5	2	5	4	6	2

sector. It must be kept in mind that these forms are to be used *only* when homogeneity is known to exist. If there is inhomogeneity, the incompressibility condition cannot be used to provide constraints in the various parity and symmetry sectors, and we must in general mix different parity objects using only the geometry of the experiment itself to eliminate any terms.

B   The $j = 1$ component in the inhomogeneous case

B.1   Antisymmetric contribution

We consider the tensor

$$T^{\alpha\beta}(\mathbf{r}) = \langle u^\alpha(\mathbf{x} + \mathbf{r}) - u^\alpha(\mathbf{x}))(u^\beta(\mathbf{x} + \mathbf{r}) + u^\beta(\mathbf{x}))\rangle. \qquad (B.1)$$

This object is trivially zero for $\alpha = \beta$. In our experimental setup, we measure at points separated in the shear direction and therefore have inhomogeneity which makes the object of mixed parity and symmetry. We cannot apply the incompressibility condition in same parity/symmetry sectors as before to provide constraints. We must in general use all 7 irreducible tensor forms. This would mean fitting for $7 \times 3 = 21$ independent coefficients plus 1 exponent $\zeta_2^{(1)}$ in the anisotropic sector, together with 2 coefficients in the isotropic sector. In order to pare down the number of parameters we are fitting for, we look at the antisymmetric part of $T^{\alpha\beta}(\mathbf{r})$,

$$\widetilde{T}^{\alpha\beta}(\mathbf{r}) = \frac{T^{\alpha\beta}(\mathbf{r}) - T^{\beta\alpha}(\mathbf{r})}{2} = \langle u^\alpha(\mathbf{x})u^\beta(\mathbf{x} + \mathbf{r})\rangle - \langle u^\beta(\mathbf{x})u^\alpha(\mathbf{x} + \mathbf{r})\rangle, \quad (B.2)$$

which will only have contributions from the antisymmetric $j = 1$ basis tensors. These are:

- Antisymmetric, odd parity

$$B_{3,1,m}^{\alpha\beta} = r^{-1}[r^\alpha\partial^\beta - r^\beta\partial^\alpha]rY_{1,m}(\hat{\mathbf{r}}). \qquad (B.3)$$

- Antisymmetric, even parity

$$
\begin{aligned}
B_{4,1,m}^{\alpha\beta} &= r^{-2}\epsilon^{\alpha\beta\mu}r_\mu r Y_{1,m}(\hat{\mathbf{r}}) \\
B_{2,1,m}^{\alpha\beta} &= r^{-2}\epsilon^{\alpha\beta\mu}\partial_\mu r Y_{1,m}(\hat{\mathbf{r}}).
\end{aligned} \tag{B.4}
$$

As with the $j = 2$ case we form a real basis $r\tilde{Y}_{1,k}(\hat{\mathbf{r}})$ from the (in general) complex $rY_{1,m}(\hat{\mathbf{r}})$ in order to obtain real coefficients in our fits:

$$
\begin{aligned}
r\tilde{Y}_{1,k=0}(\hat{\mathbf{r}}) &= rY_{1,0}(\hat{\mathbf{r}}) = r\cos\theta = r_3 \\
r\tilde{Y}_{1,k=1}(\hat{\mathbf{r}}) &= r\frac{Y_{1,1}(\hat{\mathbf{r}}) + Y_{1,1}(\hat{\mathbf{r}})}{2i} \\
&= r\sin\theta\sin\phi = r_2 \\
r\tilde{Y}_{1,k=-1}(\hat{\mathbf{r}}) &= r\frac{Y_{1,-1}(\hat{\mathbf{r}}) - Y_{1,1}(\hat{\mathbf{r}})}{2} \\
&= r\sin\theta\cos\phi = r_1.
\end{aligned} \tag{B.5}
$$

And the final forms are

$$
\begin{aligned}
B_{3,1,0}^{\alpha\beta}(\hat{\mathbf{r}}) &= r^{-1}[r^\alpha n^\beta - r^\beta n^\alpha] \\
B_{4,1,0}^{\alpha\beta}(\hat{\mathbf{r}}) &= r^{-2}\epsilon^{\alpha\beta\mu}r_\mu(\mathbf{r}.\mathbf{n}) \\
B_{2,1,0}^{\alpha\beta}(\hat{\mathbf{r}}) &= r^{-2}\epsilon^{\alpha\beta\mu}n_\mu \\
B_{3,1,1}^{\alpha\beta}(\hat{\mathbf{r}}) &= r^{-1}[r^\alpha p^\beta - r^\beta p^\alpha] \\
B_{4,1,1}^{\alpha\beta}(\hat{\mathbf{r}}) &= r^{-2}\epsilon^{\alpha\beta\mu}r_\mu(\mathbf{r}.\mathbf{p}) \\
B_{2,1,1}^{\alpha\beta}(\hat{\mathbf{r}}) &= r^{-2}\epsilon^{\alpha\beta\mu}p_\mu \\
B_{3,1,-1}^{\alpha\beta}(\hat{\mathbf{r}}) &= r^{-1}[r^\alpha m^\beta - r^\beta m^\alpha] \\
B_{4,1,-1}^{\alpha\beta}(\hat{\mathbf{r}}) &= r^{-2}\epsilon^{\alpha\beta\mu}r_\mu(\mathbf{r}.\mathbf{m}) \\
B_{2,1,-1}^{\alpha\beta}(\hat{\mathbf{r}}) &= r^{-2}\epsilon^{\alpha\beta\mu}m_\mu.
\end{aligned} \tag{B.6}
$$

We now have 9 independent terms and cannot apply incompressibility in order to reduce them further. Instead, we use the geometrical constraints of the experiment to do this:

- $\phi = 0$ (vertical separation), $\alpha = 3, \beta = 1$

$$
\begin{aligned}
B_{3,1,0}^{31}(r,\theta,\phi=0) &= -\sin\theta \\
B_{2,1,1}^{31}(r,\theta,\phi=0) &= 1 \\
B_{3,1,-1}^{31}(r,\theta,\phi=0) &= \cos\theta.
\end{aligned} \tag{B.7}
$$

There are no contributions from the reflection-symmetric terms in the $j = 0$ isotropic sector since these are symmetric in the indices. The helicity term in $j = 0$ also does not contribute because of the geometry. So, to lowest order, we have

$$
\begin{aligned}
\widetilde{T}^{31}(\mathbf{r}) &= \widetilde{T}^{31}_{j=1}(\mathbf{r}) \\
&= a_{3,1,0} r^{\zeta_2^{(1)}}(-\sin\theta) + a_{2,1,1} r^{\zeta_2^{(1)}} + a_{3,1,-1} r^{\zeta_2^{(1)}} \cos\theta. \quad \text{(B.8)}
\end{aligned}
$$

As always, we have written the scale dependent prefactors $a_{qjm}(r)$ as having a power law dependence $a_{qjm} r^{\zeta_2^{(1)}}$. We have 3 unknown independent coefficients and 1 unknown exponent to fit for in the data.

B.2   Symmetric contribution

We consider the structure function

$$
S^{\alpha\beta}(\mathbf{r}) = \langle (u^\alpha(\mathbf{x}+\mathbf{r}) - u^\alpha(\mathbf{x}))(u^\beta(\mathbf{x}+\mathbf{r}) - u^\beta(\mathbf{x})) \rangle \quad \text{(B.9)}
$$

in the case where we have homogeneous flow. This object is symmetric in the indices by construction. It is easily seen that homogeneity implies even parity in $\mathbf{r}$:

$$
\begin{aligned}
S^{\alpha\beta}(\mathbf{r}) &= S^{\beta\alpha}(\mathbf{r}) \\
S^{\alpha\beta}(-\mathbf{r}) &= S^{\alpha\beta}(\mathbf{r}). \quad \text{(B.10)}
\end{aligned}
$$

We reason that this object cannot exhibit a $j = 1$ contribution from the $SO(3)$ representation in the following. Homogeneity allows us to use the incompressibility condition

$$
\begin{aligned}
\partial_\alpha S^{\alpha\beta} &= 0 \\
\partial_\beta S^{\alpha\beta} &= 0 \quad \text{(B.11)}
\end{aligned}
$$

separately on the basis tensors of a given parity and symmetry in order to give relationships between their coefficients. For even parity, symmetric case we have for general $j \geq 2$ just two basis tensors and they must occur in some linear combination with incompressibility providing a constraint between the two coefficients. However, for $j = 1$ we only have one such tensor in the even parity, symmetric group. Therefore, by incompressibility, its coefficient must vanish. Consequently, we cannot have a $j = 1$ contribution for the even parity (homogeneous), symmetric structure function.

Now, we consider the case of an experiment when $\mathbf{r}$ has some component in the inhomogeneous direction. Now, it is no longer true that $S^{\alpha\beta}(\mathbf{r})$ is of even parity. Moreover it is not possible to use incompressibility as above to exclude the existence of a $j = 1$ contribution. We must look at all $j = 1$ basis tensors that are symmetric, but not confined to even parity.

These are:

- Odd parity, symmetric

$$B_{1,1,k}^{\alpha\beta}(\hat{\mathbf{r}}) \equiv r^{-1}\delta^{\alpha\beta}r\tilde{Y}_{1k}(\hat{\mathbf{r}})$$
$$B_{7,1,k}^{\alpha\beta}(\hat{\mathbf{r}}) \equiv r^{-1}[r^{\alpha}\partial^{\beta} + r^{\beta}\partial^{\alpha}]r\tilde{Y}_{1k}(\hat{\mathbf{r}})$$
$$B_{9,1,k}^{\alpha\beta}(\hat{\mathbf{r}}) \equiv r^{-3}r^{\alpha}r^{\beta}r\tilde{Y}_{1k}(\hat{\mathbf{r}})$$
$$B_{5,1,k}^{\alpha\beta}(\hat{\mathbf{r}}) \equiv r\partial^{\alpha}\partial^{\beta}r\tilde{Y}_{1k}(\hat{\mathbf{r}}) \equiv 0. \qquad (B.12)$$

- Even parity, symmetric

$$B_{8,1,k}^{\alpha\beta}(\hat{\mathbf{r}}) \equiv r^{-2}[r^{\alpha}\epsilon^{\beta\mu\nu}r_{\mu}\partial_{\nu} + r^{\beta}\epsilon^{\alpha\mu\nu}r_{\mu}\partial_{\nu}]r\tilde{Y}_{1k}(\hat{\mathbf{r}})$$
$$B_{6,1,k}^{\alpha\beta}(\hat{\mathbf{r}}) \equiv [\epsilon^{\beta\mu\nu}r_{\mu}\partial_{\nu}\partial_{\alpha} + \epsilon^{\beta\mu\nu}r_{\mu}\partial_{\nu}\partial_{\beta}]r\tilde{Y}_{1k}(\hat{\mathbf{r}}) \equiv 0. \quad (B.13)$$

We use the real basis of $r^{-1}\tilde{Y}_{1k}(\hat{\mathbf{r}})$ which are formed from the $r^{-1}Y_{1m}(\hat{\mathbf{r}})$. Both $B_{5,1,k}^{\alpha\beta}(\hat{\mathbf{r}})$ and $B_{6,1,k}^{\alpha\beta}(\hat{\mathbf{r}})$ vanish because of the taking of the double derivative of an object of single power in r. We thus have 4 different contributions to symmetric $j = 1$ and each of these is of 3 dimensions ($k = -1, 0, 1$) giving in general 12 terms in all:

$$B_{1,1,0}^{\alpha\beta}(\hat{\mathbf{r}}) = r^{-1}\delta^{\alpha\beta}(\mathbf{r}\cdot\mathbf{n})$$
$$B_{7,1,0}^{\alpha\beta}(\hat{\mathbf{r}}) = r^{-1}[r^{\alpha}n^{\beta} + r^{\beta}n^{\alpha}]$$
$$B_{9,1,0}^{\alpha\beta}(\hat{\mathbf{r}}) = r^{-3}r^{\alpha}r^{\beta}(\mathbf{r}\cdot\mathbf{n})$$
$$B_{8,1,0}^{\alpha\beta}(\hat{\mathbf{r}}) \equiv r^{-2}[(r^{\alpha}m^{\beta} + r^{\beta}m^{\alpha})(\mathbf{r}\cdot\mathbf{p}) - (r^{\alpha}p^{\beta} + r^{\beta}p^{\alpha})(\mathbf{r}\cdot\mathbf{m})]$$
$$B_{1,1,1}^{\alpha\beta}(\hat{\mathbf{r}}) = r^{-1}\delta^{\alpha\beta}(\mathbf{r}\cdot\mathbf{p})$$
$$B_{7,1,1}^{\alpha\beta}(\hat{\mathbf{r}}) = r^{-1}[r^{\alpha}p^{\beta} + r^{\beta}p^{\alpha}]$$
$$B_{9,1,1}^{\alpha\beta}(\hat{\mathbf{r}}) = r^{-3}r^{\alpha}r^{\beta}(\mathbf{r}\cdot\mathbf{p})$$
$$B_{8,1,1}^{\alpha\beta}(\hat{\mathbf{r}}) \equiv r^{-2}[(r^{\alpha}m^{\beta} + r^{\beta}m^{\alpha})(\mathbf{r}\cdot\mathbf{n}) - (r^{\alpha}n^{\beta} + r^{\beta}n^{\alpha})(\mathbf{r}\cdot\mathbf{m})]$$
$$B_{1,1,-1}^{\alpha\beta}(\hat{\mathbf{r}}) = r^{-1}\delta^{\alpha\beta}(\mathbf{r}\cdot\mathbf{m})$$
$$B_{7,1,-1}^{\alpha\beta}(\hat{\mathbf{r}}) = r^{-1}[r^{\alpha}m^{\beta} + r^{\beta}m^{\alpha}]$$
$$B_{9,1,-1}^{\alpha\beta}(\hat{\mathbf{r}}) = r^{-3}r^{\alpha}r^{\beta}(\mathbf{r}\cdot\mathbf{m}) \qquad (B.14)$$
$$B_{8,1,-1}^{\alpha\beta}(\hat{\mathbf{r}}) \equiv r^{-2}[(r^{\alpha}p^{\beta} + r^{\beta}p^{\alpha})(\mathbf{r}\cdot\mathbf{n}) - (r^{\alpha}n^{\beta} + r^{\beta}n^{\alpha})(\mathbf{r}\cdot\mathbf{p})].$$

These are all the possible $j = 1$ contributions to the symmetric, mixed parity (inhomogeneous) structure function.

For our experimental setup II, we want to analyze the inhomogeneous structure function in the case $\alpha = \beta = 3$, and azimuthal angle $\phi = 0$

Table 7. The number of free coefficients in the symmetric $j = 1$ sector for inhomogeneous turbulence and for different geometries.

k	$\phi = 0, \alpha = \beta = 3$		$\phi = 0, \alpha = \beta = 1$		$\phi = 0, \alpha = 3, \beta = 1$	
	$\theta \neq 0$	$\theta = 0$	$\theta \neq 0$	$\theta = 0$	$\theta \neq 0$	$\theta = 0$
0	3	3	2	1	2	0
1	1	0	1	0	0	0
-1	2	0	3	0	2	1
Total	6	3	6	1	4	1

(which corresponds to vertical separation). For that case, we obtain the basis tensors to be

$$B^{33}_{1,1,0}(\theta) = \cos\theta$$
$$B^{33}_{7,1,0}(\theta) = 2\cos\theta$$
$$B^{33}_{9,1,0}(\theta) = \cos^3\theta$$
$$B^{33}_{8,1,1}(\theta) = -2\cos\theta\sin\theta$$
$$B^{33}_{1,1,-1}(\theta) = \sin\theta$$
$$B^{33}_{9,1,-1}(\theta) = \cos^2\theta\sin\theta. \tag{B.15}$$

Table 7 gives the number of free coefficients in the symmetric $j = 1$ sector in the fit to the inhomogeneous structure function for various geometrical configurations.

C   Tests of the robustness of the interpolation formula

In order to test the robustness of the interpolation formula equation (6.2), we performed the following additional calculations. We considered the data from the probe at the height of 0.54 m. For each order n of the structure function, we defined a "window" of data extending over two decades of the separation scale, r. We first placed the lower edge of the window well inside the dissipation range and fit the interpolation formula to the data in the first window. We then moved the lower edge of the window by half a decade and fit the formula to the data in the next window. In this manner, we proceeded until the upper edge of the last window corresponded to the largest value of r. The entire range of r yields five windows. We thus obtained five values of the parameter C_n and calculate the scaling exponent $\zeta_n^{(2)} = n - 2C_n$ in each case, giving some indication of the robustness of our result.

Tables 8–10 present the results of performing these checks on structure functions of the second, third and fourth order. The mean and standard

Table 8. Second order: $\zeta_2^{(2)} = 1.25 \pm 0.05$.

C_2	0.35 ± 0.1	0.35 ± 0.05	0.39 ± 0.02	0.38 ± 0.05	0.38 ± 0.07
$\zeta_2^{(2)}$	1.31 ± 0.2	1.30 ± 0.10	1.21 ± 0.04	1.24 ± 0.10	1.23 ± 0.14

Table 9. Third order: $\zeta_3^{(2)} = 1.14 \pm 0.11$.

C_3	0.99 ± 0.03	0.95 ± 0.04	0.88 ± 0.07	0.91 ± 0.04	0.96 ± 0.08
$\zeta_3^{(2)}$	1.01 ± 0.06	1.10 ± 0.08	1.3 ± 0.14	1.2 ± 0.08	1.1 ± 0.16

Table 10. Fourth order: $\zeta_4^{(2)} = 1.61 \pm 0.13$.

C_4	1.21 ± 0.07	1.12 ± 0.09	1.15 ± 0.03	1.29 ± 0.1	1.21 ± 0.08
$\zeta_4^{(2)}$	1.58 ± 0.14	1.76 ± 0.18	1.7 ± 0.06	1.42 ± 0.2	1.58 ± 0.16

deviation of the exponent values are given in the caption for each table. It is found that the mean value in each case is in close agreement to the value of the exponents presented in the main text which were obtained by a fit to the entire range of data. This gives us greater confidence in the use of the interpolation formula.

References

[1] R.A. Antonia, B.R. Pearson and T. Zhou, *Phys. Fluids* **12** (2000) 2954-2964.

[2] I. Arad, L. Biferale, I. Mazzitelli and I. Procaccia, *Phys. Rev. Lett.* **82** (1999) 5040.

[3] I. Arad, L. Biferale and I. Procaccia, *Phys. Rev. E* **61** (2000) 2654.

[4] I. Arad, B. Dhruva, S. Kurie, V.S. L'vov, I. Procaccia and K.R. Sreenivasan, *Phys. Rev. Lett.* **81** (1998) 5330.

[5] I. Arad, V.S. L'vov and I. Procaccia, *Phys. Rev. E* **59** (1999) 6753.

[6] B. Dhruva, *An Experimental Study of High Reynolds Number Turbulence in the Atmosphere*, Ph.D. Thesis (Yale University, 2000).

[7] U. Frisch, *Turbulence: The Legacy of A.N. Kolmogorov* (Cambridge University Press, Cambridge, 1996).

[8] R.J. Hill (2000) (submitted).

[9] J. Jimenez, A.A. Wray, P.G. Saffman and R.S. Rogallo, *J. Fluid. Mech.* **255** (1993) 65.

[10] J.C. Klewicki, M.M. Metzger, E. Kelner and E.M. Thurlow, *Phys. Fluids* **7** (1995) 857.

[11] A.N. Kolmogorov, *Dokl. Akad. Nauk SSSR* **30** (1941) 301.

[12] S. Kurien, V.S. L'vov, I. Procaccia and K.R. Sreenivasan, *Phys. Rev. E* **61** (1998) 407.

[13] S. Kurien and K.R. Sreenivasan, *Phys. Rev. E* **62** (2000) 2206.

[14] D. Lohse and A. Muller–Groeling, *Phys. Rev. Lett.* **74** (1995) 1747.

[15] J.L. Lumley, *Phys. Fluids* **10** (1967) 855.

[16] V.S. L'vov, *Physics Reports* **207** (1991) 1.

[17] V.S. L'vov, E. Podivilov and I. Procaccia, *Phys. Rev. Lett.* **79** (1997) 2050.

[18] V.S. L'vov, A. Pomyalov and I. Procaccia, *Phys. Rev. E* **60** (1999) 4175.

[19] A.S. Monin and A.M. Yaglom, *Statistical Fluid Mechanics*, Vol. 2 (MIT, Cambridge, 1971).

[20] M. Nelkin, *Am. J. Phys.* **68** (2000) 310.

[21] S. Saddoughi and S. Veeravalli, *J. Fluid Mech.* **268** (1994) 333-372.

[22] X. Shen and Z. Warhaft, *Phys. Fluids* **11** (2000) 2976.

[23] K.R. Sreenivasan, *Proc. Roy. Soc. Lond. Ser. A* **434** (1991) 165.

[24] K.R. Sreenivasan and R.A. Antonia, *Annu. Rev. Fluid Mech.* **29** (1997) 435.

[25] K.R. Sreenivasan and B. Dhruva, *Prog. Theo. Phys.* **130** (1998) 103.

[26] K.R. Sreenivasan, A. Prabhu and R. Narasimha, *J. Fluid Mech.* **137** (1983) 251-272.

[27] G. Stolovitzky, *The Statistical Order of Small Scales in Turbulence*, Ph.D. Thesis (Yale University, 1994).

[28] G. Stolovitzky, K.R. Sreenivasan and A. Juneja, *Phys. Rev. E* **48** (1993) R3217.

COURSE 3

LARGE-EDDY SIMULATIONS OF TURBULENCE

O. MÉTAIS

*Laboratoire des Écoulements
Géophysiques et Industriels (LEGI),
Domaine Universitaire, BP. 53,
38041 Grenoble Cedex 9, France*

Contents

LARGE-EDDY SIMULATIONS OF TURBULENCE

O. Métais

1 Introduction

Direct-numerical simulations of turbulence (DNS) consist in solving explicitly all the scales of motion, from the largest $l_{\rm I}$ to the Kolmogorov dissipative scale $l_{\rm D}$. It is well known from the statistical theory of turbulence that $l_{\rm I}/l_{\rm D}$ scales like $R_{\rm l}^{3/4}$, where $R_{\rm l}$ is the large-scale Reynolds number $u'l_{\rm I}/\nu$ based upon the rms velocity fluctuation u'. Therefore, the total number of degrees of freedom necessary to represent the whole span of scales of a three-dimensional turbulent flow is of the order of $R_{\rm l}^{9/4}$ in three dimensions. In the presence of obstacles, around a wing or a fuselage for instance, and if one wants to simulate three-dimensionally all motions ranging from the viscous thickness $\delta_{\rm v} = \nu/v_* \approx 10^{-6}$ m up to 10 m, it would be necessary to put 10^{21} modes on the computer. Right now, the calculations done to the expense of not excessive computing times on the biggest machines take about 2×10^7 grid points, which is very far from the above estimation. Even with the unprecedented improvement of scientific computers, it may take several tenths of years (if it becomes ever possible) before DNS permit to simulate situations at Reynolds numbers comparable to those encountered in natural conditions.

Statistical modelling based on Reynolds Averaged Navier–Stokes (RANS) equations are particularly designed to deal with statistically steady flows or with flows whose statistical properties vary "slowly" with time, that is to say of characteristic time scale much larger than a characteristic turbulent time scale. The application of phase averaging constitutes another alternative which allows for the modelling of time periodic flows. With the RANS approach all the turbulent scales are modelled. First order as well as second order RANS models involve many adjustable constants and it is therefore impossible to design models which are "universal" enough to be applicable to various flow configurations submitted to diverse external forces (rotation, thermal stratification, etc.). However, since RANS models compute statistical quantities, they do not require temporal or spatial discretizations as fine as the ones necessary for DNS or even LES. They are therefore applicable to flows in complex geometries.

Large-Eddy Simulations (LES) techniques constitute intermediate tec-
niques between DNS and RANS in the sense that the large scales of the flow
are deterministically simulated and only the small scales are modelled but
statistically influence the large-scale motion. LES then explicitly resolve the
large-scales inhomogeneity and anistropy as well as the large-scales unsteadiness. This is important from an engineering point of view since the large
scales are responsible for the major part of turbulent transfers of momentum
or heat for example. Most subgrid-scale models which parameterized the
action of the small-scales are based upon "universal" properties of small-
scales turbulence: those can therefore be applied to various flows submitted
to various external effects without being modified. In this respect, they constitute "universal" models directly applicable to various flow configurations.
However, they require much finer spatial and temporal discretizations than
RANS and lie inbetween DNS and RANS as far as CPU time consump-
tion is concerned. Once confined to very simple flow configurations such as
isotropic turbulence or periodic flows, the field is evolving to include spatially growing shear flows, separated flows, pipe flows, riblet walls, and bluff
bodies, among others. This is due to the tremendous progress in scientific
computing and in particular of parallel computing. As will be seen in the
few examples presented below, LES are extremely useful in particular towards the understanding of the dynamics of coherent vortices and structures
in turbulence. We will show below that this is of special importance for flow
control problems, for detached flows and their aeroacoustics predictions and
for flows submitted to compressibility effects and density differences.

1.1   LES and determinism: Unpredictability growth

From a mathematical viewpoint, the LES problem is not very well posed.
Indeed, let us consider the time evolution of the fluid as the motion of a
point in a sort of phase space of extremely large dimension (*e.g.* $\sim 10^{21}$
around a wing, as seen above). At some initial instant, the flow computed
with LES will differ from the actual flow, due to the uncertainty contained
in the subgridscales. This initial difference between the actual and the
computed flow will grow, due to nonlinear effects, as in a dynamical system
having a chaotic behaviour. Therefore, the two points will separate in phase
space, and, as time goes on, the LES will depart from reality. However, as
will be seen below, LES permit to predict the statistical characteristics of
turbulence, as well as the dynamics of coherent vortices and structures.

Note that chaos in dynamical systems with a low number of degrees of
freedom is generally characterized by a positive Lyapounov exponent, with
exponential growth of the distance between two points initially very close
in phase space. In isotropic turbulence, one introduces for predictability
studies the error spectrum $E_\Delta(k, t)$, characterizing the spatial-frequency

distribution associated to the energy of the difference between two random fields $\vec{u}_1$ and $\vec{u}_2$ with same statistical properties:

$$\frac{1}{4} \left\langle [\vec{u}_1^2(\vec{x}, t) - \vec{u}_2^2(\vec{x}, t)] \right\rangle = \int_0^{+\infty} E_\Delta(k, t) \, \mathrm{d}k, \qquad (1.1)$$

the energy spectrum $E(k, t)$ being such that

$$\frac{1}{2} \left\langle \vec{u}_1^2 \right\rangle = \frac{1}{2} \left\langle \vec{u}_2^2 \right\rangle = \int_0^{+\infty} E(k, t) \, \mathrm{d}k. \qquad (1.2)$$

The error rate

$$r(t) = \frac{\int_0^{+\infty} E_\Delta(k, t) \mathrm{d}k}{\int_0^{+\infty} E(k, t) \mathrm{d}k} \qquad (1.3)$$

is zero when the two fields are completely correlated, and one when they are totally uncorrelated. In predictability studies, one takes generally an initial state such that complete unpredictability ($E(k) = E_\Delta(k)$) holds above $k_E(0)$, while $E_\Delta(k)$ is 0 for $k < k_E(0)$. Two-point closures of the EDQNM type (see [67] for details) show (in three or two dimensions) an inverse cascade of error, where the wavenumber $k_E(t)$ characterizing the error front decreases (see [79]). Thus, the error rate can be approximated by

$$r(t) \approx \frac{\int_{k_E(t)}^{\infty} E(k, t) \mathrm{d}k}{\int_0^{+\infty} E(k, t) \mathrm{d}k}.$$

We assume that the turbulence is forced by external forces, so that the kinetic energy arising at the denominator of equation (1.3) is fixed. In three-dimensional turbulence, and if a $k^{-5/3}$ spectrum is assumed for $k > k_E$, the error rate will be proportional to $k_E^{-2/3}$. In fact, closures (see [67, 79]) show that k_E^{-1} follows a Richardson's law ($k_E^{-1} \propto t^{3/2}$), so that the error rate grows linearly with time. This is in fact a slow increase compared with the exponential growth of chaotic dynamical systems, and quite encouraging concerning the potentialities of large-eddy simulations for three-dimensional turbulent flows.

2   Vortex dynamics

As will be seen, large-eddy simulations deal with energetic structures of the flow with a characteristic scale or wavelength larger than a given cutoff scale Δx. These so-called *large scales* may be spatially organized or not, and sometimes correspond to coherent vortices of recognizable shape. It is therefore important to be able to identify these coherent vortices.

2.1   Coherent vortices

2.1.1   Definition

Coherent vortices in turbulence are defined by Lesieur [67] as regions of the
flow satisfying three conditions:

  (i) the vorticity concentration ω, modulus of the vorticity vector, should
      be high enough so that a local roll up of the surrounding fluid is
      possible;

  (ii) they should keep approximately their shape during a time T_c longer
       enough in front of the local turnover time ω^{-1};

 (iii) they should be unpredictable.

In this context, high ω is a possible candidate for coherent-vortex identification.

2.1.2   Pressure

With such a definition, the core of the coherent vortices should be pressure
lows. Indeed, a fluid parcel winding around the vortex will be (in a frame
moving with the parcel) in approximate balance between centrifugal and
pressure-gradient effects. We are talking here of the static pressure p. The
reasoning may be made more quantitative by considering Euler equation (in
a flow of uniform density ρ_0) in the form

$$\frac{\partial \vec{u}}{\partial t} + \vec{\omega} \times \vec{u} = -\frac{1}{\rho_0}\vec{\nabla}P \qquad (2.1)$$

where $P = p + \rho_0 \vec{u}^2/2$ is now the dynamic pressure. In a frame moving with
the coherent vortex and supposed locally Galilean, the ratio of the second
to the first term in the l.h.s. of equation (2.1) is of the order of $T_c\,\omega$. Then
the equation reduces to the cyclostrophic balance

$$\vec{\omega} \times \vec{u} \approx -\frac{1}{\rho_0}\vec{\nabla}P \qquad (2.2)$$

if condition (ii) above is fulfilled. If one supposes that the coherent vortex
is a vortex tube tangent to the velocity vector, it follows that this tube is a
low for the dynamic pressure.

2.1.3   The Q-criterion

We recall now the so-called Q-criterion. Let

$$S_{ij} = \frac{1}{2}\left(\frac{\partial u_i}{\partial x_j} + \frac{\partial u_j}{\partial x_i}\right), \quad \Omega_{ij} = \frac{1}{2}\left(\frac{\partial u_i}{\partial x_j} - \frac{\partial u_j}{\partial x_i}\right) \qquad (2.3)$$

be respectively the symmetric and antisymmetric parts of the velocity-gradient tensor $\partial u_i / \partial x_j$. It is well known that the second invariant of this tensor

$$Q = \frac{1}{2}(\Omega_{ij}\Omega_{ij} - S_{ij}S_{ij}) = \frac{1}{4}(\vec{\omega}^2 - 2S_{ij}S_{ij}) , \qquad (2.4)$$

is equal to $\nabla^2 p / 2$. Indeed, the Poisson equation for the pressure in a flow of uniform density writes

$$
\begin{aligned}
-\frac{1}{\rho_0}\nabla^2 p &= \frac{\partial^2}{\partial x_i \partial x_j} u_i u_j = \frac{\partial}{\partial x_i}\left[u_j \frac{\partial u_i}{\partial x_j}\right] = \frac{\partial u_i}{\partial x_j}\frac{\partial u_j}{\partial x_i} \\
&= \left(S_{ij} + \frac{1}{2}\epsilon_{ij\lambda}\omega_\lambda\right)\left(S_{ji} + \frac{1}{2}\epsilon_{ji\mu}\omega_\mu\right) = S_{ij}S_{ij} - \frac{1}{2}\vec{\omega}^2 = -2Q.
\end{aligned}
$$

Let us consider a low-pressure tube of small section. Let $\Delta\Sigma$ be its lateral surface, assumed isobaric and convex. Let Σ_1 and Σ_2 be two cross sections of the tube, supposed normal to its axis, and ΔV the volume of the tube portion comprised between Σ_1 and Σ_2. The pressure gradient on $\Delta\Sigma$ is normal to it and directed towards the exterior. The pressure gradient on the two cross sections is tangent to them. Then, the flux of the pressure gradient getting out of the tube is equal to the flux through $\Delta\Sigma$, and is positive. Due to the divergence theorem, this is equal to the integral over ΔV of $\nabla^2 p$, which is positive, such as the integral of Q. If we suppose that the size of ΔV is small enough so that Q does not vary appreciably within it, this implies that Q is positive in ΔV. Since this reasoning may be repeated all along the length of the tube, the Q-criterion ($Q > 0$) is therefore a necessary condition for the existence of such thin convex low-pressure tubes. This is the reason which motivates Hunt *et al.* [46] to propose the Q-criterion as a way to characterize the vortices.

We have thus shown that the Q-criterion is valuable to help characterizing the convex low-pressure tubes, which are generally associated to coherent vortices. Notice however that the relation $Q = \nabla^2 p / 2\rho_0$ implies that vortex-identification criteria based upon Q involve much more small-scale activity than thosed based on the pressure, as will be verified in the simulations.

2.2 Vortex identification

Let us present a comparison of some of these vortex-identification methods (low dynamic pressure, high ω, positive Q) applied to incompressible DNS of isotropic turbulence and LES of a backward-facing step, done by Delcayre [28]. Other exemples will be provided in the rest of the chapter. More specifically, we consider isosurfaces at a given threshold of ω, p and Q. The

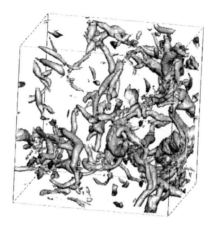

Fig. 1. Low-pressure isosurfaces in DNS of isotropic turbulence (from Delcayre [28]).

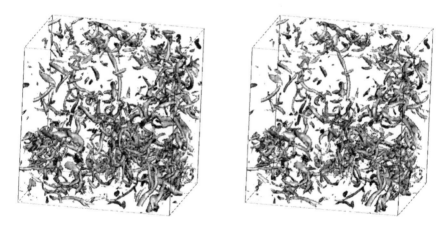

Fig. 2. High vorticity (left) and positive Q (right) isosurfaces in DNS of isotropic turbulence (from Delcayre [28]).

choice of the threshold is justified by what gives visually the best vortices, or with respect to what we know of the flow dynamics from former simulations or laboratory experiments.

2.2.1   Isotropic turbulence

For isotropic turbulence, we consider a DNS at low Reynolds number (freely-decaying case). It is well known that coherent vortices exist in such a flow, in the form of thin tubes randomly orientated, of length the turbulence integral scale (Siggia [102], Vincent and Ménéguzzi [111], Métais and Lesieur [80], Jimenez and Wray [47]). Comparison of Figures 1 and 2 (left) show that the isobaric surfaces are more fat than the vorticity surfaces, but represent the same events, in good agreement with the observations of Brachet [13] for Taylor–Green vortices and Métais and Lesieur [80] for LES of isotropic turbulence. This is confirmed by the present DNS. Figure 2 (right), showing the iso-Q maps, is close to the vorticity map, althought slightly less dense. A last point concerns the dimension of these tubes: everybody agrees on the average length, which is integral scale l. It is right now not decided yet whether the diameter scales on the dissipative scale or the Taylor microscale. If we interpret the vortices as resulting from the roll up of local vortex sheets, it is this last scale which should prevail. As a matter of fact, this strongly anisotropic vortex topology is very far from the quite naive circular eddies considered in the popular folklore of Richardson–Kolmogorov cascade.

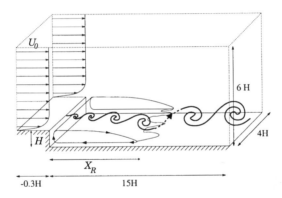

Fig. 3. Schematic view of the backward-facing step (from Delcayre [28]).

2.2.2 Backward-facing step

We present now LES results of a uniform-density flow above a straight backward-facing step. The model used is the selective structure-function model (SSF, see Sect. 5.2). Figure 3 shows a schematic view of the flow. The step-height is H, the expansion ratio 1.2, and the Reynolds number $U_0 H/\nu = 5100$, as in the configuration studied experimentally (Jovic and Driver [49]) and numerically (Le *et al.* [62], Akselvoll and Moin [2]) at Stanford. A free-slip boundary condition is used on the upper boundary, well justified with respect to the laboratory experiment consisting in a double-expansion channel. At the inlet, we impose Spalart's [106] mean turbulent boundary-layer velocity profile, to which a small three-dimensional white-noise perturbation regenerated at each time step is superposed. One assumes periodicity in the spanwise direction, and there is an outflow boundary condition of the Sommerfeld type, where the quantities are transported following a fictitious "tangential" wave phase velocity. We have checked that the latter is very good for letting the coherent vortices get out of the computational domain without any distorsion.

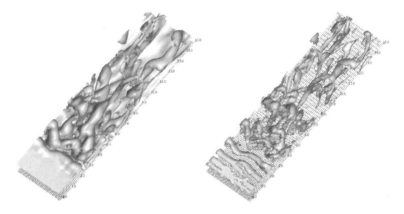

Fig. 4. Backward-facing step, visualization of coherent vortices using high-vorticity modulus (left) and positive Q (right) isosurfaces (from Delcayre [28]).

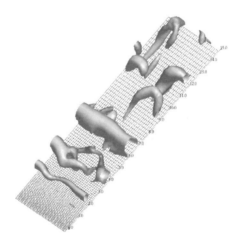

Fig. 5. Backward-facing step, low-pressure isosurfaces (from Delcayre [28]).

Flow animations show the following vortex dynamics: quasi two-dimensional Kelvin–Helmholtz type vortices are shed behind the step, resulting from the instability of the upstream vortex sheet. Then they are subject to dislocations (helical pairings) and transform into a field of big staggered arch-like vortices which impact the lower wall and are carried away downstream. Figure 4 (left), presenting iso-vorticity maps, does show the breakdown of the vortex sheet into big staggered Λ-vortices. Figure 4 (right) presents iso-Q maps, and indicates the same vortex events as for the vorticity, but the vortices are thinner, and the upstream vortex sheet has been erased. Finally, isobaric surfaces (Fig. 5) are misleading in this case, since they seem to indicate a large quasi two-dimensional vortex at the level of reattachment, whereas it is simply an erroneous reconnection of the tips of the big Λ's.

Such vortical structure has important implications in terms of pressure spectra and aeroacoustics. Figure 6 indeed shows the frequency pressure spectra at four positions in the flow: (1) just behind the step, (2) just before reattachment, (3) just behind reattachment and (4) much further downstream. Frequencies f are expressed in units U_0/H and correspond in fact to Strouhal numbers $S_t = fH/U_0$. Position (1) is marked by a peak at $S_t = 0.23$, corresponding to the shedding of Kelvin–Helmholtz vortices. At position (2), a second peak of higher amplitude is present at the subharmonic Strouhal number 0.12, corresponding physically to helical pairing. At positions (3) and (4), the two previous Strouhal numbers are still there, but a third peak forms at a Strouhal $S_t = 0.07$, corresponding to the well-known flapping of the recirculation bubble. These different Strouhal numbers associated with the different unsteady phenomena are in good agreement with those previously found by other authors (see e.g. Le et al. [62], Arnal and Friedrich [6]). Such informations regarding the pressure spectra and how they relate to the vortex dynamics is very important for acoustical studies and noise control in particular.

3 LES formalism in physical space

This chapter deals with an incompressible flow, whose density is conserved with the fluid motion, which implies the continuity equation $\vec{\nabla}\vec{u} = 0$. Then ρ may be either uniform, or have a mean variation taken into account through Boussinesq's approximation.

3.1 LES equations for a flow of constant density

To begin with, let us consider a simulation of Navier–Stokes equations with constant density ρ_0 carried out in physical space, using finite-difference or finite-volume methods. Let Δx be a scale characteristic of the grid mesh. In order to eliminate the subgridscales, we introduce a filter of width Δx.

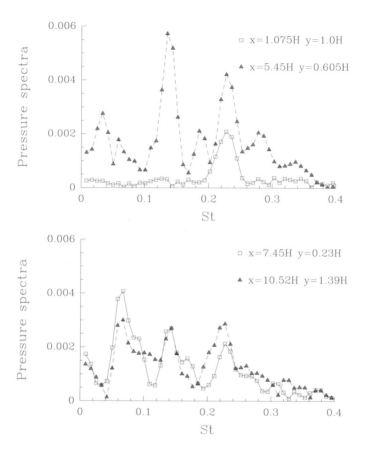

Fig. 6. Backward-facing step, spectra of spanwise-averaged pressure fluctuations at different positions (courtesy Delcayre).

Mathematically, the filtering operation corresponds to the convolution of any quantity $f(\vec{x}, t)$ of the flow by the filter function $G_{\Delta x}(\vec{x})$, in the form

$$\bar{f}(\vec{x}, t) = \int f(\vec{y}, t) G_{\Delta x}(\vec{x} - \vec{y}) \mathrm{d}\vec{y} = \int f(\vec{x} - \vec{y}, t) G_{\Delta x}(\vec{y}) \mathrm{d}\vec{y}, \qquad (3.1)$$

and the subgridscale field is the departure of the actual flow with respect to the filtered field:

$$f = \bar{f} + f'. \qquad (3.2)$$

Since Δx is here assumed constant, it is easy to show that the space and time derivatives commute with the filtering operator.

We assume that we use a Cartesian system or coordinates. Let us first write themomentum equations as

$$\frac{\partial u_i}{\partial t} + \frac{\partial}{\partial x_j}(u_i u_j) = -\frac{1}{\rho_0}\frac{\partial p}{\partial x_i} + \frac{\partial}{\partial x_j}(2\nu S_{ij}), \qquad (3.3)$$

where S_{ij} is the deformation tensor already defined by equation (2.3). The filtered momentum equations write exactly

$$\frac{\partial \bar{u}_i}{\partial t} + \frac{\partial}{\partial x_j}(\bar{u}_i \bar{u}_j) = -\frac{1}{\rho_0}\frac{\partial \bar{p}}{\partial x_i} + \frac{\partial}{\partial x_j}(2\nu \bar{S}_{ij} + T_{ij}), \qquad (3.4)$$

where

$$T_{ij} = \bar{u}_i \bar{u}_j - \overline{u_i u_j} \qquad (3.5)$$

is the subgrid-stresses tensor, responsible of momentum exchanges between the subgrid- and the filtered scales. The filtered continuity equation writes

$$\frac{\partial \bar{u}_j}{\partial x_j} = 0. \qquad (3.6)$$

Let us consider now the mixing of a passive scalar (such as temperature or density) of molecular diffusivity κ transported by the flow, and satisfying the equation

$$\frac{\partial \rho}{\partial t} + \frac{\partial}{\partial x_j}(\rho u_j) = \frac{\partial}{\partial x_j}\left\{\kappa\frac{\partial \rho}{\partial x_j}\right\}. \qquad (3.7)$$

The filtered scalar equation writes

$$\frac{\partial \bar{\rho}}{\partial t} + \frac{\partial}{\partial x_j}(\bar{\rho}\bar{u}_j) = \frac{\partial}{\partial x_j}\left\{\kappa\frac{\partial \bar{\rho}}{\partial x_j} + T_j^{(\rho)}\right\}, \qquad (3.8)$$

where

$$T_j^{(\rho)} = \bar{\rho}\bar{u}_j - \overline{\rho u_j} \qquad (3.9)$$

is the subgrid scalar flux. T_{ij} and $T_j^{(\rho)}$ can be written as:

$$T_{ij} = -\left(\overline{u_i' u_j'} + \overline{\bar{u}_i u_j'} + \overline{u_i' \bar{u}_j} + \overline{\bar{u}_i \bar{u}_j} - \bar{u}_i \bar{u}_j\right), \qquad (3.10)$$

$$T_j^{(\rho)} = -\left(\overline{\rho' u_j'} + \overline{\bar{\rho} u_j'} + \overline{\rho' \bar{u}_j} + \overline{\bar{\rho} \bar{u}_j} - \bar{\rho} \bar{u}_j\right). \qquad (3.11)$$

In equation (3.10), $-\overline{u_i' u_j'}$ is a Reynolds-stress like term, $-(\overline{\bar{u}_i u_j'} + \overline{u_i' \bar{u}_j})$ is called the Clark term (Clark *et al.* [21]), and $\bar{u}_i \bar{u}_j - \overline{\bar{u}_i \bar{u}_j}$ is the Leonard's tensor (Leonard [65]). The latter is explicit in the sense that it is defined in terms of the filtered field, and has been used in scale-similarity models to provide information on the subgrid stresses (see Sect. 7.3). Leonard's stresses are also a major ingredient of the so-called Germano's identity for the dynamic approach in physical space (see Sect. 6).

These subgridscale tensors and fluxes need of course to be modelled.

3.2 LES Boussinesq equations in a rotating frame

We give now the LES equations corresponding to Navier–Stokes equations within the Boussinesq approximation in a Cartesian frame of reference rotating with a constant angular velocity Ω about the x_3 axis:

$$
\frac{\partial \bar{u}_i}{\partial t} + \frac{\partial}{\partial x_j}(\bar{u}_i \bar{u}_j) = -\frac{1}{\rho_0}\frac{\partial \bar{p}}{\partial x_i} + \frac{\partial}{\partial x_j}(2\nu \bar{S}_{ij} + T_{ij})
$$
$$
+2\epsilon_{ij3}\Omega \bar{u}_j - g_i \delta_{i3}\frac{\bar{\rho}}{\rho_0}, \tag{3.12}
$$

$$
\frac{\partial \bar{\rho}}{\partial t} + \frac{\partial}{\partial x_j}(\bar{\rho}\bar{u}_j) = \frac{\partial}{\partial x_j}\left\{ \kappa \frac{\partial \bar{\rho}}{\partial x_j} + T_j^{(\rho)} \right\}. \tag{3.13}
$$

This comes from the filtering of a particular version of Boussinesq equations, valid both for a liquid and a perfect gas, where ρ is the total density, p the static pressure, and ρ_0 an average density on the height of the fluid layer. T_{ij} and $T_j^{(\rho)}$ are defined as above. Here, the density is still a scalar transported by the flow (as in Eq. (3.7)). But it is not passive since it reacts through gravity in the momentum equation. In fact, this system of equations is very useful to study stably-stratified or thermally convective rotating flows.

3.3 Eddy-viscosity and diffusivity assumption

By analogy with what is done in the framework of Reynolds equations for the ensemble-averaged equations, the subgridscale tensors are in most of the cases expressed in terms of eddy-viscosity and diffusivity coefficients in the form

$$
T_{ij} = 2\nu_t(\vec{x}, t)\, \bar{S}_{ij} + \frac{1}{3}T_{ll}\, \delta_{ij}; \quad T_j^{(\rho)} = \kappa_t(\vec{x}, t)\frac{\partial \bar{\rho}}{\partial x_j}. \tag{3.14}
$$

Then LES equations for a flow of uniform density write

$$
\frac{\partial \bar{u}_i}{\partial t} + \frac{\partial}{\partial x_j}(\bar{u}_i \bar{u}_j) = -\frac{1}{\rho_0}\frac{\partial \bar{P}}{\partial x_i} + \frac{\partial}{\partial x_j}\left\{ (\nu + \nu_t)\left(\frac{\partial \bar{u}_i}{\partial x_j} + \frac{\partial \bar{u}_j}{\partial x_i} \right) \right\} \tag{3.15}
$$
$$
\frac{\partial \bar{\rho}}{\partial t} + \frac{\partial}{\partial x_j}(\bar{\rho}\bar{u}_j) = \frac{\partial}{\partial x_j}\left\{ (\kappa + \kappa_t)\frac{\partial \bar{\rho}}{\partial x_j} \right\}, \tag{3.16}
$$

where

$$\bar{P} = \bar{p} - \frac{1}{3}\rho_0 T_{ll} \qquad (3.17)$$

is a modified pressure which can be determined with the aid of the filtered continuity equation.

Two questions are in fact posed: the first one is how to determine the eddy-viscosity ν_t and the corresponding turbulent Prandtl number

$$Pr_t = \frac{\nu_t}{\kappa_t}, \qquad (3.18)$$

and the second one concerns the validity of the eddy-viscosity assumption itself. Indeed, it is based on an analogy with Newtonian fluids, which is certainly not fulfilled here. Let us discuss briefly this point: molecular viscosity ν characterizes for a "macroscopic" fluid parcel the momentum exchanges with the surrounding fluid due to molecular diffusion across its interface. Here, one assumes a wide separation between macroscopic and microscopic scales and it is this separation which allows to calculate these molecular exchange coefficients using kinetic theories of liquids or gases where molecules are assumed to follow some sort of Gaussian random walk. No such scale separation exists in the LES problem, where one observes in general a distribution of energy (kinetic-energy spectrum) continuously decreasing from the energetic to the smallest dissipative scales, even in inflexional shear flows with vigorous coherent vortices. Since the cutoff scale Δx is in the middle of this spectrum, there is obviously no spectral gap at this level. On the other hand, trajectories of fluid parcels are very far from a random walk, since they may be either trapped around a vortex or strained in stagnation regions between vortices.

We believe therefore that the lack of spectral gap is the major drawback of the eddy-viscosity assumption, responsible for the fact that numerous numerical and even experimental *a priori* tests (see *e.g.* Clark *et al.* [21], Liu *et al.* [77]) invalidate relation (3.14): when a low-pass filter is for instance applied to DNS results, one can calculate explicitly the subgrid-stresses tensors, and correlate them to the filtered deformation. The correlation found is very poor, of the order of 0.1 instead of 1. This justifies the development of models going beyond the classical eddy-viscosity concept: it is the case of the spectral eddy viscosity (see Sect. 4.1), and also of models presented in Section 5. However, LES results based on classical eddy-viscosities in physical space derived from Smagorinsky or structure-function models may give very good results, as will be seen below, from the point of view of vortex dynamics and statistical predictions.

3.4   Smagorinsky's model

As already pointed out, the most widely used eddy-viscosity model was
proposed by Smagorinsky [105]. He introduced an eddy-viscosity which was
supposed to model subgridscale dissipation through a Kolmogorov $k^{-5/3}$
cascade. Smagorinsky's model is an adaptation of Prandtl's mixing-length
theory to subgridscale modelling. Prandtl assumes that the eddy viscosity
arising in Reynolds-averaged Navier–Stokes equations is proportional to a
scale characteristic of turbulence (the mixing length) times a characteristic
turbulent velocity. In the same way, Smagorinsky assumes that the LES
eddy-viscosity is proportional to the subgrid-scale characteristic length scale
Δx, and to a characteristic subgrid-scale velocity

$$v_{\Delta x} = \Delta x\, |\bar{S}|, \tag{3.19}$$

based on the second invariant of the filtered-field deformation tensor

$$|\bar{S}| = \sqrt{2\bar{S}_{ij}\bar{S}_{ij}}. \tag{3.20}$$

Thus Smagorinsky's eddy viscosity writes

$$\nu_t = (C_S \Delta x)^2 |\bar{S}|. \tag{3.21}$$

If one assumes that $k_C = \pi/\Delta x$, the cutoff wavenumber in Fourier space,
lies within a $k^{-5/3}$ Kolmogorov cascade $E(k) = C_K \epsilon^{2/3} k^{-5/3}$ (C_K is the
Kolmogorov constant), one can analytically determine the constant C_S. It
is then found (see Lilly [75]):

$$C_S \approx \frac{1}{\pi} \left(\frac{3 C_K}{2} \right)^{-3/4}. \tag{3.22}$$

It yields $C_S \approx 0.18$ for a Kolmogorov constant of 1.4. This value proves
to give acceptable results for LES of isotropic turbulence. However, most
researchers prefer $C_S = 0.1$ (which represents a reduction by nearly a factor of 4 of the eddy-viscosity), a value for which Smagorinsky's model behaves reasonably well for free-shear flows and for channel flow[1] (Moin and
Kim [85]). This clearly indicates that C_S is not a universal constant, and
that assuming k_C within a $k^{-5/3}$ Kolmogorov cascade is too much a constraint. In fact, Smagorinsky's model is obviously too dissipative in the
presence of a wall, and does not work in particular for transition in a
boundary-layer developing upon a flat plate: it artificially relaminarizes
the flow if the upstream perturbation is not high enough. This is due to

---

[1] These channel flow computations require wall laws.

the heavy influence in the eddy-viscosity expression of the velocity gradient in the direction normal to the wall (see Meneveau and Katz [82]). Furthermore, the subgrid-scale dissipation $P_{sm} = T_{ij}\overline{S}_{ij}$ is on the average positive corresponding to a global flow of kinetic energy from the large-scale towards the subgrid scale (forward scatter). However, backscatter, associated with a locally negative P_{sm} and with a reversed energy transfer, occurs in some flow regions (Piomelli *et al.* [95], Liu *et al.* [77]). Classical eddy-viscosity models like Smagorinsky's assume that P_{sm} is positive everywhere. This justifies Smagorinsky's dynamic approach which will be presented in Section 6, and where the constant is dynamically adjusted to the flow conditions.

4   LES in Fourier space

4.1   Spectral eddy viscosity and diffusivity

We assume that Navier–Stokes is written in Fourier space. This requires statistical homogeneity in the three directions of space, but we will see below how to handle flows with only one direction of inhomogeneity. Let $\hat{u}_i(\vec{k}, t)$ and $\hat{\rho}(\vec{k}, t)$ be the spatial Fourier transform of the velocity and passive-scalar fields where $\vec{k}$ is the wavenumber. The filter consists in a sharp cut-off filter, simply clipping all the modes larger than k_C, where $k_C = \pi/\Delta x$ is still the cut-off wavenumber.

We write Navier–Stokes in Fourier space as

$$\frac{\partial}{\partial t}\hat{u}_i(\vec{k}, t) + [\nu + \nu_t(\vec{k}|k_C)]k^2\hat{u}_i(\vec{k}, t)$$
$$= -ik_m P_{ij}(\vec{k})\int_{|\vec{p}|,|\vec{q}|<k_C}^{\vec{p}+\vec{q}=\vec{k}} \hat{u}_j(\vec{p}, t)\hat{u}_m(\vec{q}, t)d\vec{p} \qquad (4.1)$$

where $P_{ij}(\vec{k}) = \delta_{ij} - (k_i k_j/k^2)$ is the projector on the plane perpendicular to $\vec{k}$, which allows in particular to eliminate the pressure. The spectral eddy viscosity $\nu_t(\vec{k}|k_C)$ is defined by

$$\nu_t(\vec{k}|k_C)k^2\hat{u}_i(\vec{k}, t) = ik_m P_{ij}(\vec{k})\int_{|\vec{p}|\,\text{or}\,|\vec{q}|>k_C}^{\vec{p}+\vec{q}=\vec{k}} \hat{u}_j(\vec{p}, t)\hat{u}_m(\vec{q}, t)d\vec{p}. \qquad (4.2)$$

The r.h.s. of equation (4.1) corresponds to a resolved transfer.

A spectral eddy-diffusivity for the passive scalar may be defined in the same way, by writing the passive-scalar equation in Fourier space

$$\frac{\partial}{\partial t}\hat{\rho}(\vec{k}, t) + [\kappa + \kappa_t(\vec{k}|k_C)]k^2\hat{\rho}(\vec{k}, t) = -ik_j\int_{|\vec{p}|,|\vec{q}|<k_C}^{\vec{p}+\vec{q}=\vec{k}} \hat{u}_j(\vec{p}, t)\hat{\rho}(\vec{q}, t)d\vec{p} \qquad (4.3)$$

with

$$\kappa_t(\vec{k}|k_C)k^2\hat{\rho}(\vec{k},t) = ik_j \int_{|\vec{p}|\text{or}|\vec{q}|>k_C}^{\vec{p}+\vec{q}=\vec{k}} \hat{u}_j(\vec{p},t)\hat{\rho}(\vec{q},t)\mathrm{d}\vec{p}. \qquad (4.4)$$

4.2 EDQNM plateau-peak model

Expressions (4.2) and (4.4) give exact expressions of the eddy coefficients. They are however useless since they involve subgrid quantities. In fact, they can be evaluated at the level of kinetic-energy and passive-scalar spectra evolution equations obtained with the aid of two-point closures of three-dimensional isotropic turbulence. It is in this context that the concept of k-dependent eddy-viscosity was first introduced by Kraichnan [54]. The spectral eddy-diffusivity for a passive scalar was introduced by Chollet and Lesieur [19, 20].

Kraichnan used the so-called Test–Field Model. We work using a slightly different closure called the Eddy-Damped Quasi-Normal Markovian (EDQNM) theory introduced by Orszag [89, 90] (see also André and Lesieur [3], and Lesieur [67], for details). In this theory, which is easily handable only in the case of isotropic turbulence, the fourth-order cumulants in the hierarchy of moments equations are supposed to relax linearly the third-order moments, in the same qualitative way as does the molecular viscosity. The EDQNM provides for isotropic turbulence a closed equation of evolution for the kinetic energy spectrum $E(k,t)$. In a LES approach, we split the transfers between interactions involving only modes smaller than k_C, and the others. The equations for the supergrid-scale velocity, $\bar{E}(k,t)$, and scalar, $\bar{E}_\rho(k,t)$ spectra then write

$$\left(\frac{\partial}{\partial t} + 2\nu k^2\right)\bar{E}(k,t) = T_{<k_C}(k,t) + T_{>k_C}(k,t) \qquad (4.5)$$

$$\left(\frac{\partial}{\partial t} + 2\kappa k^2\right)\bar{E}_\rho(k,t) = T^\rho_{<k_C}(k,t) + T^\rho_{>k_C}(k,t), \qquad (4.6)$$

where $T_{<k_C}(k,t)$ and $T^\rho_{<k_C}(k,t)$ are the spectral transfers corresponding to resolved triads such that $k, p, q \le k_C$, and $T_{>k_C}$ (resp. $T^\rho_{>k_C}$) to modes such that $k < k_C, p$ and (or) $q > k_C$.

We first assume that k_C lies within a $k^{-5/3}$ inertial range. For $k \ll k_C$, both modes being larger than k_i the kinetic-energy peak, expansions in powers of the small parameter k/k_C yield to the lowest order

$$T_{>k_C}(k,t) = -2\nu_t^\infty k^2 \bar{E}(k,t) \qquad (4.7)$$

$$T^{\rho}_{>k_{\mathrm{C}}}(k,t) = -2\kappa_{\mathrm{t}}^{\infty} \, k^2 \, \bar{E}_{\rho}(k,t) \tag{4.8}$$

with

$$\nu_{\mathrm{t}}^{\infty} = 0.441 \, C_{\mathrm{K}}^{-3/2} \left[\frac{E(k_{\mathrm{C}})}{k_{\mathrm{C}}} \right]^{1/2} \tag{4.9}$$

$$\kappa_{\mathrm{t}}^{\infty} = \frac{\nu_{\mathrm{t}}^{\infty}}{P_{\mathrm{r}}^{(t)}}; \quad P_{\mathrm{r}}^{(t)} = 0.6. \tag{4.10}$$

Here, $E(k_{\mathrm{C}})$ is the kinetic-energy spectrum at the cutoff k_{C}. The 0.6 value for the Prandtl number is in fact the highest one permitted by the choice of two further adjustable constants arising in the EDQNM passive-scalar equation (see Lesieur [67]). When k is close to k_{C}, the numerical evaluation of the EDQNM transfers yields

$$T_{>k_{\mathrm{C}}}(k,t) = -2\nu_{\mathrm{t}}(k|k_{\mathrm{C}}) \, k^2 \, \bar{E}(k,t) \tag{4.11}$$
$$T^{\rho}_{>k_{\mathrm{C}}}(k,t) = -2\kappa_{\mathrm{t}}(k|k_{\mathrm{C}}) \, k^2 \, \bar{E}_{\rho}(k,t), \tag{4.12}$$

with

$$\nu_{\mathrm{t}}(k|k_{\mathrm{C}}) = K \left(\frac{k}{k_{\mathrm{C}}} \right) \nu_{\mathrm{t}}^{\infty}; \quad \kappa_{\mathrm{t}}(k|k_{\mathrm{C}}) = C \left(\frac{k}{k_{\mathrm{C}}} \right) \kappa_{\mathrm{t}}^{\infty} \tag{4.13}$$

where $\nu_{\mathrm{t}}^{\infty}$ and $\kappa_{\mathrm{t}}^{\infty}$ are the asymptotic values given by equations (4.9) and (4.10), and $K(x)$ and $C(x)$ nondimensional functions equal to 1 for $x = 0$. As shown also by Kraichnan [54], $K(x)$ has a plateau-value at 1 up to $k/k_{\mathrm{C}} \approx 1/3$. Above, it displays a strong peak (cusp-behaviour). Let us mention that Kraichnan did not point out the scaling of the eddy viscosity against $[E(k_{\mathrm{C}})/k_{\mathrm{C}}]^{1/2}$, which turns out to be essential for LES purposes. Indeed, when the energy spectrum decreases rapidly at infinity (for instance during the initial stage of decay in isotropic turbulence, see below), the eddy viscosity will be very low and inactive. On the other hand, we have $[E(k_{\mathrm{C}})/k_{\mathrm{C}}]^{1/2} \sim \epsilon^{1/3} k_{\mathrm{C}}^{-4/3}$ in an inertial-range expression, which may be important even before the establishment of the $k^{-5/3}$ range. Furthermore, we will show below that the plateau-peak model may be generalized to spectra different from Kolmogorov at the cutoff (spectral-dynamic model).

It was shown by Chollet and Lesieur [19, 20] that $C(x)$ behaves qualitatively as $K(x)$ (plateau at 1 and positive peak), and that the spectral turbulent Prandtl number $\nu_{\mathrm{t}}(k|k_{\mathrm{C}})/\kappa_{\mathrm{t}}(k|k_{\mathrm{C}})$ is approximately constant, and thus equal to 0.6 as given by equation (4.10).

It is clear that the plateau part corresponds to the usual eddy-coefficients assumption when one goes back to physical space, so that the "peak" part goes beyond the scale-separation assumption inherent to the classical eddy-viscosity and diffusivity concepts. The peak is mostly due to semi-local interactions across k_C: near the cutoff wavenumber, the main nonlinear interactions between the resolved and unresolved scales involve the smallest eddies of the former and the largest eddies of the latter (such that $p \ll k \sim q \sim k_C$).

At the level of kinetic-energy exchanges, this formulation of the spectral eddy-viscosity includes all backscatter effects in the following sense: when kinetic energy is injected around a particular wavenumber k_i, and for the decaying case, it may be shown with the aid of expansions in terms of the small parameter $k/k_i \ll 1$ that the transfer is proportional to k^4, and hence a spectrum proportional to k^4 is produced at low wavenumbers $k \ll k_i$ (see Lesieur and Schertzer [72]). Such a backscatter transfer is due to nonlinear resonance between two energetic modes in the neighborhood of k_i. This was checked in an LES in which k_i was close to k_C by Lesieur and Rogallo [71]. Considering the eddy-viscosity (4.13), it may be shown that, for $k \ll k_C$ (both modes being larger than k_i and in the inertial range), the backscatter due to subgrid-scale modes is negligible. Indeed, its relative importance in terms of transfers is, according to EDQNM theory, $(k/k_C)^2[E(k_C)/E(k)]$, which is very small since $E(k_C) \ll E(k)$ (see Lesieur [67]). The cusp results from the difference between a "drain", which sends energy to the subgrid scales, and a "backscatter" which injects energy back to the supergrid scales, so that the net effect is a positive eddy-viscosity.

As shown by Chollet [18], the plateau-peak behaviour of $K(x)$ can approximately be expresssed with the following analytical expression:

$$K(x) = 1 + 34.5 \, e^{-3.03/x}. \tag{4.14}$$

We will see below another analytic expression of this spectral eddy viscosity in terms of hyper-viscosities (see Sect. 7.1).

The plateau-peak model consists in using these eddy-viscosities in the deterministic equations ((4.1), (4.3)). One advantage of using such a subgridscale modelling is that they are correct from an energetic-transfer viewpoint.

4.2.1   The spectral-dynamic model

Another drawback of the plateau-peak model is that it is restricted to the case where k_C lies within a $k^{-5/3}$ Kolmogorov cascade. Fortunately, this can be cured with the introduction of the spectral-dynamic model. One assumes now that the kinetic-energy spectrum is $\propto k^{-m}$ for $k > k_C$, whith m not necessarily equal to 5/3. We modify the spectral eddy viscosity as

$$\nu_t(k|k_C) = 0.31 \, C_K^{-3/2} \sqrt{3-m} \frac{5-m}{m+1} K\left(\frac{k}{k_C}\right)\left[\frac{E(k_C)}{k_C}\right]^{1/2}, \tag{4.15}$$

for $m \leq 3$. This expression is exact for $k \ll k_C$ within the same nonlocal expansions of the EDQNM theory, as shown in Métais and Lesieur [80]. We retain the peak shape through $K(k/k_C)$ in order to be consistent with the Kolmogorov spectrum expression of the eddy viscosity. For $m > 3$, the scaling is no more valid, and the eddy viscosity will be set equal to zero. Indeed, we are very close to a DNS for such spectra. In the spectral-dynamic model, the exponent m is determined through the LES with the aid of least-squares fits of the kinetic-energy spectrum close to the cutoff. On may also check that the turbulent Prandtl number is given by:

$$P_r^t = 0.18 \, (5 - m) \tag{4.16}$$

(see Métais and Lesieur [80]). This value does not depend on the Kolmogorov and model constants. It is a great advantage in LES of heated or variable-density flows to have the possibility of a variable turbulent Prandtl number. This possibility exists also for the dynamic models in physical space which will be presented in the following sections.

4.2.2   Existence of the plateau-peak

The spectral LES of decaying isotropic turbulence and associated scalar mixing performed by Lesieur and Rogallo [71], together with those of Métais and Lesieur [80], have been used to compute directly the spectral eddy-viscosity and diffusivity. The method is the same as that employed by Domaradzki *et al.* [29] for a direct numerical simulation: one defines a fictitious cutoff wavenumber $k_C' = k_C/2$, across which the kinetic-energy transfer T and scalar transfer T^ρ are evaluated. Since we deal with a large-eddy simulation, the latter correspond to triadic interactions such that $k < k_C'$, p and (or) $q > k_C'$ and $p, q < k_C$: they are termed $T^{<k_C}_{>k_C'}(k,t)$ and $T^{\rho <k_C}_{>k_C'}(k,t)$. They correspond to resolved transfers, and satisfy energetic equalities of the type

$$T^{<k_C}_{>k_C'}(k,t) = T_{>k_C'}(k,t) - T_{>k_C}(k,t) \tag{4.17}$$

where $T_{>k_C'}$ and $T_{>k_C}$ are the total kinetic energy transfers across k_C' and k_C. It is important to note that equation (4.17) is the exact energetic equivalent in spectral space of Germano's identity (Germano [41], see Sect. 6). A similar relation holds for $T^{\rho <k_C}_{k_C'}$. Once divided by $-2k^2 E(k,t)$ and $-2k^2 E_\rho(k,t)$, they give the resolved spectral eddy-viscosity and diffusivity. Figure 7 shows these functions normalized by $[E(k_C')/k_C']^{1/2}$, taken from Métais and Lesieur [80]. Similar results had been found in Lesieur and Rogallo [71]. It demonstrates that the plateau-peak behaviour does exist for the eddy viscosity, but is questionable for the eddy diffusivity. This

anomalous scalar range still exists in a DNS of decaying isotropic turbulence: in this case, the double filtering yields a plateau-peak eddy viscosity with a plateau value approximately 0, as was discovered by Domaradzki *et al.* [29]. The eddy-diffusivity on the contrary still behaves as in the LES. In fact, Métais and Lesieur [80] have checked that the anomaly disappears when the temperature is no more passive and coupled with the velocity within the frame of Boussinesq approximation (stable stratification). It is possible that the same holds for compressible turbulence, which would legitimate the use of the plateau-peak eddy diffusivity in this case. Note that the plateau-peak behaviour was recently confirmed by the experimental data from the turbulent wake experiment performed by Cerutti *et al.* [17]

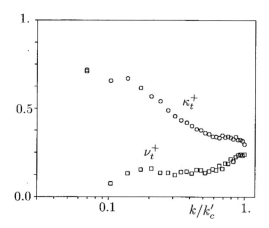

Fig. 7. Resolved eddy-viscosity and diffusivity evaluated through a double filtering in LES of isotropic decaying turbulence (from Métais and Lesieur [80]).

The plateau-peak spectral eddy-viscosity and diffusivity coefficients have been used successfully to study freely-decaying three-dimensional turbulence at high Reynolds number. LES at a resolution of 128^3 Fourier modes have been carried out by Lesieur and Rogallo [71]. It was checked that, at first, Kolmogorov and Corrsin–Oboukhov $k^{-5/3}$ cascades establish. Afterwards, the kinetic-energy spectrum decays self-similarly, with a spectral slope comprised between $-5/3$ and -2. The scalar spectrum exhibits a very short inertial-convective range close to the cutoff, and a very wide range shallower than k^{-1} in the large scales. Here, the scalar decays much faster than the temperature. This anomalous range was explained by Métais and Lesieur [80] as due to the quasi two-dimensional character of the scalar diffusion in the large scales, leading to large-scale intermittency of the scalar. More precisely, the scalar diffusion seems to be dominated by the effect of coherent vortices already considered in Section 2.1, and tending to form scalar discontinuities in the stagnation regions between the vortex tubes.

The plateau-peak model been applied to various temporally growing free-shear flows like wakes, jets and mixing-layers for which the flow is assumed to be periodic in the streamwise direction. Silvestrini ([104], see also Lesieur and Métais [69]) showed that LES, even at moderate resolution are able to reproduce the complexity of the mixing-layer dynamics. With three-dimensional initial noise, the helical pairing arrangement, previously found in the mixing-layer DNS of Comte et al. [24], was recovered: vortex filaments oscillate out-of-phase in the spanwise direction, and reconnect, yielding a vortex-lattice structure. On the other hand, and if the perturbation is quasi two-dimensional, the mixing layer was shown to evolve into a set of big quasi two-dimensional Kelvin–Helmholtz vortices which both undergo pairing and stretch intense longitudinal hairpin vortices in the stagnation regions between them. This stretching of longitudinal vortices had been observed experimentally for a long time (see e.g. Bernal and Roshko [11]). It is remarkable that LES, done at a quite low resolution, are indeed able to capture the longitudinal vortices, which are at quite small scales. The DNS of Comte et al. [24] at an equivalent resolution were unable to find organized intense longitudinal vortices, because their molecular Reynolds number is too low. It is only at a much higher resolution that DNS can capture these vortices (Rogers and Moser [97]). This is an example where large-eddy simulations are an excellent tool to capture not only large- but also small-scale vortices.

However, a spectral eddy-viscosity is difficult to employ when the geometry of the problem obliges one to work in physical space. This has motivated the development of new models directly inspired from the previous spectral formulation but which can be utilized in physical space (see Sect. 5.1).

4.3 Incompressible plane channel

We show now how the spectral dynamic model may be applied to an incompressible turbulent Poiseuille flow between two infinite parallel flat plates. A schematic view of the channel is presented in Figure 8. A rotation axis oriented in the spanwise direction is indicated for further applications, but rotation is inactive right now. The channel has a width $2h$, and we define the macroscopic Reynolds number by $Re = 2hU_{\mathrm{m}}/\nu$, where U_{m} is the bulk velocity. We assume periodicity in the streamwise and spanwise directions. Calculations are carried out at constant U_{m}. They are initiated by a parabolic laminar profile perturbed by a small three-dimensional random noise, and pursued up to complete statistical stationarity.

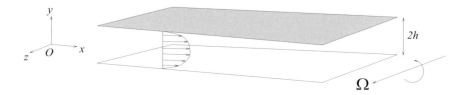

Fig. 8. Schematic view of a plane channel.

4.3.1   Wall units

Let us recall the so-called wall units, very useful when turbulence has developed. The friction velocity v_* is defined by

$$v_*^2 = \nu \, \frac{d\langle u \rangle}{dy} \Big|_{y=0}, \qquad (4.18)$$

with the aid of the mean-velocity derivative at the wall. The velocities will be normalized by v_* and noted $u_i^+ = u_i/v_*$. One defines also a viscous thickness

$$l_v = \frac{\nu}{v_*}, \qquad (4.19)$$

characteristic of motions very close to the wall which are dominated by viscosity, so that the spatial scales will be normalized by l_v and noted $x_i^+ = x_i/l_v$. Let y be the distance perpendicular to the wall. Substituting Taylor-series expansions of the velocity components in powers of y^+ close to the wall, together with the use of the continuity equation, leads to the result that u^+ and w^+ scale like y^+, while v^+ scales likes y^{+2}.

A consequence is that the mean longitudinal velocity profile scales also like y^+, behaviour which persists up to about $y^+ = 4 \approx 5$, which characterizes the width of the viscous region. DNS and LES show that this region is certainly not laminar, and strongly marked by the system of high and low longitudinal velocity streaks.

4.3.2   Streaks and hairpins

These coherent structures have been discovered experimentally in a turbulent boundary layer by Kline $et\ al.$ [52], and had been observed previously by Klebanoff $et\ al.$ [51] in a celebrated paper related to transition in a boundary layer forced upstream by a vibrating ribbon. Klebanoff associated the streaks (which he could detect with anemometers) to a system of longitudinal hairpins travelling downstream in phase and pumping between their legs fluid lower fluid slowed by the wall: this model explains the formation of low-speed streaks in the "peaks" of the hairpins, and high-speed streaks in the "valleys". The system of streaks in a turbulent channel was recovered numerically in the LES of Moin and Kim [85] already mentioned before and using Smagorinsky's model with wall laws.

4.3.3 Spectral DNS and LES

We will present turbulent channel DNS and LES, taken from the work of Lamballais [57] and Lamballais *et al.* [61] (see also [58–60]). It is interesting to see that $h^+ = v_* h / \nu$ defines a microscopic Reynolds number based on the friction velocity. The numerical code used combines pseudo-spectral methods in the streamwise and spanwise directions, and compact finite-difference schemes of sixth order (see [64]) in the transverse direction with grid refinement close to the walls. The subgrid model is the spectral-dynamic eddy viscosity, computed thanks to two-dimensional kinetic-energy spectra calculated at each time step by spatial averages in planes parallel to the wall. Therefore, the exponent m in the eddy viscosity depends of y and t. This spectral eddy viscosity is implemented spectrally in the directions parallel to the wall, and in physical space in the transverse direction. This is a very precise code of accuracy comparable to a spectral method at equivalent resolution as shown by the comparison of DNS at $h^+ = 162$ with spectral DNS of Kuroda [56] at $h^+ = 150$. We reproduce this picture. We see in Figure 9a that the logarithmic range starts at $y^+ = 30$. Figure 9b presents the rms velocity profiles as a function of y^+. It confirms the strong u' production close to the wall, with a peak at $y^+ = 12$, and which is obviously the signature of high- and low-speed streaks discussed before. Figures 9e (Reynolds stresses) and 9d (rms pressure fluctuations) have a peak higher ($y^+ \approx 30$). It might correspond to the tip of hairpin vortices ejected above the low-speed streaks. Figure 9f corresponds to rms vorticity fluctuations. It indicates that the maximum vorticity produced is spanwise and at the wall[2]. The vorticity perpendicular to the wall is about 40% higher than the longitudinal vorticity in the region $5 < y^+ < 30$, which shows only a weak longitudinal vorticity stretching by the ambient shear.

We next present two LES using the spectral-dynamic model, at $Re = 6666$ ($h^+ = 204$, case A) and $Re = 14\,000$ ($h^+ = 389$, case B). They are respectively subcritical and supercritical with respect to the linear-stability analysis of the Poiseuille profile. In the two simulations there is a grid refinement close to the wall, in order to simulate accurately the viscous

---

[2]It corresponds in fact to a steepening of du/dy at the wall under the high-speed streaks, resulting from a sort of squashing of the boundary layer upon the wall consecutive to the fact that the fluid is descending, and inducing an increase of the friction-coefficient.

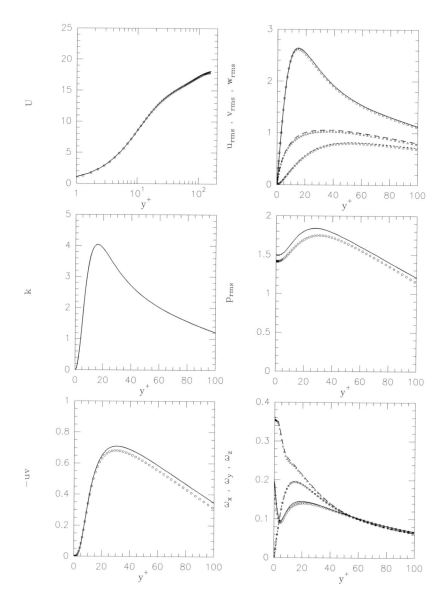

Fig. 9. Statistical data obtained in DNS of a turbulent channel flow by
Lamballais (straight line) and Kuroda (symbols); from left to right and top to
bottom, a) mean velocity, b) rms velocity fluctuations (respectively from top to
bottom, longitudinal, spanwise, vertical), c) kinetic energy, d) rms pressure fluctuation, e) Reynolds stresses, f) rms vorticity (from top to bottom, spanwise,
vertical, longitudinal). Courtesy Lamballais.

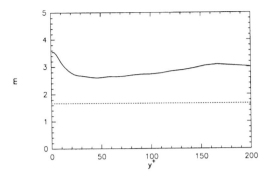

Fig. 10. Spectral-dynamic LES of the channel flow (case A), time-averaged exponent $m(y^+)$ of the kinetic-energy spectrum at the cutoff.

sublayer. Figure 10 shows for case A the time-averaged exponent m arising in the energy spectrum at the cutoff, as a function of the distance to the wall y^+. Regions where $m > 3$ correspond to a zero eddy viscosity and hence a direct-numerical simulation. This is the case in particular close to the wall, up to $y^+ \approx 12$ where we know that longitudinal velocity fluctuations are very intense, due to the low- and high-speed streaks. Therefore, and since the first point is very close to the wall ($y^+ = 1$), such LES have the interesting property of becoming DNS in the vicinity of the wall, enabling to capture events which occur in this region.

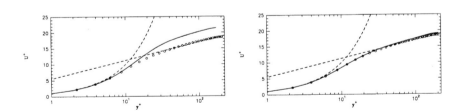

Fig. 11. Mean velocity profiles in wall units. Lines: spectral eddy-viscosity based simulations ($Re = 6666$); symbols: Piomelli ([94]; $Re = 6500$). Left: $m = 5/3$ ($h^+ = 181$); right: dynamic evaluation of $m(y,t)$ ($h^+ = 204$). The dashed straight line corresponds to the universal logarithmic mean velocity profile $\langle u \rangle = 2.5 \ln y^+ + 5.5$.

Figure 11 shows (in semi-logarithmic coordinates) the mean velocity profile in case A, compared with the LES of Piomelli [94] using the dynamic model of Germano [41]. The latter is known to agree very well with experiments at these low Reynolds numbers. The simulation using the spectral dynamic model (right part of the figure) coincides with the DNS, yielding a correct value of the additive constant 5.5 in the logarithmic velocity profile. On the other hand, the LES using the classical spectral-cusp model with $m = 5/3$ (left of figure) gives an error of 100% for this constant. The dashed parabola corresponds to the linear profile at the wall, which is exact up to 4 wall units. In case A, it was ckecked (see Lamballais *et al.* [61]) that the mean velocity and rms velocity fluctuations, compared very well with the dynamic-model predictions of Piomelli [94]. The agreement of rms velocities is still very good, with a correct prediction of the longitudinal velocity fluctuations peak.

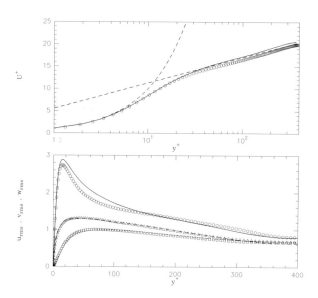

Fig. 12. Turbulent channel flow, comparisons of the spectral-dynamic model (solid lines, $h^+ = 389$) with the DNS of Antonia *et al.* ([4], symbols, $h^+ = 395$); top, mean velocity, bottom, rms velocity components.

Concerning the supercritical case, the LES of case B are in very good agreement with a DNS at $h^+ = 395$ carried out in Antonia *et al.* [4], both for the mean velocity and the rms velocity components. They are shown in Figure 12. Notice that the LES allows in this case to reduce the computational cost by a factor of the order of 100, which is huge. Notice also that the extent of the linear-velocity profile range close to the wall has slightly increased (from 4 to 5) with the Reynolds number. We present finally in Figure 13 a map of the vorticity modulus at the same threshold for cases A

Fig. 13. Turbulent plane channel, vorticity modulus; a) DNS ($h^+ = 165$), b) LES using the spectral-dynamic model ($h^+ = 389$), from Lamballais [57].

and B. The flow goes from left to right. It is clear that the large-eddy simulation does reproduce features expected from turbulence at higher Reynolds number, and displays much more vortical activity in the small scales that the DNS.

5 Improved models for LES

We have clearly shown in the former chapter the advantages of the spectral eddy-viscosity models, with in particular the possibility of accounting for local or semi-local effects in the neighbourhood of the cutoff. However, we have already stressed that in most of industrial or environmental applications, the complexity of the computational domain forbids the use of spectral methods. One has thus to deal with numerical codes written in physical space and employing finite-volume or finite-differences methods, often with unstructured grids. This last point will not be considered here, although it is crucial for practical applications. We will present however simulations on orthogonal grids of mesh size varying in direction and location, sometimes in curvilinear geometry. The present chapter will be devoted to models of the structure-function family, with applications to isotropic turbulence, free-shear flows and boundary layers.

5.1 Structure-function model

5.1.1 Formalism

The Structure-Function model is an attempt to go beyond Smagorinsky, while keeping in physical space the same scalings as the spectral eddy viscosity. The original Structure-Function (SF) model is due to Métais and Lesieur [80] (see also Lesieur and Métais [69]). It consists in building in physical space an eddy viscosity normalized by $\sqrt{E_{\bar{x}}(k_C)/k_C)}$, still with $k_C = \pi/\Delta x$. The spectrum $E_{\bar{x}}(k_C, t))$ is a local kinetic-energy spectrum at a given point $\vec{x}$, which has to be properly determined assuming that turbulence is locally isotropic. This allows to take into account the spatial intermittency of turbulence. One first discards the peak behaviour[3] of $K(x)$ in equation (4.14) and adjust the constant as proposed by Leslie and Quarini [66], by balancing in a $k^{-5/3}$ inertial range extending from zero to k_C the subgridscale flux $2\int_0^{k_C} \nu_t k^2 \bar{E}(k) \mathrm{d}k$ with the kinetic energy flux ϵ. This yields

$$\nu_t(k_C) = \frac{2}{3} C_K^{-3/2} \left[\frac{E(k_C)}{k_C}\right]^{1/2}. \tag{5.1}$$

We keep in mind that $E(k_C)$ is now a local kinetic-energy spectrum, which has to be evaluated in terms of physical-space quantities. The best candidate for that is the second-order velocity structure function

$$F^{\mathrm{is}}(r) = \left\langle [\vec{u}(\vec{x}, t) - \vec{u}(\vec{x} + \vec{r}, t)]^2 \right\rangle, \tag{5.2}$$

where the label "is" stands for isotropic turbulence, and the brackets correspond to ensemble averaging. The equivalence of Kolmogorov's $\epsilon^{2/3} k^{-5/3}$ spectrum is the $\langle \delta v(r)^2 \rangle \sim (\epsilon\, r)^{2/3}$ structure-function. This was the original formulation of Kolmogorov's law (Kolmogorov [53]). We recall also Batchelor's relation in isotropic turbulence

$$F^{\mathrm{is}}(\Delta x) = 4 \int_0^\infty E(k, t) \left(1 - \frac{\sin(k\Delta x)}{k\Delta x}\right) \mathrm{d}k. \tag{5.3}$$

For the subgrid-modelling problem, we consider the following local structure-function

$$F_2(\vec{x}, \Delta x) = \left\langle [\bar{\vec{u}}(\vec{x}, t) - \bar{\vec{u}}(\vec{x} + \vec{r}, t)]^2 \right\rangle_{\|\vec{r}\| = \Delta x}. \tag{5.4}$$

The difference with relation (5.2) is that F_2 is calculated with a local statistical average of square (filtered) velocity differences between $\vec{x}$ and the six closest points surrounding $\vec{x}$ on the computational grid. In some cases, the average may be taken over four points parallel to a given plane[4]. The equivalent Batchelor's formula is

$$F_2(\vec{x}, \Delta x) = 4 \int_0^{k_C} \bar{E}(k, t) \left(1 - \frac{\sin(k\Delta x)}{k\Delta x}\right) \mathrm{d}k, \tag{5.5}$$

---

[3]It will be shown later on how to reintroduce the peak in terms of hyperviscosity.
[4]In a channel, for instance, the plane is parallel to the boundaries.

since the filtered field has no energy at modes larger than k_C. Assuming again a $k^{-5/3}$ spectrum extending from zero to k_C, one obtains

$$\nu_t^{SF}(\vec{x}, \Delta x) = 0.105 \, C_K^{-3/2} \, \Delta x \, [F_2(\vec{x}, \Delta x)]^{1/2}. \tag{5.6}$$

In fact, this derivation of the SF model equation is different and simpler than the one proposed in the original paper of Métais and Lesieur [80], and may be found in Ducros [31].

5.1.2 Non-uniform grids

Interpolations of equation (5.6) based on Kolmogorov's 2/3 law for the above structure function may be proposed if the computational grid is not regular (but still orthogonal). Let Δc be a mean mesh in the three spatial directions[5]. We have (in the six-point formulation)

$$F_2(\vec{x}, \Delta c) = \frac{1}{6} \sum_{i=1}^{3} F_2^{(i)} \left(\frac{\Delta c}{\Delta x_i} \right)^{2/3}, \tag{5.7}$$

with

$$F_2^{(i)} = [\vec{u}(\vec{x}) - \vec{u}(\vec{x} + \Delta x_i \, \vec{e}_i)]^2 + [\vec{u}(\vec{x}) - \vec{u}(\vec{x} - \Delta x_i \, \vec{e}_i)]^2, \tag{5.8}$$

where $\vec{e}_i$ is the unit vector in direction $\vec{x}_i$.

5.1.3 Structure-function *versus* Smagorinsky models

We have found a relation between Smagorinsky's and the structure-function models, by replacing the velocity increments in the latter by first-order spatial derivatives. One finds for the six-point formulation

$$\nu_t^{SF} \approx 0.777 \, (C_S \Delta x)^2 \sqrt{2\bar{S}_{ij}\bar{S}_{ij} + \bar{\omega}_i\bar{\omega}_i} \, , \tag{5.9}$$

where $\vec{\bar{\omega}}$ is the vorticity of the filtered field, whereas C_S is Smagorinsky's constant defined by equation (3.22) in terms of Kolmogorov's constant C_K. Then the SF model appears, within this crude first-order approximation, to be a combination of Smagorinsky model in a strain and vortical version. Suppose as an example that we are in the stagnation regions between two quasi two-dimensional vortices (in a mixing layer, or a wake, or in a round jet...) when there is a low residual vorticity which is going to be stretched longitudinally. At this initial stage, and since vorticity in the stagnation

---

[5]It may be geometric, or of another type.

region is low, the vortical term will be small in front of the strain term, so that the SF model will be about 80% less dissipative than Smagorinsky's, which will favour the eventual stretching of longitudinal vortices. On the other hand, SF should be more dissipative than Smagorinsky within the core of the large vortices.

5.1.4 Isotropic turbulence

It has been shown in Métais and Lesieur [80] that, for $C_K = 1.4$, the SF model gives a quite good $k^{-5/3}$ energy spectrum[6], whereas Smagorinsky[7] has more like a k^{-2} inertial range.

5.1.5 SF model, transition and wall flows

It was at first quite a disapointment to realize that the SF model was, like Smagorinsky's, too dissipative for transition in a boundary layer (yielding again relaminarization) or in a channel. One might have thought that at least the four-point formulation in planes parallel to the wall would have eliminated the effect of the mean shear at the wall on the eddy viscosity. In fact, it turns out that the isotropic relation (5.5) introduces spurius inhomogeneous effects in the eddy-viscosity, which increase the latter, and the SF model is too dissipative for quasi two-dimensional or transitional situations. This is of course a real concern, specially for turbulent boundary layers or channel flows, and has motivated the development of two improved versions of the SF model: the selective structure-function model (SSF), and the filtered structure-function model (FSF), for which turbulence gets rid of large-scale inhomogeneities before the SF model is applied.

5.2 *Selective structure-function model*

In the selective structure-function (SSF) model (David [27]), the eddy-viscosity is switched off when the flow is not three-dimensional enough. We need for that a criterion of three-dimensionalization, defined as follows: one considers at a given time the angle between the vorticity vector at a given grid point and the arithmetic mean of vorticity vectors at the six closest neighbouring points (or the four closest points in the four-point formulation). If one carries out LES of isotropic turbulence at a resolution of $32^3 \sim 64^3$, one finds that the p.d.f. peaks for an angle of 20^0, which is thus the most probable value. Then, the eddy viscosity will be cancelled at

---

[6]With a nearly flat compensated $k^{5/3}E(k)$ spectrum.
[7]With C_S still given by equation (3.22).

points where this angle is smaller than 20^0. We will give various applications of this model to various heated or compressible flows throughout the book.

5.3   Filtered structure-function model

5.3.1   Formalism

The filtered structure-function model was developed by Ducros ([31], see also Ducros et al. [33]). The filtered field $\bar{u}_i$ is now submitted to a high-pass filter $\widetilde{(.)}$ consisting in a Laplacian operator discretized by second-order centered finite differences and iterated three times. We first apply relation (5.5) to the high-pass filtered field

$$\tilde{F}_2(\vec{x}, \Delta x) = 4 \int_0^{k_C} \tilde{E}(k) \left(1 - \frac{\sin(k\Delta x)}{k\Delta x}\right) \mathrm{d}k, \qquad (5.10)$$

where $\tilde{F}_2(\vec{x}, \Delta x)$ is the second-order structure function of the high-pass filtered field $\widetilde{\bar{u}}_i$, and $\tilde{E}(k)$ its spectrum. This allows (for isotropic turbulence) to relate $\tilde{F}_2$ to $\tilde{E}(k_C)$, and hence to $E(k_C)$ thanks to the transfer function of the "tilde" operator, determined with the aid of isotropic test fields. Using equation (5.1) yields for the eddy viscosity

$$\nu_t^{\mathrm{FSF}}(\vec{x}, \Delta x) = 0.0014 \, C_K^{-3/2} \, \Delta x \, [\tilde{F}_2(\vec{x}, \Delta x)]^{1/2}. \qquad (5.11)$$

A further advantage of the FSF model is that it does not contain adjustable constants. We will show below very satisfactory applications of this model to mixing layers and boundary layer on a flat plate.

5.4   A test case for the models: The temporal mixing layer

We present now in Figure 14 a comparison between Smagorinsky's model, the plain spectral plateau-peak model (not dynamic)) and the various structure function models (original, selective and filtered versions). The comparison is carried out in the case of a temporally-growing mixing layer. Pseudo-spectral numerical methods are here used. We take a three-dimensional initial isotropic perturbation, but the domain contains now only two fundamental longitudinal most-unstable wavelengths, so that no helical pairing develops. Instead, we see two big rollers oscillating in phase[8], and stretching longitudinal haipins exactly as in the model of Bernal and Roshko [11], with

---

[8]Such a configuration corresponds to "translative instability", from the work of Pierrehumbert and Widnall [93] on secondary instabilities (Floquet-type analysis) of Stuart vortices.

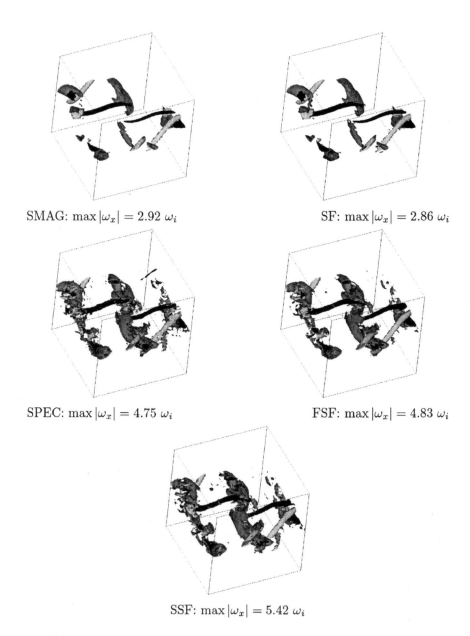

SMAG: max $|\omega_x| = 2.92\,\omega_i$

SF: max $|\omega_x| = 2.86\,\omega_i$

SPEC: max $|\omega_x| = 4.75\,\omega_i$

FSF: max $|\omega_x| = 4.83\,\omega_i$

SSF: max $|\omega_x| = 5.42\,\omega_i$

Fig. 14. Comparison of various SGS models (Smagorinsky, Structure-Function, Plateau-Peak, FSF, SSF) applied to a temporal mixing layer, visualized by iso-surfaces $\omega_x = \omega_i$ (black), $\omega_x = -\omega_i$ (light grey) and $\omega_z = \omega_i = -2U/\delta_i$ (dark grey).

very neat alternate longitudinal vortices. Notice the strong resemblance between the results obtained with the plateau-peak, filtered structure-function and selective structure-function models. They give bigger spanwise and longitudinal vortices than Smagorinsky and the SF models, and much more small-scale variability. This confirms that both modifications of the original structure-function model go in the right direction, since the primary and secondary instabilities are less damped with the new models. Note also that the SF model seems to be here only slightly less dissipative that Smagorinsky's model.

5.5 Spatially growing mixing layer

The temporal approximation is only a crude approximation of a mixing layer spatially developing, where one works in a frame traveling with the average velocity between the two layers. We consider now an incompressible mixing layer spatially-developing between two streams of velocity U_1 and U_2 ($U_1 > U_2$). Further details can be found in Comte *et al.* [25]. The inflow consists of an hyperbolic-tangent velocity profile

$$\bar{u}(y) = \frac{U_1 + U_2}{2} + \frac{U_1 - U_2}{2} \tanh \frac{2y}{\delta_i}, \tag{5.12}$$

where δ_i is the upstream vorticity thickness. The Reynolds number is here built on δ_i and half the velocity difference $U = (U_1 - U_2)/2$. A weak random perturbations is superimposed onto the mean profile. The same mixed spectral-compact code already discussed for the channel and the wake is used here. Periodicity is assumed in the spanwise direction z. Sine/cosine expansions are used in the transverse direction y, enforcing free-slip boundary conditions. Non-reflective outflow boundary conditions are approximated by a multi-dimensional extension of Orlansky's discretization scheme, with limiters on the phase velocity (see Gonze [43] for a detailed description of the numerical code). We first compare a DNS at low Reynolds number ($Re = 100$) with a LES (without molecular viscosity) using the FSF model. The upstream forcing consists here in a quasi-twodimensional random perturbation.

Figure 15 (top) shows an isosurface of the vorticity modulus obtained in the DNS. The vortex sheet undergoes oscillations leading to a first roll-up further downstream. Subsequently, various pairings of Kelvin–Helmholtz vortices are observed. Again, thin intense longitudinal vortices are stretched as in Bernal and Roshko's [11] experiment. For the DNS, the vorticity magnitude during the run peaks at $2\omega_i$, where $\omega_i = 2U/\delta_i$ is the maximal vorticity magnitude introduced at the inlet. Although less computational points are used in the LES, the LES (Fig. 15, bottom) is obviously much more turbulent than the DNS, and has also a lot of oblique waves propagating

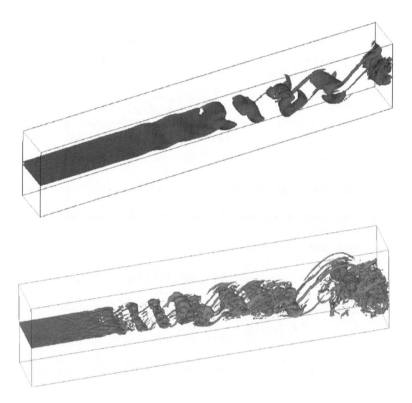

Fig. 15. Perspective views of isovorticity surface: top, DNS, $\|\vec{\omega}\| = \omega_i/3$; bottom, run LES, $\|\omega\| = (2/3)\omega_i$.

along the upstream vortex sheet. The latter breaks down much faster, and the longitudinal vortices are stretched much more efficiently. Indeed, the maximal vorticity magnitude is now $\approx 4\ \omega_i$ for the whole run. Roll-up and pairing events occur much faster than in the DNS. Notice the complexity of the dynamics with a cluster of three fundamental Kelvin–Helmholtz vortices undergoing a first pairing and, at its downstream end, a billow made of 4 fundamental KH vortices whose second pairing is in progress.

Similar simulations but with a domain of spanwise size doubled (still with a quasi-twodimensional random forcing) show that the vortical structure changes quite radically when the spanwise direction is increased. Figure 16, taken from Comte *et al.* [25], indeed shows respectively the low-pressure and vorticity fields in that case. It is clear at least on the pressure that helical pairing develops, as in the experiments of Browand and Troutt [15].

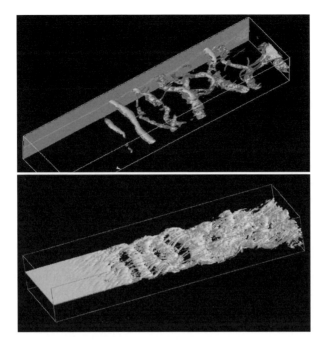

Fig. 16. Wide domain, LES. Top, low pressure; bottom, high vorticity.

5.6   Vortex control in a round jet

Our goal here is to demonstrate the ability of the LES to properly reproduce
the coherent-vortex dynamics in the transitional region of the jet. We also
show the possibility of controlling the jet behaviour by manipulating the inflow conditions. The control of the turbulent jets find numerous industrial
applications in thermohydraulics, aeronautics, industrial processes or even
the dispersion of pollutants. For these applications, it is particularly interesting to control certain flow characteristics such as the mixing efficiency,
the acoustic generation, etc. We will show below that an efficient control requires a precise knowledge of the spatial and temporal flow organization to
manipulate the three-dimensional coherent vortices. In the last ten years,
the progress in the experimental methods of detection and identification
has made possible a detailed investigation of the complex three-dimensional
coherent vortices imbedded within this flow. For instance, the influence of
the entrainment of the secondary streamwise vortices has been studied by
Liepmann and Gharib [74]. On the numerical side, several simulations of
two-dimensional or temporally evolving jets have been performed. Very few
have however investigated the three-dimensionnal spatial development of the

round jet. We here show how LES can be used to perform a precise study of the spatial growth of the round jet from the nozzle up to several diameters downstream. Round jets and there control have been numerically studied in details by Urbin [108], Urbin and Métais [109] and Urbin et al. [110]. We here briefly present recent results obtained by Silva [103] with the mixed compact/spectral numerical code previously described. The use of LES techniques allow us to reach high values of the Reynolds number. We here use the Selective Structure Function Model which is well adapted for transitional flows. We consider a computational domain starting at the nozzle and extending up to several jet diameters downstream. We successively consider two jets configurations: the "natural" jet ($Re = 25\,000$) which is forced upstream by the top-hat profile to which is superposed a weak 3D white noise; the "excited" jet development ($Re = 10\,000$) is controlled with the aid of a given deterministic inflow forcing (plus a white noise) designed to trigger a specific type of three-dimensional coherent structures.

Experimental studies by Michalke and Hermann [83] have clearly pointed out the capital effect of the inflow momentum boundary layer thickness θ and of the ratio R/θ (R: jet radius) on the jet downstream development. It was shown that the detailed shape of the mean velocity profile strongly influences the nature of the coherent vortices appearing near the nozzle: either axisymmetric structures (vortex rings) or helical structure can indeed develop. Here, the flow inside the nozzle was not simulated, but a mean axial velocity profile of top-hat shape in accordance with the experimental measurements was imposed:

$$W(r) = \frac{1}{2}W_{\mathrm{o}}\left[1 - \tanh\left(\frac{1}{4}\frac{R}{\theta}\left(\frac{r}{R} - \frac{R}{r}\right)\right)\right] \tag{5.13}$$

where W_{o} is the velocity on the axis. Here, $R/\theta = 20$. For such an inlet profile, linear stability analysis predicts a slightly higher amplification rate for the axisymmetric (varicose) mode than for the helical mode (see Michalke and Hermann [83]).

We first consider the "natural" jet which is forced upstream by the top-hat profile given by equation (5.13), to which is superposed a weak three-dimensional white noise. The frequency spectra revealed the emergence of a predominant vortex-shedding Strouhal number (normalized by D and W_0), $Str_{\mathrm{D}} = 0.375$, in good agreement with the experimental value. The LES shown in Figure 17 shows that the Kelvin–Helmholtz instability along the jet edge yields further downstream vortices having mainly an axisymmetric toroidal shape. One sees in Figure 17b that an original vortex arrangement can be observed subsequent to the varicose mode growth: the "alternate pairing". Such a vortex interaction was previously observed by Fouillet [36] and Comte et al. [23] in the DNS of a temporally evolving round jet at low Reynolds number ($Re = 2000$). The direction normal to the toroidal vortices symmetry plane tends, during their advection downstream, to differ from the jet axis. The inclination angle of two consecutive vortices appears to be of opposite sign, eventually leading to local vortex reconnections with an alternate arrangement. Note that vortex loop's inclination at the end of the potential core was experimentally observed by Petersen [92]. Experimental evidence of "alternate pairing" was given by Broze and Hussain [16]. This alternate-pairing mode corresponds to the growth of a subharmonic perturbation (of wavelength double of the one corresponding to the rings) developing after the formation of the primary rings. It therefore presents strong analogies with the helical-pairing mode observed in plane mixing layers (see above).

We here now show how a deterministic inflow perturbation can trigger one particular flow organization. We apply a periodic fluctuation associated with a frequency corresponding to $Str_D = 0.375$ for which the jet response is known to be maximal. The inflow excitation is here chosen such that alternate-pairing mode previously described is preferentially amplified. The resulting structures are analogous to Figure 17 except that the alternatively inclined vortex rings now appear from the nozzle (see Fig. 18). One of the striking features is the very different spreading rates in different directions. Note that the present jet exhibits strong similarities with the "bifurcating" jet of Lee and Reynolds [63]. One of the important technological application of this peculiar excitation resides in the ability to polarize the jet in a preferential direction.

5.7 LES of spatially developing boundary layers

The standard Structure-Function model permits to go beyond transition in a temporal (periodic in the flow direction) compressible boundary layer upon an adiabatic wall at Mach 4.5 (see Ducros *et al.* [32]). But it does not work for transition in a boundary layer at low Mach (or incompressible) where, like Smagorinsky, it is too dissipative and prevents small perturbations to degenerate into turbulence. Conversely, it has been used with success in its filtered version (FSF model) for the simulation of a quasi-incompressible ($M_\infty = 0.5$) boundary layer of an ideal gas developing spatially over an adiabatic flat plate with a low level of upstream forcing (Ducros *et al.* [33]). Although it gives interesting qualitative information on the structure of turbulent boundary layers, the above LES did not have a sufficient resolution close to the wall (first point at $y^+ = 5 \approx 6$) for good predictions of average quantities such as the friction coefficient at the wall or the shape factor. Here, we present new results with a finer resolution at the wall ($y^+ = 1$ or 2), at a lower Mach number (0.3). The computations are performed with the COMPRESS numerical code briefly described in Section 10.

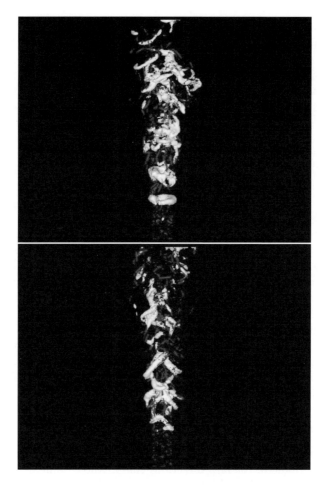

Fig. 17. LES of the natural jet at $Re = 25\,000$: instantaneous visualizations at two different times. White: low pressure isosurface; dark: regions of $Q > 0$ (courtesy Silva).

The details of this computations are presented in Briand [14]. It is known that transition in the boundary layer on a flat plate depends upon the type of perturbations exerted upstream on the flow (see Lesieur [67]). In Klebanoff *et al.* [51], the boundary layer was forced upstream with a thin metal ribbon parallel to the wall and stretched in the spanwise direction, which vibrates two-dimensionally close to the wall. In this experiment, the 3D forcing was harmonic. This corresponds to what is referred to as the K-mode, where the crests of the TS waves oscillate in phase in the spanwise

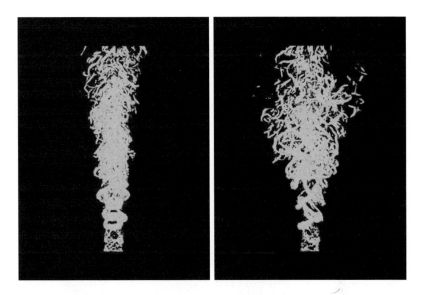

Fig. 18. Bifurcation of the jet with alternate-pairing excitation: forced jet at $Re = 10\,000$. Instantaneous vizualisations of regions of $Q > 0$ at the same instant but seen from two different perpendicular directions (courtesy Silva).

direction. The resulting vortex filaments are therefore aligned in the flow direction. On the other hand, if the perturbation is subharmonic, the crests oscillate out of phase. This is called H-mode, from Herbert [45], and corresponds to a staggered organization of vortex filaments. Herbert could show for the temporal problem[9], that the staggered mode was more amplified than the aligned mode. This should favour the emergence of H-mode during transition in natural situations, and explain why the transition above seems to be of subharmonic type.

We now return to the LES of the spatially-developing boundary layer over a flat plate. It is started with a set of upstream conditions (harmonic K-mode or subharmonic H-mode) obtained with the aid of nonlinear parabolized stability expansion (PSE) calculations (Bertolotti and Herbert [12], Airiau [5]). To the upstream state corresponding with a Reynolds number $R_{\delta_i} = 1000$ (δ_i being the upstream displacement thickness), one superposes a 3D white-noise of amplitude 0.2 the amplitude of the PSE perturbation. In the K-case, one sees in the transitional region formation of big longitudinal Λ-shaped vortices lying on the wall, and in phase in the spanwise

---

[9]Using a secondary-instability analysis where a perturbation is superposed on a TS wave of finite amplitude.

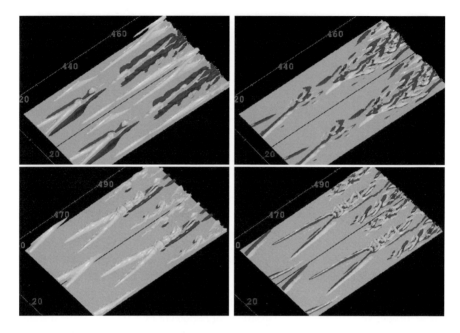

Fig. 19. LES of a spatial boundary layer at Mach 0.3; top and bottom, K- and H-transition respectively; the l.h.s. and r.h.s. correspond respectively to velocity and vorticity fluctuation components (dark, positive, light grey, negative); dark grey marks isosurfaces of positive Q (courtesy Briand).

direction (see Fig. 19, top). In the H-case, the vortices are staggered (see Fig. 19, bottom). The figures show at the end of transition the longitudinal components of velocity and vorticity, and also positive Q. One sees that the Λ vortices are very well correlated with a system of induced high and low-speed streaks[10]. Remark also on the vorticity plots that the big Λ's induce "antivorticity" close to the wall, due to the zero velocity condition at the wall. Downstream of $\approx$440 δ_i, the streaks become purely longitudinal. This is accompanied by the fast shedding of small arch vortices ejected from the tip of the Λ's, as indicated by Q-isosurfaces.

Figure 20 shows for the K-transition the downstream evolution of the friction coefficient at the wall, with comparison against the theoretical predictions of Van Driest[11] and Barenblatt and Prostokishin [7]. One sees a good agreement of the LES with these predictions, a resolution of $y^+ = 1$ improving the result. It is even better in the H-case. The peak in the

---

[10]This is not apparent on the figure for the H-case, due to an ill-chosen threshold.

[11]Discussed in Cousteix [26].

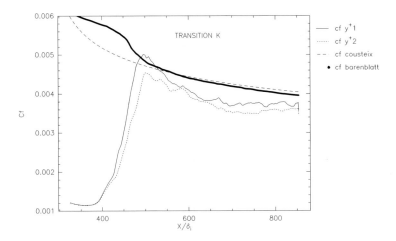

Fig. 20. LES of a spatial boundary layer at Mach 0.3: friction coefficients against downstream distance, compared with theoretical predictions of Cousteix and Barenblatt (courtesy Briand).

friction coefficient is at 490 δ_i, much further than the change of regime of the velocity streaks, and might be associated to an event such as the localized creation of a big hairpin vortex observed in the simulations of Ducros *et al.* [33]. Figure 21 shows for the K-case (but results are very close in the H-case) the rms longitudinal velocity component u' at a downstream distance such that $R_{\delta_1} = 1670$ (δ_1 local displacement thickness), compared with Spalart's [106] DNS at Reynolds numbers of 1000 and 2000. Again, the agreement is good, since the results are inbetween Spalart's predictions. If one looks at developed turbulence further downstream, various plots of vorticity components and pressure, as well as Q, accompanied with animations, show the very long longitudinal velocity streaks (about 1000 wall units for the low speeds). Above these streaks are ejected hairpins through what resembles a secondary Kelvin–Helmholtz instability occuring at a height of about $30 \approx 40$ wall units. The hairpins first creep at the wall, then rise, due to self-induction effects. Their length is about 300 wall units, so that there are several hairpins (about 3) above a single low-speed streak. In this sense, we have no more the perfect correlation hairpins-streaks which we observed during the transitional stage. It is therefore difficult to associate in the developed region the streaks to a system of purely longitudinal alternate vortices at the wall.

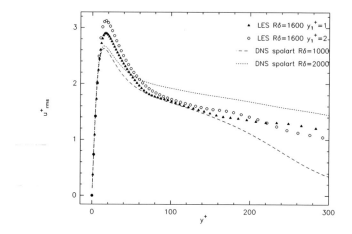

Fig. 21. Spatial boundary layer at Mach 0.3; rms velocity fluctuations compared with Spalart's DNS (courtesy Briand).

6   Dynamic approach in physical space

6.1   Dynamic models

We have shown already in Section 4.1 how a valuable information concerning a given eddy viscosity (here the spectral plateau-peak eddy viscosity) could be obtained thanks to a double filtering through the cutoff k_C and a fictitious cutoff $k_C' = k_C/2$, and the calculation of explicit resolved transfers across k_C'. This is the underlying philosophy of the dynamic model in physical space (Germano [41]). The basic method applies to a LES using an eddy-viscosity model. Most of the historical developments have been done with Smagorinsky's model, but the procedure may be applied to any eddy-viscosity models. One starts with a regular LES corresponding to a filter *bar* of width Δx, operator associating to a function[12] $f(\vec{x}, t)$ the function $\bar{f}(\vec{x}, t)$. One then defines a second "test filter" *tilde* of larger width $\alpha \Delta x$ (for instance $\alpha = 2$), associating $\tilde{f}(\vec{x}, t)$ to $f(\vec{x}, t)$. We then have two filter operators *bar* and *tilde* which apply on the functions, the product of them[13] being *tilde* o *bar*. This product, applied to $f(\vec{x}, t)$, means that we apply first to f the *bar* filter (to yield $\bar{f}$) then the *tilde* filter to obtain $\widetilde{\bar{f}}$. Let us first apply this filter product to the Navier–Stokes equation (with constant density): the subgrid-scale tensor of the field $\widetilde{\bar{u}}_i$ is readily obtained from

---

[12]Scalar, or vector, or tensor.
[13]In the sense of product of operators.

equation (3.5) with the replacement of the filter *bar* by the double filter, that is:

$$\mathcal{T}_{ij} = \widetilde{\bar{u}_i}\widetilde{\bar{u}_j} - \widetilde{\overline{u_i u_j}}. \tag{6.1}$$

We consider now the field $\bar{u}_i$ per se[14], and evaluate the resolved turbulent stresses obtained by application of the *tilde* filter. They write:

$$\mathcal{L}_{ij} = \widetilde{\bar{u}_i}\widetilde{\bar{u}_j} - \widetilde{\bar{u}_i \bar{u}_j}. \tag{6.2}$$

We now apply the filter "tilde" to equation (3.5), which leads to

$$\widetilde{T_{ij}} = \widetilde{\overline{\bar{u}_i \bar{u}_j}} - \widetilde{\overline{u_i u_j}}. \tag{6.3}$$

Adding equations (6.2) and (6.3), yields, with the aid of (6.1)

$$\mathcal{L}_{ij} = \mathcal{T}_{ij} - \widetilde{T_{ij}}. \tag{6.4}$$

This expression is called Germano's identity. In the r.h.s., $\mathcal{T}_{ij}$ and $\widetilde{T_{ij}}$ have to be modelled, while the l.h.s. $\mathcal{L}_{ij}$ (the resolved stresses) can be explicitly calculated by applying the *tilde* filter to $\bar{u}_i$.

We use Smagorinsky's model expression defined by equation (3.21) and "tilde-filter" it, to get

$$\widetilde{T_{ij}} - \frac{1}{3}\widetilde{T_{ll}}\,\delta_{ij} = 2\widetilde{\mathcal{A}_{ij}C}, \tag{6.5}$$

whith $C = C_S^2$ and

$$\mathcal{A}_{ij} = (\Delta x)^2\,|\bar{S}|\bar{S}_{ij}.$$

We have now to determine $\mathcal{T}_{ij}$, stress resulting from the filter product. This is again obtained using Smagorinsky for the largest filter, which yields

$$\mathcal{T}_{ij} - \frac{1}{3}\mathcal{T}_{ll}\,\delta_{ij} = 2\mathcal{B}_{ij}, \tag{6.6}$$

whith

$$\mathcal{B}_{ij} = \alpha^2(\Delta x)^2\,|\widetilde{\bar{S}}|\,\widetilde{\bar{S}}_{ij}.$$

Substracting equation (6.5) from equation (6.6) yields with the aid of Germano's identity

$$\mathcal{L}_{ij} - \frac{1}{3}\mathcal{L}_{ll}\,\delta_{ij} = 2\mathcal{B}_{ij}C - 2\widetilde{\mathcal{A}_{ij}C}.$$

---

[14]As if it were the instantaneous field.

This is a nice result relating the model coefficient (unknown) to the resolved stresses. However there are some difficulties. First, one removes C from the filtering as if it were constant[15], leading to

$$\mathcal{L}_{ij} - \frac{1}{3}\mathcal{L}_{ll}\,\delta_{ij} = 2CM_{ij}, \qquad (6.7)$$

with

$$M_{ij} = \mathcal{B}_{ij} - \widetilde{\mathcal{A}_{ij}}.$$

All the terms of equation (6.7) may now be determined with the aid of $\bar{u}$. But there are five independent equations for only one variable C, so that the problem is overdetermined. A first solution proposed by Germano [41] is to multiply tensorially equation (6.7) by $\bar{S}_{ij}$, to get

$$C = \frac{1}{2}\frac{\mathcal{L}_{ij}\bar{S}_{ij}}{M_{ij}\bar{S}_{ij}} \qquad (6.8)$$

(indeed, and due to incompressibility, $\bar{S}_{ii} = 0$). This provides finally a dynamical evaluation of $C(\vec{x}, t)$, which can be used in the LES of the *bar* field $\bar{u}$. But problems still arise: in tests using channel flow data obtained from DNS, Germano [41] could show that the denominator in equation (6.8) could locally vanish or become sufficiently small to yield computational instabilities. To get rid of this problem, Lilly [76] chose to determine the value of C in equation (6.7) by minimizing the error using a least squares approach, which gives

$$C = \frac{1}{2}\frac{\mathcal{L}_{ij}M_{ij}}{M_{ij}^2} \qquad (6.9)$$

and removes the undeterminacy of equation (6.7). However, the analysis of DNS data reveals that the C field predicted by the models (6.8) or (6.9) varies strongly in space and contains a significant fraction of negative values, with a variance which may be ten times higher than the square mean. So, the removal of C from the filtering operation is not really justified and the model exhibits some mathematical inconsistencies. The possibility of negative C is an advantage of the model since it allows a sort of backscatter in physical space, but very large negative values of the eddy viscosity is a destabilizing process in a numerical simulation, yielding a non-physical growth of the resolved scale energy. The cure which is often adopted to avoid excessively large values of C consists in averaging the numerators and denominators of (6.8) and (6.9) over space and/or time, thereby losing some of the

---

[15]This is in some way contradictory with the original aim of having a dynamic evaluation of C depending on space and time.

conceptual advantages of the "dynamic" local formulation. Averaging over direction of flow homogeneity has been a popular choice, and good results have been obtained by Germano [42] and Piomelli [94], who took averages in planes parallel to the walls in their channel-flow simulation. Remark that the same thing has been done, with success, when averaging the dynamic spectral eddy viscosity in the channel-flow LES presented above. It can be shown that the dynamic model gives a zero subgrid-scale stress at the wall, where L_{ij} vanishes, which is a great advantage with respect to the original Smagorinsky model; it gives also the proper asymptotic behavior near the wall.

As already stressed, the use of Smagorinsky's model for the dynamic procedure is not compulsory, and any of the models described in the present paper can be a candidate. As an example, El Hady and Zang [34] have applied the dynamic structure-function model applied to a compressible boundary layer above a long cylinder.

7   Alternative models

7.1   Generalized hyperviscosities

One of the drawback of the structure-function model given by equation (5.6) is the absence of a cusp near k_C. However, EDQNM data show that the exponential form given in equation (4.14) can be correctly approximated by a power law of the type:

$$\nu_t^* \left(\frac{k}{k_C} \right) = \left(1. + \nu_{tn}^* \left(\frac{k}{k_C} \right)^{2n} \right), \tag{7.1}$$

with $2n \approx 3.7$. Lesieur and Métais [69] have shown that ν_{tn}^* can be determined by considering the energy balance between explicit and subgrid-scale transfers. This yields:

$$\nu_{tn}^* = 0.512 \left(\frac{3n}{2} + 1 \right). \tag{7.2}$$

In fact, the EDQNM value of $2n = 3.7$ is not so far from the exponent $2n = 4$ which would be obtained with a Laplacian operator iterated twice. Therefore, Lesieur and Métais [69] proposed a physical-space turbulent dissipative operator based upon the structure-function model and taking into account the "cusp" behaviour:

$$2 \frac{\partial}{\partial x_j} \left[\nu_t^{\mathrm{SF}} \bar{S}_{ij} \right] + \nu_t^{(2)} \left(\frac{\Delta x}{\pi} \right)^4 \left(\frac{\partial^2}{\partial x_j^2} \right)^3 \bar{u}_i, \tag{7.3}$$

where $\bar{S}_{ij}$ is the deformation tensor of the field $\bar{u}_i$. $\left(\partial^2/\partial x_j^2\right)^3$ designates the Laplacian operator iterated three-times. ν_t^{SF} is given by the r.h.s. of equation (5.6) multiplied by $0.441/(2/3)$, $\nu_t^{(2)} = \nu_t^{\mathrm{SF}} \times \nu_{t2}^*$, and ν_{t2}^* is given by equation (7.2) with $n = 2$. The expression (7.3) is interesting in the sense that it provides an eddy dissipation combining the structure function model with a hyperviscosity $(\nabla^2)^3\bar{u}_i$. The latter represents in physical space the action of the cusp in Kraichnan's spectral eddy viscosity.

7.2 Hyperviscosity

The model given by equation (7.3) bears some resemblance to hyperviscosity models which are widely used in the study of geophysical flows because of their simplicity. Indeed, the hyperviscosity consists in replacing the molecular dissipative operator $\nu\nabla^2$ by $(-1)^{\alpha-1}\nu_\alpha(\nabla^2)^\alpha$, where α is a positive integer. As opposed to equation (7.3), ν_α is here a constant (positive) coefficient which has to be adjusted. This has been widely used in two-dimensional isotropic turbulence (see [10]), with $\alpha = 2$ or $\alpha = 8$, as a way to shift the dissipation to the neighbourhood of k_{C}. This allows for a reduction of the number of scales strongly affected by viscous effects, and has rendered possible in the case of two-dimensional turbulence to demonstrate the existence of coherent vortices.

In three-dimensional turbulence, it was used by Bartello *et al.* [9] to study the influence of a solid-body rotation, with surprisingly good results.

7.3 Scale-similarity and mixed models

The lack of correlation between the subgrid-scale stress and the large-scale strain rate tensors has led Bardina *et al.* [8] to propose an alternative subgrid-scale model called the scale similarity model. This is based upon a double filtering approach and on the idea that the important interactions between the resolved and unresolved scales involve the smallest eddies of the former and the largest eddies of the latter. They suggest that the real subgrid tensor is similar to the stress tensor constructed from the resolved velocity field. One then writes:

$$T_{ij} = \bar{\bar{u}}_i\bar{\bar{u}}_j - \overline{\bar{u}_i\bar{u}_j}. \tag{7.4}$$

The analysis of DNS and experimental data [8, 77] have shown that the modelled subgrid-scale stress deduced from (7.4) exhibits a good correlation with the real (measured) stress. However, when implemented in LES calculations, the model hardly dissipates any energy. It is therefore necessary to combine it with an eddy-viscosity type model such as Smagorinsky's model to produce the "mixed" model (see *e.g.* Meneveau and Katz [82]).

In the line of Bardina *et al.* model, new formulations have been proposed to correct this lack of dissipation. Liu *et al.* [77] have proposed the following model:

$$T_{ij} = C_{\mathrm{L}} \left(\tilde{\bar{u}}_i \tilde{\bar{u}}_j - \widetilde{\bar{u}_i \bar{u}_j} \right), \tag{7.5}$$

where C_{L} is a dimensionless coefficient. The operator $\sim$ consists in a second filter of different width [77].

7.4 Anisotropic subgrid-scale models

As stressed above, the subgrid-scale tensor and buoyancy flux given by (3.14) are assumed to be strictly proportional to the grid-scale strain rate tensor and buoyancy flux, respectively. Abbà *et al.* [1] have proposed an anisotropic formulation of (3.14) using eddy-viscosity and eddy-diffusivity tensors instead of scalar ones:

$$T_{ij} - \frac{1}{3} T_{ll}\, \delta_{ij} = 2 \sum_{r,s} \nu^{\mathrm{t}}_{ijrs} \bar{S}_{rs} - \frac{2}{3} \delta_{ij} \sum_{l,r,s,} \nu^{\mathrm{t}}_{llrs} \bar{S}_{rs}. \tag{7.6}$$

This formulation allows for a better description of the small-scale anisotropy. This model in conjunction with the dynamic procedure previously described has been used successfully in Large-Eddy Simulations of turbulent natural convection [1]. The reader is referred to the book by Sagaut [98] for a presentation of other anisotropic models.

8 LES of rotating flows

Our purpose here is to show the ability of LES and DNS to accurately reproduce the detailed vorticity dynamics and flow statistics even in the presence of external forces like solid-body rotation. Rotating flows are extremely important in engineering for studies related to turbo-machinery of turbines, pumps, or air-intakes of jet engines. They are also crucial in internal geophysics[16], oceanography, meteorology, planetary or stellar physics. We will first consider rotating shear flows (free or wall bounded), where the effects of rotation are extremely spectacular in terms of modification of the vortical structure and of the statistics. Then we will review studies concerning homogeneous turbulence submitted to rotation.

---

[16]To understand the Earth magnetic-field generation.

8.1   Rotating shear flows

We assume a purely incompressible flow submitted to solid-body rotation.
We consider a parallel basic (or mean[17]) velocity $\langle u \rangle \, (y)$, and assume that
the axis of rotation is parallel to the spanwise direction, such as for the
channel of Figure 8 We work in a relative rotating frame of angular-rotation
vector $\vec{\Omega}$. Coriolis acceleration $-2\vec{\Omega} \times \vec{u}$ is added to Navier–Stokes equations,
while centrifugal effects are incorporated in the pressure gradient. Let

$$R_{\mathrm{o}}(y, t) = -\frac{1}{f} \frac{\mathrm{d} \langle u \rangle}{\mathrm{d} y} \qquad (8.1)$$

(with $f = 2\Omega$) be the local Rossby number. It characterizes the ratio of
the local relative basic vorticity upon the entrainment vorticity f associated
to the solid-body rotation. Regions with a positive (resp. negative) local
Rossby will be called cyclonic (resp. anticyclonic). We recall also that the
absolute vorticity vector is $\vec{\omega}_{\mathrm{a}} = \vec{\omega} + f\vec{z}$, and satisfies Helmholtz theorem
in its conditions of applicability, within which absolute-vortex elements are
material.

8.1.1   Free-shear flows

We synthesize first results concerning free-shear layers coming both from
3D linear-stability studies (Yanase et al. [112]), and DNS or LES (Lesieur
et al. [73], Métais et al. [78]). As in instability studies, we start with a basic
parallel velocity profile, weakly perturbed. There is a critical local Rossby
number of -1 such that:

- in regions where initially $R_{\mathrm{o}}(y) \geq -1$, the shear layer is two-
  dimensionalized. In a mixing layer for instance, 3D perturbations
  are damped, and straight Kelvin–Helmholtz billows form[18]. This result agrees in particular with Proudman–Taylor's theorem when the
  Rossby number modulus is small;

- for $R_{\mathrm{o}}^{\min} < R_{\mathrm{o}}(y) < -1$ ("weak" anticyclonic rotation), where
  $R_{\mathrm{o}}^{\min} \approx -10 \sim -20$ decreases as the Reynolds increases, the flow
  is highly three-dimensionalized, with production of intense Görtler-
  like alternate longitudinal rolls. This flow three-diemnsionalization
  results from the development of the so-called "Shear-Coriolis instability". Examination of the vorticity fields shows that they correspond
  in fact to the condensation of absolute-vortex lines into very long

---

[17]In the sense of an ensemble average.
[18]Without stretching of longitudinal vortices nor helical pairing.

hairpins which are oriented in a purely longitudinal direction. As a result, their spanwise vorticity component is zero, which implies that the mean velocity gradient becomes constant and equal to f, so that the local Rossby number uniformizes to the value -1. This has been clearly shown in the mixing layer DNS by Métais *et al.* [78] as well as in the anticyclonic region of a wake. There is in fact universality of this result for all shear flows (free or wall bounded), since this law is also found for the channel as shown below.

8.1.2   Wall flows

As already stressed, and due to their numerous applications in turbo-machinery and also in oceanography, the turbulent flow in a rotating channel of spanwise rotation axis has been subject to extensive studies. Experimentally, it is not easy to cover a wide range of rotation regimes. Conversely, the introduction of the Coriolis acceleration is rather straightforward in numerical codes simulating the three-dimensional Navier–Stokes equations. This may explain why more numerous numerical studies based either on DNS (see *e.g.* Kristoffersen and Andersson [55]) or LES (see *e.g.* Kim [50], Miyake and Kajishima [84], Tafti and Vanka [107], and Piomelli and Liu [96]) have been devoted to this topic than experimental ones (see *e.g.* Johnston *et al.* [48], Nakabayashi and Kitoh [86]). These previous works have mainly focussed on weak-rotation regimes, and the analysis restricted to statistics directly issued from the velocity field, or studies of the large-scale flow organization. We here recall the main results obatined by Lamballais *et al.* [57,61]. We precise again the notations. $\vec{\Omega}$ is oriented along the spanwise direction z, and may be positive or negative. For the channel flow, the vorticity vector associated with the mean velocity profile $\langle \vec{\omega} \rangle = (0, 0, -\mathrm{d}\langle u \rangle /\mathrm{d}y)$ is parallel to $\vec{\Omega}$ near one wall and antiparallel near the opposite wall: we refer to the two particular walls as cyclonic and anticyclonic. Various other terms are currently used. The names suction and pressure sides originate from the pressure gradient due to the Coriolis force, and the terms trailing and leading sides are borrowed from turbo-machinery. The initial Rossby number (already defined above for the channel) turns out to be equal to $Ro_{\mathrm{g}} = \dfrac{3\,U_{\mathrm{m}}}{2\Omega\,h}$ where U_{m} is the bulk velocity.

The previous studies have clearly shown that, due to the action of moderate rotation, the flow becomes very asymmetric with respect to the channel center, with a turbulent activity much reduced on the cyclonic side as compared with the anticyclonic side.

We first show a LES of a rotating channel flow based upon the spectral-dynamic model with the same characteristic parameters than Piomelli and Liu [96]) in their LES using a localized version of the dynamic model:

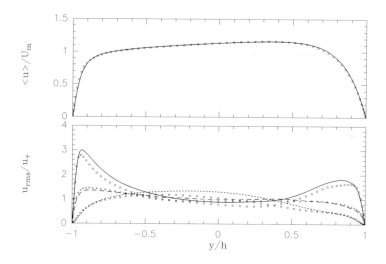

Fig. 22. Mean velocity and turbulence intensities ($Re = 5700$, $Ro_g = 21$). —, ○, $\sqrt{\langle \overline{u'^2} \rangle}/u_\tau$; ..., ⊔, $\sqrt{\langle \overline{v'^2} \rangle}/u_\tau$; - - -, △, $\sqrt{\langle \overline{w'^2} \rangle}/u_\tau$. Lines: spectral-dynamic LES; symbols: LES of Piomelli and Liu [96]; courtesy Lamballais.

$Ro_g = 21$ and $Re = 5700$. Figure 22 shows that the mean and rms velocity profiles predicted by the two models are in excellent agreement.

We next investigate rotation regimes for which anticyclonic destabilization is achieved: this corresponds to rotation rates such that $Ro_g > 1$. For these rotation rates, the shear-Coriolis instability should be at hand, as in the free-shear case, at least if the Rossby number is not too high. The computations are here performed at $Ro_g = \infty, 18, 6, 2$. We also study Reynolds number effects by comparing spectral-dynamic model based LES at $Re = 14\,000$ with DNS at $Re = 5000$.

We here examine the three-dimensional flow structure. Figure 23 clearly shows that rotation strongly modifies the vortex organization. We observe the following trends:

- the turbulent activity is gradually reduced near the cyclonic wall as the rotation rate is increased. For $Ro_g = 2$, the very flat isosurfaces indicate an almost complete flow relaminarization. This will be confirmed by the statistics;

- on the anticyclonic region, the flow presents a strong turbulence activity. We have checked in the DNS (see Lamballais *et al.* [61]) the existence of large-scale longitudinal roll cells similar to those already observed in the laboratory experiments of Johnston *et al.* [48] and in the numerical simulations of Kristoffersen and Andersson [55] and Piomelli and Liu [96]. The roll cells are no longer present for $Ro_g = 2$;

- the vortical structures are more and more organized as rotation is increased, and their inclination with respect to the wall is reduced. This is clearly demonstrated by considering the statistics of the inclination angle of the vorticity vector (see Lamballais *et al.* [61]).

It is important to note that the LES are capable to reproduce all the characteristic features of the flow organization. The subgridscale model is indeed able to capture cyclonic relaminarization with inactive turbulent motions as well as detailed turbulent flow organization on the anticyclonic side (Fig. 23).

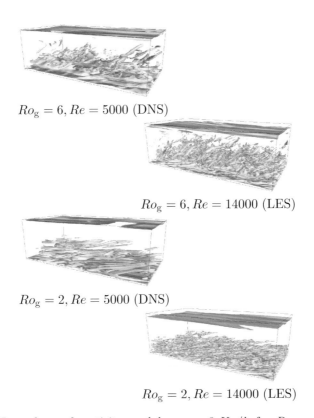

$Ro_g = 6, Re = 5000$ (DNS)

$Ro_g = 6, Re = 14000$ (LES)

$Ro_g = 2, Re = 5000$ (DNS)

$Ro_g = 2, Re = 14000$ (LES)

Fig. 23. Isosurfaces of vorticity modulus $\omega = 3\,U_m/h$ for $Ro_g = 6$ or $\omega = 2.25\,U_m/h$ for $Ro_g = 2$ (for the DNS results, only a quarter of the computational domain is presented).

As far as the statistical quantities are concerned, Figure 24, taken from Lamballais *et al.* [61] clearly shows that the important physical result corresponding to the establishment of the $R_o(y) = -1$ plateau, predicted by the DNS, is well reproduced by the LES. Furthermore, Lamballais *et al.* [61] have thoroughfully checked various statistical quantities based upon the mean velocity field, the fluctuating velocity field or the fluctuating vorticity field. An excellent agreement with the DNS is obtained and the LES are able to correctly reproduce all the anisotropy characteristics of the flow. It is important to note that the rotating channel is quite a challenging test case for the one-point closure models based of a Reynolds Averaged Navier–Stokes approach and quite sophisticated models have to be designed to obtain satisfactory results. Subgrid-scale models turn out to be much more universal, since identical models can be used for different flows submitted to various external forces.

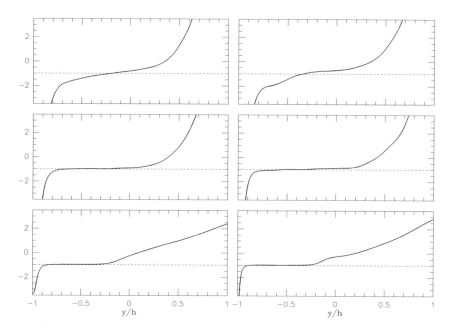

Fig. 24. Final local Rossby in the DNS (left, global Reynolds number 5000) and LES (right, Reynolds 14 000) of a rotating channel. From top to bottom, initial Rossby at the wall: 18, 6 and 2.

8.1.3   Homogeneous turbulence

Bartello *et al.* [9] have performed numerical simulations investigating the formation and stability of quasi-twodimensional coherent vortices in rotating homogeneous three-dimensional flow. Isolated coherent two-dimensional vortices obtained from a purely two-dimensional decay simulation, were superposed with a low-amplitude three-dimensional perurbation, and used to initialize a first set of simulations. In the non-rotating case, a three-dimensionalization of all vortices was observed. Conversely, when $2\Omega \approx [\vec{\omega} \cdot \vec{\Omega}]_{rms}$, a rapid destablization of the anticyclones was observed to occur, whereas the initial two-dimensional cyclonic vortices persisted throughout the simulation. At larger Ω, both cyclones and anticyclones remained two-dimensional, consistent with the Taylor–Proudman theorem. A second set of simulations starting from isotropic three-dimensional fields was initialized by allowing a random field to evolve with $\Omega = 0$ to a fully-developed state. When the simulation were continued with $2\Omega \approx [\vec{\omega} \cdot \vec{\Omega}]_{rms}$, the three-dimensional flow was observed to organize into two-dimensional cyclonic vortices. At large Ω, two-dimensional anticyclones also emerged from the initially-isotropic flow.

9   LES of flows of geophysical interest

Most of the flows encountered in the atmosphere or the ocean are composed of interacting waves and turbulence. Geophysical eddies and turbulence often originate from the development of instabilities resulting from the combined effects of density gradients and rotation, and these strongly affect the dynamics over a large range of scales. We summarize here the results of DNS and LES aimed at investigating the effects of stable or unstable density stratification and/or solid-body rotation on turbulence and coherent vortices, and we particularly focus our study on three-dimensional processes.

9.1   Baroclinic eddies

The baroclinic instability results from the combined effects of horizontal temperature gradients and fast rotation on a stably-stratified fluid. It corresponds to a very efficient mechanism of conversion of potential energy into horizontal kinetic energy. When one considers horizontal scales of the order of the internal Rossby radius of deformation (≈ 1000 km in the atmosphere and ≈ 50 km in the ocean, at mid-latitude), this instability becomes very active and gives rise to "baroclinic" eddies. Garnier [37], Garnier *et al.* [38, 39] have performed direct and large-eddy simulations of baroclinic jet flows instabilities with the goal to study the nature of the coherent vortices and in particular the asymmetry between cyclonic and anticyclonic eddies. Garnier *et al.* [38,39] have considered a stably-stratified medium associated with a constant vertical mean density gradient characterized by a constant Brunt–Vaissala frequency N. The initial basic state consits in an horizontal density front oriented in the meridional direction $\vec{y}$. The rotation vector $\vec{\Omega}$ is oriented in the z direction. Let $f = 2\Omega$ be the Coriolis parameter. In the limit of fast rotation and strong stratification, it can be shown that the density front has to be associated with a basic velocity profile. Indeed, the geostrophic equilibrium corresponding to a balance between the Coriolis force and the pressure gradient and the hydrostatic balance imply that this basic state has to satisfy the thermal wind equation:

$$\frac{\partial \vec{u}_{\mathrm{H}}}{\partial z} = -\frac{g}{\rho_0 f}\vec{z} \times \vec{\nabla}_{\mathrm{H}}\rho, \qquad (9.1)$$

where $\vec{a}_{\mathrm{H}}$ stands for the horizontal projection of the vector $\vec{a}$ on the horizontal plane. The meridional density gradient then give rise to a mean velocity corresponding to a jet, sheared along the vertical direction and directed to the east at the top of the domain and to the west at the bottom. When a small random perturbation is superposed to this basic state the flow becomes unstable. The nature of the instability is however very different depending upon the characteristic parameters. The two non-dimensionalized parameters are the Rossby (Ro) and Froude (Fr) numbers (see Garnier *et al.* [38,39]).

First, Garnier *et al.* [38] have used direct numerical simulations to carry out linear stability studies. They have shown the existence of a critical value of the ratio $Ro/Fr = 1.5$ constituting the threshold between to distinct regimes:

1) $Ro/Fr > 1.5$: the instability is weak and mainly barotropic: it is of Kelvin–Helmholtz type and is associated with the inflexional nature of the mean velocity profile;

2) $Ro/Fr \leq 1.5$: the baroclinic instability corresponding to a conversion of potential energy associated with the horizontal density gradient into horizontal kinetic energy can develop. The amplification of the perturbations is much stronger than in the barotropic case.

We now concentrate on the second regime $Ro/Fr \leq 1.5$: here $Ro/Fr = 0.5$. The Rossby number is fixed to 0.1. The numerical code is similar to the channel flow study previously described except that compact differences schemes are here used into two spatial directions. The Reynolds number is low in the DNS ($Re = 400$) and much higher in the LES ($Re = 10\,000$).

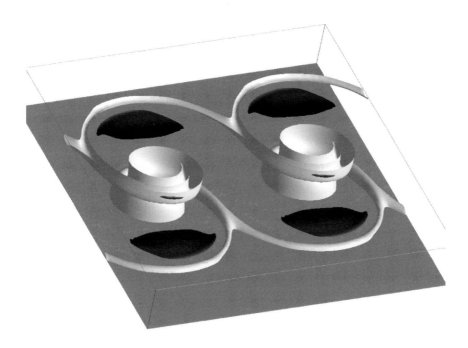

Fig. 25. Iso-surfaces of vertical vorticity; n light-gray: cyclonic vorticity, black: anticyclonic vorticity (courtesy Garnier).

9.1.1 Synoptic-scale instability

Figure 25 shows the vorticity structure obtained by DNS once the instability has fully developed. We observe the formation of cyclonic eddies of strong intensity, composed of nearly two-dimensional cores between which braids of very high cyclonic vorticity are formed. The vorticity maxima are observed within these braids and correspond here to ≈ 8 times the vorticity maximum of the initial mean velocity profile. The vorticity intensification in the anticyclonic eddies is weaker (3 times the initial vorticity): we have checked that those are far more three-dimensional than the cyclonic eddies and strongly stretched by them. The asymmetry cyclones/anticyclones is clear and the vorticity maxima are cyclonic and are localized in very concentrated regions of the space. Contrarily, the anticyclonic vorticity is much more spread.

9.1.2   Secondary cyclogenesis

In the previous DNS, the late stage of the cyclogenesis was dominated by dissipative and diffusive effects. We have thus performed high Reynolds number LES at $Re = 10\,000$ to study the flow development subsequent to the occlusion process. The subgrid-scale structure function model with a cusp given by equation (7.3) have been used. Here the grid is non-isotropic $\Delta x \neq \Delta y \neq \Delta z$ and the formulation for non-uniform gird, given by equations (5.7) and (5.8), is used.

Figure 26 shows a time evolution of the vorticity contours of a cyclonic eddies. As compared with the DNS presented in the preceeding section, one may notice that the spiralling of the vorticity contours inside the core of the cyclonic eddies is much more pronounced. Due to viscous effects, the vorticity was indeed homogenized in the DNS. We have checked that the frontal region are much steeper in the LES indicating more energy near the wavenumber cut-off. The steepening of the fronts is associated with the appearance of a secondary instability resulting in a local intensification of the vertical vorticity. This instability seems to take place in regions where the local values of the Rossby and Froude number $Ro(\vec{x})$ and $Fr(\vec{x})$ verify the criterion $Ro(\vec{x})/Fr(\vec{x}) \leq 1.5$. The potential energy associated with local horizontal fronts is then converted into horizontal kinetic energy and gives rise to vertical vorticity intensification. It is important to notice that if the structure model without cusp is used excessive accumulation of energy is observed at the smallest scales eventually leading to numerical divergence. This demonstrates the importance of the cusp-like behaviour and the feasibility of the subgrid-scales previously described for the LES of geophysical flows with quasi-twodimensional regions and sharp frontal regions.

Note that the present results have been compared by Lesieur *et al.* [70] with satellite observations corresponding to the severe storm of 26 Dec. 1999. This storm, together with its companion on 28 December, caused casualties and immense damage in France and neighbouring countries. Lesieur *et al.* [70] discuss possible analogies and differences, as well as some consequences in terms of numerical weather forecasting.

Note that the same subgrid-scale models and the same numerical code has also been successfully used to study the combined effects of unstable density stratification and rotation. Padilla–Barbosa and Métais [91] have indeed performed LES of rotating turbulent convection with an application to oceanic deep-water formation.

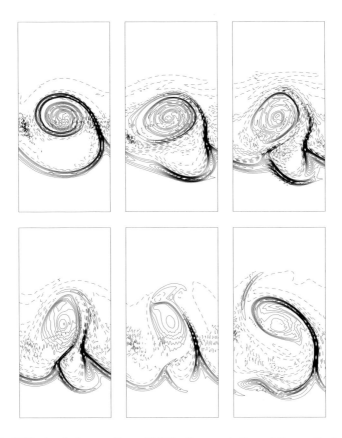

Fig. 26. LES simulation at $Re = 10\,000$: time evolution of the vertical vorticity isocontours at the top of the computational domain. Solid-lines cyclonic vorticity; dashed lines anticyclonic vorticity.

10 LES of compressible turbulence

Compressible turbulence has extremely important applications in subsonic, supersonic and hypersonic aerodynamics. More generally, and even at low Mach numbers, strong density differences due to an intense heating (in combustion for instance) may have profound consequences on the flow structure and the associated mixing. Heating a wall may, for instance, completely destabilize a boundary layer, due to the generalized inflexion-point instability. Examples and details are given below.

10.1 Compressible LES equations

We here briefly recall the specific difficulties attached to the filtered compressible Navier–Stokes equations. Details may be found in Lesieur and Comte [68] and Métais *et al.* [81]. The momentum equation with variable density, $\rho(\vec{x}, t)$, and for a newtonian fluid can be written in the so-called conservative form as:

$$\frac{\partial}{\partial t}(\rho u_i) + \frac{\partial}{\partial x_j}(\rho u_j u_i) = -\frac{\partial}{\partial x_j}\left[p\,\delta_{ij} + \frac{2\mu}{3}\left(\vec{\nabla}\cdot\vec{u}\right)\delta_{ij} - 2\mu D_{ij}\right]. \quad (10.1)$$

Similarly, the continuity equation can be expressed as:

$$\frac{\partial \rho}{\partial t} = -\frac{\partial}{\partial x_j}(\rho u_j). \quad (10.2)$$

These equation have to be completed by an equation representing the evolution of the total energy defined by, for an ideal gas,

$$\rho e = \rho\, C_{\mathrm{v}}\, T + \tfrac{1}{2}\rho(u_1^2 + u_2^2 + u_3^2) \quad (10.3)$$

as well as the equation of state.

The application of the filter operator $\overline{}$ to these equations yields quantities of the following form:

$$\overline{\rho\Phi}$$

where $\Phi(\vec{x}, t) = u_i(\vec{x}, t)$, $e(\vec{x}, t)$, $T(\vec{x}, t)$, etc. These are not easily expressable as a simple function of $\overline{\rho}$ and $\overline{\Phi}$. To overcome this difficulty it is customary (see Favre [35]) to introduce the density-weighted Favre average (here density-weighted filter) denoted as $\widetilde{}$:

$$\widetilde{\Phi} = \frac{\overline{\rho\Phi}}{\overline{\rho}}. \quad (10.4)$$

One then tries to write a closed system of equations for the variables $\widetilde{\Phi}$ and for $\overline{\rho}$. For instance, the equation for $\overline{\rho}$ is obtained through the filtered continuity equation:

$$\frac{\partial \overline{\rho}}{\partial t} = -\frac{\partial}{\partial x_j}(\overline{\rho u_j})$$

$$\frac{\partial \overline{\rho}}{\partial t} = -\frac{\partial}{\partial x_j}(\overline{\rho}\,\widetilde{u}_j). \quad (10.5)$$

The application of the filter $\overline{}$ to the momentum equation gives rise to the subgrid-scale stress tensor, whose form, in the compressible case, is:

$$T_{ij} = \overline{\rho}\,\widetilde{u}_i\widetilde{u}_j - \overline{\rho u_i u_j}. \quad (10.6)$$

The filtered momentum equation may be closed through a classical turbulent viscosity assumption. The closure of the energy equation may be resolved by introducing an eddy-conductivity $k_t(\vec{x}, t)$:

$$k_t(\vec{x}, t) = C_p \frac{\nu_t(\vec{x}, t)}{Pr_t} \tag{10.7}$$

where $\nu_t(\vec{x}, t)$ is the turbulent viscosity. Note that the difficulties associated with the trace of the subgrid-scale tensor T_{ll} may be nicely overcome by the introduction of a "macro-pressure"

$$\varpi = \bar{p} - \frac{1}{3}T_{ll} \tag{10.8}$$

and of a "macro-temperature"

$$\vartheta = \widetilde{T} - \frac{1}{2C_v\bar{\rho}}T_{ll} \tag{10.9}$$

as proposed by Lesieur and Comte [68]. Except for these differences, the formulation of the subgrid-scale model is identical to the incompressible case but $\nu_t(\vec{x}, t)$ is now determined from the density-weighted filtered velocity field $\widetilde{u}_i$. Another important difference with the incompressible case is that the dynamic viscosity varies with temperature through the classical Sutherland empirical law.

10.2   Heated flows

The compressible results presented here are based upon the COMPRESS code developed in Grenoble. The details on the numerical procedure can be found in [22]. The numerical code uses curvilinear co-ordinates. The system is solved in the transformed grid by means of a extension of the fully-explicit McCormack scheme, second order in time and fourth in space, devised by Gottlieb and Turkel [44]. High-Mach number boundary layer simulations have been performed by Normand et al. [87, 88] which constitute a validation of the numerics and SGS model. It showed in particular the ability of the code to reproduce the effect of strong heating on Reynolds stresses. It thus can be considered as a suitable tool for prediction of heat fluxes in situations for which experimental data are absent or sparse. Indeed, the understanding of the dynamics of turbulent flows submitted to strong temperature gradients is still an open challenge for numerical and experimental research. It is of vital importance due to the numerous industrial applications such as the heat exchangers, the cooling of turbine blades, the cooling of rocket engines, etc.

176 New Trends in Turbulence

10.2.1 The heated duct

We want here to show the ability for LES to adequately reproduce the effects of an asymetric heat flux in a square duct flow. The details of the computations are reported in [99] and [100]. We solve the three-dimensional compressible Navier–Stokes equations with the *COMPRESS* code previously mentioned. The subgrid-scale model is the selective structure function model. We have successively considered the isothermal duct, at a Reynolds number $Re_b = 6000$ (based on the bulk velocity), with the four wall at the same temperature and the heated duct for which the temperature of one of the walls is imposed to be higher than the temperature of the three other walls ($Re_b = 6000$). It is important to note that moderate resolutions are used: the grid consists of $32 \times 50 \times 50$ nodes in the isothermal case and of $64 \times 50 \times 50$ nodes in the heated case along x (streamwise), y and z (transverse) directions. This moderate resolution renders the computation very economical compared with a DNS. One crucial issue in LES is to have a fine description of the boundary layers. In order to correctly simulate the near-wall regions, a nonuniform (orthogonal) grid with a hyperbolic-tangent stretching is used in the y and z directions: the minimal spacing near the walls is here 1.8 wall units. The Mach number is $M = 0.5$ based upon the bulk velocity and the wall temperature.

We have first validated our numerical procedure by comparing our results, for the isothermal duct, with previous incompressible DNS results [40]: a very good agreement was obtained at a drastically reduced computer cost. The flow inside a duct of square cross section is characterized by the existence of secondary flows (Prandtl's flow of second kind) which are driven by the turbulent motion. The secondary flow is a mean flow perpendicular to the main flow direction. It is relatively weak ($2 - 3\%$ of the mean streamwise velocity), but its effect on the transport of heat and momentum is quite significant. If a statistical modelling approach is employed, elaborate second-order models have to been utilized to be able to accurately reproduce this weak secondary flow. Figure 27a shows the contours of the streamwise vorticity in a quarter of a cross section. The secondary flow vectors reveal the existence of two streamwise counter-rotating vortices in each corner of the duct. The velocity maximum associated with this flow is 1.169% of the bulk velocity: this agrees very well with experimental measurements. It shows the ability for LES to accuratly reproduced statistical quantities. Figure 27b shows the instantaneous flow field for the entire duct cross-section. As compared Figure 27a, it clearly indicates a very pronounced flow variability with an instantaneous field very distinct from the mean field. The maximum for the transverse fluctuating velocity field is of the order of ten times the maximum for the corresponding mean velocity field. As far as the vorticity is concerned, the transverse motions are

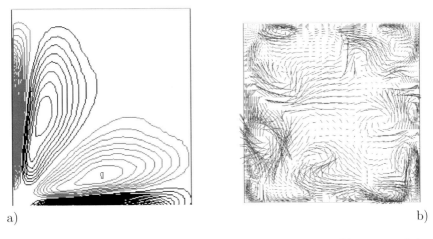

a)   b)

Fig. 27. a) Ensemble averaged streamwise vorticity contours; b) vectors of the instantaneous velocity field (courtesy Salinas–Vasquez).

associated with streamwise vorticity generation, whose maximum is about one third of the transverse vorticity maximum.

In the heated case, Salinas and Métais [100] have investigated the effect of the heating intensity by varying the temperature ratio between the hot wall and the other walls. When the heating is increased, an amplification of the mechanism of ejection of hot fluid from the heated wall is observed. Figure 28 shows temperature structures near the heated wall of the duct. Only one portion of the duct is here represented. As shown in Figure 28, these ejections are concentrated near the middle plane of the heated wall. This yields a strong intensification of the secondary flow. It is also shown that the turbulent intensity is reduced near the heated wall with strong heating due to an increase of the viscous effect in that region.

10.2.2   Towards complex flow geometries

Several applications of LES to compressible flows in geometries of inductrial interest are presented in Lesieur and Comte [68] and Métais *et al.* [81]. We here briefly mentioned two of these applications.

The first application is a LES of the detached boundary layer over a curved compression ramp at Mach 2.5 modelling the wind-side region of the body-flap of HERMES during its projected re-entry. Note that the simulated Mach number is lower than for real situations. Indeed, the external Mach number relevant to the shuttle is about 10 (altitude 50 km, incidence 30°, flap extension angle $\alpha_0 = 20°$). The whole computational domain is

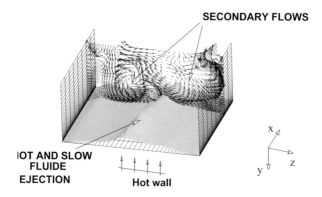

Fig. 28. Large scale motion over the hot wall in a heated duct ($T_h/T_w = 2.5$). Instantaneous transversal vector field and a isosurface of temperature ($T/T_w = 2.1$) (courtesy Salinas–Vasquez).

contained within the bow shock. The grid used is shown, upside down, in Figure 29. The simulation is therefore performed at the maximal Reynolds number permitted by our resolution, that is $Re_{\delta_i} = 280$, where δ_i is the upstream displacement thickness of the boundary layer. The first part of the boundary (up to 13.6 m away from the nose) is curved. It corresponds to the wind side of the body. The ramp corresponds to the body flap, assumed to be flat. For computational reasons, it is prolonged by a fictitious horizontal surface introducing a cut-off with the lee-side of the flap and the after-body. This enables the prescription of well-posed boundary conditions at the exit of the domain. The wall temperature is $T_w = 290$ K and the "external" (outside of the boundary layer, but inside the bow shock) temperature is $T_\infty = 460$ K. The adiabatic recovery temperature, defined by

$$T_{ad} = T_\infty \left(1 + \sqrt{Pr}\ \frac{\gamma - 1}{2}\ M_\infty^2 \right), \qquad (10.10)$$

is $T_{ad} = 1047$ K, yielding $T_w/T_{ad} = 0.277$. The ramp is therefore very cool with respect to the fluid, which models the radiative balance of the true shuttle during its re-entry.

Experimental evidence of (streamwise counter-rotating) Görtler vortices in a similar case was brought in particular by [101], but the consequence of these vortices on the wall heat flux has remained an open question. Figure 30 shows such Görtler vortices, obtained from a 3D simulation performed with the selective structure-function model in a domain of spanwise extension equal to $4.5\ \delta_i$. One clearly sees two large structures, cross-cuts of which show that each of them corresponds to a pair of counter-rotating Görtler

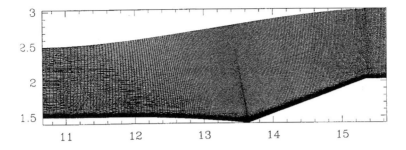

Fig. 29. Transverse section of the $220\times140\times25$ – point grid used for the simulation of the transition on a the curved ramp (angle $20°$). The axes are graded in metres, counted from the nose of the full-size shuttle. The spanwise size of the domain is 4.5 the displacement thickness δ_i of the boundary layer prescribed at the upstream boundary.

vortices. We have checked that the extreme values of the temperature fluctuations close to the wall associated with these structures are ±90 K. These 30% of temperature fluctuations induce huge fluctuations of the Stanton number (normalized heat-flux), between 2×10^{-3} and 14×10^{-3}, with an average of about 6×10^{-3}. The rms of the Stanton-number fluctuations is thus 133%. The same trend is observed for the skin-friction coefficient C_f. Note that LES is the only available tool able to reproduce such a strong variability for high-Reynolds number flows. Time-averaged plots also prove that the Görtler vortices are, in this simulation, fairly stable in time. This is likely to enhance considerably their destructive effects on the material of the body flap.

The second example is taken from the recent work by Dubief and Delcayre [30]. It consists in the LES of a transonic flow past a rectangular cavity. It required the implementation of domain decomposition in the COMPRESS code. The Reynolds number is here 1.25×10^6 based upon the external velocity and the depth of the cavity and the Mach number is 0.95. The Figure 31 displays, through the Q criterion, the vortices which are shed by the cavity. The main interest of this flow is related with aeroacoustical aspects and with the noise generated by the various eddies. It has been checked that the present LES is able to correctly reproduce the characteristics frequencies which are experimentally measured. This makes the LES a precious tool for aeroacoustics studies.

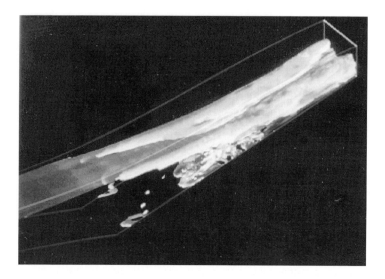

Fig. 30. Ramp flow. Zoom on the hinge and body-flap region showing an isosurface of the vorticity magnitude. This surface is coloured by temperature. This shows clearly that hot fluid in the outer part of the boundary layer is being downwashed to the wall, which brings about wall-heat-flux fluctuations.

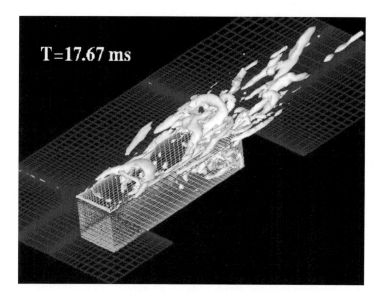

Fig. 31. LES of a transonic flow past a rectangular cavity at $Re = 1.25 \times 10^6$ and $Mach = 0.95$. White: vortices identified through the Q-criterion.

11 Conclusion

Turbulence plays a major role in the aerodynamics of cars, trains and planes, combustion in engines, acoustics, cooling of nuclear reactors, dispersion of pollution in the atmosphere and the oceans, or magnetic-field generation in planets and stars. Applications of turbulence, industrial in particular, are thus immense. Since the development of computers in the sixties, so-called industrial numerical models have been created. These models solve Reynolds ensemble-averaged equations of motions (RANS), and they require numerous empirical closure hypotheses which need to be adjusted on given particular experimentally-documented cases. RANS are widely used in the industry. However, it has become clear than RANS models suffer from a lack of universality and require specific adjustments when dealing with a flow submitted to such effects as separation, rotation, curvature, compressibility, or strong heat release.

Classical turbulence modelling, based on one-point closures and a statistical approach allow computation of mean quantities. In many cases, it is however necessary to have access to the fluctuating part of the turbulent fields such as the pollutant concentration or temperature: LES is then compulsory. Large-eddy simulations (LES) of turbulent flows are extremely powerful techniques consisting in the elimination of small scales by a proper low-pass filtering, and the formulation of evolution equations for the large scales. The latter have still an intense spatio-temporal variability. History of large-eddy simulations (LES) started also at the beginning of the sixties with the introduction of the famous Smagorinsky's [105] eddy viscosity. Due to the tremendous progress in scientific computing and in particular of parallel computing, LES, which were first confined to very simple flow configurations, are able to deal with more and more complex flows. We have here shown several examples of applications showing that LES are an invaluable tool to decipher the vortical structure of turbulence. Together with DNS, LES is then able to perform deterministic predictions (of flows containing coherent vortices, for instance) and to provide statistical information. The last is very important for assessing and improving one-point closure models, in particular for turbulent flows submitted to external forces (stratification, rotation, ...) or compressibility effects. The ability to deterministically capture the formation and ulterior evolution of coherent vortices and structures is very important for the fundamental understanding of turbulence and for designing efficient turbulent flow control.

The complexity of problems tackled by LES is continuously increasing, and this has nowadays a decisive impact on industrial modelling and flow control. Among the current challenges for LES in dealing with very complex geometries (like the flow around an entire car) are the development of efficient wall functions, the use of unstructured meshes and the use of adaptative meshes. However, the design of efficient industrial turbulence models will necessarily require an efficient coupling of LES and RANS techniques.

The results presented have greatly benefitted from the contributions of E. Briand, P. Comte, E. David, F. Delcayre, Y. Dubief, E. Garnier, E. Lamballais, M. Lesieur, M. Salinas-Vasquez, C. Silva, G. Silvestrini, G. Urbin. We are indebted to P. Begou for the computational support. Some of the computations were carried out at the IDRIS (Institut du Développement et des Ressources en Informatique Scientifique, Paris).

References

[1] A. Abbà, C. Cercignani, L. Valdettaro and P. Zanini, LES of turbulent thermal convection, in *Direct and Large Eddy Simulation II*, edited by J.P. Chollet, P.R. Voke and L. Kleiser (Kluwer Academic Publishers, 1997) pp. 147-156.

[2] K. Akselvoll and P. Moin, *Large Eddy Simulation of turbulent confined coannular jets and turbulent flow over a backward facing step*, TF 63 (Stanford University, 1995).

[3] J.C. André and M. Lesieur, *J. Fluid Mech.* **81** (1977) 187-207.

[4] R.A. Antonia, M. Teitel, J. Kim and L.W.B. Browne, *J. Fluid Mech.* **236** (1992) 579-605.

[5] C. Airiau, *Stabilité linéaire et faiblement non-linéaire d'une couche limite laminaire incompressible par un système d'équations parabolisé (PSE)*, Ph.D. Thesis (Toulouse University, 1994).

[6] M. Arnal and R. Friedrich, Large-eddy simulation of a turbulent flow with separation, in *Turbulent Shear Flows 8*, edited by F. Durst, R. Friedrich, B.E. Launder *et al.* (Springer, 1993) p. 169.

[7] G. Barenblatt and V.M. Prostokishin, *J. Fluid Mech.* **248** (1993) 513-529.

[8] J. Bardina, J.H. Ferziger and W.C. Reynolds, *Improved subgrid model for large-eddy simulation*, AIAA paper Nos. 80-1357 (1980).

[9] P. Bartello, O. Métais and M. Lesieur, *J. Fluid Mech.* **273** (1994) 1-29.

[10] C. Basdevant and R. Sadourny, *J. Mec. Theor. Appl.*, Numéro Spécial (1983) 243-269.

[11] L.P. Bernal and A. Roshko, *J. Fluid Mech.* **170** (1986) 499-525.

[12] P. Bertolotti and T. Herbert, *Theoret. and Comp. Fluids Dynamics* **3** (1991) 117-124.

[13] M. Brachet, *C.R. Acad. Sci. Paris II* **311** (1990) 775-780.

[14] E. Briand, *Dynamique des structures cohérentes en couche limite transitionnelle et turbulente étudiée par simulation des grandes échelles*, Ph.D. Thesis (National Polytechnic Institute, Grenoble, 1999).

[15] F.K. Browand and T.R. Troutt, *J. Fluid Mech.* **93** (1980) 325-336.

[16] G. Broze and F. Hussain, *J. Fluid Mech.* **311** (1996) 37-71.

[17] S. Cerutti, C. Meneveau and O.M. Knio, *J. Fluid Mech.* (2000) (in press).

[18] J.P. Chollet, *Turbulence tridimensionnelle isotrope : modélisation statistique des petites échelles et simulations numérique des grandes échelles*, Thèse de Doctorat d'État (Grenoble, 1984).

[19] J.P. Chollet and M. Lesieur, *J. Atmos. Sci.* **38** (1981) 2747-2757.

[20] J.P. Chollet and M. Lesieur, Modélisation sous maille des flux de quantité de mou-
      vement et de chaleur en turbulence tridimensionnelle isotrope, *La Météorologie*,
      29-30 (1982) pp. 183-191.

[21] R.A. Clark, J.H. Ferziger and W.C. Reynolds, *J. Fluid Mech.* **91** (1979) 1-16.

[22] P. Comte, Numerical methods for compressible flows, in *Computational Fluid
      Dynamics*, Les Houches Session LIX, 1993, edited by M. Lesieur, P. Comte and J.
      Zinn-Justin (Elsevier Science, 1996).

[23] P. Comte, Y. Fouillet and M. Lesieur, Simulation numérique des zones de mélange
      compressibles. *Revue scientifique et technique de la defense*, 3e trimestre (1992)
      pp. 43-63.

[24] P. Comte, M. Lesieur and E. Lamballais, *Phys. Fluids A* **4** (1992) 2761-2778.

[25] P. Comte, J.H. Silvestrini and P. Bégou, *Eur. J. Mech. B/Fluids* **17** (1998) 615-637.

[26] J. Cousteix, *Turbulence et couche limite*, CEPADUES (1989).

[27] E. David, *Modélisation des Écoulements Compressibles et Hypersoniques : une
      Approche Instationnaire*, Ph.D. Thesis (National Polytechnic Institute, Grenoble,
      1993).

[28] F. Delcayre, *Étude par simulation des grandes échelles d'un écoulement décollé :
      la marche descendante*, Ph.D. Thesis (National Polytechnic Institute, Grenoble,
      1999).

[29] J.A. Domaradzki, R.W. Metcalfe, R.S. Rogallo and J.J. Riley, *Phys. Rev. Lett.* **58**
      (1987) 547-550.

[30] Y. Dubief and F. Delcayre, *J. Turbulence* **1** (2000) 011.

[31] F. Ducros, *Simulation numérique directe et des grandes échelles de couches limites
      compressibles*, Ph.D. Thesis (National Polytechnic Institute, Grenoble, 1995).

[32] F. Ducros, P. Comte and M. Lesieur, Direct and large-eddy simulations of a super-
      sonic boundary layer, in *Selected Proceedings of Turbulent Shear Flows 9* (Springer,
      1995) pp. 283-300.

[33] F. Ducros, P. Comte and M. Lesieur, *J. Fluid Mech.* **326** (1996) 1-36.

[34] N.M. El-Hady and T.A. Zang, *Theoret. Comput. Fluid Dynamics* **7** (1995) 217-240.

[35] A. Favre A., *J. de Mécanique* **4** (1965) 361.

[36] Y. Fouillet, *Contribution à l'étude par expérimentale numérique des écoule-
      ments cisaillés libres. Effets de compressibilité*, Ph.D. Thesis (National Polytechnic
      Institute, Grenoble, 1992).

[37] E. Garnier, *Étude numérique des instabilités de jets beroclines*, Ph.D. Thesis (Natl.
      Polytech. Inst., Grenoble, 1996).

[38] E. Garnier, O. Métais and M. Lesieur, *J. Atmos. Sci.* **55** (1998) 1316-1335.

[39] E. Garnier, O. Métais and M. Lesieur, *C.R. Acad. Sci. Paris Sér. II B* **323** (1996)
      161-168.

[40] S. Gavrilakis, *J. Fluid Mech.* **244** (1986) 101.

[41] M. Germano, *J. Fluid Mech.* **238** (1992) 325-336.

[42] M. Germano, U. Piomelli, P. Moin and W. Cabot, *Phys. Fluids A* **3** (1991) 1760-
      1765.

[43] M.A. Gonze, *Simulation Numérique des Sillages en Transition à la Turbulence*,
      Ph.D. Thesis (National Polytechnic Institute, Grenoble, 1993).

[44] D. Gottlieb and E. Turkel, *Math. Comp.* **30** (1976) 703.

[45] T. Herbert, *Ann. Rev. Fluid Mech.* **20** (1988) 487-526.

[46] J.C.R. Hunt, A.A. Wray and P. Moin, Eddies, stream, and convergence zones in
      turbulent flows. Center for Turbulence Research Rep., **CTR-S88** (1988) p. 193.

[47] J. Jimenez and A.A. Wray, *J. Fluid Mech.* **373** (1998) 255-285.

[48] J.P. Johnston, R.M. Halleen and D.K. Lezius, *J. Fluid Mech.* **56** (1972) 533-557.

[49] S. Jovic and M. Driver, *Backward-facing step measurement at low Reynolds number* Re_h = 5000, Ames Research Center, NASA Technical Memorandum, 108807 (1994).

[50] J. Kim, The effect of rotation on turbulence structure, in *Proc. 4th Symp. on Turbulent Shear Flows, Karlsruhe*, (1983) pp. 6.14-6.19.

[51] P.S. Klebanoff, K.D. Tidstrom and L.M. Sargent, *J. Fluid Mech.* **12** (1962) 1-34.

[52] S.J. Kline, W.C. Reynolds, F.A. Schraub and P.W. Runstadler, *J. Fluid Mech.* **30** (1967) 741-773.

[53] A.N. Kolmogorov, *Dokl. Akad. Nauk. SSSR* **30** (1941) 301-305.

[54] R.H. Kraichnan, *J. Atmos. Sci.* **33** (1976) 1521-1536.

[55] R. Kristoffersen and H.I. Andersson, *J. Fluid Mech.* **256** (1993) 163-197.

[56] A. Kuroda, *Direct-numerical simulation of Couette-Poiseuille flows*, Ph.D. Thesis (University of Tokyo, 1990).

[57] E. Lamballais, *Simulations numériques de la turbulence dans un canal plan tournant*, Ph.D. Thesis (National Polytechnic Institute, Grenoble, 1996).

[58] E. Lamballais, M. Lesieur and O. Métais, *Int. J. Heat and Fluid Flow* **17** (1996) 324-332.

[59] E. Lamballais, M. Lesieur and O. Métais, *C. R. Acad. Sci. Sér. II B* **323** (1996) 95-101.

[60] E. Lamballais, O. Métais and M. Lesieur, Influence of a spanwise rotation upon the coherent-structure dynamics in a turbulent channel flow, in *Direct and Large Eddy Simulation II*, edited by J.P. Chollet, P.R. Voke and L. Kleiser (Kluwer Academic Publishers, 1996) pp. 225-236.

[61] E. Lamballais, O. Métais and M. Lesieur, *Theoret. Comput. Fluid Dynamics* **12** (1998) 149-177.

[62] H. Le, P. Moin and J. Kim, *J. Fluid Mech.* **330** (1997) 349-374.

[63] M. Lee and W.C. Reynolds, Bifurcating and blooming jets at high Reynolds number, in *Fifth Symp. on Turbulent Shear Flows* (Ithaca, New York, 1985) pp. 1.7-1.12.

[64] S.K. Lele, *J. Comput. Phys.* **103** (1992) 16-42.

[65] A. Leonard, *Adv. Geophys. A* **18** (1974) 237-248.

[66] D.C. Leslie and G.L. Quarini, *J. Fluid Mech.* **91** (1979) 65-91.

[67] M. Lesieur, *Turbulence in Fluids, Third Revised and Enlarged Edition* (Kluwer Academic Publishers, Dordrecht, 1997).

[68] M. Lesieur and P. Comte, Large-eddy simulations of compressible turbulent flows, AGARD-VKI course *Turbulence in compressible flows* (Belgique, 2-5 June and USA 20-24 October, AGARD report 819, 1997) ISBN 92-836-1057-1.

[69] M. Lesieur and O. Métais, *Annu. Rev. Fluid Mech.* **28** (1996) 45-82.

[70] M. Lesieur, O. Métais and E. Garnier, *J. Turbulence* **1** (2000) 002.

[71] M. Lesieur and R. Rogallo, *Phys. Fluids A* **1** (1989) 718-722.

[72] M. Lesieur and D. Schertzer, *J. Mécanique* **17** (1978) 609-646.

[73] M. Lesieur, S. Yanase and O. Métais, *Phys. Fluids A* **3** (1991) 403-407.

[74] D. Liepmann and M. Gharib, *J. Fluid Mech.* **245** (1992) 643-668.

[75] D.K. Lilly, in *Lecture Notes on Turbulence*, edited by J.R. Herring and J.C. McWilliams (World Scientific, 1987) pp. 171-218.

[76] D.K. Lilly, *Phys. Fluids A* **4** (1992) 633-635.

[77] S. Liu, C. Meneveau and J. Katz, *J. Fluid Mech.* **275** (1994) 83-119.

[78] O. Métais, C. Flores, S. Yanase, J.J. Riley and M. Lesieur, *J. Fluid Mech.* **293** (1995) 41-80.

[79] O. Métais and M. Lesieur, *J. Atmos. Sci.* **43** (1986) 857-870.

[80] O. Métais and M. Lesieur, *J. Fluid Mech.* **239** (1992) 157-194.

[81] O. Métais, M. Lesieur and P. Comte, Large-eddy simulations of incompressible and compressible turbulence, in *Transition, Turbulence and Combustion Modelling*, edited by A. Hanifi *et al.*, ERCOFTAC Series (Kluwer Academic Publishers, 1999) pp. 349-419.

[82] C. Meneveau and J. Katz, *Annu. Rev. Fluid Mech.* **32** (2000) 1-32.

[83] A. Michalke and G. Hermann, *J. Fluid Mech.* **114** (1982) 343-359.

[84] Y. Miyake and T. Kajishima, *Bull. JSME* **29** (1986) 3347-3351.

[85] P. Moin and J. Kim, *J. Fluid Mech.* **118** (1982) 341-377.

[86] K. Nakabayashi and O. Kitoh, *J. Fluid Mech.* **315** (1996) 1-29.

[87] X. Normand, *Transition à la turbulence dans les écoulements cisaillés libres et pariétaux*, Ph.D. Thesis (National Polytechnic Institute, Grenoble, 1990).

[88] X. Normand and M. Lesieur, *Theor. and Comp. Fluid Dyn.* **3** (1992) 231-252.

[89] S.A. Orszag, *J. Fluid Mech.* **41** (1970) 363-386.

[90] S.A. Orszag, Statistical theory of turbulence, in *Fluid Dynamics 1973*, Les Houches Summer School of Theoretical Physics, edited by R. Balian and J.L. Peube (Gordon and Breach, 1977) pp. 237-374.

[91] J. Padilla–Barbosa and O. Métais, *J. Turbulence* **1** (2000) 009.

[92] R.A. Petersen, *J. Fluid Mech.* **89** (1978) 469-495.

[93] R.T. Pierrehumbert and S.E. Widnall, *J. Fluid Mech.* **114** (1982) 59-82.

[94] U. Piomelli, *Phys. Fluids A* **5** (1993) 1484-1490.

[95] U. Piomelli, W.H. Cabot, P. Moin and S. Lee, *Phys. Fluids A* **3** (1991) 1766-1771.

[96] U. Piomelli and J. Liu, *Phys. Fluids A* **7** (1995) 839-848.

[97] M. Rogers and R. Moser, *Phys. Fluids A* **6** (1994) 903.

[98] P. Sagaut, *Introduction à la simulation des grandes échelles pour les écoulements de fluide incompressible* (Springer-Verlag, 1998).

[99] M. Salinas–Vazquez, *Simulations des grandes échelles des écoulements turbulents dans les canaux de refroidissement des moteurs fusée*, Ph.D. Thesis (National Polytechnic Institute, Grenoble, 1999).

[100] M. Salinas–Vazquez and O. Métais, Large-eddy simulation of the turbulent flow in a heated square duct, in *Direct and Large Simulation III*, edited by P.R. Voke *et al.* (Kluwer Academic Publishers, 1999) pp. 13-24.

[101] G.S. Settles, T.J. Fitzpatrick and S.M. Bogdonoff, *AIAA J.* **17** (1979) 579-585.

[102] E.D. Siggia, *J. Fluid Mech.* **107** (1981) 375-406.

[103] C.B. da Silva, Ph.D. Thesis (National Polytechnic Institute, Grenoble, 2000).

[104] J.H. Silvestrini, *Simulation des grandes échelles des zones de mélange : application à la propulsion solide des lanceurs spatiaux*, Ph.D. Thesis (National Polytechnic Institute, Grenoble, 1996).

[105] J. Smagorinsky, *Mon. Weath. Rev.* **91** (1963) 99-164.

[106] P.R. Spalart, *J. Fluid Mech.* **187** (1988) 61-98.

[107] D.K. Tafti and S.P. Vanka, *Phys. Fluids A* **3** (1991) 642-656.

[108] G. Urbin, *Étude numérique par simulation des grandes échelles de la transition à la turbulence dans les jets*, Ph.D. Thesis (National Polytechnic Institute, Grenoble, 1998).

[109] G. Urbin and O. Métais, Large-eddy simulation of three-dimensional spatially-developing round jets, in *Direct and Large-Eddy Simulation II*, edited by J.P. Chollet, L. Kleiser and P.R. Voke (Kluwer Academic Publishers, 1997) pp. 35-46.

[110] G. Urbin, C. Brun and O. Métais, Large-eddy simulations of three-dimensional spatially evolving roud jets, in *11th symposium on Turbulent Shear Flows* (Grenoble, September 8-11, 1997) pp. 25-23/25-28.

[111] A. Vincent and M. Meneguzzi, *J. Fluid Mech.* **258** (1994) 245-254.

[112] S. Yanase, C. Flores, O. Métais and J.J. Riley, *Phys. Fluids A* **5** (1993) 2725-2737.

COURSE 4

STATISTICAL TURBULENCE MODELLING FOR THE COMPUTATION OF PHYSICALLY COMPLEX FLOWS

M.A. LESCHZINER

Department of Engineering,
Queen Mary & Westfield College,
University of London, Mile End Road,
London E1 4NS, U.K.

Contents

STATISTICAL TURBULENCE MODELLING FOR THE COMPUTATION OF PHYSICALLY COMPLEX FLOWS

M.A. Leschziner

Abstract

Turbulence modelling is the cornerstone of any Reynolds-averaged computational method for predicting turbulent flows. The nature and properties of the model used often dictate the method's predictive accuracy and breadth of applicability. While Large Eddy Simulation is beginning to be applied to practical flows, modelling will continue to be the dominant approach for complex industrial applications for many years to come. This article discusses the formulation and application of statistical turbulence models at various levels of complexity and rational foundation, placing particular emphasis on advanced non-linear eddy-viscosity models and second-moment closure. Alternative modelling approaches are reviewed and their rationale, formulation and properties are discussed. The predictive performance of major model forms are then illustrated by computational solutions for nine two-dimensional and three-dimensional flows, some incompressible and others compressible. These applications serve to demonstrate the sensitivity to modelling in complex strain and the predictive advantages that can be gained from attention to modelling and the use of advanced model forms.

1 Approaches to characterising turbulence

Turbulence modelling is a statistical artifact born out of the need to characterise, in quantitative terms, the effects of turbulence on practical flows, in the face of the impossibility of resolving the details of the chaotic motion that characterises the phenomenon. Turbulence, once established by strain-induced instability and a still ill-understood process of repeated bifurcation, was eloquently described by Richardson in 1922 [1] as follows:

Great whirls have little whirls that feed on their velocity and little whirls have lesser whirls and so on to viscosity.

While this sentence tells only a small part of the story, it captures a key aspect of turbulence: large-scale, energetic eddies are continuously generated as a consequence of strain-induced non-linear instabilities; these are then broken down into smaller eddies by mutual interaction – referred to as *vortex stretching*; once the eddies are sufficiently small, viscous dissipation sets in destroying these eddies. The above process is accompanied by a transfer of turbulence energy – referred to as the *energy cascade* – across the eddy-size range, terminating with an irreversible conversion of this energy into heat by viscous destruction.

Recourse to the concept of *turbulence energy* is a first reflection of the need to describe turbulence in statistical terms, based on time- or ensemble-averaging of the components of the turbulent velocity-fluctuation vector $u_i \equiv (u_1, u_2, u_3)$. With the averaging operation denoted by $\langle \rangle$, the turbulence energy arises as:

$$k \equiv \frac{1}{2} \left(\langle u_1^2 \rangle + \langle u_2^2 \rangle + \langle u_3{}^2 \rangle \right). \tag{1}$$

A second important statistical quantity is the rate at which this energy is dissipated, by viscous destruction; the precise form of this term – a correlation of gradients of turbulent velocity fluctuations – will be dealt with later. Here, we merely note that dissipation rate is characterised by the symbol ε. Accepting the notion that dissipation arises by the interaction of the smallest eddies with viscosity, a dissipative length scale, η – the *Kolmogorov scale* – may be derived from dimensional reasoning:

$$\eta = \left(\nu^3 / \varepsilon \right)^{0.25}. \tag{2}$$

In effect, it is at this eddy size that dissipation takes place.

The above process of energy transfer, associated with Richardson's description, is shown schematically in Figure 1, in the form of the spectral distribution of the turbulence energy across the wave-number range, k_l, which is inversely proportional to the eddy-length scale (L representing the large, energetic eddies and l the smallest eddies in the dissipative range). The *inertial subrange* is that range in which energy *cascades* down the size range as eddies are broken down by vortex stretching.

The fact that a turbulent flow is non-repeatably time-varying and involves motion of scales ranging from the overall size of the flow down to the Kolmogorov scale, with a time-scale range to match, poses the severe fundamental problem of how to describe it in quantitative terms. A seemingly straightforward approach is to solve numerically the Navier–Stokes equations in time and space, to obtain a full description of the velocity $U(x_i, t)$ and any other relevant flow properties. This approach is referred to *Direct Numerical Simulation* (DNS). While this has become possible over the past

Table 1. Numerical grid and approximate CPU requirements for DNS.

$Re =$	6600	2×10^4	10^5	10^6
$N =$	2×10^6	40×10^6	3×10^8	15×10^{12}
T at 150 MFlops $=$	37 h	740 h	6.5 y	3000 y
T at 1 TFlops $=$	20 s	400 s	8.3 h	4000 h

few year, with the advent of powerful super-computers, the resource implications are daunting for any but relatively simple flows at low Reynolds numbers. There is also the problem of how to handle the enormous data sets arising from DNS and how to process these in a form that allows useful statements to be derived.

Order-of-magnitude considerations can be used to demonstrate that the ratio of the largest-to-smallest eddy-length scales, L/η, present in a turbulent flow scales with the turbulent Reynolds number according to:

$$L/\eta \propto Re_{\mathrm{t}}^{3/4}. \tag{3}$$

This means that the number of nodes in the numerical grid needed to resolve all turbulence scales in a flow rises as $(L/\eta)^3 = O\left(Re_{\mathrm{t}}^{2.25}\right)$. Based on this deduction and the fact that the shortest time scale, corresponding to the Kolmogorov scale, that needs to be resolved is $\tau = (\nu/\varepsilon)^{0.5}$, an estimate of the computer-resources requirements for a fully-resolved simulation may be derived as a function of the mean-flow Reynolds number, assumed to be $O(10Re_{\mathrm{t}})$. This is given in Table 1 and demonstrates that DNS is untenable in practical, high Reynolds-number flows.

In reality, the resource problem is not quite as severe as implied by Table 1, for a simulation need not resolve scales lower than $O(5\eta)$. However, a dramatic as well as realistic illustration of the enormous resource requirements is provided by Moin and Kim [2] in an article in the Scientific American. They estimate that to fully simulate the flow around an airliner at cruise conditions would require at least 10^{16} nodes, and that the computation of one second of flight time would require several thousands of CPU years using a Teraflop computer. It is now generally accepted that DNS, even if possible one day for practical flows, will not be used beyond the scope of studying fundamental aspects of turbulence in relatively simple, low-Reynolds-number flows. The most useful outcome of such simulations is insight into the dynamics of turbulence, enabled by the

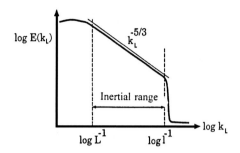

Fig. 1. Schematic representation of the spectral distribution of turbulence energy across the wave-number spectrum.

application of *structure-identification* methods, and data for statistical properties and correlations of turbulence (for example k as defined in (1)). The reader interested in DNS is referred to a recent review article by Moin and Mahesh [3].

Of particular practical interest are the statistical correlations of turbulent quantities that arise when the Navier–Stokes equations and any other transport equations for scalar properties (*e.g.* energy, species concentration) are time- or ensemble-averaged to give corresponding equations for the mean velocity and scalar properties. This type of averaging was first introduced, formally, by Osborne Reynolds in 1894 [4] and is generally referred to as *Reynolds averaging*[1]. Time-averaging is based on the decomposition[2],

$$\Phi(t) = \overline{\Phi} + \phi(t) \tag{4}$$

where Φ is any scalar flow property. Insertion of this decomposition into the parent transport equation for $\Phi(t)$ and time-averaging gives (for constant density and zero generation rate):

$$\frac{\partial \rho \overline{U}_j \overline{\Phi}}{\partial x_j} = \frac{\partial}{\partial x_j} \left(\Gamma_\Phi \frac{\partial \overline{\Phi}}{\partial x_j} \right) - \frac{\partial \overline{\rho u_j \phi}}{\partial x_j} \tag{5}$$

which is written here in Cartesian tensor notation and in which $\overline{\rho u_j \phi}$ are the *turbulent fluxes*. An analogous decomposition for velocity and pressure, insertion into the NS equations and time-averaging give rise to the set of

---

[1] Reynolds used ensemble-averaging. While there are certain fundamentally important differences between ensemble- and time-averaging, the formal result is the same, and both types of averaging are now collectively referred to as Reynolds-averaging.

[2] While this is attributed to Reynolds, Boussinesq was the first to at least imply it in 1872 (see Boussinesq [5]).

Reynolds stresses $\rho\overline{u_j u_j}$. Both sets of correlations represent the mixing action of turbulence, and their gradients may be interpreted as turbulence transport, akin to diffusive transport associated with Brownian motion in non-turbulent flow. From a practical point of view, these correlation are of critical importance, because they encapsulate the effects of turbulence on the time-averaged flow properties – momentum, temperature, concentration of chemical compounds etc., which are of primary interest to the engineer wishing to predict the mean operational characteristics of fluid-flow processes and related equipment.

While the solution of the Reynolds-averaged equations is a far more economical proposition than extracting the mean-flow quantities from simulations, the problem is that the turbulent stresses and fluxes are unknown, reflecting the loss of information associated with the averaging (integration) process. It is the determination of these stresses and fluxes from auxiliary equations that constitutes *turbulence modelling*, which is the subject of much of what is to follow. While turbulence modelling is necessarily an approximation process, in that it is simply impossible to devise a closed mathematical framework, the *holy grail* of turbulence modelling is to formulate a model that is applicable to a wide range of flow features – free and wall-bounded shear, separation, stagnation, swirl, density stratification, compressibility etc. – and which is sufficiently simple and robust to be used routinely in an industrial CFD environment. Few would claim that we are anywhere near this goal. All indications are, however, that the increased fundamental rigour and physical realism of a model goes hand-in-hand with greater mathematical complexity and numerical difficulties. Hence, turbulence modelling is and will probably always remain an exercise of compromise. A flavour of this tortuous path will be conveyed in what is to follow.

Before attention is turned to statistical modelling issues, mention ought to be made of *Large Eddy Simulation* – a *half-way house* between DNS and RANS. This technique (covered extensively in this book by Metais) is, in essence, a coarse-grid DNS, augmented by a statistical *sub-grid-scale* (SGS) model which aims to describe the turbulence processes that have been filtered out within the mesh intervals (cells) that are larger than the smallest turbulent scales. The underlying rationale is that all important turbulence-transport processes arise from the large-to-medium-size eddies, with the typical lower bound being of order 1% of the scale of the whole flow, while smaller eddies are principally responsible for dissipating the turbulence energy. The relative contribution of the SGS model depends on the Reynolds number. At increasingly large Reynolds numbers, the disparity between the largest and the smallest scales increases as well, according to (3). If the numerical mesh is kept invariant, a progressively wider range of scales needs to be represented by the SGS model, and the simulation is increasingly sensitive to modelling errors.

While LES shows considerable promise (for a review, see Lesieur and Metais [6]) and performs well when stochastic turbulence coexists with periodic coherent structures and in flows that are dominated by large-scale structures in free shear layers separating from sharp edges, it also faces a number of not unimportant limitations. Like DNS, it is resource-intensive, albeit to a far lesser extent. A typical simulation on a 1M-node mesh requires of order $5 - 10$ K CPU hours on a high-performance workstation processor, e.g. SGI R12000 or DEC Alpha, for time-integrated turbulence data to be obtained. Perhaps even more serious are resolution limitations that arise, especially in high Reynolds numbers, from the need to use coarse grids. Thus, in most circumstances, the subgrid-scale eddies are not simply dissipative, but can contribute significantly to turbulent mixing. Indeed, small-scale eddies can transfer energy to larger ones – a process termed *backscatter*. Moreover, the "large" eddies become small as any bounding wall is approached, requiring the use of extremely fine grids in this region. This does not merely aggravate the problem of high resource requirements, but can also provokes serious errors due to high levels of mesh aspect ratio, unless the grid is refined locally in all directions simultaneously as the wall is approached. This practice is possible, in principle, but poses challenges to accuracy and also means that the computational time steps between successive spatial realisations must be reduced in proportion to the degree of refinement to avoid instability and/or serious numerical inaccuracies. Yet another challenge facing LES is its reliance on spectral information as part of the boundary conditions at inflow boundaries of the simulation domain.

Two examples illustrating problematic flow conditions are shown in Figures 2 and 3. The former shows solutions by Temmerman and Leschziner [7] for a periodic segment of a channel with an infinite number of "hills" on the lower wall, which provoke separation from the leeward side of the hills. A favourable feature of streamwise periodicity is that no inlet-boundary conditions are required; the flow is driven by a pressure gradient adjusted so as to achieve the requisite flow rate. The flow is at the relatively low Reynolds number of 10 000, based on the mean velocity and hill height. One solution arises from a highly resolved simulation obtained with almost 5M nodes, in which SGS effects are insignificant and for which the viscous near-wall layer is resolved in detail. The other solution arises from a simulation with 0.66M – the kind of mesh density which might be considered tenable for practical simulations performed as an alternative to RANS computations. For this coarser grid, an approximate near-wall (*wall-function*) treatment was required because the wall-nearest grid plane was too far from the wall to resolve the viscous sublayer in sufficient detail. As seen, the solutions differ greatly, and this is a reflection of the strong sensitivity of the

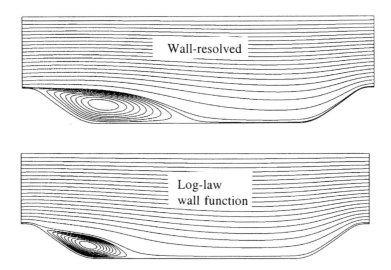

Fig. 2. LES solutions for a separated flow in a periodic channel segment with hill on the lower wall, by Temmerman and Leschziner [7]: nearly fully resolved simulation with 6M nodes vs. a simulation with 0.66M nodes and wall-function near-wall treatment.

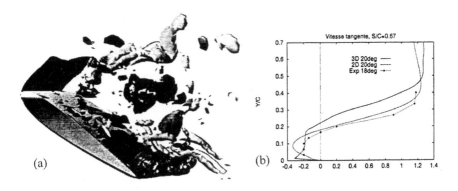

Fig. 3. LES solution of the flow around a NACA 12 aerofoil at $20°$ incidence, by Lardat [8]; (a) iso-vorticity surfaces; (b) velocity profiles above suction side.

simulation to the near-wall treatment, in combination with errors arising from the numerical approximation and the SGS model.

The second example, shown in Figure 3, is a separated flow around a NACA 12 aerofoil at $20°$ incidence and a chord Reynolds number of 10^5. The figure contains, on the left, an instantaneous realisation of the flow and, on the right, a comparison of time-averaged velocity profiles predicted by Lardat [8] relative to corresponding experimental data. An added problem in separated flows of the above type is the large disparity in the turbulence scales that must be resolved to achieve a credibly accurate simulation, even if the semi-viscous sublayer is bridged by some kind of semi-analytical *wall function*. Thus, to resolve the important streaky transverse structure of the boundary layer on the aerofoil's suction side requires a transverse grid distance of order 10^{-3} of the chord, while the large vortical structure ejected from the surface following separation might require a spanwise domain of $1 - 2$ chords. This is not economically tenable.

In summary then, Direct and Large Eddy Simulations are becoming increasingly important approaches to the computational characterisation of turbulence. The role of DNS is likely to remain one of being a tool for fundamental investigations. LES, in contrast, offers the prospect of some classes of flows being resolved in a practical environment, in preference to RANS approaches, which rely greatly on the quality of the turbulence model. However, in the large majority of practical circumstances, turbulence modelling within the RANS framework will remain, for some years to come, the principal approach to characterising turbulence effect. Therein lies the importance of the topic and the merit of continuing efforts to improve turbulence models.

2   Some basic statistical properties of turbulence and associated implications

In any turbulent flow subjected to strain, the turbulence structure is anisotropic – that is, the autocorrelations or kinematic normal stresses in the principal directions, $\overline{u_\alpha^2}$, are unequal (and always positive, *i.e. realisable*). Indeed, the cross-correlations or kinematic shear stresses, $\overline{u_i u_j}(i \neq j)$ cannot be finite in isotropic turbulence, *i.e.* when $\overline{u_\alpha^2}$ are equal. If, for the sake of argument, turbulence is assumed to be Gaussian[3], *the joint probability-density function* (PDF) for the fluctuations u_1 and u_2 has the typical shape shown in Figure 4. Isotropy gives rise to circular contours in the PDF, in which case u_1 and u_2 are uncorrelated and the shear stress $\overline{u_1 u_2}$ is zero. The level of anisotropy is especially high near walls and close to density jumps, because the fluctuations normal to the wall or jump are severely damped relative to fluctuations in other directions. This is well illustrated in Figure 5 which shows the variations of the turbulent intensities (rms values) in a channel flow. The damping of wall-normal fluctuations is responsible for the fact that near-wall flows spread considerably more slowly than free

---

[3]This is a gross oversimplication, in that it implies that higher moments are zero.

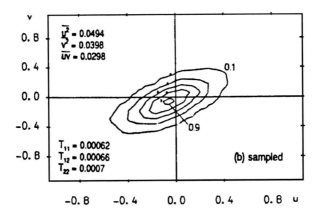

Fig. 4. Contours of Gaussian joint PDF representative of correlated u and v turbulent-velocity fluctuation.

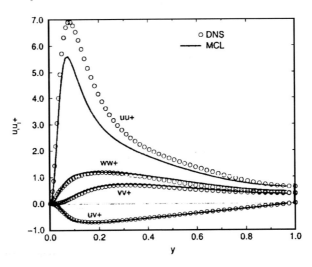

Fig. 5. Near-wall anisotropy of turbulence intensities in channel flow.

flows (jets, wakes), and this also implies that the shear stress is closely linked to the cross-flow turbulence intensity, normal to the shear strain.

A coordinate-independent, scalar representation of turbulence aniso-tropy

$$a_{ij} \equiv \frac{\overline{u_i u_j}}{k} - \frac{2}{3}\delta_{ij} \tag{6}$$

is necessary for general considerations, and this relies on the definition of

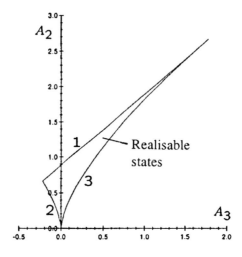

Fig. 6. Lumley's realisability map.

scalar *invariants* of the anisotropy tensor. Thus, the *second* and *third invariants* are, respectively:

$$A_2 = a_{ij}a_{ij}, A_3 = a_{ij}a_{jk}a_{ki}. \tag{7}$$

Lumley [9] has shown that particular functional relationships between A_2 and A_3 represent distinct physical states of turbulence. These are shown as lines in the $A_2 - A_3$ map in Figure 6.

The triangle in Figure 6 identifies *realisable* (*i.e.* realistic) states of turbulence, characterised by the positivity of the normal stresses, in any orientation, and the limit on the cross-correlations (*Schwarz inequality*),

$$\frac{\overline{u_i u_j}}{\sqrt{\overline{u_i^2}}\sqrt{\overline{u_j^2}}} < 1. \tag{8}$$

Adherence to the above realisability constraints is one important fundamental test of the quality of any turbulence model. The origin $A_2 = A_3 = 0$ represents isotropic turbulence, while the line defined by:

$$A = 1 - \frac{9}{8}(A_2 - A_3) = 0 \tag{9}$$

represents *two-component turbulence* – a state present at any wall or sharp density interface. The derived invariant A (sometimes called *flatness parameter*) characterises, in a qualitative sense, the flatness of eddies as turbulence moves from isotropy, $A = 1$ to the two-component limit $A = 0$.

The outermost vertex of the triangle represents the one-component limit. Finally, the two curved sides represent the change in conditions arising from isotropic turbulence being subjected to axi-symmetric compressive straining and axi-symmetric expansion, respectively.

The ability of a model to reproduce the correct response of the turbulent stresses to different types of straining is central to its fundamental strength and prospective generality. Hence, this response is often the subject of considerable attention, and newer modelling proposals are frequently formulated and calibrated by reference to tests in which turbulence is subjected to homogeneous shear, compression and expansion. This testing is substantially aided by DNS simulations which adhere closely to Lumley's criteria. Many models, especially those based on the *eddy-viscosity* concept, return a very poor response to normal straining. Such models were designed for flows which are dominated by a single shear stress and associated shear strain, as arising in thin shear flow. In such circumstances, normal straining is weak and the normal stresses are of sub-ordinate importance to the mean-flow behaviour. In complex strain, however, these models show serious defects – a topic considered later in detail. More advanced models, referred to as *second-moment closures*, are based on the solution of transport equations for all stress components, and these give a much better response to different strain types. Two such recent models are those of Jakirlić and Hanjalić [10] and Craft and Launder [11]. Figure 7 shows their response to different types of homogeneous straining, by reference to DNS data. Model solutions have been taken from Batten *et al.* [12] who have adapted and extended Craft and Launder's model to compressible flows. The plots show how the Reynolds stresses (a_{ij} is the normal-stress anisotropy) respond, in time t, when isotropic turbulence is subjected to the strain rate S of the relevant type. As seen, there are significant variations in model performance, even among models in the most complex *second-moment closure* category.

Apart from including substantial directional differences in turbulence intensity, strained turbulent flows feature significant structural properties which depend greatly on mean-flow (strain) characteristics and the proximity of boundaries. For example, eddies embedded in a flow impinging on a wall will tend to be deformed into "pancake"-like structures, while a shear flow along a wall will give rise to elongated "hair-pin"- or "sausage"-like structures. The level of anisotropy, characterised by the components of the Reynolds-stress tensor, might be similar in both types of flow, yet the turbulence structure will be different, and this difference must be expected to manifest itself by variations in the sensitivity of the all-important turbulent stresses to the strain components. The implication here is thus that a general turbulence model aught to be able to account in some way for structural features of turbulence. In particular, a distinction is needed

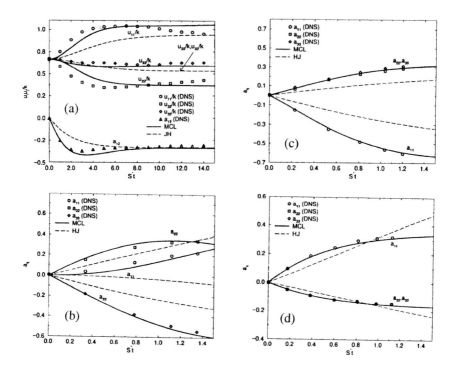

Fig. 7. Response of isotropic turbulence to homogeneous shear (a), plane strain (b), axisymmetric contraction (c) and axisymmetric expansion (d) as predicted by two second moment closure models relative to DNS data (taken from Batten *et al.* [12]).

between the *componentality* of turbulence, characterised by the normal-stress components, and its *dimensionality*, characterising the dimensional shape of coherent structures (Reynolds and Kassinos [13]).

Although the above remarks pertain, principally, to the large, energetic eddies, the componentality of the smallest eddies, associated with dissipation, might also require consideration. Away from walls, in fully-developed, high-Reynolds-number turbulence, the large separation in the scales of the large and small eddies means that the latter can be assumed to be isotropic in terms of both componentality and dimensionality. However, near walls and sharp fluid-fluid interfaces, this separation of scales is greatly reduced, and the severe anisotropy of the turbulent motion also affects the dissipative process. The implication is thus that dissipation cannot generally be described by a single scalar quantity, but must be regarded as a tensorial set equivalent to the set of Reynolds stresses. There is, as yet, no turbulence

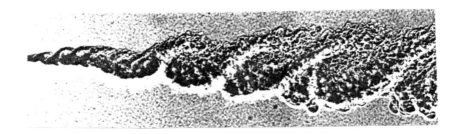

Fig. 8. Coherent motion super-imposed on random turbulence in a *young* shear layer (from Van Dyke, An Album of Fluid Motion, Courtesy Rebollo).

model which accounts for structural features of turbulence and is generally applicable to complex flows.

The decomposition (4) and subsequent time-averaging is based on the notion that the mean-flow is not a function of time and that the correlations arising from the averaging process represent the effects of the stochastic turbulent fluctuations. There are many turbulent flows, however, which involve organised, periodic (non-turbulent) motion, in addition to stochastic turbulence, and which cannot, therefore, be described in terms of conventional Reynolds averaging. Any shear layer with *young* turbulence, shortly after natural transition, contains significant coherent structures, as indicated in Figure 8. Most flows behind bluff bodies, at least in the near-wake portion, contain dominant periodic motions associated with the shedding of vortices at distinct frequencies. This is especially so when the body is long in the span-wise direction – say, a long cylinder or thick blade with a blunt or rounded trailing edge. An aerofoil in pitching motion or flutter can give rise to dynamic stall, with large time-dependent vortices separating from the suction side and giving rise to strong hysteresis effects in lift and drag. A turbine-blade row in a cascade sweeping across the wakes of the preceding row of stator blades will be subjected to strong periodic motion, dictated by the relative speed of the blade rows.

The most appropriate computational approach to such flows is by way of *Large Eddy Simulation*. A statistical framework may be acceptable as an alternative, but its adequacy greatly depends on the nature of the coherent motion, especially its time scale relative to that of the truly turbulent motion, and the nature of the closure model.

In a flow which contains a periodic component, the appropriate decomposition is:

$$\Phi(t) = \bar{\Phi} + \tilde{\Phi}(t) + \phi(t) \tag{10}$$

$$\langle \Phi(t) \rangle = \bar{\Phi} + \tilde{\Phi}(t) \tag{11}$$

where the second component is the periodic fluctuation. The sum (11) is the ensemble or phase-averaged value which varies periodically with time. There are now two optional routes. One involves the substitution of the *triple decomposition* (10) into the NS equation and time-averaging. This results in equations containing various correlations of turbulent and periodic fluctuations $\left(\overline{u_i u_j}, \overline{\tilde{U}_i \tilde{U}_j}, \overline{\tilde{U}_i u_j} ... \right)$, the first of which is the familiar set of Reynolds stresses. This framework is much more complex than conventional Reynolds averaging, and no suitable closure exists at present to approximate the correlations involving the periodic fluctuations, although pertinent proposals have been made by Hussain and Reynolds [14] and Drikakis [15]. The other route involves a substitution of (10) and (11) into the NS equations and time-averaging (with the rate-of-change term $\partial \langle \Phi \rangle / \partial t$ retained) over a period significantly larger than the turbulent time scale and significantly smaller than the time scale associated with the periodic fluctuation $\langle \Phi \rangle$. This then gives rise to the conventional form of the RANS equations, except that the subjects of the equations are the ensemble-averaged, time-dependent properties $\langle \Phi \rangle$. This is the usual modelling framework used for computing time-dependent turbulent flows (as opposed to *Large Eddy Simulation!*), with conventional closures providing the (time-dependent) Reynolds stresses. To a large extent, this approach relies on the closure assumptions, derived for and calibrated by reference to steady flow, carrying over to unsteady conditions. The likelihood of this being an adequate framework increases with decreasing frequency of the coherent motion and with the mathematical rigour with which stress-transport is represented. A related *Semi-Deterministic-Modelling* approach advocated by Haminh and Kourta [16] is claimed to "filter out" the dominant periodic motion and to includes only the stochastic element in the modelling framework. In reality, however, this approach is closely akin to that based on the conventional phase-averaged formulation.

We finally return to the fact that turbulent eddies occupy a wide range of length scales and that the turbulence energy is, correspondingly, spread over a wide wave-number spectrum (Fig. 1). This fact suggests that it might be beneficial or even necessary to devise closures that devide the turbulence energy and turbulent stresses into several spectral segments or contributions, each govered by its own closure relations and each linked to its spectral neighbours by transfer functions. This idea has been pursued (though not vigorously) for many years (*e.g.* Schiestel [17] and Wilcox [18]), but has not yielded demostrably superior closure forms, mainly because of the difficulty of devising general models linking the spectral partitions.

3 Review of "simple" modelling approaches

3.1 The eddy-viscosity concept

The key task of a turbulence model is to relate the unknown Reynolds stresses and/or scalar fluxes to known or determinable flow properties. The large majority of models used in practical computational schemes for engineering flows are based on the linear Boussinesq stress-strain relationships:

$$-\overline{u_i u_j} = \nu_t \left(\frac{\partial U_i}{\partial x_j} + \frac{\partial U_j}{\partial x_i} \right) - \frac{1}{3} \overline{u_k u_k} \delta_{ij} \tag{12}$$

in conjunction with the isotropic (kinematic) eddy viscosity which, on dimensional grounds, must be of the form:

$$\nu_t \propto \text{turbulent velocity scale} \times \text{turbulent length scale.} \tag{13}$$

The designation "simple" arises from the simplicity of the linear relationships (12) for physical interactions which are known to be non-linear and much more complex than the linkage represented by equation (12).

Equation (12) may be viewed as a generalisation of the thin-shear-flow relation:

$$-\overline{uv} = \nu_t \frac{\partial U}{\partial y} \cdot \tag{14}$$

Although equation (14) was first proposed on the basis of the analogy with the viscous process, a justification on physical grounds is provided by the observation that the exact rate of production of the shear stress in thin shear flow $U(y)$ (see Sect. 4) is:

$$P_{\overline{uv}} = -\overline{v^2} \frac{\partial U}{\partial y} \cdot \tag{15}$$

To a first approximation, it may be supposed that the shear stress is proportional to its rate of generation multiplied by a turbulent time scale[4]:

$$-\overline{uv} \propto \overline{v^2} \frac{\partial U}{\partial y} \times \frac{k}{\varepsilon} \tag{16}$$

where k is the turbulence energy and ε is its dissipation rate. It is evident that the above may be interpreted as an eddy-viscosity relation with:

$$\nu_t \propto \overline{v^2} \frac{k}{\varepsilon} \cdot \tag{17}$$

---

[4]...the banking analogy being that wealth is proportional to interest rate multiplied by the period over which saving takes place.

As an aside, it is noted that use of

$$\overline{v^2} = f\,k \tag{18}$$

(with f being an appropriately calibrated function) makes equation (17) part of the widely-used $k - \varepsilon$ model of turbulence, wherein k and ε are determined from their own transport equations. The fact that the normal stress, $\overline{v^2}$ occurs in equation (16) is a first indication of the importance of resolving the anisotropy of turbulence. While the relation (18) appears to circumvent this requirement, its usefulness hinges on the form of f, and it is illusory to assume that this can be determined empirically for general flow conditions.

It is pertinent to point out here that equation (12) predicts a state of isotropy, $\overline{u_i^2} = 2/3k$, in a fully-developed channel flow or slowly developing boundary layer, while experiments show that the maximum wall-normal intensity is only about 30% of that in the streamwise direction. This defect reflects the fundamentally flawed relationship established by equation (12) between strain and stress. As will be shown later, the production rates of the normal stresses in simple shear $U(y)$ are:

$$P_{\overline{u^2}} = -2\overline{uv}\,\frac{\partial U}{\partial y}, \qquad P_{\overline{v^2}} = 0. \tag{19}$$

This demonstrates, with striking clarity, that the streamwise normal stress is driven by the shear strain, rather than – as is implied in (12) – by the normal strain $\partial U/\partial x$. There is no generation of the transverse normal stress $\overline{v^2}$, which explains, in part, why it is lower than $\overline{u^2}$. In addition, near a wall, v-fluctuations are damped much more strongly than u-fluctuations by the blockage effect of the wall, which adds to the disparity. That $\overline{v^2}$ is at all finite, despite zero generation, is due to an energy-redistribution process which will be considered in Section 4.

3.2   Model categories

The $k - \varepsilon$ model (or rather group of models), mentioned in the previous section, is but one of numerous approaches to determining the eddy viscosity, and will be considered later in more detail. In fact, this type of closure – based on the solution of related evolution equations – is the most advanced currently used for modelling most industrial flows in aero-mechanical engineering. At the same time, it is the minimum level of closure offering any scope for generally in complex flow conditions involving complex strain and separation. Yet, much simpler algebraic, mixing-length-type models are still in widespread use. Among them, the models of Baldwin and Lomax [19] and Cebeci and Smith [20] are the most popular, especially in external aero-dynamics and in turbomachine flows. The range of applicability of algebraic models is essentially limited to fully turbulent boundary layers in weak pressure gradient (*i.e.* close to turbulence equilibrium). They display serious weaknesses in adverse pressure gradient, and do not apply to separated flow. Their inability to account for non-equilibrium and history effects also make them unsuitable for modelling unsteady flows, unless extended to at least include a differential equation for the streamwise evolution of the maximum shear stress across the boundary layer. Such an extension has been pursued by Hytopoulos *et al.* [21], leaning on elements of the Johnson and King model [22] for steady attached and separated boundary layers.

The next category comprises *one-equation models* which include a transport equation for the turbulence energy – the square of the turbulent velocity scale – and an algebraic equation for the length scale in equation (13). Typical representatives of this group are the models of Wolfshtein [23] and Norris and Reynolds [24]. Other variants, such as those of Baldwin and Barth [25] and Spalart and Allmaras [26], are based a transport equation for the turbulent viscosity itself, derived *via* the dependence expressed in equation (13) and assumed variations for the length scale (usually in relation to the distance from the nearest wall). These models are applied predominantly in computations of external aerodynamic flows, and offer the advantage of accounting for some aspects of history in non-equilibrium flows, such as accelerating and decelerating boundary layers. However, like algebraic model, one-equation formulations are restricted to attached flows, do not take important history effects on the turbulent length scale into account and rely on explicit transition models. These restrictions have discouraged their use in conditions in which algebraic models can be made to work adequately, if carefully calibrated. An exception to this general rule or trend has been the use of one-equation models for resolving the thin semi-viscous near-wall layer in calculations employing so-called *high-Re two-equation models* in the fully-turbulent field away from the wall $(y^+ > 50)$. The principal motivation for this hybrid-modelling approach lies in the facts that *low-Re two-equation models*, applicable right down to the wall, are relatively complex, require especially dense near-wall grids and tend to return excessive levels of near-wall length scale (and hence turbulent viscosity) in adverse pressure gradient. There are several computational studies employing such hybrid models, among them those of Franke and Rodi [27] and Cho *et al.* [28]. One-equation models have also been used in conjunction with *high-Re Reynolds-stress models* – for example, by Leschziner *et al.* [29], Lien and Leschziner [30,31] and Apsley *et al.* [32].

Two-equation models may, as already claimed, be regarded as the minimum level of closure offering any scope for generality in complex strain.

This is rooted in the fact that both scales to which the turbulent viscosity is related are determined from transport equations which account for history effects and which relate the scales to local conditions as they evolve in the computational solution. Although displaying important predictive defects in complex strain fields, as will be demonstrated later, two-equation models apply, in principle, to separated flows and are capable (if appropriately constructed) of resolving (to some degree) bypass transition and relaminarisation, without the use of specific transition models.

There are numerous two-equation-model variants documented in the literature. Most involve transport equations of the turbulence energy and rate of dissipation (e.g. Jones and Launder [33], Launder and Sharma [34], Lam and Bremhorst [35], Chien [36], Myong and Kasagi [37], Nagano and Hishida [38], Orszag et al. [39], Lien and Leschziner [40, 41]), but there are also a number of others in which the ratios k/ε (representing a time scale τ, Kalitzin et al. [42]) or ε/k (representing a turbulent vorticity ω, Wilcox [43, 44]) or k^2/ε (representing the turbulent Reynolds number, Goldberg [45]) or $\varepsilon/k^{0.5}$ (Gibson and Daffa'Alla [46]) are used as subjects of the length-scale-determining equation. There are even hybrid variants, such as that of Menter [47], which blends the $k - \varepsilon$ model in the outer flow with the $k - \omega$ model near the wall. The arguments for moving away from ε are not always transparent, but are justified, in some cases, on the grounds of gentler near-wall variation of the alternative length-scale parameters proposed, allowing more stable or economical numerical resolution, and a physically more realistic representation by the associated transport equation of the response of the length scale to adverse pressure gradients – a general problem area for $k - \varepsilon$ models. To a degree, the multiplicity of two-equation models betrays the frailties of the linear eddy-viscosity concept as a general approach to representing turbulence effects in complex flows: each model variant reflects yet another attempt to plug yet another hole in yet another class of flows. We will return to this topic later.

With attention restricted to the $k - \varepsilon$ framework, all model variants may be written collectively as:

$$\nu_{\text{t}} = c_\mu f_\mu \frac{k^2}{\tilde{\varepsilon}} \tag{20}$$

$$\frac{Dk}{Dt} = \frac{\partial}{\partial x_i} \left[(\nu + \nu_{\text{t}}) \frac{\partial k}{\partial x_i} \right] + P_k - \varepsilon \tag{21}$$

$$\frac{D\tilde{\varepsilon}}{Dt} = \frac{\partial}{\partial x_i} \left[\left(\nu + \frac{\nu_{\text{t}}}{\sigma_\varepsilon} \right) \frac{\partial \tilde{\varepsilon}}{\partial x_i} \right] + \frac{\tilde{\varepsilon}}{k} (c_{\varepsilon 1} P_k - c_{\varepsilon 2} \tilde{\varepsilon}) + E \tag{22}$$

with each variant having different forms for f_μ, E, $c_{\varepsilon 1}$ and $c_{\varepsilon 2}$. In the above, the tilde denotes the rate of homogeneous dissipation $\tilde{\varepsilon} = \varepsilon - 2\nu \left(\frac{\partial \sqrt{k}}{\partial x_i} \right)^2$

which is used (mainly for reasons of numerical stability) in some models in preference to the inhomogeneous ε which approaches a constant value at the wall. Typically, f_μ, E, $c_{\varepsilon 1}$ and $c_{\varepsilon 2}$ are made functions of the viscosity, *via* y^+ or $y^* = k^{0.5}y/\nu$ or $R_t = k^{3/2}/\varepsilon\nu$, on the basis of empirical observations and/or DNS data for low-Re near-wall flows (for a review of low-Re models, see Patel *et al.* [48]).

3.3 Model applicability

The large majority of turbulence models have been calibrated by reference to key steady flows and are expected, therefore, to apply to steady conditions. For example, $c_{\varepsilon 1}$ and $c_{\varepsilon 2}$ in equation (22) have been derived, respectively, by reference to experimental data for near-wall turbulence equilibrium in a steady, zero-pressure-gradient boundary layer and decaying grid turbulence. Most models thus perform reasonably well in flows which are not far removed from the range of calibration, namely attached boundary layers and other thin shear flows[5]. More or less serious weaknesses come to light, however, when the strain field becomes complex due to strong departure from equilibrium (acceleration/deceleration), strong curvature, impingement, separation, reattachment, swirl, strong 3D skewness etc. This will be discussed in detail in Sections 4–6 when models more elaborate than those based on the linear eddy-viscosity concept are considered.

Low-Re eddy-viscosity models which include transport equations for the velocity and length scales also perform reasonably well, up to certain limits, in accelerating and decelerating boundary layers. In fact, most predict relaminarisation in strong acceleration and bypass transition in simple boundary layers, although performance can vary greatly, depending on the precise closure approximations used. As an example, Figure 9, taken from Henkes *et al.* [49], shows a typical range of variability of model performance for a zero-pressure-gradient boundary layer subjected to 3% free-stream turbulence intensity. The standard $k - \varepsilon$ model does not apply to low-Re conditions and fails to represent transition in any shape or form. The other models, however, include viscosity-related closure approximations and are able to capture transition, albeit not well.

Of particular relevance to turbomachine and pitching-aerofoil aerodynamics is the adequacy of turbulence models for unsteady flow. It is instructive to consider first the model problem of a fully-developed pipe flow subjected to periodic variations in pressure gradient and hence mass flow. In laminar conditions, at all Reynolds numbers, the flow field depends only on

---

[5]Although even in this group problems are encountered. For example, two-equation models predict the round jet to spread faster than the plane jet, while the reverse is true in reality.

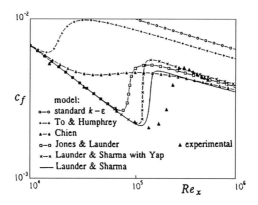

Fig. 9. Prediction of bypass transition in zero-pressure-gradient flat-plate boundary layer at 3% free-stream turbulence (taken from Henkes *et al.* [49]).

the Stokes parameter $\omega D^2/\nu$. This suggests that, in turbulent conditions, the structure will depend on $\omega D^2/\nu_t$. In steady flow, the turbulent viscosity rises in proportion to $U_b D$ (U_b is the bulk velocity). Hence, the flow will depend on $\omega D/U_b^*$ (U_b^* being the phase-averaged velocity), and it follows that the flow structure will depend on the amplitude of the motion. It is further known, also from steady-state experiments, that a flow subjected to strong acceleration tends to laminarise. The controlling criterion is

$$K_t = \frac{\nu}{U^3}\frac{dU}{dt} \geq 3 \times 10^{-6} \tag{23}$$

where U is, typically, the free-stream velocity in a boundary layer or the maximum velocity in a passage or pipe. Hence, the amplitude of the motion and its rate of change may cause periodic relaminarisation.

Returning to the Stokes parameter $\omega D^2/\nu_t$, we conclude that, since ν_t varies rapidly with position across the boundary layer (or radius), the effects of unsteadiness on the turbulent structure will vary significantly across the flow. A measure of the variation in the response of turbulence to periodic disturbances is provided by the ratio of time scales $\omega k/\varepsilon$. At the wall, k/ε is small (the asymptotic decay being quadratic), and then rises to a maximum by a factor of $100 - 500$. Hence, we expect to see significant variations in lag in the adjustment of the turbulence structure to mean-flow periodicity. In particular, near-wall turbulence responds faster to changes in the mean flow than does the turbulence in the outer part of the flow, because the time scale of turbulence is small, allowing turbulence to adjust itself quickly.

As ω increases, the turbulent stresses in the outer boundary-layer flow (or the flow in the centre of a pipe) fail to respond to increased rates of

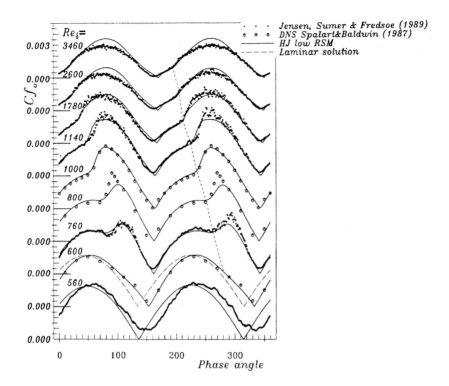

Fig. 10. Skin-friction in a sinusoidally oscillating boundary layer predicted with a low-Re Reynolds-stress-transport model (taken from Jakirlić [59]).

mean-flow changes $\partial U / \partial t$, and the stresses have a negligible effect on the phase-averaged mean flow. Ramaprian and Tu [50] define four levels of unsteadiness which relate to ω:

(i) Quasi-steady: the instantaneous state of the flow is the same as that of the steady flow at the same Reynolds number, ie the effect of unsteadiness on both the mean and turbulence fields is negligible. Turbulence models which do not take history and unsteadiness into account are expected to be adequate (provided convective effects associated with spatial rates of change are also small).

(ii) Slow: $\partial U / \partial t$ is significant, except near the wall. Thus, unless the boundary layer is subjected to strong pressure gradients, local equilibrium prevails at the wall. The rate of change of shear stress $\partial \overline{uv} / \partial t$ is small, and "steady-state" turbulence models which include some aspects of history effects (*e.g.* through Dk/Dt and $D\varepsilon/Dt$) are expected to apply.

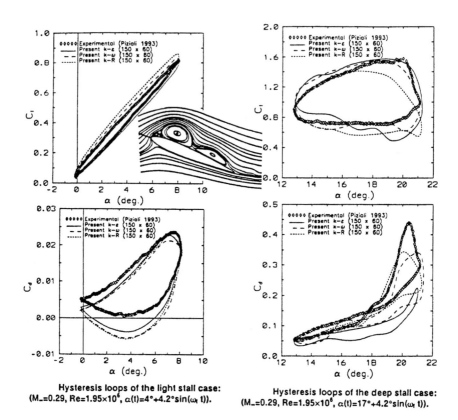

Fig. 11. Hysteresis of lift and drag in a dynamically-stalled pitching aero-foil, predicted by three low-Re eddy-viscosity models (taken from Barakos and Drikakis [61]).

(iii) Moderate: $\partial U/\partial t$ becomes important everywhere, including the low-Re region, and $\partial \overline{uv}/\partial t$ is significant (although the stresses are of little importance in the outer part of the boundary layer or in the core of the pipe). The law of the wall no longer applies, and turbulence models which do not fully account for history effects by including transport of stresses are likely to seriously misrepresent the turbulence structure.

(iv) Fast: turbulence affects only the near-wall sublayer. Conventional models are unlikely to give a realistic representation.

To represent unsteady interactions with a reasonable level of realism for *slow* processes, we require, as a minimum, the application of low-Re two-equation models, and there have been a number of studies which aimed to investigate the performance of such models, among them those of Jackson and He [51], Fan et al. [52] and Fan and Lakshminarayana [53].

A particularly interesting fundamental case is one in which an outer stream is made to oscillate, sinusoidally about a zero-velocity, parallel to a stationary flat plate according to $U_0 \cos(\omega t)$. This creates an boundary layer which oscillates from a positive to a negative velocity maximum, moving through periods of very low velocity. A DNS study of this flow was performed by Spalart and Baldwin [54], and corresponding experiments were performed by Jensen et al. [55]. A number of computational studies of this flow with different low-Re models were undertaken by Jutesen and Spalart [56], Haminh et al. [57], Shima [58] and Jakirlić [59], the last three with low-Re Reynolds-stress-transport models. Figure 10 is taken from Jakirlić [59] and shows predictions of a low-Re Reynolds-stress model for the variation of skin friction at different boundary-layer Reynolds numbers ($U_0 \delta / \nu$, where U_0 is the amplitude and δ is the Stokes thickness $(2\nu/\omega)^{0.5}$) over one oscillation period. As seen, the model returns a very credible representation of the skin friction at all Reynolds number, the lowest two corresponding, in effect, to laminar conditions prevailing at low U_0 and/or high ω values. A somewhat less satisfactory performance is returned by two low-Re two-equation models examined by Jutesen and Spalart [56].

A significant amount of information on the global performance of turbulence models in unsteady conditions has been derived from studies on dynamic stall (unsteady separation) from oscillating aerofoils at high incidence. This type of flow is, of course, considerably more complex than the unsteady boundary layers considered earlier. In particular, the separated flow shows substantial hysteresis effects, manifesting itself in drastically different lift and drag polars during the upward and downward strokes of the oscillating aerofoil. A further complicating feature is natural transition, associated with the low-turbulence approach stream and the laminar boundary layer preceding separation. Several studies, among them those by Guilmineau et al. [60], Barakos and Drikakis [61], Wu et al. [62], Srinivasan et al. [63] and Ekaterinaris and Menter [64] demonstrate that two-equation eddy-viscosity models are able to return a credible representation of the hysteresis of drag, lift and moments in attached conditions and light stall, while deep stall presents the models with much greater challenges. Figure 11, taken from Barakos and Drikakis [61], shows the hysteresis of lift and drag coefficients predicted with three transport models of turbulence for light-stall and deep-stall conditions in comparison to experiments by Piziali [65]. Unfortunately, dynamic-stall studies provide little information on fundamental issues and processes associated with the interaction of unsteadiness and turbulence.

The above review leads to the general conclusion that low-Re turbulence-transport models, which take history effects into account, are capable, in principle, of representing the interaction between mean-flow and the turbulence structure in boundary layers subjected to periodic oscillations. However, this applies only to relatively slow transients at low frequencies. Also, different models return different levels of predictive quality, and considerable care is therefore called for in extrapolating from observations arising from specific tests or applications.

4 Second-moment equations and implied stress–strain interactions

Models reviewed in Section 3 are based on the linear eddy-viscosity hypothesis (12). These perform well in relatively simple strain, but often do badly in the presence of high curvature, separation, recirculation, impingement and swirl. Predictive defects repeatedly observed in computations of complex flows include excessive shear stress, particularly in curved shear layers and in the presence of adverse pressure gradient; suppression of separation along curved walls; grossly excessive levels of turbulence in regions of stagnation and impingement; wrong response to swirl; insensitivity of turbulence transport to density stratification; grossly excessive heat transfer in reattachment regions; and suppression of periodic motions. The cause of many problems is that equation (12) gives a seriously erroneous linkage between the stresses and strain components and fails to represent the substantial directional misalignment of the stress and strain tensors. The physically correct linkage is implicit in the equations governing the evolution of the Reynolds stresses, to which attention is directed next.

Laborious but otherwise straightforward manipulations of the NS and RANS equations lead to the following exact transport equations for the (kinematic) Reynolds-stresses, applicable in the form given to incompressible flow:

$$\frac{D\overline{u_i u_j}}{Dt} = \underbrace{\left\{ \overline{u_i u_k}\frac{\partial U_j}{\partial x_k} + \overline{u_j u_k}\frac{\partial U_i}{\partial x_k} \right\}}_{P_{ij}} + \underbrace{\left(\overline{f_i u_j} + \overline{f_j u_i} \right)}_{F_{ij}}$$

Strain production   Body-force production

$$+ \underbrace{\overline{\frac{p}{\rho}\left(\frac{\partial u_i}{\partial x_j} + \frac{\partial u_j}{\partial x_i}\right)}}_{\Phi_{ij}} - \underbrace{2\nu \overline{\frac{\partial u_i}{\partial x_k}\frac{\partial u_j}{\partial x_k}}}_{\varepsilon_{ij}}$$

Redistribution   Dissipation   (24)

$$-\frac{\partial}{\partial x_k} \underbrace{\left[\overline{u_i u_j u_k} + \frac{\overline{pu_j}}{\rho}\delta_{ik} + \frac{\overline{pu_i}}{\rho}\delta_{jk} - \nu\frac{\partial \overline{u_i u_j}}{\partial x_k} \right]}_{d_{ij}}.$$

Diffusion

It is instructive to consider the interaction between some of the terms of equation (24) by reference to the simple example of a boundary layer in which, as shown in Figure 12, the normal stresses differ greatly among one another – an observation which the eddy-viscosity relations are entirely unable to represent.

For this flow, equation (24) reduce to the following (nearly) exact set[6]:

$$\frac{D\overline{uv}}{Dt} = -\overline{v^2}\frac{\partial U}{\partial y} \quad + \overline{\frac{p}{\rho}\left(\frac{\partial u}{\partial y}+\frac{\partial v}{\partial x}\right)} \quad - \frac{\partial}{\partial y}\left(\overline{uv^2}+\frac{\overline{pu}}{\rho}\right) \quad - \quad 0$$

$$\frac{D\overline{v^2}}{Dt} = 0 \quad + \overline{\frac{2p}{\rho}\frac{\partial v}{\partial y}} \quad - \frac{\partial}{\partial y}\left(\overline{v^3}+\frac{2\overline{pv}}{\rho}\right) \quad - \frac{2}{3}\varepsilon$$

$$\frac{D\overline{u^2}}{Dt} = -2\overline{uv}\frac{\partial U}{\partial y} + \overline{\frac{2p}{\rho}\frac{\partial u}{\partial x}} \quad - \frac{\partial}{\partial y}\overline{(u^2 v)} \quad - \frac{2}{3}\varepsilon \qquad (25)$$

$$\frac{D\overline{w^2}}{Dt} = 0 \quad + \overline{\frac{2p}{\rho}\frac{\partial w}{\partial z}} \quad - \frac{\partial}{\partial y}\overline{(w^2 v)} \quad - \frac{2}{3}\varepsilon$$

$$\underbrace{\qquad}_{P_{ij}} \qquad \underbrace{\qquad}_{\Phi_{ij}} \qquad \underbrace{\qquad}_{d_{ij}} \qquad \underbrace{\qquad}_{\varepsilon_{ij}\approx}$$

For enhanced clarity, $\overline{u_i u_j}$ have been replaced above by $\overline{u^2}$, $\overline{v^2}$, $\overline{w^2}$ and $\overline{uv}$. The "cycle of turbulence", shown in Figure 13, starts with the generation of streamwise stress $\overline{u^2}$ by the (exact!) production rate $2\overline{uv}\partial U/\partial y$. The other normal stresses $\overline{u^2}$ and $\overline{w^2}$ are not generated, but receive a proportion of the energy contained in $\overline{u^2}$ by a redistribution process effected through and interaction between pressure and velocity (strain) fluctuations and represented by Φ_{ij}. Although this redistribution steers turbulence towards isotropy and, consistently, reduces the magnitude of the shear stress $\overline{uv}$, anisotropy remains high. In particular, $\overline{v^2}$ is quite low due to inertial damping by the wall – a process which tends to steer turbulence towards a two-component state as the wall is approached (see Fig. 12).

All normal stresses are attenuated, at different rates close to the wall, by viscous dissipation, with turbulence energy being transferred to heat

---

[6]The dissipation rate is assumed to be isotropic, i.e. $\varepsilon_{ij} = 2/3\varepsilon\delta_{ij}$.

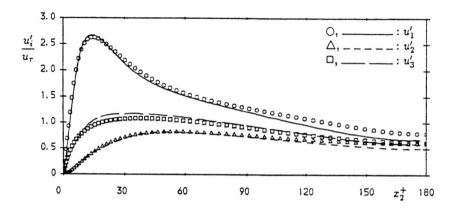

Fig. 12. Normal-stress anisotropy in the near-wall layer of a fully-developed channel flow (taken from Launder and Tselepidakis [66]).

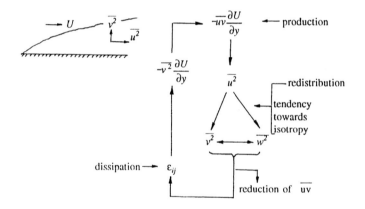

Fig. 13. The *cycle of turbulence* which sustains turbulence in a simple shear flow.

through ε_{ij}. The stress $\overline{v^2}$ plays a key role in the cycle by contributing to the generation of the shear stress uv through the (exact!) production term $-\overline{v^2}\partial U/\partial y$. This last link in the chain identifies one of the important challenges of turbulence modelling: even if stress production is represented exactly – as it is in the modelling framework advocated later – the level of the wall-normal stress is only weakly related to this production and is highly sensitive to the manner in which Φ_{ij} and ε_{ij} are approximated.

The above simple example highlighted two important facts: first, determining normal-stress anisotropy is of crucial importance to the correct representation of the generation of shear stress; second; the exact stress-production terms contains only strains and stresses, and this carries over to all flow conditions, however complex. The latter fact, coupled with the expectation that the level of any stress should be closely linked to the rate of its production, may be used to argue the merits of a modelling framework which is based directly on the set of equations (24) and which yields solutions for the individual Reynolds-stress components.

In view of the complexity of interactions inherent in equations (24) and the absence of any obvious connection to the eddy-viscosity concept, the question might be posed as to why eddy-viscosity models are observed to return a credible representation of a fair number of flows, especially those which fall into the category of thin shear flows. The answer is given, in effect, by equation (16) in Section 3.1 for the shear stress, together with the observation that, in the case of simple shear, this stress is the only one which has any relevance to the behaviour of the mean flow. However, this simple stress-strain relationship does not extend to complex strain fields and cannot be generalized in the form (12) without serious conflict with reality. A demonstration of the fact that equation (12) fails to represent the directional misalignment of the stress and strain tensors in complex conditions is given in Figure 14. This compares the directions of the Eigenvectors of the strain and stress-anisotropy tenors in the case of a flow across a sub-channel of a heat-exchanger tube-assembly, as computed by Archambeau and Laurence [67]. If relations (12) were to apply, the Eigenvectors of the strains and stresses would be aligned. However, closer inspection shows that this is far from valid in several regions of the sub-channel.

A key to understanding the relationship between stresses and strains and to identifying the origin of several phenomena which the eddy-viscosity concept is unable to represent is the exact stress-production terms P_{ij} arising in the stress-transport equations (24). Since the level of generation has, as one would expect, a dominant influence on the associated stresses, it follows that P_{ij} is crucial to the resolution of *stress anisotropy*. To gain some appreciation of the role of stress anisotropy and its importance to mean-field characteristics, it is instructive to consider the response of generation to different flow features or processes, each viewed in isolation. This is done below for the examples of near-wall shear, flow curvature, normal strain and heat transfer coupled with buoyancy.

4.1   Near-wall shear

It has been noted in Section 3.1 that the production of the shear stress in a near-wall layer implies that a sensible Ansatz for the shear stress is:

$$-\overline{u_1 u_2} \propto \overline{u_2^2}\frac{\partial U_1}{\partial x_2} \times \frac{k}{\varepsilon} \qquad (26)$$

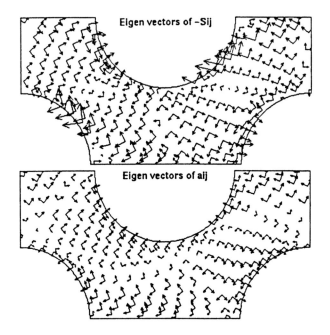

Fig. 14. Misalignment of Eigenvectors of strain and stress-anisotropy tensors in a separated flow across a tube-bank sub-channel (Archambeau and Laurence [67]).

which suggests that the decay of wall-normal intensity, turbulence energy (k) and turbulence dissipation (ε) govern the near-wall variation of the shear stress (if the interaction between turbulence and viscosity very close to the wall is ignored). With y denoting the distance from the wall, it is readily shown–on the basis of simple Taylor-series expansions of the form:

$$u_k = a(t)y + b(t)y^2 + c(t)y^3...\tag{27}$$

followed by their insertion into $\overline{u_i u_j}$, k and ε, the application of kinematic constraints and time-averaging–that ε tends to a constant value, while k and $\overline{u_2^2}$ decay quadratically and quartically, respectively; that is, $\overline{u_2^2}$ decays much faster than k. This demonstrates that turbulence tends towards a two-component state at the wall. In eddy-viscosity models, however, the turbulence energy k is used to represent the normal stress $\overline{u_2^2}$, which then necessitates the introduction of a Reynolds-number-dependent "damping function" multiplying k and compensating for the inconsistency between k and the wall-normal stress. Because this inconsistency arises purely from kinematic constraints, sensitising the shear stress to the viscosity (usually *via* the turbulent Reynolds number) in the manner indicated above is untenable on fundamental grounds and must be viewed as an artifact mimicking the effects of the strong wall-induced stress anisotropy.

4.2 Streamline curvature

Flow curvature coexistent with shear is encountered in the large majority of engineering flows. A boundary layers along a curved wall, a shear layer bordering a separation bubbles and a swirling shear flow are examples in which curvature induces significant secondary strain rates additional to the primary shear strain. Even if the curvature strain is weak, its effect on the turbulence structure can be profound; this interaction is shown below to be closely associated with anisotropy.

In 2D flow described within a Cartesian framework, the exact production of shear stress is given by:

$$P_{12} = -\overline{u_2^2}\frac{\partial U_1}{\partial x_2} - \overline{u_1^2}\frac{\partial U_2}{\partial x_1} . \tag{28}$$

Streamwise curvature is essentially expressed by the secondary strain $\partial U_2/\partial x_1$. In any shear layer, normal-stress anisotropy is high, since the only normal stress generated by shear is that aligned with the streamwise direction. In a wall-bounded shear layer, the wall-normal stress is only a small fraction (typically 25%) of the one in the streamwise direction. It is thus evident that curvature strain in a boundary layer has a disproportionately large influence on the level of shear-stress production and hence on the shear stress itself. In the case of a boundary layer on a convex wall, the curvature strain is negative, and the overall result is a considerable attenuation in the shear stress. This attenuation is further accentuated by the fact that convex curvature tends to reduce $\overline{u_2^2}$ relative to $\overline{u_1^2}$. This emerges from a consideration of the production of $\overline{u_2^2}$:

$$P_{22} = -2\overline{u_1 u_2}\frac{\partial U_2}{\partial x_1} . \tag{29}$$

Since, here, both the shear stress and curvature strain are negative, the result is a negative normal-stress production and hence a reduction in $\overline{u_2^2}$ (though this stress remains positive due to the transfer of energy from $\overline{u_1^2}$ effected by the pressure-strain-interaction process). Of course, this subtle interaction can only be captured by a model returning a realistic level of anisotropy, and its importance is obvious in the context of curved boundary layers approaching separation as well as free shear layers bordering regions of recirculation.

The impact which the interaction between curvature and turbulence can have is illustrated in Figure 15. This shows two computational solutions for

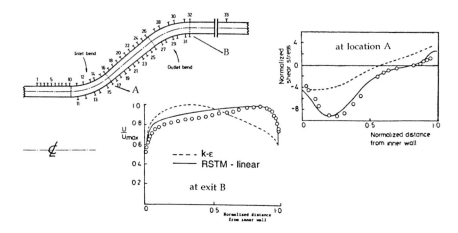

Fig. 15. Effects of curvature on the velocity and shear stress in an annular diffuser; from Jones and Manners [68].

the attached flow in an annular diffuser, one obtained with a linear k-ε eddy-viscosity model and the other with a Reynolds-stress transport model (see later), relative to experimental data. The computations have been performed by Jones and Manners [68]. As seen, the predicted velocity profile at the diffuser exit is strongly dependent on the turbulence model used. Only the Reynolds-stress model gives a realistic prediction of the velocity, and this reflects its ability to account for the interactions elucidated above. In particular, the shear-stress profiles in Figure 15, which relate to location "A" just downstream of the first bend, demonstrate that the origin of the effect on the velocity is the curvature-induced amplification of turbulence on the inner diffuser wall and its attenuation on the outer wall.

4.3 Separation and recirculating flow

A sharp corner will almost invariably induce separation, irrespective of how turbulence is modelled. A much more challenging form of separation is that provoked by a continuous adverse pressure gradient or a shock acting on a curved boundary layer – say, that developing on the suction side of an aerofoil or blade. In this case, the turbulence structure of the boundary layer, especially the distribution of the shear stress, will have a decisive effect on the point of separation. Hence the interaction between curvature and the stresses, identified in the previous section, is of considerable importance to the prediction of separation from continuous curved surfaces. This interaction is especially influential at walls because of the particularly large difference between the streamwise and wall-normal turbulence intensities, which increases the relative influence of the second, curvature-strain-related term in equation (28).

Curvature is also important in the separated, curved shear layer bordering any recirculation zone, whether emerging from a sharp corner or a continuous surface. There are two aspects to this interaction. First, curvature tends to diminish the shear stress in the shear layer, inhibiting entrainment into the layer and thus delaying reattachment. Second, since the typical computational mesh covering the recirculation zone in a general-flow RANS code is not generally aligned with the curved shear layer, the normal stresses (and hence the level of anisotropy) affect the predicted structure of the separated zone. For example, if a Cartesian decomposition of momentum is used in conjunction with a Cartesian coordinate system and numerical mesh to compute the recirculation zone behind a backward-facing step, the turbulent diffusion fluxes in the U- and V-momentum equations are given, respectively, by:

$$\text{Diff}_U = -\frac{\partial \overline{\rho u^2}}{\partial x} - \frac{\partial \overline{\rho uv}}{\partial y} \tag{30}$$

$$\text{Diff}_V = -\frac{\partial \overline{\rho v^2}}{\partial y} - \frac{\partial \overline{\rho uv}}{\partial x}. \tag{31}$$

This suggests that the gradients of the normal stresses might be as important as those of the shear stress – in contrast to a simple attached boundary layer which would normally be computed with a mesh following the solid boundary and in which, therefore, the only dynamically active term is the wall-normal gradient of the shear stress.

4.4   Rotation

System rotation gives rise a body force which interacts with turbulence so as to damp it in some parts of the flow and to amplify it in others, depending upon the orientation (sign) of the strain relative to the rotation vector. This interaction has obvious relevance to turbomachine aerodynamics.

It may be shown that rotation introduces into the Reynolds-stress-transport equations the additional exact body-form term:

$$P_{ij,\text{rot}} = -2\Omega_p \left(\varepsilon_{ipq} \overline{u_q u_j} + \varepsilon_{jpq} \overline{u_p u_j} \right) \tag{32}$$

where ε_{ipq} is the alternating third-rank unit tensor.

In the simple case of a fully developed flow in a channel rotating anticlockwise in orthogonal mode $\Omega_p = \Omega_3 = \Omega$, as shown in Figure 16,

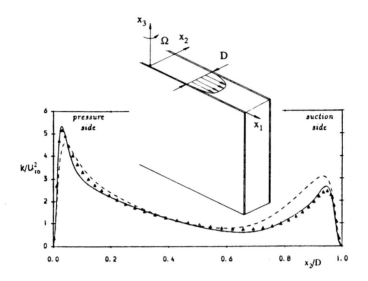

Fig. 16. Effects of rotation on turbulence energy in a plane, rotating channel flow; from Launder and Tselepidakis [66].

relation (32) (augmented by shear-induced production) simplifies to:

$$P_{22} + P_{22,\text{rot}} = -4\Omega \,\overline{uv}$$

$$P_{11} + P_{11,\text{rot}} = -2\overline{uv}\frac{\partial U}{\partial y} + 4\Omega \,\overline{uv} \tag{33}$$

$$P_{12} + P_{12,\text{rot}} = -\overline{v^2}\frac{\partial U}{\partial y} - 2\Omega \left(\overline{u^2} - \overline{v^2}\right).$$

As the shear stress is negative on the pressure side (the shear strain being positive) and positive on the suction side, relations (33) imply that rotation amplifies (destabilizes) turbulence on the pressure and damps it on the suction side, leading to an asymmetric velocity profile and higher level of skin friction on the former relative to the latter. This is precisely in accord with experimental observation and DNS data, Figure 16, and is a feature which no linear eddy-viscosity model is able to reproduce.

4.5 Irrotational strain

Turbulence responds quite differently to rotational (shear) and irrotational strain, the latter arising principally in impingement and reattachment zones, in shock waves and in compressed (*e.g.* IC engine) flows. This is of importance, for exemple, in relation to flow impingement on the leading edge of a turbine blade and can have dramatic consequences to the ability of a model to capture leading-edge separation and transition.

Eddy-viscosity models are *shear-strain oriented* in the sense that they have arisen from, and have been calibrated by reference to flows which are strongly sheared. Applying these models to flows in which compressive or extensive straining dominates tends to result in physically unrealistic behaviour. To illustrate this fact, attention is directed towards the production of turbulence energy $P_k = 0.5(P_{11} + P_{22} + P_{33})$:

$$P_k = -\overline{u_1^2}\frac{\partial U_1}{\partial x_1} - \overline{u_2^2}\frac{\partial U_2}{\partial x_2} - \overline{u_1 u_2}\left(\frac{\partial U_1}{\partial x_2} + \frac{\partial U_2}{\partial x_1}\right). \tag{34}$$

Because of the mass-continuity constraint (in incompressible flow), the first two terms, involving the normal stresses as multipliers, counteract each other. In fact, the production can easily become <u>negative</u> if the negative normal strain is multiplied by the higher of the two normal stresses, with isotropy being large. When the Boussinesq stress-strain relations are used, however – as is done in eddy-viscosity models – to remove the normal stresses, the associated terms involve quadratic strains and are additive rather than compensatory. Hence, eddy-viscosity models which include the turbulence-energy-transport equation tend to return excessive levels of energy and thus turbulence diffusion in the presence of strong compressive strains.

There are many manifestations of the above defect, among them the suppression of separation from the leading edge of aerofoils and turbomachine blades, the under-estimation of separation from sharp corners or edges of obstacles on which the flow impinges and the prediction of seriously excessive wall heat transfer at impingement points. Hence, the ability of a model to distinguish between the effects of shear and irrotational strain may be of crucial importance to the primary operational characteristics of fluid-flow equipment.

4.6   Heat transfer and stratification

In a plain shear layer featuring positive shear strain $\partial U_1/\partial x_2$ and positive temperature gradient $\partial \Phi/\partial x_2$, the exact production rates for the stresses and heat fluxes are as follows:

$$P_{11} = -2\overline{u_1 u_2}\frac{\partial U_1}{\partial x_2} \qquad\qquad P_{22} = 2g\beta\overline{u_2\phi}$$

$$P_{12} = -\overline{u_2^2}\frac{\partial U_1}{\partial x_2} + (\beta g\overline{u_2\phi}) \qquad\qquad P_{\overline{u_1\phi}} = -\overline{u_1^2}\frac{\partial \Phi}{\partial x_2} - \overline{u_2\phi}\frac{\partial U_1}{\partial x_2} \tag{35}$$

$$P_{\overline{u_2\phi}} = -\overline{u_2^2}\frac{\partial \Phi}{\partial x_2} + \left(\beta g\overline{\phi^2} - \overline{u_2\phi}\frac{\partial U_1}{\partial x_2}\right)$$

where β is the volumetric expansion coefficient. The above expressions identify a number of potentially important interactive mechanisms and processes
which are closely associated with anisotropy:

(i) the heat flux across the shear layer is very sensitive to the level of the
normal stress $\overline{u_2^2}$ and hence to the level of anisotropy;

(ii) the streamwise heat flux can be (indeed, usually *is*) larger than the
transverse flux, even if the temperature gradient in the streamwise
direction is zero; that is, Fourier's law has no validity;

(iii) stable temperature stratification results in negative production P_{22},
since $\overline{u_2\phi}$ is negative, and hence leads to a suppression of the shear
stress and lateral heat flux through the effect of P_{22} on $\overline{u_2^2}$. Analogous considerations for unstable stratification show that, in that case,
turbulence is amplified.

5   Second moment closure

The considerations in the previous section justify the claim that the most
rational approach to constructing a turbulence model intended to possess
wide-ranging applicability should proceed *via* the exact set (24), which is
the basis of *second-moment closure*. The term *closure* gives expression to
the fact that diffusion, redistribution and dissipation need to be approximated – a process substantially aided in recent years by a growing body
of accurate and detailed DNS data (Spalart [69], Spalart and Baldwin [54],
Kim *et al.* [70], Eggels *et al.* [71], Le and Moin [72]).

With attention initially restricted to high-Re flow regions, applicable
when the turbulent Reynolds number exceeds $O(100)$, stress diffusion (which
is rarely a dominant process) is usually approximated by the "Generalised
Gradient Diffusion Hypothesis" (GGDH) of Daly and Harlow [73]:

$$d_{ij} = \frac{\partial}{\partial x_k}\left(c_s \overline{u_k u_\ell}\frac{k}{\varepsilon}\frac{\partial \overline{u_i u_j}}{\partial x_\ell}\right). \tag{36}$$

More complex forms of (36) exist, but are not demonstrably superior, in
general, and have rarely been used. In the exact equations (24), the domi-
nant fragment is $\partial(\overline{u_i u_j u_k})/\partial x_k$, with pressure diffusion being sub-ordinate
(estimated by Lumley [9] at around 20%). Some attempts have thus been
made to determine the triple correlations from third-moment closure, but
this has not generally been found to be a profitable route, except in highly

stratified flow (Craft *et al.* [74]). A model for the fragment representing diffusion by pressure fluctuations has been used by Craft and Launder [11] within their "cubic" second-moment closure.

At high Reynolds numbers, dissipation is usually assumed to be isotropic, because it occurs at eddy length scales which tend to be very much smaller than the scales of the large energetic eddies which are sensitive to the mean strain (and hence to its orientation). Isotropy implies:

$$\varepsilon_{ij} = \frac{2}{3}\varepsilon\delta_{ij} \tag{37}$$

in which ε is the dissipation rate of turbulence energy. This approximation is inadequate close to the wall where length scales are generally small and anisotropy is large. Proposals have thus been made (*e.g.* by Launder and Reynolds [75] and Hanjalić and Jakirlić [76]) to sensitise, in an algebraic fashion, ε_{ij} to invariants of the stress anisotropy $a_{ij} = (\overline{u_i u_j}/k - 2/3\delta_{ij})$ or the equivalent invariants of the dissipation anisotropy $e_{ij} = (\varepsilon_{ij}/\varepsilon - 2/3\delta_{ij})$,

$$E_2 = e_{ij}\,e_{ij}; \quad E_3 = e_{ij}\,e_{jk}\,e_{ki}; \quad E = 1 - 9/8(E_2 - E_3). \tag{38}$$

A model widely used to represents the anisotropic dissipation is:

$$\varepsilon_{ij} = \frac{2}{3}f_\varepsilon\delta_{ij}\varepsilon + (1 - f_\varepsilon)\varepsilon_{ij}^* \tag{39}$$

where ε_{ij}^* are the wall-limiting values of ε_{ij},

$$\frac{\varepsilon_{11}}{\varepsilon} \to \frac{\overline{u_1^2}}{k}; \quad \frac{\varepsilon_{22}}{\varepsilon} \to 4\frac{\overline{u_2^2}}{k}; \quad \frac{\varepsilon_{33}}{\varepsilon} \to \frac{\overline{u_3^2}}{k}; \quad \frac{\varepsilon_{12}}{\varepsilon} \to 2\frac{\overline{u_1 u_2}}{k} \tag{40}$$

which can be readily obtained by kinematic arguments (Launder and Reynolds [75]). The "blending function" f_ε varies from model to model – there are at least five forms – and its subjects are A and/or R_t. Apart from securing the correct wall-limiting behaviour of ε_{ij} and introducing shear-stress dissipation, proposal (39) also ensures that dissipation of the wall-normal intensity is shut off as turbulence approaches the two-component near-wall limit. This is an important element of any model designed to satisfy *realizability*, a property which includes the unconditional satisfaction of $\overline{u_\alpha^2} \geq 0$.

Alongside dissipation, the redistribution or "pressure-strain" term Φ_{ij} presents the modeller with the biggest challenge in the context of second-moment closure. It can be shown, analytically, that this process consists of two major constituents, one involving an interaction between turbulent quantities only ($\Phi_{ij,1}$ and termed *slow*) and the other involving an interaction between mean strain and turbulence fluctuations ($\Phi_{ij,2}$ and termed *rapid*). This fact has led most modellers to make separate proposals for these two fragments. The simplest proposal forms, used for most complex-flow computations, are the linear relations by Rotta [77] and Gibson and Launder [78]:

$$\Phi_{ij1} = \frac{-c_1\varepsilon}{k}\left(\overline{u_iu_j} - \frac{1}{3}\delta_{ij}\,\overline{u_ku_k}\right)$$

$$\Phi_{ij2} = -c_2\left(P_{ij} - \frac{1}{3}\delta_{ij}P_{kk}\right). \tag{41}$$

Fu *et al.* [79] have shown, by reference to swirling flow, that the above form of Φ_{ij2} is not frame-invariant, but that invariance is assured if the body-force-related production terms F_{ij} and the convection tensor are included to give the form:

$$\Phi_{ij2} = -c_2\left(\left[P_{ij} + F_{ij} - \frac{1}{3}\delta_{ij}(P_{kk} + F_{kk})\right] - \left[C_{ij} - \frac{1}{3}\delta_{ij}C_{kk}\right]\right). \tag{42}$$

Although this combined linear model satisfies the basic requirement of steering turbulence towards isotropy, the isotropisation process is far too intense at the high levels of anisotropy prevailing near the wall. As the wall is approached, turbulence tends towards a two-component state $(A = 0)$, and redistribution must vanish to allow this state to be achieved. By intensifying the isotropisation process as the wall is approached, the linear model does not merely fail to represent the physical process correctly, but can lead to one of the principal normal stresses becoming negative – a condition violating realizability.

Correcting the above weakness, within the linear framework, relies invariably on the introduction of elaborate and influential *ad-hoc* terms (Shir [80], Gibson and Launder [78], Craft and Launder [81]) which counteract the isotropisation process in proportion to the distance from the wall, normalised by the turbulent length scale $k^{3/2}/\varepsilon$. For example, for a shear layer along a single horizontal wall, the correction terms damping the isotropisation of the wall-normal stress $\overline{v^2}$ are:

$$\Phi_{22,1}^{\mathrm{w}} = -c_1^{\mathrm{w}}\frac{4}{3}\frac{\varepsilon}{k}\,\overline{v^2}\,\frac{k^{1.5}/\varepsilon}{c_\mu^{-3/4}\kappa y}$$

$$\Phi_{22,2}^{\mathrm{w}} = 2c_2^{\mathrm{w}}\left(P_{22} - \frac{1}{3}P_{kk}\right)\frac{k^{1.5}/\varepsilon}{C_\mu^{-3/4}\kappa y}. \tag{43}$$

In addition, the redistribution process needs to be sensitized to anisotropy invariants, especially in low-*Re* forms which allow the model to be used down to the wall (So *et al.* [82], Shima [83], Ince *et al.* [84], Jakirlić and

Hanjalić [10]). An example of the latter practice is that of Jakirlić and Hanjalić who made extensive use of DNS data to calibrate their linear low-Re model and use:

$$c_1 = 2.5A \left[\min(0.6, A_2)\right]^{1/4} f + \sqrt{A}E^2$$

$$c_2 = 0.8A^{1/2} \tag{44}$$

where f is a function of the turbulent Reynolds number R_t. Similarly, c_1^w and c_2^w are sensitized to A, A_2 and R_t.

An alternative practice, recently proposed by Durbin [85], introduces the "elliptic relaxation equation",

$$L^2 \nabla^2 \frac{\Phi_{ij}^c}{k} - \frac{\Phi_{ij}^c}{k} = \frac{\Phi_{ij}}{k} \tag{45}$$

where Φ_{ij}^c is the wall-corrected form of the standard (uncorrected) Φ_{ij}, L is the turbulence length scale and ∇^2 is the elliptic operator. Equation (45) steers Φ_{ij} towards the correct wall values, prescribed as boundary conditions for Φ_{ij}^c. The rationale of this proposal is rooted in the observation that the effect of walls on Φ_{ij} occurs through pressure reflections, which is an elliptic mechanism propagated into the flow by the condition imposed on the boundary.

From a fundamental point of view, as well as on practical grounds, the use of wall corrections is unsatisfactory, not only because of their non-general nature, but also because they relies heavily on the wall distance (y in Eq. (43)). The latter is especially disadvantageous in complex geometries, where the influence of more than one wall needs to be taken into account, and when general non-orthogonal numerical grids are used. Hence, much of the recent fundamental research in the area of turbulence modelling has been concerned with constructing *non-linear* pressure-strain models which satisfy the realizability constrains and do not require wall corrections. Non-linear models or variants have been proposed by Shih and Lumley [86], Fu *et al.* [87] and Speziale *et al.* [88], Launder and Tselepidakis [66], Craft *et al.* [89] and Craft and Launder [11], the last three being extensions of Fu *et al.*'s model. The models differ in detail and in respect of the order of terms included, but all have arisen from the common approach of proposing non-linear expansions, in terms of components of the Reynolds-stress tensor $\overline{u_i u_j}$ (or rather the anisotropy tensor $a_{ij} = (\overline{u_i u_j}/k - 2/3\delta_{ij})$), to second- and fourth-rank tensors which arise in the most general Ansatz for the pressure-strain term, prior to its approximation:

$$\Phi_{ij} = \varepsilon A_{ij}(a_{ij}) + k M_{ijkl}(a_{ij}) \frac{\partial U_k}{\partial x_\ell} \tag{46}$$

with the two groups of terms corresponding, respectively, to $\Phi_{ij,1}$ and $\Phi_{ij,2}$. The coefficients of the various terms in the expansions for A_{ij} and M_{ijkl} are then determined by imposing necessary kinematic constraints (continuity, symmetry etc.). Realizability is introduced into some model forms by sensitising the pressure-strain model to invariants of the stress anisotropy. For example, the model of Speziale *et al.* [88] involves a quadratic form of $\Phi_{ij,1}$, the linear fragment of which is pre-multiplied by the coefficient:

$$c_1 = 1 + 3.1 \left(A_2 A\right)^{1/2} . \tag{47}$$

The most elaborate model is that of Craft and Launder [11] and is quadratic in $\Phi_{ij,1}$ and cubic in $\Phi_{ij,2}$, the latter containing six distinct groups of terms and associated coefficients. Unfortunately, the validity or usefulness of A as an indicator of the approach to the two-component limit turns out not to be as general as one might wish. For example, Batten *et al.* [12] have recently shown that Craft and Launder's model interprets shock waves as virtual walls, in so far as the model returns a value $A \sim 0$ as the shock is traversed, thus virtually shutting off the redistribution process and preventing turbulence from recovering downstream of the shock. Batten *et al.* have therefore modified Craft and Launder's model so as to ensure that the approach to the two-component limit is confined to the near-wall region where both A and turbulent Reynolds number R_t decline together. An application of this model to a 3D supersonic flow will be reported in Section 7.

It must be acknowledged that the above cubic forms continue to rely on wall corrections (or *"inhomogeneity"* terms), albeit much weaker than those associated with the linear models. To at least avoid reliance on the wall distance, efforts have thus been made to replace the wall-distance parameter in equation (43) by local turbulence-structure parameters which indicate the wall proximity by implication. Examples for such parameters are those proposed by Craft and Launder [11] and Jakirlić [59]:

$$f^{\mathrm{w}} = \frac{1}{c_l} \frac{\partial l}{\partial x_n}; \qquad f^{\mathrm{w}} = \frac{1}{c_l} \frac{\partial A^{1/2} l}{\partial x_n} \tag{48}$$

where $c_l = c_\mu^{-3/4} \kappa$, $l = k^{3/2}/\varepsilon$.

Determining the dissipation rate ε (and hence ε_{ij}) is another challenge in the context of second-moment closure. With few exceptions, ε is determined from a single transport equation representing, rather intuitively, a balance between transport, generation and destruction of dissipation:

$$\frac{\partial \rho U_k \varepsilon}{\partial x_k} = \frac{\partial}{\partial x_k} \left(\rho c_t \frac{\overline{u_k u_\ell}}{\varepsilon} k \frac{\partial \varepsilon}{\partial x_\ell} \right)$$

$$+ 0.5 \rho \frac{\varepsilon}{k} c_{\varepsilon 1} \left(P_{kk} + F_{kk}\right) - \rho c_{\varepsilon 2} \frac{\varepsilon^2}{k} + S_\varepsilon \tag{49}$$

in which S_ε is a model-dependent source-like term containing specific corrections and terms associated with the influence of viscosity on dissipation. Apart from associating the dissipation process with the single macro-length scale $k^{3/2}/\varepsilon$, the above equation suggests only a very weak sensitivity of dissipation to the structure of turbulence. As turbulence anisotropy increases, especially at walls, the normal components of the dissipation tensor also become anisotropic (as expressed by Eqs. (39) and (40)), and the scalar dissipation is altered. A proposal sensitising the scalar dissipation rate to anisotropy is that of Haroutunian et al. [90], in which $c_{\varepsilon 1} = 1.44$ and $c_{\varepsilon 2} = 1.92$ are replaced by:

$$c_{\varepsilon 1} = 1 \qquad c_{\varepsilon 2} = \frac{1.92}{\left(1 + 0.7 A A_2^{1/2}\right)} . \tag{50}$$

An apparently intractable defect of the dissipation-rate equation is that it returns excessive levels of turbulent length scale in boundary layers subjected to adverse pressure gradient. Among other problems, this property results in excessive near-wall shear stress and hence inappropriate suppression of separation from continuous surfaces. This defect is common to both two-equation eddy-viscosity and Reynolds-stress models. Within the former framework, Lien and Leschziner [40] have introduced constraints which drive, as the wall is approached, the length scale towards the value prescribed algebraically as part of the one-equation model of Norris and Reynolds [24]. However, in most model forms and applications the defect is addressed by introducing, via S_ε in equation (49), some variant of the ad-hoc correction of Yap [91], which forces the ε-equation to return a length scale close to the local-equilibrium value. An example is the recent variant of Jakirlić and Hanjalić [10]:

$$S_\varepsilon = \max\left(\left[\left(\frac{1}{c_l}\frac{\partial l}{\partial x_n}\right)^2 - 1\right]\left(\frac{1}{c_l}\frac{\partial l}{\partial x_n}\right)^2, 0\right) A\frac{\varepsilon\tilde{\varepsilon}}{k} \tag{51}$$

in which $\tilde{\varepsilon}$ is the homogeneous part of the dissipation ε.

Little has been said so far about accommodating the effects of viscosity in the context of low-Re modelling. Most recent models, among them those of Launder and Tselepidakis [66], So et al. [82], Shima [83], Jakirlić and Hanjalić [10] and Craft and Launder [11] are low-Re variants, allowing an integration through the viscous sublayer. This is an area in which much reliance is placed on recent DNS data for near-wall flows. In essence, different model elements, especially the dissipation equation (via $c_{\varepsilon 1}$, $c_{\varepsilon 2}$ and S_ε), are sensitized to viscosity by way of damping functions, with subjects being forms of the turbulent Reynolds number. As the near-wall structure is substantially affected by both inertial and viscous damping, the former provoking strong anisotropy *via* pressure reflections, low-*Re* extensions involve a functionalisation on anisotropy invariants (7) as well as viscosity, each expressing a different physical process. In fact, the dissipation invariants (38) can also be used, as has been done by Jakirlić and Hanjalić [10]. Because the functionalisation process is non-rigorous, essentially aiming to make the model return a phenomenological behaviour consistent with experimental or DNS data, there is a considerable amount of ambiguity in extending models to low-*Re* conditions, and thus each model features its own individual set of functions derived along different routes. Such extensions are not, therefore, considered in detail herein.

Although low-*Re* second-moment-closure models are beginning to be applied to quite complex 2D and even 3D flows (*e.g.* Craft [92], Batten *et al.* [12]), the desire for relative simplicity and the uncertainties associated with near-wall modelling have encouraged the application of somewhat simpler hybrid models which combine high-*Re* second-moment closure with low-*Re* eddy-viscosity models, the latter applied to the viscous near-wall layer (Lien and Leschziner [30]), or even with wall functions (Jakirlić [59], Hanjalić *et al.* [93]). Justification, especially for the former option, is provided by the observation that stress transport is usually uninfluential very close to the wall, and that the principal function of the near-wall model is to provide the correct level of the shear stress and wall-normal heat flux.

6   Non-linear eddy-viscosity models

Second moment closure has compelling fundamental merits, as well as yielding real predictive benefits in complex 2D and 3D flows. On the other hand, it is mathematically elaborate, numerically challenging and (often) computationally expensive – all regarded as important limitations in the context of industrial CFD. This has thus motivated efforts to construct models which combine the simplicity of the eddy-viscosity formulation with the superior fundamental strength and predictive properties of second-moment closure. These efforts have given rise to the group of *non-linear eddy-viscosity models* (NLEVMs).

NLEVMs can be traced back to Pope's [94] observation that Rodi's [95] algebraic approximation of the Reynolds-stress-transport model of Launder *et al.* [96] can be arranged in the explicit form:

$$a_{ij} = \sum G^{\lambda} T_{ij}^{\lambda} \tag{52}$$

where T_{ij}^λ is a tensorial power expansion in the strain and vorticity tensors:

$$S_{ij} \equiv \frac{1}{2}\left(\frac{\partial U_i}{\partial x_j} + \frac{\partial U_j}{\partial x_i}\right), \quad \Omega_{ij} \equiv \frac{1}{2}\left(\frac{\partial U_i}{\partial x_j} - \frac{\partial U_j}{\partial x_i}\right) \qquad (53)$$

while G^λ are coefficients which are functions of vorticity and strain invariants. This proposal received little attention until Speziale [97] first formulated and applied a workable quadratic model in which coefficients G^λ were simply taken to be powers of the time scale k/ε, so as to achieve dimensional consistency. Since then, a number of models of various complexity and derived along quite different routes have emerged (Yoshizawa [98], Shih et al. [99], Rubinstein and Barton [100], Gatski and Speziale [101], Craft et al. [102,103], Lien and Durbin [104], Lien et al. [105], Taulbee et al. [106], Apsley and Leschziner [107]). Most models are quadratic, while those of Craft et al., Lien et al. and Apsley and Leschziner are cubic and that of Gatski and Speziale is quartic. These differences in order are of considerable significance. In particular, the cubic fragments play an important role in capturing the strong effects of curvature on the Reynolds stresses. As regards model origin and derivation, an important distinction arises from the fact that some models (e.g. Shih et al., Lien et al., Craft et al.) start from a general series-expansion of the Reynolds-stress tensor in terms of strain and vorticity tensors, while others (Gatski and Speziale, Apsley and Leschziner, Taulbee et al.) start from an algebraic Reynolds-stress model. Other routes involve the direct interaction approximation adopted by Yoshizawa and the RNG approach taken by Rubinstein and Barton. Most models utilize constitutive equations which are functions of two turbulence scales (usually k and ε) as well as strain and vorticity invariants. In contrast, one variant of Craft et al.'s cubic model makes use of a transport equation for the stress invariant $A_2 = a_{ij}a_{ij}$, while Lien and Durbin's quadratic model depends on the Reynolds stress normal to the streamlines, which is also obtained from a related transport equation.

To a degree, the multiplicity of the NLEVMs published in the literature is indicative of the loss of rigour inherent in moving away from the complete second-moment framework towards a simpler closure level, which necessarily involves more empirical input and greater intuitive content. Clearly, the critical question is whether the simplification is justified in terms of the predictive performance which non-linear formulations return relative to linear models and second-moment closure. This question cannot be answered categorically, at present, because of insufficient evidence and the diversity of models, each displaying individual predictive characteristics.

Although the rational foundation and derivation of different models can differ greatly, the stress-strain/vorticity constitutive relationship for quadratic or cubic models can be written (for incompressible flow) in the following canonical form:

$$a_{ij} = -2c_\mu \frac{k}{\tilde{\varepsilon}} S_{ij}$$

$$+ c_1 \frac{k^2}{\tilde{\varepsilon}^2} \left(S_{ik} S_{jk} - \frac{1}{3} S_{kl} S_{kl} \delta_{ij} \right) + c_2 \frac{k^2}{\tilde{\varepsilon}^2} \left(S_{ik} \Omega_{jk} + S_{jk} \Omega_{ik} \right)$$

$$+ c_3 \frac{k^2}{\tilde{\varepsilon}^2} \left(\Omega_{ik} \Omega_{jk} - \frac{1}{3} \Omega_{kl} \Omega_{kl} \delta_{ij} \right) + c_4 \frac{k^3}{\tilde{\varepsilon}^3} \left(S_{ik} \Omega_{jl} + S_{jk} \Omega_{il} \right) S_{kl} \quad (54)$$

$$+ c_5 \frac{k^3}{\tilde{\varepsilon}^3} \left(\Omega_{ik} \Omega_{kl} S_{lj} + \Omega_{jk} \Omega_{kl} S_{li} - \frac{2}{3} \Omega_{kl} S_{lm} \Omega_{mk} \delta_{ij} \right)$$

$$+ c_6 \frac{k^3}{\tilde{\varepsilon}^3} S_{kl} S_{kl} S_{ij} + c_7 \frac{k^3}{\tilde{\varepsilon}^3} \Omega_{kl} \Omega_{kl} S_{ij}.$$

This expression, which is either the starting point of a non-linear model or the outcome of certain simplifications introduced into the Reynolds-stress equations, satisfies all requisite symmetry and contraction properties in incompressible flow.

The mechanism by which NLEVMs represent anisotropy emerges upon considering simple 2D shear in the (x_1, x_2) plane, in which $\sigma \equiv (k/\varepsilon)$ $(\partial U/\partial y)$ characterizes the strain, for which equation (54) yields:

$$
\begin{aligned}
a_{11} &= \frac{1}{12}(c_1 + 6c_2 + c_3)\sigma^2 \\
a_{22} &= \frac{1}{12}(c_1 - 6c_2 + c_3)\sigma^2 \\
a_{33} &= -\frac{1}{6}(c_1 + c_3)\sigma^2 \quad (55) \\
a_{12} &= -c_\mu \sigma + \frac{1}{4}(c_6 + c_7 - c_5)\sigma^3.
\end{aligned}
$$

This demonstrates that the quadratic terms are responsible for the ability of non-linear models to capture anisotropy: without these, $\overline{u_\alpha^2} = 2/3k$ for $\alpha = 1, 2, 3$. Also, by establishing a link between the normal stresses and the shear strain, relations (54) are, qualitatively, compatible with statements derived from the Reynolds-stress equations. However, as shown in Figure 18, the predictive quality with which a particular non-linear model resolves anisotropy depends significantly on the calibration of the model's coefficients.

The mechanism by which curvature effects are represented transpires from a consideration of the shear stress in 2D shear. Assuming that $c_6 + c_7 - c_5 = 0$ (so that the cubic terms play no role in simple shear), relations (54) give:

$$a_{12} = -2 \left[c_\mu + \frac{1}{4}(c_7 - c_5)(S^2 - \Omega^2) \right] \frac{k}{\varepsilon} S_{12} \quad (56)$$

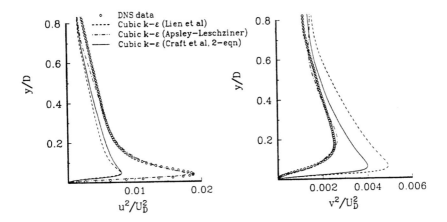

Fig. 17. Normal-stress distributions in fully-developed channel flow at $Re = 7890$ predicted by different non-linear EVMs (Apsley and Leschziner [107]).

where,

$$S = \frac{k}{\varepsilon}\sqrt{2S_{ij}S_{ij}}, \qquad \Omega = \frac{k}{\varepsilon}\sqrt{2\Omega_{ij}\Omega_{ij}}. \tag{57}$$

Result (56) highlights the fact that (part of) the cubic fragments can be assimilated into the linear term, and this allows the effects of curvature to be brought out most clearly. Thus, if $c_7 - c_5$ is chosen positive, the shear stress will increase when $S^2 - \Omega^2 > 0$ and will decrease when $S^2 - \Omega^2 < 0$. Here again, the effect is qualitatively equivalent to that predicted by the Reynolds-stress-transport equations, although in that case the sensitivity is brought about by a direct interaction between normal-stress anisotropy and curvature strain, *via* the stress-generation terms.

The sensitivity to curvature is well illustrated in Figure 18 which shows the result of applying Lien *et al.*'s [105] cubic model to the fully-developed curved-channel flow investigated experimentally by Ellis and Joubert [108] at $Re = 70\,000$. In this case it is readily shown that $(S^2 - \Omega^2)$ changes sign across the channel in a manner that attenuates the shear stress at the inner wall and amplifies it at the outer wall, thus provoking the distinct velocity asymmetry observed in the experiment.

There are other consequences that may be derived directly from the canonical form (54). First, in 2D flow, the quadratic terms (those multiplied by coefficients c_1, c_2 and c_3) have no *direct* effect on the turbulence-energy production. Second, again in 2D conditions, the cubic term associated with c_4 vanishes, whilst the remaining cubic terms are proportional to the mean-strain tensor.

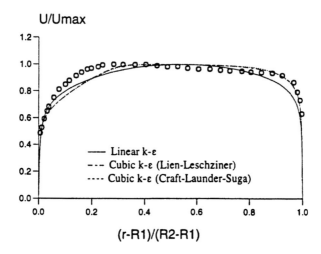

Fig. 18. Velocity distribution in a fully-developed flow in a curved channel predicted by the models of Lien *et al.* [105] and Craft *et al.* [102].

Although all cubic models are based on equation (54), they differ substantially in detail, especially in respect of the determination of the coefficients c_μ and c_1 to c_7 and the form of the dissipation equation. Specifically, one of the two models of Craft *et al.* [102] sensitizes the coefficients to A_2 which is determined from a related transport equation derived from second-moment closure. Of particular importance is the form of c_μ. In most model variants, this coefficient is sensitized to the strain and vorticity invariants, so as to avoid the excessive generation of turbulence energy in stagnation regions–a defect that is characteristic of standard linear eddy-viscosity models using the k-transport equation. In turbomachine blades, for example, this suppression is crucial for the prediction of transition following leading-edge impingement. While different models adopt different functional forms of c_μ, there is a remarkable commonality in respect of the variation at strain rates above those corresponding to the state of turbulence equilibrium. This is demonstrated in Figure 19 which shows the variation of c_μ as a function of the strain rate in simple shear, non-dimensionalised by the turbulent time scale k/ε. Thus, all models but one return a behaviour $c_\mu \propto 1/S$, at least at strain rates exceeding the equilibrium value. This commonality reflects the fact that the correct dependence of the turbulence generation on strain is $P_k \propto S$, while the eddy-viscosity formulation with constant c_μ gives $P_k \propto S^2$.

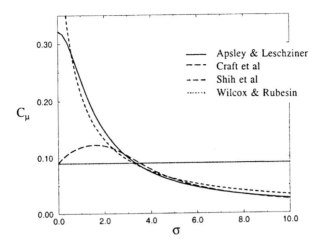

Fig. 19. Functional dependence of c_μ on non-dimensional strain rate returned by several non-linear eddy-viscosity models in simple shear (Loyau *et al.* [138]).

7   Application examples

7.1   Overview

This section reports a few selected results arising from the many validation studies that have involved the application of anisotropy-resolving models of the type summarised in Sections 5 and 6. Attention is focused, principally, on flows featuring separation and streamwise vorticity. In one flow, impingement and associated shear-layer curvature are the main features of interest. Second-moment closure also displays superior predictive qualities in swirling, rotating and buoyancy-affected flows, but space constraints prevent the inclusion of related results. The interested reader may wish to refer to papers, serving as examples, by Hogg and Leschziner [109] on swirling flow, by Cresswell *et al.* [110] on buoyant-jet flow and by Bo and Iacovides [111] on rotating flows.

Results are presented here for nine flows:

(i) moderately separated, 2D, incompressible flow in a plane asymmetric diffuser, Obi *et al.* [112];

(ii) moderately separated, 2D, incompressible flow around the high-lift, single-element "Aerospatiale-A" aerofoil at high incidence, Piccin and Cassoudessale [113];

(iii) massively separated, 2D, incompressible flow around a compressor-cascade blade operating in highly-loaded conditions, Zierke and Deutsch [114];

(iv) axisymmetric, incompressible jet impinging normally on a flat plate, Cooper *et al.* [115];

(v) separated, 3D, incompressible flow around a prolate spheroid at high angle of attack, Kreplin *et al.* [116], Meier *et al.* [117];

(vi) flow in a round-to-rectangular transition duct, Davis and Gessner [118];

(vii) flow in a 3D wing-flat-plate junction, Devenport and Simpson [119] and Fleming *et al.* [120];

(viii) supersonic, 3D, separated flow around a fin-plate junction, Barberis and Molton [121, 122];

(ix) transonic, 3D, separated flow over a rectangular jet-afterbody configuration, Putman and Mercer [123].

It must be emphasized that the present selection only provides a narrow window to a field which is populated with numerous model variants and computational solutions of variable quality for many different flows. This makes it exceedingly difficult to arrive at categorical, incontestable conclusions on the fundamental merits of any one modelling group or class; all that can be done is to make cautious judgements on the basis of broad experience.

In consonance with the general view that the use of log-law-based *wall functions* is seriously flawed in non-equilibrium flow, all calculations have involved integration through the near-wall sublayer, either with low-Re variants or with hybrid implementations, combining a high-Re model for the outer flow with a lower-order low-Re viscous-sub-layer model.

The computations have been performed with two quite different numerical schemes. For incompressible flows, single-block and multi-block versions of the general non-orthogonal, fully co-located, finite-volume algorithm "STREAM", Lien and Leschziner [124], has been used. In this, convection of mean-flow as well as turbulence quantities is approximated by the "UMIST" scheme, Lien and Leschziner [125], a second-order TVD approximation of the QUICK scheme of Leonard [126]. Mass continuity is enforced indirectly by way of a pressure-correction algorithm. For compressible flow, the Riemann-solver-based, implicit HLLC scheme of Batten *et al.* [127] has been used. This solves the conservation laws for mean-flow quantities and the turbulence-model equations simultaneously over an arbitrary structured mesh.

7.2 Asymmetric diffuser

This diffuser involves separation from the inclined plane wall and reattachment in the constant-area duct following the expansion. It offers accurate and well-resolved mean-flow and turbulence data for well-controlled 2D conditions and, importantly, it includes detailed data well removed from the diffuser section, allowing boundary conditions to be prescribed with a high level of confidence. The diffuser length is 21 times the upstream channel height H, and the overall expansion ratio is 4.7. The Reynolds number based on upstream-channel conditions is 21 200. Following grid-independence studies, the flow was computed with a second-order TVD scheme and a 272×82 grid, extending $11H$ and $40H$ upstream and downstream of the diffuser section, respectively. The y^+ value along the grid line closest to the wall was kept below 1 throughout. Results presented here have been taken from a study by Apsley and Leschziner [107].

Figure 20 shows the development of the streamwise-mean-velocity profile along the diffuser. Results have been included for one linear and two cubic low-Re eddy-viscosity models (EVMs) and the Reynolds-stress-transport model of Speziale *et al.* [88], with the near-wall sublayer resolved by the one-equation model of Norris and Reynolds [24]. The linear EVM predicts a symmetric mean-velocity profile across the diffuser, and near-isotropy amongst the normal-stress components (not shown). The addition of quadratic terms, as done by Speziale [97], distinguishes the individual normal stresses, but fails to improve the mean-velocity predictions significantly. In contrast, the cubic model of Apsley and Leschziner [107], which combines (S, Ω)-dependent coefficients and non-linear terms in the stress-strain relationship, provides a very satisfactory prediction of cross-channel asymmetry, close to that achieved with second-moment models. Good agreement is also returned in respect of normal and shear stresses which are not included here.

7.3 Aerospatiale aerofoil

This geometry and associated flow, resolved by a 354×66 C-type grid, are shown in Figure 21. The condition of principal interest is that at $Re = 2.1 \times 10^6$ (based on chord and free-stream velocity) and an angle of attack of $13.3°$, at which stall has just set in. Hence, the main computational challenge is to predict the onset of separation, which is a very sensitive function of the structure of the boundary layer ahead of the separation point. This case involves free transition at 12% and 30% of chord on the suction and pressure sides, respectively, and these locations were prescribed in the calculations. The results presented here have been taken from an extensive study by Lien and Leschziner [128] and have been obtained with the linear EVM of Lien and Leschziner [40], the quadratic EVM of

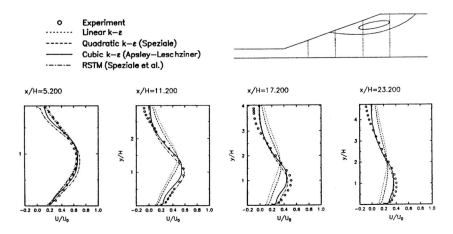

Fig. 20. Plane asymmetric diffuser: mean-streamwise-velocity profiles at four diffuser stations (taken from Apsley and Leschziner [108]).

Shih *et al.* [99] and the Reynolds-stress model of Gibson and Launder [80], the last combined with the one-equation eddy-viscosity model of Wolfshtein [23], which is applied in the viscous sub-layer. In a later study, reported in Apsley *et al.* [32], the cubic model of Apsley and Leschziner [107] is shown to return results qualitatively consistent with those obtained by Lien and Leschziner [128].

Figure 21 shows boundary-layer profiles along the suction surface. The best performance by far is returned by the "truncated" Reynolds-stress model and the "modified" non-linear model, both giving similar predictive quality and both (just) returning the relatively weak separation observed in the experiment. The linear model gives a much too weak response of the boundary layer to the adverse pressure gradient and fails to predict even the tendency towards separation.

The nature of the "truncation" and "modification" are explained in detail in Lien and Leschziner [128]. Suffice it to say here that the "modification" is designed to mimic the effects of the important cubic terms (accounting for curvature effects – see Sect. 6) which Shih *et al.*'s model [99] lacks, while the "truncation" refers to a removal of a particular wall-related correction term which is part of the original model, but which causes the near-wall shear stress to be higher than it should be. A recent study by Apsley and Leschziner [107] with their own cubic EVM gives solutions for the present aerofoil geometry in close agreement with those of the "modified" quadratic model of Shih *et al.* [99]. Figure 21 also contains variations of pressure, skin friction and suction-side displacement-thickness

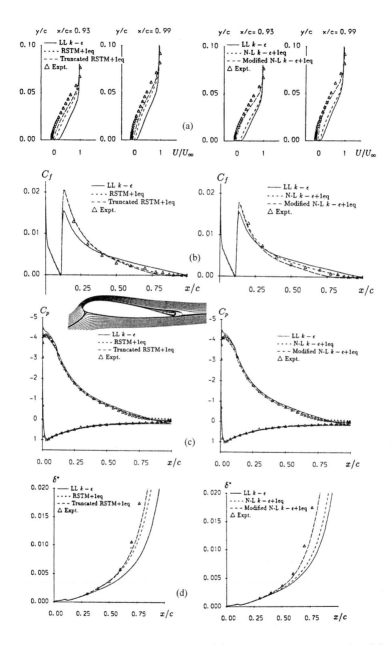

Fig. 21. High-lift *Aerospatiale* aerofoil: (a) velocity on suction side; (b) skin friction on suction side; (c) pressure coefficient; (d) displacement thickness on suction side (taken from Lien and Leschziner [128]).

distributions which confirm comments made above on the relative performance of the models examined. Particular attention is drawn to the fact
that both the non-linear EVM and the Reynolds-stress model correctly predict the pressure variation at the leading edge, and this appears to be sensi-
tively linked (*via* the circulation constraint) to the correspondingly correct
representation of the pressure variation in the trailing-edge region.

7.4 Cascade blade

Two cascade flows, one around a controlled-diffusion (CD) blade and the
other around a double-circular-arc (DCA) blade, have recently been investigated by Chen *et al.* [129,130], using Craft *et al.*'s [103] cubic eddy-viscosity
model and Gibson and Launder's [78] Reynolds-stress model, the latter coupled with a low-Re linear two-equation eddy-viscosity model which was applied to the near-wall region. Of the two flows, the DCA blade is more
challenging, because of its intense curvature and the fact that massive separation is provoked at the rear part of the suction side, even when the flow
is close to the design state. On the suction side there is a small leading-
edge separation bubble, and this causes the flow to become turbulent within
$2 - 3\%$ of the chord. Hence, although the turbulence structure in the impingement zone of the leading edge obviously influences the behaviour of the
boundary layers on the blade surface, the details of the transition process
are not of great importance in the real flow. At about 80% of the chord,
the turbulent boundary layer separates, forming a substantial recirculation
zone which extends well beyond the trailing edge.

Figure 22 shows the multiblock grid used, and gives selected results for
an incidence angle which departs by -1.5 deg from the design state. The
pressure distributions demonstrate that the second-moment closure performs much better than both the linear and cubic EVMs, especially near
the trailing edge where separation occurs. The streamwise velocity profiles
confirm that only the second-moment closure resolves the trailing-edge separation. The success of this model may again be attributed to its inherent
sensitivity to streamline curvature and adverse pressure gradient, the latter
causing a rapid deceleration of the near-wall flow and eventually separation.
The cubic model, whilst also capable of representing this sensitivity to some
extent, fails to capture the separation process. The main reason appears
to be that the model overestimates the leading-edge separation bubble, associated with a substantial delay in the onset of transition, thus returning a relatively thin boundary layer upstream of the leading edge, which
then resists separation. This delay in transition is readily identified from
the streamwise turbulence-intensity profiles at 7.3% of chord. The linear
EVM predicts the initial boundary layer fairly well. However, it also fails to

capture the trailing-edge separation because of excessive turbulent mixing in the suction-side boundary layer.

The variations of integral boundary-layer quantities indicate that the small leading-edge separation bubble on the suction side is barely resolved by the measurement, its existence being reflected by the slightly elevated momentum thickness and shape factor. The steep rise in shape factor beyond 75% of chord signifies the presence of the trailing-edge separation. Only the second-moment closure gives a credible prediction of the boundary-layer quantities on the suction side, consistent with that of the velocity profiles. The other two models inevitably result in a far too low displacement thickness. A significant weakness of the second-moment closure can only be observed in relation to the shape factor near the leading edge on the suction side, reflecting a failure of the model to capture the small leading-edge separation bubble.

7.5   Axisymmetric impinging jet

This case allows a searching test of the ability of models to capture the interaction between turbulence and curvature in the outer shear layer of the deflecting jet, and of the response to normal straining in the centre of the jet as it decelerates. The jet is discharged at twice the pipe diameter above the plate, at a Reynolds number of 23 000, based on bulk velocity and pipe diameter. Computations were performed by Batten *et al.* [12] on a 180×180 mesh with various models, including a modified version of Craft and Launder's [11] low-Re cubic Reynolds-stress model. Computations for the same geometry with the original Craft–Launder model are also reported in Craft and Launder [11]. In all cases, the grid line closest to the wall was located at $y^+ < 1$.

Figure 23 presents profiles of mean (absolute) velocity at two radial positions, a normal-stress profile at one radial position and the Nusselt-number distribution along the impingement plate, predicted by four models: the linear $k - \varepsilon$ model of Launder and Sharma [34], the "Shear Stress Transport" (SST) variant of Menter [47,131], the Jakirlić–Hanjalić [10] Reynolds-stress model (JH) (see also Hanjalić [132] and Hanjalić and Jakirlić [76]) and the Craft–Launder model, modified by Batten *et al.* (MCL). The most noticeable defects of linear EVMs occur in the stagnation region which is characterized by high rates of normal straining, where both models show a serious over-prediction of the normal stresses (and hence, turbulence energy) and heat transfer. The SST model is actually a little worse than the $k - \varepsilon$ model on the stagnation line, but it recovers remarkably once the flow turns and the turbulent boundary layer on the plate starts to develop. The $k - \varepsilon$ model is also seen to seriously misrepresent the curved shear layer. The Jakirlić–Hanjalić model gives only a modest improvement over the $k-\varepsilon$

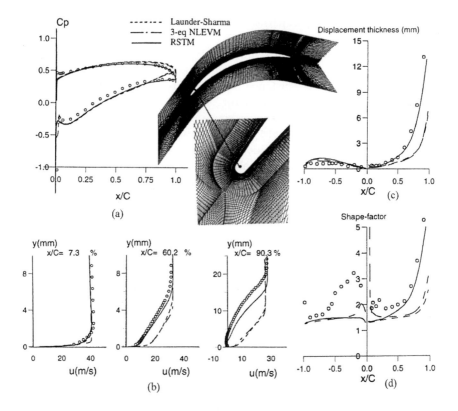

Fig. 22. Double-circular-arc cascade blade: (a) pressure coefficient; (b) velocity on suction side; (c) displacement thickness; (d) shape factor (taken from Chen *et al.* [129, 130]).

model. This is due, in part, to weaknesses of the conventional wall-reflection terms used in this model (see, for example, Craft and Launder [81]), which have especially deleterious effects in impinging flows. The predictions of the cubic model are generally good, and this applies also to other radial positions and to the shear stress, which are not included herein.

7.6 Prolate spheroid

This is a flow around an elliptical body of axes ratio 6:1 and inclined at $10°$ or $30°$ to an oncoming, uniform stream. It was one of several test cases in the European-Commission-funded international validation exercise ECARP (Haase *et al.* [133]). The geometry represents the group of external flows around streamlined bodies which feature vortical separation that arise from

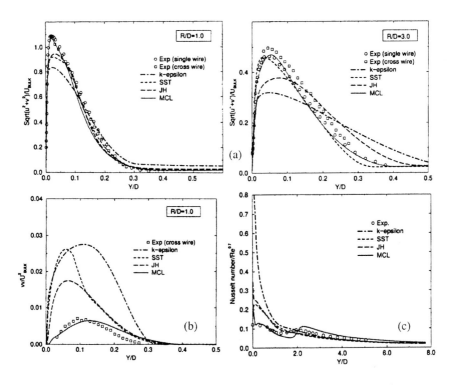

Fig. 23. Normally-impinging round jet: (a) velocity magnitude; (b) normal stresses perpendicular to the wall (c) Nusselt number along wall (taken from Batten *et al.* [12]).

an oblique "collision" and subsequent detachment of boundary layers on the body's leeward side. Of the two flows, that at 30° and $Re = 6.5 \times 10^6$ (based on chord) is much more challenging, but poses significant uncertainties due to a complex pattern of natural transition on the windward surface. Experimental data have been obtained by Kreplin *et al.* [116] and Meier *et al.* [117] for pressure, skin friction and mean-velocity, the last with a 5-hole probe. Corresponding computations have been performed by Lien and Leschziner [41] with the low-Re linear eddy-viscosity model of Lien and Leschziner [40,41], a low-Re adaption, by Lien *et al.* [105], of Shih *et al.*'s [99] model and the second-moment closure of Gibson and Launder [78], coupled to the above linear eddy-viscosity model in the viscous sublayer. A second-order TVD scheme was used on a high-quality conformal mesh of $98 \times 82 \times 66$ nodes, with the y^+-value closest to the wall being kept to $0.5-1$ across the entire surface. Test calculations with the non-linear eddy-viscosity model on a 128^3 grid have shown only the skin friction to change slightly at this level of grid refinement.

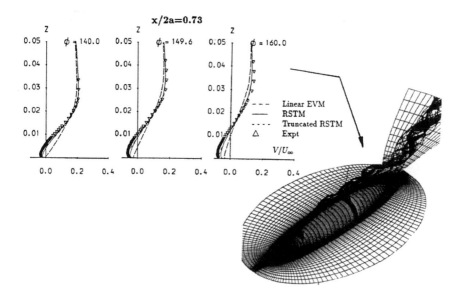

Fig. 24. prolate spheroid at $10°$ incidence: azimuthal velocity profiles above lee-ward side at one axial position (taken from Lien and Leschziner [41]).

Figure 24 contains comparisons of azimuthal velocity profiles at one streamwise location at $10°$ incidence, while Figure 25 gives, for $30°$ incidence, one azimuthal pressure distribution, one skin-friction distribution and one velocity field, the last showing the lee-ward separation and the associated transverse vortex. In general, the non-linear eddy-viscosity model and the second-moment closure give similar results which are closer to the experimental data that those obtained with the linear EVM. However, the improvement is not uniformly pronounced across all flow properties, and the uncertainties associated with transition in the $30°$ case do not warrant a categorical statement on model performance in this very complex case.

7.7   Round-to-rectangular transition duct

While this flow, shown in Figure 26, is attached, it involves strong trans-verse motion which is induced by the lateral squeezing and the streamwise realignment of the vortex lines in the boundary layer. The convective trans-port associated with the transverse motion creates a strong distortion in the streamwise velocity which persists far downstream. The main challenge thus posed to the turbulence models is to represent the evolution and decay of

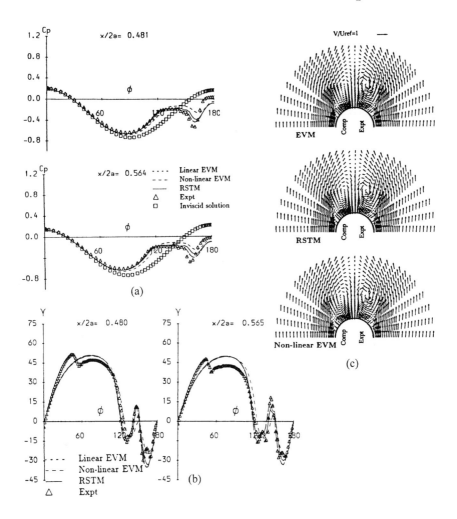

Fig. 25. Prolate spheroid at 30° incidence: (a) circumferential pressure variations; (b) circumferential variations of skin-friction direction; (c) structure of vortices above rear lee-ward side (taken from Lien and Leschziner [41]).

the streamwise vortices and to predict the complex stress pattern which goes with them.

Computations for this case were performed by Lien and Leschziner [41] with a multi-block grid containing 530 000 nodes, using Lien and Leschziner's [40] low-Re linear $k - \varepsilon$ model and Gibson and Launder's [78] Reynolds-stress model, the latter coupled to the former to bridge the semi-viscous sublayer.

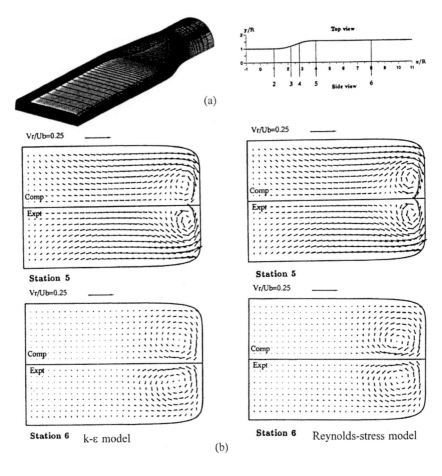

Fig. 26. Round-to-rectangular transition duct: (a) geometry; (b) transverse velocity (taken from Lien and Leschziner [41]).

The results shown in Figures 26 and 27 indicate – and this is supported by solutions for pressure and friction factor – that the most important process to resolve is the skewness of the flow very close to the wall, arising from the transverse motion which reaches its peak just outside the viscous sublayer. Thus, in relation to mean-flow features, second-moment closure offers here only minor benefits relative to those arising from a detailed resolution of the near-wall mean-flow structure. It is noted, however, that only this closure is able to return a credible representation of the stress fields, especially in respect of anisotropy.

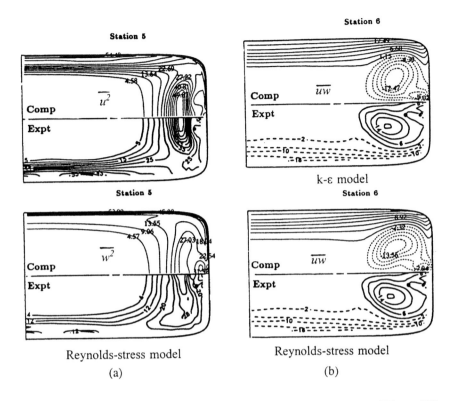

Fig. 27. Round-to-rectangular transition duct: (a) normal stresses $\overline{u^2}$ and $\overline{w^2}$; (d) shear stress $\overline{uw}$ (taken from Lien and Leschziner [41]).

7.8   Wing/flat-plate junction

This generic geometry, shown in Figure 28a, is representative of wing/aircraft-body, blade/hub, appendage/ship-hull and bridge-pillar/river-bed junctions. It consists of a 3:2 semi-elliptic nose connected to a NACA 0020 tail, for which detailed experimental data have been obtained by Devenport and Simpson [119] and Fleming et al. [120]. Numerical solutions were obtained by Apsley and Leschziner [134] with several non-linear eddy-viscosity and second-moment models, and comparisons with exprimental data are reported in the original reference for the around the wing and on the plane surface, streamwise velocity in the upstream symmetry plane, transverse velocity in cross-flow planes, and streamwise normal stress and the shear stress in cross-flow planes. The flow is characterised by the formation of a recirculation region upstream of the wing nose and a horshoe vortex wrapped around the wing in the junction. The transverse motion in this vortex causes, among other effects, a convective redistribution of streamwise momentum and turbulence quantities in the junction region, and the details of this process are sensitive to turbulence modelling.

Figure 28(b) gives, contours of streamwise normal stress in the upstream symmetry plane bisecting the aerofoil. These contours are highly sensitive indicators of the structure of the upstream vortex, and also identify its strength in terms of the turbulence-generating strain within it. The contours indicate the predictive benefits arising from second-moment closure. This model returns correctly, in contrast to the eddy-viscosity model, the significant elevation of the streamwise normal-stress in the intensely sheared central portion of the recirculating vortex – reflecting the ability of the model to represent the fact that shear feeds, preferentially, the streamwise normal stress through generation, while that generation is suppressed in the surrounding regions in which either shear-straining is low or normal straining is dominant. In effect, this behaviour signifies the ability of the model to resolve normal-stress anisotropy and its interaction with stress generation.

Comparisons of contours of streamwise velocity and shear stress, at one cross-flow plane close to the trailing edge, are given in Figure 29. Here again, the Reynolds-stress model gives a broadly satisfactory behaviour due to the intensity of the transverse motion it predicts. The linear eddy-viscosity model returns large levels of stress (and turbulence energy) over most of the cross-sectional area, signifying high upstream generation and downstream advection of turbulence. The Reynolds-stress model returns a field which contains the foot-print of the vortex, specifically the rotational advection of energy generated in the junction and along the walls.

7.9 Fin-plate junction

This is amongst the most complex flows predicted, so far, with second-moment closure (the most complex case is probably the transonic flow around a whole aircraft model computed by Batten *et al.* [135]). The geometry, shown in Figure 30, was the subject of a recent Europe-US workshop on high-speed flows. A Mach 2 flat-plate boundary layer collides with the rounded normal fin, producing a complex shock/boundary-layer interaction and multiple horse-shoe vortices.

Experimental data are available for surface pressure, LDA velocity, skin-friction patterns and Reynolds stresses (Barberis and Molton [121, 122]). While the geometry and flow are well controlled, some minor uncertainties arise because of a lack of detail in the measured boundary layer well upstream of the fin (only its thickness was given) and the presence of leakage between the fin tip and one wall of the windtunnel. The latter poses some uncertainty about the boundary conditions on the computational boundary plane above the lower flat plate along which the interaction takes place.

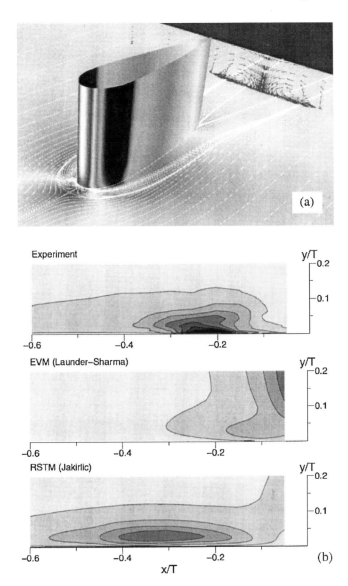

Fig. 28. Wing/plate junction: (a) overall flow field; (b) contours of streamwise normal stress in symmetry plane (taken from Apsley and Leschziner [134]).

Computations were performed by Batten *et al.* [12] with Menter's [47] SST linear eddy-viscosity model, the Jakirlić–Hanjalić [10] linear Reynolds-stress model and the Craft–Launder cubic model, modified by

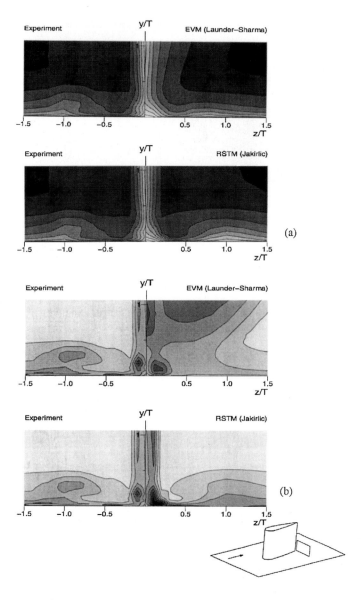

Fig. 29. Wing/plate junction: (a) contours of streamwise velocity in cross-flow plane close to trailing edge; (b) contours of shear stress in same plane as (a) (taken from Apsley and Leschziner [134]).

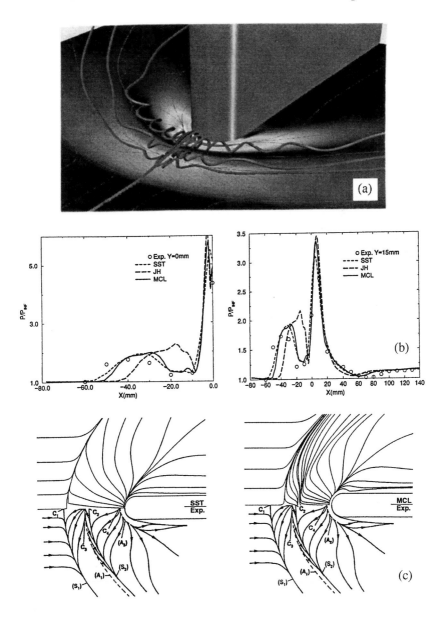

Fig. 30. Supersonic fin-plate-junction flow: (a) overview and iso-Mach contours in centre-plane; (b) plate-pressure distributions at two spanwise (y) stations; (c) surface streamlines (taken from Batter *et al.* [12]).

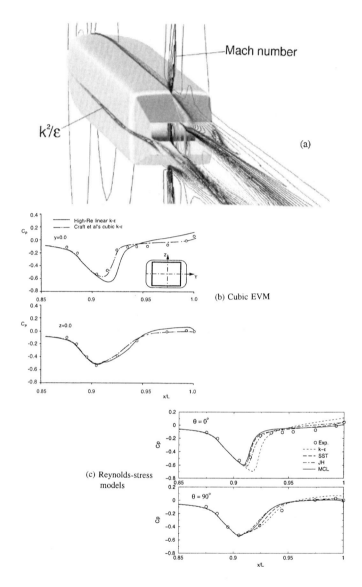

Fig. 31. Jet-afterbody: a) geometry and overall flow field; b,c) pressure distributions on lines formed by intersection of the vertical and horizontal centre-planes with the afterbody at $M = 0.94$ (taken from Leschziner *et al.* [136] and Batten *et al.* [135]).

Batten *et al.* [12]. A fin-adapted $80 \times 80 \times 70$ C-type grid was used, with the y^+ closest to the wall being of order 0.5.

The results shown in Figure 30 illustrate that only the second-moment closure is able to reproduce the multiple separations/reattachments ahead of the fin, which are observed in the experiment, although the patterns are not identical. However, the size of the separated zone and the associated pressure field, which are closely linked to the predicted strength of the shock/boundary-layer interaction, are strongly dependent on which model variant is used. As seen, the Jakirlić–Hanjalić model significantly underestimates the interaction, thus predicting a delayed boundary-layer separation. In contrast, the cubic model does significantly better, returning pressure distributions close to the experimental variations.

7.10   Jet–afterbody combination

This geometry, shown in Figure 31, is a rectangular afterbody with rounded edges, which discharges a supersonic, under-expanded jet. Experimental data have been obtained by Putman and Mercer [123] for free-stream Mach numbers of 0.6 and 0.94, the latter generating a substantial shock-induced separation from the boat-tail surface as a consequence. While the jet itself is not a primary region of interest, it plays an important role, in that it affects the boat-tail shock by the blockage effect it introduces. Corresponding computations were performed by Batten *et al.* [135] and Apsley (see Ref. [136]), the former with the cubic EVM of Craft *et al.* [102, 103] and the latter with the second-moment models of Jakirlić and Hanjalić [10] and Craft and Launder [11] (modified by Batten *et al.* [12] for compressible flow), over a multi-block mesh containing $600\,000 - 987\,000$ nodes. Figure 31 gives afterbody pressure distributions for $M = 0.94$. At the lower Mach number, for which there are also experimental velocity traverses available, there is no afterbody shock and no separation, and sensitivity to turbulence modelling is found to be weak. Hence, no results have been included for that condition. At the higher value, however, there is a substantial separation, and sensitivity is seen to be high. The cubic eddy-viscosity model and the Craft–Launder Reynolds-stress model return this separation, and predict pressure distributions along the major axes of the afterbody close to the experimental variations. In contrast, the linear eddy-viscosity model fails to predict separation and gives a far too rapid increase in afterbody pressure.

8   Concluding remarks

Much effort has gone over the past decade or two into the construction, calibration and validation of improved forms of turbulence closure. This is an especially difficult area of CFD, and progress has been much slower than that in the field of numerical methods, grid generation, and solvers. Turbulence modelling, at whatever level, is inevitably an exercise of compromise between predictive realism and computational effort. There are several sets of circumstances in which turbulence modelling will probably never provide a satisfactory representation of the flow, a typical example being a flow which is dominated by periodic, coherent structures. In flows in which transport is dominated by true turbulence and for which only statistical flow properties are sought, turbulence modelling, rather than simulation, is likely to remain the preferred option for many years to come. With this premise accepted, considerable care is required in choosing the appropriate turbulence model, for this choice may dictate whether the prediction is quantitatively meaningful.

In terms of computing effort, there is very little to choose between conventional linear eddy-viscosity models and modern non-linear variants. While it is far from clear, at present, which model variant offers the best overall predictive performance in complex strain, it is clear that the better models are fundamentally superior to their linear counterparts, as well as being preferable on fundamental grounds. In particular, low-Re cubic models, which are derived *via* simplifications to second-moment closure and carefully calibrated by reference to DNS data, return a fair representation of normal-stress anisotropy, sensitivity to curvature and a broadly correct response to irrotational straining. Thus, there is no rational justification for not adopting non-linear models in physically complex flows involving strong curvature and separation.

Second-moment closure is computationally more challenging and can be costly in terms of both memory and CPU time, especially in 3D flows and if flux-transport equations are included for the thermal or species-concentration fields. However, a careful implementation, including some special stabilisation practices (*e.g.* Lien and Leschziner [124]), allow the CPU time for second-moment computations in 3D flow to be depressed down to a factor $1.5 - 2$ of the time taken by eddy-viscosity models. The key advantage of second-moment closure is its fundamental strength, rooted in the formally exact representation of stress generation. The main challenge of second-moment closure lies in the fact that the redistribution and dissipation processes are highly influential and difficult to model, especially in the near-wall region in which the turbulence structure experiences very steep spatial rates of change. Thus, different closure variants display different performance characteristics. However, all model variants return the anisotropic nature of strained turbulence and a realistic sensitivity to body forces, curvature and normal straining.

Difficulties with stable and accurate model implementation, the fine grids and high CPU resources required, the lack of sufficiently accurate and resolved data derived from well-controlled experiments, and the small number of groups engaged/collaborating in advanced modelling (especially that involving 3D flows) makes validation very challenging and its outcome subject to uncertainty. Nevertheless, the considerable number of validation studies reported in the open literature justify the overall observation that anisotropy-resolving closures offer not insubstantial predictive advantages over simpler closures in complex strain fields. In general, the advantages in 3D flow appear to be less pronounced than in 2D ones (at least in terms of the dynamic state). While this might initially appear odd, a possible explanation lies in the fact that turbulent transport in 3D flows tends to be less dominant, relative to inviscid contributions, than in 2D flows. This is linked to higher levels of flow curvature present in many complex 3D flows, with which convective transport and pressure gradients are associated.

On a somewhat pessimistic note, it is unrealistic to expect that the future will see the evolution of a turbulence-modelling framework that will offer a dramatic widening of the generality and range of applicability of the most complex model forms currently available. Practical constraints dictate that truncation and closure need to effected at a level not higher than third or even second order, with the current type of closure equations perhaps supplemented by additional equations for quantities related to turbulence structure or processes in different spectral partitions. While this increases the motivation to switch to LES, there are still not insubstantial obstacles to overcome in this area too, especially in complex industrial flows. All these challenges are aptly expressed by the title of a recent paper by Bradshaw [137]: "Turbulence: the chief outstanding difficulty of our subject".

The author gratefully acknowledges the contributions made to this article by his collaborators Dr. D.D. Apsley, Dr. P. Batten, Dr. W.-L. Chen, Dr. T.J. Craft, Prof. F.-S. Lien and Dr. H. Loyau. Computational results included in the article originate from research funded by EPSRC, BAE Systems, DERA, Rolls-Royce and the EU. The author is grateful for their financial support.

References

[1] L.F. Richardson, *Weather prediction by numerical process* (University Press, Cambridge, 1992).

[2] P. Moin and J. Kim, *Scient. Amer.* **276** (1997) 62.

[3] P. Moin and K. Mahesh, *Annu. Rev. Fluid Mech.* **30** (1998) 539.

[4] O. Reynolds, *Trans. Roy. Soc. A.* **186** (1894) 123.

[5] J. Boussinesq, *Mem. Acad. Sci. β* (1877) 1-680.

[6] M. Lesieur and O. Metais, *Annu. Rev. Fluid Mech.* **28** (1996) 45.

[7] L. Temmerman and M.A. Leschziner, Large Eddy Simulation of separated flow in a streamwise periodic channel constriction, in *Proc. 2nd Symp. on Turbulence and Shear Flow Phenomena* (Stockholm, 2001) (to appear).

[8] R. Lardat, *Simulations numériques d'écoulements externes instationnaires décolls autour d'une aile avec des modèles de sous maille*, Ph.D. Thesis, University Paris 6 (1997).

[9] J.L. Lumley, *Adv. Appl. Mech.* **18** (1978) 123.

[10] S. Jakirlić and K. Hanjalić, A second-moment closure for non-equilibrium and separating high- and low-*Re*-number flows, in *Proc. 10th Symp. on Turbulent Shear Flows* (Pennsylvania State University, 1995) 23-25.

[11] T.J. Craft and B.E. Launder, *Int. J. Heat Fluid Flow.* **17** (1996) 245.

[12] P. Batten, T.J. Craft, M.A. Leschziner and H. Loyau, *J. AIAA J.* **37** (1999) 785.

[13] W.C. Reynolds and S.C. Kassinos, One-point modelling of rapidly deformed homogeneous turbulence, in *Proc. Osborne Reynolds Centenary Symposium*, Manchester. *Proc. The Royal Society of London, Series A* **451** (1994) 87.

[14] A.K.M.F. Hussain and W.C. Reynolds, *J. Fluid Mech.* **54** (1972) 263.

[15] D. Drikakis, The equations for the coherent structure dynamics in turbulent flows, in *Proc. FEDSM '98* (Washington DC, 1998).

[16] H. Ha Minh and A. Kourta, Semi-deterministic turbulence modelling for flows dominated by strong organised structures, in *9th Symp. on Turbulent Shear Flows* (Kyoto, 1994) 10.5.1.

[17] R. Schiestel, *Phys. Fluids* **30** (1987) 722.

[18] D.C. Wilcox, *AIAA J.* **26** (1988) 1311.

[19] B.S. Baldwin and H. Lomax, *AIAA Paper* 78-0257 (1978).

[20] T. Cebeci and A.M.O. Smith, *Analysis of turbulent boundary layers (Ser. in Appl. Math. & Mech* **XV** (Academic Press, London, 1974).

[21] E. Hytopoulos, J.A. Schetz and R.L. Simpson, *J. Fluids Engrg.* **119** (1997) 541.

[22] D.A. Johnson and L.S. King, *AIAA Paper* 84-0175 (1985).

[23] M. Wolfshtein, *Int. J. Heat Mass Transfer.* **12** (1969) 139.

[24] L.H. Norris and W.C. Reynolds, Turbulence channel flow with a moving wavy boundary, *Rep. FM-10*, Dept. of Mech. Engrg. (Stanford University, 1975).

[25] B.W. Baldwin and T.A. Barth, *AIAA Paper* 91-0610 (1991).

[26] P.R. Spalart and S.R. Allmaras, *AIAA Paper* 92-0439 (1992).

[27] R. Franke and W. Rodi, Calculation of vortex shedding past a square cylinder with various turbulence models, in *Proc. 8th Shear Flows Symp.* (Munich, 1991) 20-1-1.

[28] H.-H. Cho, X. Liu, W. Rodi and B. Schönung, 1993, *J. Turbomachinery* **115** (1993) 675.

[29] M.A. Leschziner, K.P. Dimitriadis and G. Page, *The Aeronautical J.* **97** (1993) 43.

[30] F.S. Lien and M.A. Leschziner, *J. Fluids Engrg.* **115** (1993) 717.

[31] F.-S. Lien and M.A. Leschziner, Modelling 2D and 3D separation from curved surfaces with variants of second-moment closure combined with low-*Re* near-wall formulations, in *Proc. 9th Symposium Turbulent Shear Flows* (Kyoto, 1993) 13.1.1.

[32] D. Apsley, W.L. Chen, M.A. Leschziner and F.S. Lien, *IAHR J. Hydraulic Res.* **35** (1998) 723.

[33] W.P. Jones and B.E. Launder, *Int. J. Heat Mass Transfer.* **15** (1972) 301.

[34] B.E. Launder and B.I. Sharma, Letters in *Heat Mass Transfer* **1** (1974) 131.

[35] C.K.G. Lam and K. Bremhorst, *J. Fluids Engrg.* **103** (1981) 456.

[36] K.Y. Chien, *AIAA J.* **20** (1982) 33.

[37] H.K. Myong and N. Kasagi, *Japanese Society Mechanical Engineering Int. J.*, Series II **33** (1990) 63.

[38] Y. Nagano and M. Hishida, *Trans. ASME* **109** (1987) 156.

[39] S.A. Orzag, V. Yakhot, W.S. Flannery, F. Boysan, D. Choudhury, J. Maruzewski and B. Patel, Renormalisation group modelling and turbulence simulations. *Near-Wall Turbulent Flows*, edited by R.M.C. So, C.G. Speziale and B.E. Launder (Elsevier, 1993) 1031.

[40] F.S. Lien and M.A. Leschziner, Modelling the flow in a transition duct with a non-orthogonal FV procedure and low-*Re* turbulence-transport models. ASME Summer *Meeting, Symposium on Advances in Computational Methods in Fluid Dynamics* (1994) 93.

[41] F.S. Lien and M.A. Leschziner, *Computers and Fluids* **25** (1995) 237.

[42] G. Kalitzin, A.R.B. Gould and J.J. Benton, *AIAA Paper* 96-0327 (1996).

[43] D.C. Wilcox, *AIAA J.* **26** (1988) 1299.

[44] D.C. Wilcox, *AIAA J.* **32** (1994) 247.

[45] U.C. Goldberg, *J. Fluids Engrg.* **116** (1994) 72.

[46] M.M. Gibson and A.A. Dafa'Alla, The $q - \zeta$ model for turbulent wall flow. Fluid Dynamics Division, American Physical Society, in *Proc. 47th Annual Meeting* (Atlanta, Georgia, 1994).

[47] F.R. Menter, *AIAA J.* **32** (1994) 1598.

[48] C.V. Patel, W. Rodi and G. Scheuerer, *AIAA J.* **23** (1985) 1308.

[49] R.A.W.M. Henkes, B.F.A. van Hest and D.M. Passchier, *ERCOFTAC Bulletin* **24** (1995) 34.

[50] B.R. Ramaprian and S.W. Tu, Study of periodic turbulent pipe flow, Inst. of Hydr. Res. (University of Iowa Report IJHR 238, 1982).

[51] J.D. Jackson and S. He, Simulations of transient turbulent flow using various two-equation low-reynolds number turbulence models, in *Proc. 10th Symp. on Turbulent Shear Flows* (Penn-State University, 1995) 11.19.

[52] S. Fan, B. Lakshminarayana and M. Barnett, *AIAA J.* **31** (1993) 1777.

[53] S. Fan and B. Lakshminarayana, *J. Turbomachinery* **118** (1996) 96.

[54] P.R. Spalart and B.S. Baldwin, Direct numerical simulation of a turbulent oscillating boundary layer. *Turbulent Shear Flows* 6, edited by J.C. André *et al.* (Springer, 1989) 417.

[55] B.L. Jensen and B.M. Sumer, *J. Fluid Mech.* **206** (1989) 265.

[56] P. Justesen and P.R. Spalart, *AIAA Paper* 90-0496 (1990).

[57] H. Ha Minh, J.R. Viegas, M.W. Rubesin, D.D. Vandromme and P. Spalart, Physical analysis and second-order modelling of an unsteady turbulent flow: The oscillating boundary layer on a flat plate, in *Proc. 7th Symp. on Turbulent Shear Flows* (Stanford, 1989) 11.5.1.

[58] N. Shima, *J. Fluids Engrg.* **115** (199) 356.

[59] S. Jakirlić, *Reynolds-spannungs-modellierung komplexer turbulenter Stroemungen*, Ph.D. Thesis (University of Erlangen–Nuernberg, 1997).

[60] E. Guilmineau, J. Piquet and P. Queutey, *Int. J. Num. Meths. in Fluids* **25** (1997) 315.

[61] G. Barakos and D. Drikakis, *ASME paper* FEDSM97-3651 (1997).

[62] J.-C. Wu, D.L. Huff and L.N. Sankar, *J. Aircraft* **27** (1990) 382.

[63] G.R. Srinivasan, J.A. Ekaterinaris and W.J. McCroskey, *Computers and Fluids* **24** (1995) 883.

[64] J.A. Ekaterinaris and F.R. Menter, *AIAA J.* **32** (1994) 2359.

[65] R.A. Piziali, *An experimental investigation of 2D and 3D osciallating wing aerodynamics for a range of angle of attach including stall*, NASA Technical Memorandum 4632 (1993).

[66] B.E. Launder and D.P. Tselepidakis, Contribution to the modelling of near-wall turbulence. *Turbulent Shear Flows 8*, edited by F. Durst *et al.* (Springer-Verlag, 1993) 81.

[67] F. Archambeau and D. Laurence, Description of numerical methodology for tube bank, in *Proc. 2nd ERCOFTAC-IAHR Workshop on Refined Flow Modelling* (Manchester, 1993) UMIST.

[68] W.P. Jones and A. Manners, The calculation of the flow through a two-dimensional faired diffuser. *Turbulent Shear Flows 6*, edited by J.C. André *et al.* (Springer, 1989) 18.

[69] P.R. Spalart, *J. Fluid Mech.* **187** (1988) 61.

[70] J. Kim, P. Moin and R. Moser, *J. Fluid Mech.* **177** (1987) 133.

[71] J.G.M. Eggels, F. Unger, M.H. Weiss, J. Westerweel, R.J. Adrian, R. Friedrich and F.T.M. Nieuwstadt, *J. Fluid Mech.* **268** (1994) 175.

[72] H. Le and P. Moin, Direct numerical simulation of turbulent flow over a backward-facing step, *Report No. TF-58*, Dept. of Mech. Eng. (Stanford University, 1994).

[73] B.J. Daly and F.H. Harlow, *Phys. Fluids* **13** (1970) 2634.

[74] T.J. Craft, J.W. Kidger and B.E. Launder, Importance of third-moment modelling in horizontal, stably-stratified flows, in *Proc. 11th Shear Flows Symposium* (Grenoble, 1997) 20.13.

[75] B.E. Launder and W.C. Reynolds, *Phys. of Fluids* **26** (1983) 1157.

[76] K. Hanjalić and S. Jakirlić, *Appl. Scient. Res.* **51** (1993) 513.

[77] J.C. Rotta, *Zeitschrift der Physik* **129** (1951) 547.

[78] M.M. Gibson and B.E. Launder, *J. Fluid Mech.* **86** (1978) 491.

[79] S. Fu, M.A. Leschziner and B.E. Launder, Modelling strongly swirling recirculating jet flow with Reynolds-stress transport closure, in *Proc. 6th Symposium on Turbulent Shear Flow* (Toulouse, 1987) 17.6.1.

[80] C.C. Shir, *J. Atmos. Sci.* **30** (1973) 1327.

[81] T.J. Craft and B.E. Launder, *AIAA J.* **30** (1992) 2970.

[82] R.M.C. So, Y.G. Lai, H.S. Zhang and B.C. Hwang, *AIAA J.* **29** (1991) 1819.

[83] N. Shima, A Reynolds-stress redistribution model applicable up to the wall, in *Proc. Int. Symp. on Mathematical Modelling of Turbulent Flows* (Tokyo, 1995) 309.

[84] N.Z. Ince, P.L. Betts and B.E. Launder, Low Reynolds number modelling of turbulent buoyant flows, in *Proc. EUROTHERM Seminar 22*, Turbulent Natural Convection in Cavities, Delft, Editions européenes Thermique et Industrie, edited by R.A.W.M. Henkes and C.J. Hoogendoorn (Paris, 1994) 76.

[85] P.A. Durbin, *J Fluid Mech.* **249** (1993) 465.

[86] T.H. Shih and J.L. Lumley, Modelling of pressure correlation terms in Reynolds-stress and scalar-flux equations. *Report FDA-85-3*, Sibley School of Mech. and Aerospace Eng. (Cornell University, 1985).

[87] S. Fu, B.E. Launder and D.P. Tselepidakis, Accommodating the effects of high strain rates in modelling the pressure-strain correlation. *Report TFD/87/5* Mechanical Engineering Dept., UMIST (Manchester, 1987).

[88] C.G. Speziale, S. Sarkar and T.B. Gatski, *J. Fluid Mech.* **227** (1991) 245.

[89] T.J. Craft, L.J.W. Graham and B.E. Launder, *Int. J. Heat Mass Transfer* **36** (1993) 2685.

[90] V. Haroutunian, N. Ince and B.E. Launder, A new proposal for the e equation, in *Proc. 3rd UMIST CFD Colloquium*, UMIST (Manchester, 1988) Paper 1.3.

[91] C.R. Yap, *Turbulent heat and momentum transfer in recirculating and impinging flows*. Ph.D. Thesis (University of Manchester, 1987).

[92] T.J. Craft, *Int. J. Heat Fluid Flow* (1998) (in press).

[93] K. Hanjalić, S. Jakirlić and I. Hadžić, Computation of oscillating turbulent flows at transitional Re-numbers. *Turbulent Shear Flows* 9, edited by F. Durst *et al.* (Springer, 1995) 323.

[94] S.B. Pope, *J. Fluid Mech.* **72** (1975) 331.

[95] W. Rodi, *Math. Mech.* **56** (1976) 219.

[96] B.E. Launder, G.J. Reece and W. Rodi, *J. Fluid Mech.* **68** (1975) 537.

[97] C.G. Speziale, *J. Fluid Mech.* **178** (1987) 459.

[98] A. Yoshizawa, *Phys. Fluids* **27** (1987) 1377.

[99] T.-H. Shih, J. Zhu and J.L. Lumley, A realisable Reynolds stress algebraic equation model. *NASA TM105993* (1993).

[100] R. Rubinstein and J.M. Barton, *Phys. Fluids A* **2** (1990) 1472.

[101] T.B. Gatski and C.G. Speziale, *J. Fluid Mech.* **254** (1993) 59.

[102] T.J. Craft, B.E. Launder and K. Suga, A non-linear eddy-viscosity model including sensitivity to stress anisotropy, in *Proc. 10th Symp. on Turbulent Shear Flows*, Pennstate **3** (1995) 23.19.

[103] T.J. Craft, B.E. Launder and K. Suga, *Int. J. Heat Fluid Flow* **18** (1997b) 15.

[104] F.S. Lien and P.A. Durbin, Non-linear $k-\varepsilon-v^2$ modelling with application to high-lift, in *Proc. Summer Prog.*, Centre For Turbulence Research (Stanford University, 1996) 5.

[105] F.S. Lien, W.L. Chen and M.A. Leschziner, Low-Reynolds-number eddy-viscosity modelling based on non-linear stress-strain/vorticity relations. *Engineering Turbulence Modelling and Measurements* – 3, edited by W. Rodi and G. Bergeles (Elsevier, 1996) 91.

[106] D.B. Taulbee, J.R. Sonnenmeier and K.M. Wall, Application of a new non-linear stress-strain model to axisymmetric turbulent swirling flows. *Engineering Turbulence Modelling and Experiments* 2, edited by W. Rodi and F. Martelli (Elsevier, 1996) 103.

[107] A.D. Apsley and M.A. Leschziner, *Int. J. Heat Fluid Flow* **19** (1997) 209.

[108] L.B. Ellis and P.N. Joubert, *J. Fluid Mech.* **62** (1974) 65.

[109] S. Hogg and M.A. Leschziner, *AIAA J.* **27**, (1989) 57.

[110] R. Cresswell, V. Haroutunian, N.Z. Ince, B.E. Launder and R.T. Szczepura, Measurement and modelling of buoyancy-modified, elliptic turbulent shear flows, in *Proc. 7th Symp. on Turbulent Shear Flows* (Stanford, 1989) 12.41.

[111] T. Bo and H. Iacovides, *Int. J. Num. Meths. Heat Fluid Flow* **6** (1996) 47.

[112] S. Obi, K. Aoki and S. Masuda, Experimental and computational study of turbulent separated flow in an asymmetric diffuser, in *Proc. 9th Symp. on Turbulent Shear Flows*, 3 (Kyoto, 1993) 305.

[113] O. Piccin and D. Cassoudesalle, Etude dans la soufflerie F1 des profils AS239 et AS240. *ONERA Technical Report, PV 73/1685 AYG* (1987).

[114] W.C. Zierke and S. Deutsch, The measurement of boundary layers on a compressor blade in a cascade. *NASA CR-185118* (1989).

[115] D. Cooper, D.C. Jackson, B.E. Launder and G.X. Liao, *Int. J. Heat and Mass Transfer* **36** (1993) 2675.

[116] H.P. Kreplin, H. Vollmers and H.U. Meier, Wall shear stress measurements on an inclined prolate spheroid in the DFVLR 3 m × 3 m low speed wind tunnel, Goettingen. *DFVLR Report IB 222-84 A 33* (1985).

[117] H.U. Meier, H.P. Kreplin, A. Landhauser and D. Baumgarten, Mean velocity distribution in 3D boundary layers developing on a 1:6 prolate spheroid with artificial transition. *DFVLR Report IB 222-84 A11* (1984).

[118] D.O. Davis and F. Gessner, Experimental investigation of turbulent flow through a circular-to-rectangular duct. *AIAA Paper 90-1505* (1990).

[119] W.J. Devenport and R.L. Simpson, *J. Fluid Mech.* **210** (1990) 23.

[120] J.L. Fleming, R.L. Simpson and W.J. Devenport, *Experiments in Fluids* **14** (1993) 366.

[121] D. Barberis and P. Molton, Shock wave-turbulent boundary layer interaction in a three dimensional flow – laser velocimeter results. *Technical Report, ONERA TR.31/7252AY* (1992).

[122] D. Barberis and P. Molton, Shock wave-turbulent boundary layer interaction in a three dimensional flow. *AIAA Paper 95-0227* (Reno, Nevada, 1995).

[123] L.E. Putnam and C.E. Mercer, Pitot-pressure measurements in flow fields behind a rectangular nozzle with exhaust jet for free-stream Mach numbers of 0, 0.6 and 1.2. *NASA TM 88990* (1986).

[124] F.-S. Lien and M.A. Leschziner, *Comp. Meths. Appl. Mech. Engrg.* **114** (1994) 123.

[125] F.-S. Lien and M.A. Leschziner, *Int. J. Num. Meths. in Fluids* **19** (1994) 527.

[126] B.P. Leonard, *Comp. Meths. Appl. Mech. Engrg.* **19** (1979) 59.

[127] P.Batten, M.A. Leschziner and U.C. Goldberg, *J. Comput. Phys.* **137** (1997) 38.

[128] F.S. Lien and M.A. Leschziner, *The Aeronautical J.* **99** (1995) 125.

[129] W.L. Chen, F.S. Lien and M.A. Leschziner, *Int. J Heat Fluid Flow* **19** (1997) 307.

[130] W.L. Chen, F.S. Lien and M.A. Leschziner, in *Proc. 11th Symp. on Turbulent Shear Flows* (Grenoble, 1997) 1.13.

[131] F.R. Menter, Zonal two equation $k - \omega$ turbulence models for aerodynamic flows. *AIAA Paper 93-2906* (1993).

[132] K. Hanjalić, *Int. J Heat Fluid Flow* **15** (1994) 178.

[133] W. Haase, E. Chaput, E. Elsholz, M.A. Leschziner and U.R. Müller (eds.) ECARP: European Computational Aerodynamics Research Project. II: Validation of CFD Codes and Assessment of Turbulent Models, *Notes on Numerical Fluid Mechanics* (Vieweg Verlag, 1996) 58.

[134] D.D. Apsley and M.A. Leschziner, Advanced turbulence modelling in a generic wing-body junction. *Flow Turbulence and Combustion* (2000) (to appear).

[135] P. Batten, M.A. Leschziner and T.J. Craft, Reynolds-stress modelling of transonic afterbody flows. *Aeronautical Journal* (2001) (to appear). (see also Proc. 1st Symp. on Turbulence and Shear Flow Phenomena, edited by S. Banerjee and J.K. Eaton (Begell) 215.

[136] M.A. Leschziner, H. Loyau and D.D. Apsley, Prediction of shock/boundary-layer interaction with non-linear eddy-viscosity models, *CDROM Proc. European Congress on Computational Methods in Applies Sciences and Engineering*, ECCOMAS 2000, Barcelona.

[137] P. Bradshaw, *Exp. Fluids* **16** (1994) 203.

[138] H. Loyau, P. Batten and M.A. Leschziner, *Flow, Turbulence and Combustion* **60** (1998) 257.

COURSE 5

COMPUTATIONAL AEROACOUSTICS

R. MANKBADI

Embry-Riddle Aeronautical University,
Daytona Beach, FL, U.S.A., and
ICOMP, NASA Glenn Research
Center, Cleveland, OH, U.S.A.

Contents

COMPUTATIONAL AEROACOUSTICS

R. Mankbadi

1 Fundamentals of sound transmission

Sound propagates as a wave that evolves as it travels away from the source. Far from the source, its curvature diminishes and it resembles a one-dimensional plane wave propagating radically outward with constant speed, c (Fig. 1).

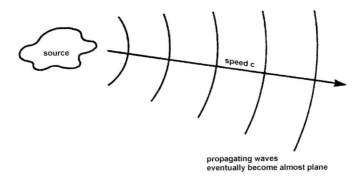

Fig. 1. Sound waves propagating away from a source.

Consider a fluid at rest with uniform pressure p_0 and density ρ_0. When this is perturbed by a sound wave p and ρ at position x and time t, they change to $p_0 + p'(x,t)$ and $\rho_0 + \rho'(x,t)$. The ratios (p'/p) and (ρ'/ρ) are usually $\ll 1$. If this is the case, the products of the perturbations involved in acoustic waves can be neglected. Consequently, the acoustic field response can usually be linearized.

A sound wave transports energy with it that energizes our eardrums to vibrate. Tones of frequencies ranging between 20–20 000 Hz (cycles per second) are audible. The human ear is most sensitive to sound in 1–5 kHz range.

1.1 One-dimensional wave analysis

Far from their source all sound waves become virtually plane waves. The flow parameters p, ρ, v are functions of a single space variable, x, and time, t. Since there are no pressure gradients in either the y or z directions, the accelerations in these directions are zero and $v(x,t) = (u(x,t), 0, 0)$.

Consider the mass conservation in control volume $(\delta x, 1, 1)$ (in x, y, z directions):

$$\frac{\partial \rho'}{\partial t} \delta x = [(\rho_0 + \rho')u](x,t) - [(\rho_0 + \rho')u](x + \delta x, t).$$

Neglecting products of small quantities gives the linearized equation:

$$\frac{\partial \rho'}{\partial t} + \rho_0 \frac{\partial u}{\partial x} = 0. \tag{1.1}$$

In a sound field the pressure represents a greater stress than that induced by viscosity. The ratio of the two stresses $2\pi c\lambda/v \gg 1$, where $\omega =$ angular frequency and $\lambda =$ wavelength. The second has to travel $\omega\lambda^2/v$ wavelengths before viscous effects are important. Thus, the momentum conservation can be written, neglecting viscosity, as:

$$\rho_0 \frac{\partial u}{\partial t} \delta x = p'(x,t) - p'(x + \delta x, t).$$

Which, upon linearization becomes:

$$\rho_0 \frac{\partial u}{\partial t} + \frac{\partial p'}{\partial x} = 0. \tag{1.2}$$

Manipulating the conservation and momentum equations gives:

$$\frac{\partial^2 \rho'}{\partial t} - \frac{\partial^2 p'}{\partial x^2} = 0. \tag{1.3}$$

The relationship between p' and ρ' depends on the fluid and heat exchange:

$$p = p_0 + (\rho - \rho_0) \frac{\mathrm{d}p}{\mathrm{d}\rho}(\rho_0) + \cdots$$

Linearization gives:

$$p' = \rho' \frac{\mathrm{d}p}{\mathrm{d}\rho}(\rho_0)$$

$$c^2 = \frac{\mathrm{d}p}{\mathrm{d}\rho}(\rho_0).$$

Then substituting $p' = c^2\rho'$, gives:

$$\frac{1}{c^2}\frac{\partial^2 p'}{\partial t^2} - \frac{\partial^2 p'}{\partial x^2} = 0. \qquad (1.4)$$

This is a 1-D equation for waves traveling at speed, c. The small pressure perturbations are organized into sound waves. The same equation describes the propagation of electromagnetic waves in free space.

1.1.1   General solution of the wave equation

$$p'(x, t) = f(x - ct) + g(x + ct) \qquad (1.5)$$

where $p'(x, t) = f(x + ct)$ represents waves traveling in the direction of increasing x and $p'(x, t) = g(x+ct)$ represents waves traveling in the direction of decreasing x. The pressure wave maintains its form but travels at speed, c (Fig. 2).

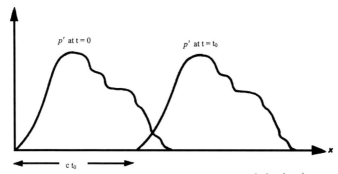

- Development of pressure perturbations in the traveling wave $p'(x,t) = f(x - ct)$.

Fig. 2.

A particular form of waves when f and g are harmonic functions is:

$$p'(x, t) = Ae^{i\omega\left(t - \frac{x}{c}\right)} + Be^{i\omega\left(t + \frac{x}{c}\right)} \qquad (1.6)$$

where ω is the frequency in rad/s, and the wave period is $2\pi/\omega$ s with wavelength $\lambda = 2\pi c/\omega$.

1.1.2   The particle velocity

We show herein that ρ', u can be evaluated from p' by considering the simple wave traveling to the right: $p'(x, t) = f(x - ct)$.

Since
$$\rho' = p'/c^2.$$

Then
$$\rho' = c^{-2}f(x - ct).$$

Thus,
$$\frac{\partial \rho'}{\partial t} = \frac{1}{c^2}\frac{\partial f(x - ct)}{\partial t}.$$

Let $(x - ct) = X$, then

$$\frac{\partial \rho'}{\partial t} = -\frac{1}{c^2}\frac{df(X)}{dX}\frac{\partial X}{\partial t}$$

$$\frac{\partial \rho'}{\partial t} = -\frac{1}{c}\frac{df}{dX}.$$

Using the linearized mass equation, $\dfrac{\partial \rho'}{\partial t} + \rho_0\dfrac{\partial u}{\partial x} = 0$, gives:

$$\frac{\partial u}{\partial x} = \frac{1}{\rho_0 c}\frac{df(X)}{dX} = \frac{1}{\rho_0 c}\frac{\partial f(x - ct)}{\partial x}.$$

Integrating and noting that u has zero mean value gives:

$$u = \frac{1}{\rho_0 c}f(x - ct).$$

Note that all flow quantities are functions of x and t in $(x - ct)$ combination only and that $p' = \rho_0 cu$. For the wave propagating in the negative direction, $p'(x,t) = g(x + ct)$, $p' = -\rho_0 cu$. The acoustic impedance is defined as $p'/u = \rho_0 c$.

1.2   Three-dimensional sound waves

In unconfined spaces, all sound fields are three-dimensional. The flow parameters will be functions of the three space coordinates and time. The velocities of the particles displaced by the sound waves will be a vector field $\mathbf{v}(\mathbf{x},t)$ or $(\mathbf{v}_1(\mathbf{x},t), \mathbf{v}_2(\mathbf{x},t), \mathbf{v}_3(\mathbf{x},t))$. Consider the mass conservation for a small volume δV enclosed by a surface s.

$$\delta V\frac{\partial \rho'}{\partial t} = -\int_S \rho_0\mathbf{v}\cdot\mathbf{n}ds.$$

According to Gauss's theorem, $\displaystyle\int_S \rho_0\mathbf{v}\cdot\mathbf{n}ds = \rho_0\text{div}\cdot\mathbf{v}\delta V$

$$\frac{\partial \rho'}{\partial t} + \rho_0\text{div}\mathbf{v} = 0. \tag{1.7}$$

In Cartesian coordinates,

$$\frac{\partial \rho'}{\partial t} + \rho_0 \frac{\partial v_i}{\partial x_i} = 0. \tag{1.8}$$

The momentum equations can be written as:

$$\rho_0 \frac{\partial \mathbf{v}}{\partial t} + \mathrm{grad} \cdot p' = 0. \tag{1.9}$$

In Cartesian coordinates,

$$\rho_0 \frac{\partial v_i}{\partial t} + \frac{\partial p'}{\partial x_i} = 0. \tag{1.10}$$

The wave equation can be obtained as:

$$\frac{\partial^2 \rho'}{\partial t^2} - \nabla^2 p' = 0.$$

Use $p' = c^2 \rho'$ to obtain:

$$\frac{1}{c^2} \frac{\partial^2 p'}{\partial t^2} - \nabla^2 p' = 0. \tag{1.11}$$

Since $\rho' = p'/c^2$ then ρ' satisfies the wave equation. Differentiation of equation (1.11) shows that $\partial p'/\partial x_i$ satisfies the wave equation. Hence, we can deduce from equation (1.10) that the velocity field also satisfies the wave equation.

$$\frac{1}{c^2} \frac{\partial^2 \mathbf{v}}{\partial t^2} - \nabla^2 \mathbf{v} = 0. \tag{1.12}$$

Taking the curl of equation (1.9), we get $\partial(\mathrm{curl} \cdot \mathbf{v})/\partial t = 0$. Since the fluid is perturbed from an initial state of rest, the flow is irrotational and curl $V = 0$.

1.3 Sound spectra

The total composition of noise can be investigated by spectral analysis using Fourier transform. According to Fourier's theorem, a function, $s(t)$, can be represented as an integral of harmonic elements:

$$s(t) = \frac{1}{2\pi} \int_{-\infty}^{+\infty} F(\omega) e^{i\omega t} d\omega$$

$$F(\omega) = \int_{-\infty}^{+\infty} s(t) e^{-i\omega t} dt$$

F is the spectral decomposition of $s(t)$ or the strength of each harmonic element. If $s(t)$ is the sound pressure perturbation, the above equation indicates that it is made up of harmonic waves, each with time dependence $e^{i\omega t}$, with an amplitude, $F = \delta\omega/2\pi$, of the waves with frequencies in the band between ω and $\omega + \delta\omega$.

1.3.1 Spectral composition of a square pulse

If $s(t) = s$ between t_0 and $(t_0 + \tau)$, but zero at all other times.

$$F(\omega) = s \int_{t_0}^{t_1+\tau} e^{-i\omega t}dt = -\frac{is}{\omega}e^{i\omega t_0}(1 - e^{-i\omega\tau})$$

$$F(\omega) = s\tau e^{i\omega\left(t_0 + \frac{\tau}{2}\right)} \frac{\sin(\omega\tau/2)}{\omega\tau/2}.$$

1.3.2 Spectral composition of a harmonic signal

If $s(t)$ is a harmonic signal $S(t) = Ie^{i\omega_0 t}$

$$F(\omega) = I \int e^{i(\omega_0 - \omega)t}dt$$

$$F(\omega) = 2\Pi I\delta(\omega_0 - \omega).$$

The spectrum of a function can be continuous or discrete.

Most real noise has continuous spectra, but if pure tones are present, the spectrum will have spikes. Experimentally pressure spectrum is found by measuring Sound Pressure Level (SPL) in a special frequency band over the frequencies of interest. Most sound meters have bandwidths wider than 1 Hz (of 1 and 1/3 octave). An octave is a bandwidth over which the frequency doubles (center frequency is $\sqrt{2}$ bigger than lowest frequency in the octave). A one third octave is the interval between two frequencies having a ratio of $(2)^{1/3}$. Commonly used octaves are 37.5–75 Hz, 75–150 Hz, etc.

1.4 Logarithmic scales for rating noise

A logarithmic scale is used to measure the sound levels because of the wide range involved. (10^{-10} in human whisper, 10^{-5} in human shout, 10^5 W emitted from large jet aircraft at take off, 10^7 W acoustic energy emitted from large rocket launch.)

1.4.1 The Sound Power Level (PWL)

PWL (in dB decibels) $= 10 \log_{10}$ (sound power output/10^{-12} W)dB.
$\qquad\qquad\qquad\qquad = 120$ dB $+ 10 \log_{10}$ (sound power output in watts).

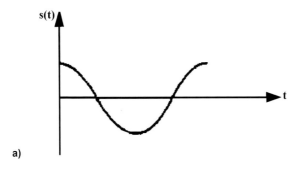

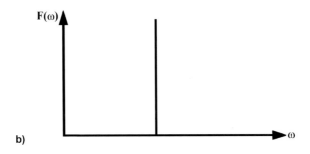

Fig. 3. Shows the Fourier transform of the pure tone in (a), and *vice versa*.

1.4.2   Sound Pressure Levels (SPL)

Is a measure of mean square level of pressure fluctuation on a logarithmic scale:

$$\begin{aligned} \text{SPL (in dB)} &= 20 \log_{10} (\text{prms} \,/\, 2 \times 10^{-5} \text{ N/m}^2) \\ &= 20 \log_{10} \text{prms} + 94 \\ &= 20 \log_{10} (\text{prms} \,/\, 2 \times 10^{-4} \,\mu\text{bar}). \end{aligned}$$

1.4.3   Pressure Band Level (PBL)

Is the SPL in decibels in a given bandwidth Δf (Fig. 4).

1.4.4   Pressure Spectral Level (PSL) per unit frequency

The sound pressure level, SPL, in the frequency band of 1 Hz centered at ω

$$\text{PSL} = \text{PBL} - 10 \log_{10}(\Delta f).$$

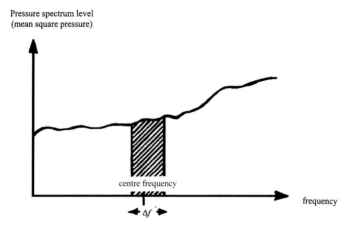

Fig. 4. The hatched area indicates the pressure band level, PBL, in the frequency band of width Δf and the mean pressure spectrum level per unit frequency in the band is $= \text{PBL} - 10 \log_{10}(\Delta f)$.

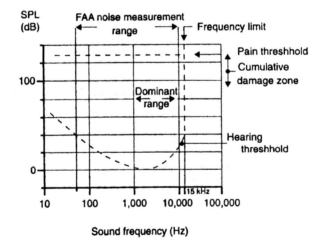

Fig. 5.

(Obtained by PBL by bandwidth if Δf is small enough for PSL to be constant within each frequency band.)

Noise Range.

Sound Pressure, p (Pa)	Sound Intensity, I (Wm^{-1})	SPL (dB)	Noise Source
200	10^{-6}	140	Painful; close to jet exhaust
62	10^{-7}	130	Saturn V at 2 nautical miles
20	10^{-8}	120	Turbojet at .3 nautical miles
2	10^{-10}	100	Jackhammer at 3 m (10 ft)
.2	10^{-12}	80	Suburban Street
.02	10^{-14}	60	Speech at 1 m (3 ft)
.002	10^{-16}	40	Interior Domestic
2×10^{-4}	10^{-18}	20	Dormitory
6×10^{-5}	10^{-19}	10	Leaves Rustling
$P_{\text{ref}} = 2 \times 10^{-5}$	10^{-20}	0	Hearing Threshold

1.4.5   Acoustic intensity

The acoustic power carried per unit area of the wave front. For a plane wave, the intensity $I = pu$ (w/m^2), where $u =$ rms velocity associated with rms pressure.

$$u = p^2/\rho c \text{ (since } u = p/\rho_{0c})$$

for air at $(0\ °\text{C and } 101 \text{ kPa})$ $\rho_{0c} = 428$ kgm^{-2} s^{-1}

$$I_{\text{ref}} = (4 \times 10^{-10})/428 \approx 1e - 12 \text{ w/m}^2.$$

In a plane wave, the wave front is flat and perpendicular to the direction of travel. For more complicated waves, such spiraling waves in ducts or in the near field close to the source, the intensity, power and pressure levels are not conveniently related as in a plane wave.

1.4.6   Overall pressure levels

Pressure or power signals have a frequency spectrum. The SPL or PWL referred to may be the total signal level (spectrum integrated over a frequency range), or the component of the spectrum at a particular frequency. If the overall signal is used to obtain the rms pressure or power, the levels are known as the overall sound pressure level OAPWL. The results depend on the bandwidth over which the signal is summed.

1.4.7   Subjective noise measures

Unlike pressure and intensity, which can be measured by a meter, loudness is purely subjective since it depends on the response of the listener. The human ear responds over a range of about 120 dB (10^6 in P_{rms}), and can feel changes in the sound level ≥ 3 dB. The response is stronger to sound at some frequencies than others, and sounds of the same intensity but of different frequencies will not appear to be equally loud. There is a marked fall in human perceived noisiness below 300 Hz and above 10 kHz.

Subjective noise measures are used to rate the annoyance produced on a group of people. The simplest ratings weigh the overall SPL in a manner related to the sensitivity of the human ear. The most common is the A weighted SPL expressed in dBA or dB(A). (reduces the effect of noise outside the range of 1–5 kHz). There is also B, C, D Scales.

1.4.8 Perceived Noise Level (PNL)

The civil noise standard, FAR 36 (U.S. FAA 1969) adopted the concept of perceived noise level (PNL) to evaluate noise intrusion. Its numerical value was intended to represent SPL of the octave band of "white noise" centered at 1 kHz that would be judged as equally noisy as the sound to be rated. The noise is first analyzed into third octave SPL before applying appropriate correlation for amplitude and frequency. When spectrum is broad band PNL is a good indicator of annoyance. When the noise spectrum contains a strong tone, PNL is not a good indicator of annoyance.

1.4.9 Tone Corrected Perceived Noise Level (PNLT)

Third octave spectrum is examined to determine if tones are presented evident by protrusion of one band above its neighbor by a prescribed ratio. If so, a correction is added to PNL depending on the frequency and extent of protrusion.

1.4.10 Effective Perceived Noise Level (EPNL)

Noise caused annoyance is also related to the time over which the noise level is high. FAR 36 specifies 24 one-third octave bands in the frequency spectrum between 50 and 10 000 Hz and correct the PNL derived in a specified way. (A scale) from the SPL in each band for tone and duration.

PNLT at 0.5 s intervals are summed to produce EPNL as follows:

$$\text{EPNL} = 10 \log \sum_{i=0}^{d} \text{anti} \log \frac{PNLTi}{10} - B.$$

The summation is over the time for which the PNLT is within 10 dB of the peak level.

2 Aircraft noise sources

Flow-generated noise is an important issue in several technical applications including turbomachinery, rotorcraft, inlets, nozzles, ducts, mixer-ejectors, and jet noise. The recent growing interest in turbulence noise is largely due to efforts to develop quieter and faster airplanes. The success of this new technology is contingent upon reducing its jet exhaust noise. This noise is generated by the time-dependent turbulence fluctuations in the near field, which are associated with pressure fluctuations that propagate to the Fairfield producing the radiated sound. The emphasis of this chapter is on the aerodynamically generated sound associated with airplanes.

2.1 Noise regulations

Noise-caused annoyance is dependent on its level as well as the time over which the noise level is high. FAR 36 specifies 24 one-third octave bands in the frequency spectrum between 50 and 10 000 Hz and correct the PNL derived in a specific way (A scale) from the SPL in each band for tone and duration. Take off and landing are the two most important operating points relevant to community noise. Figure 6 shows the Perceived Noise Level (PNL) for three engine conditions typical of take off and landing. Figure 7 shows the EPNL for various aircrafts, and Figure 8 shows the stages 2 and 3 of noise regulation.

2.2 Contribution from various components

Aerodynamic noise of airctrafts is attributed to various components: Fan, Jet, Turbine, Airframe, Core, Wheel Wells, etc. Figure 9 shows the relative contribution of each component at take off and approach.

2.3 Engine noise

As the figure indicates, the power of noise can be classified into engine noise [1–3], airhose noise [4–11], and jet noise [12, 13]. Engine noise will be discussed herein.

The two operating points of interest for community noise, approach and takeoff correspond to subsonic and supersonic tip Mach numbers. As shown by narrow band in Figure 13. Subsonially, blade passage frequency and its harmonics are superimposed on a broadband component, while supersonically; all multiples of shaft frequency prominently radiates from turbofan inlets during takeoff.

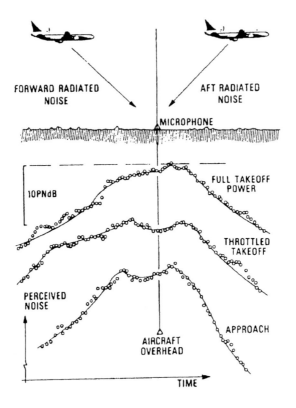

Fig. 6. Three flyovers for an aircraft with high bypass ratio.

2.3.1   Elements of the generation process

The four processes – blade unsteady aerodynamic response; acoustic coupling to the duct; propagation in the duct, which may contain other blade rows and have acoustically treated walls; and acoustic coupling (radiation) to the far field – have each been studied and modeled separately as convenient elements of overall problem.

2.3.2   Noise measurements

The noise characteristics of turbomachinery components in an aircraft engine are usually quantified in terms of several noise measurement parameters. These include the following:

1. overall sound power level (OAPWL);

2. sound power level spectrum (PWL (f));

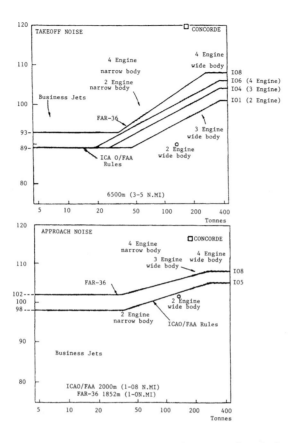

Fig. 7. Take-off and approach noise levels regulations and typical measurements, abscissa shows all up weight.

3. overall sound pressure level directivity (OASPL (θ));

4. sound pressure level spectrum (SPL (θ, f));

5. an appropriate subjective noise level, *e.g.*, perceived noise level (PNL (θ)).

These parameters are obtainable from measurements of sound pressure made with microphones placed at strategic locations around the component (or engine) during a test. The overall sound power level (OAPWL) produced by a turbomachinery is generally a function of aerodynamic and performance-related parameters such as air flow rate, tip speed, pressure ratio and/or shaft horsepower, and geometric design parameters.

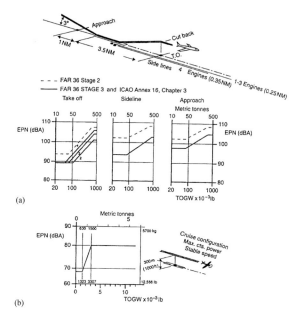

(a)

(b)

Fig. 8. Take-off and approach noise levels regulations and typical measurements, abscissa shows all up weight.

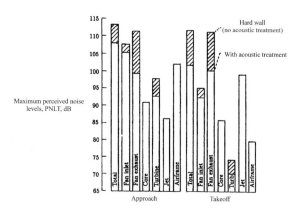

Fig. 9. Component fly-over noise levels for advanced turbofan.

The sound power level spectrum (PWL (f)) is the distribution of the generated sound over a range of audible frequencies. The directivity characteristics of the noise generated at a given frequency describe how the internally generated sound power is distributed in the radiation field at some

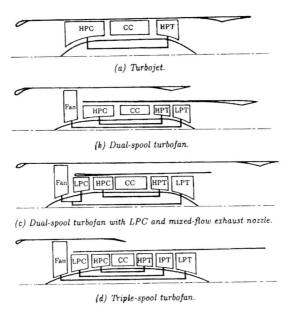

(a) Turbojet.

(b) Dual-spool turbofan.

(c) Dual-spool turbofan with LPC and mixed-flow exhaust nozzle.

(d) Triple-spool turbofan.

Fig. 10. Typical turbojet-turbofan engine schematic arrangements.

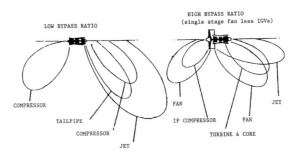

Fig. 11. Typical Sound Intensity distribution arising from various components.

distance away from the turbomachinery component or engine in terms of sound pressure (SPL) measured by microphones or heard by the ear. Sound Pressure level can have both azimuthal directivity and polar directivity. Azimuthal directivity describes the variation of sound pressure azimuthally, or around the machine axis. In most cases, azimuthal variations are small and can be neglected, especially for the broadband components of noise. In certain special cases, the azimuthal variation in sound pressure can be significant for discrete tones. Polar directivity refers to the variation in sound pressure from inlet centerline to exhaust centerline on a constant radius in

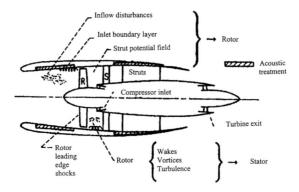

Fig. 12. Schematic cross-section turbofan engine with turbomachinery noise-generating mechanisms.

a fixed azimuthal plane. The polar directivity of broadband noise is usually smooth with maximum levels occurring near the engine (or component) inlet and/or exhaust exit. The polar directivity of discrete tones can be highly irregular with several peaks and valleys referred to as lobes, the number and size of which depend on the type of tone and the source mechanism which produces it. Typical examples of polar directivity patterns are shown in Figure 15.

Subjective noise levels refer to appropriately integrated or summed sound pressure levels which best represent human ear annoyance to the generated sound field. Summing is performed in the frequency domain and, often, also in the time domain. For example, perceived noise level (PNL) refers to a summation, over all 1/3-octave frequency bands, of the sound pressure levels at a given observer polar angle, with level in each band weighted by a factor which represents the degree of annoyance to noise observed at that particular frequency. Effective perceived noise level (EPNL) refers to a time integration of PNL received by an observer as the noise source passes by, such as that which would occur during an aircraft flyover. It represents effects of time duration over which a given PNL must be "endured".

An example of a typical high-bypass-ratio turbofan engine noise filed is shown in Figure 16. The component contributions of fan noise, combustor noise, LPT noise, and jet noise are shown in Figure 15 to demonstrate the eminent sources which typically control the noise in the various regions of the spectrum and directivity patterns. We see, for example, that the fan noise usually contributes the highest levels in the forward arc at midrange and high frequencies and in the aft arc at high frequencies. The turbine (LPT) only contributes in the aft arc at high frequencies. The jet dominates the low

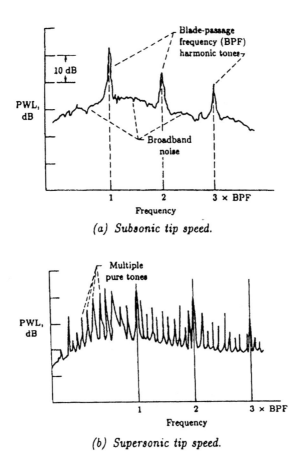

(a) *Subsonic tip speed.*

(b) *Supersonic tip speed.*

Fig. 13. Typical turbomachinery sound power spectra.

frequencies over most of the directivity arc, while the combustor significantly around the sideline angles to 120°, mostly at low to mid frequencies.

The trends shown in Figures 13–16 are typically for bypass from about 3–8. For low-bypass-ratio engines (mass flow ratio less than about 1.0), the jet noise is the greatest contributor to the overall noise and may actually control the total noise in the aft arc during takeoff conditions.

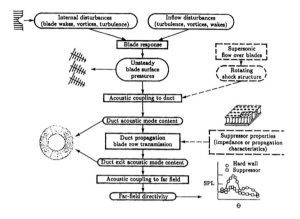

Fig. 14.

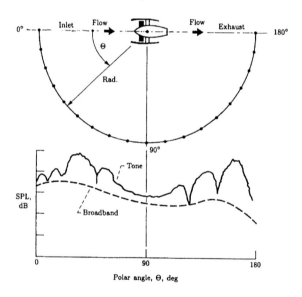

Fig. 15. Typical turbomachinery sound pressure level polar directivity patterns.

3   Methodology for jet noise

3.1   Jet noise physics

Consider a round jet issuing from a nozzle of diameter D (Fig. 17). Due to entrainment the jet spreads in the axial direction x. Three flow-regimes can be identified [14–23]. In the first, a potential core is firmed wherein

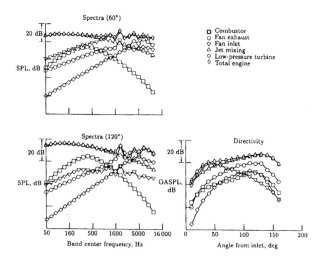

Fig. 16. Typical turbofan engine component noise source contributions at takeoff power.

the mean flow velocity is constant. The nozzle-lip's shear layer grows and reaches the centerline. This marks the end of the potential core. This is followed by the transitional regime leading to the fully developed regime. In the fully developed regime the flow is elf similar and the mean flow velocity decays like the square route of the inverse of x. The initial region of the jet is characterized by large-scale coherent structure. The fully developed region of the jet is dominated by small-scale random turbulence.

The unsteady structure in the jet radiates noise with a peak on the forward direction as Figures 18 and 19 show (Refs. [24–29]). At high speeds, the radiation pattern is characterized by Mach-wave emission, which can be detected in Figure 20. The structure radiates at various frequencies with a peak corresponding to Strouhal number of 0.2–0.6 (Figs. 5–7).

3.2 CAA for jet noise

In order to appreciate the numerical difficulties associated with jet simulation, let us briefly examine the physics involved [21]. The streamwise development of the jet can be split into three regimes. In the potential core regime, the shear-layer, formed at the nozzle lip, spreads and reaches the centerline of the jet marking the end of the potential core. The mean flow centerline velocity is constant within this core. This is followed by the transitional regime until the fully developed regime is reached. In measuring the acoustic field, the microphone is usually placed at a circle centered at

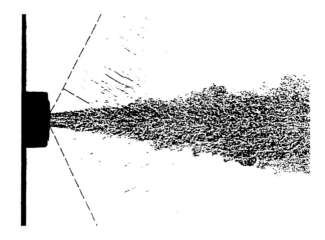

Fig. 17. Directional acoustic radiation from supersonic jet generated by instability waves.

the jet exit of a least $40D$ radius. Thus, the computational domain must extend radically to about $40D$. A computation domain needs to extend $60D$ downstream before the centerline velocity has considerably decayed. Here, D is the nozzle diameter. Though the mean flow of the jet may be axisymmetric, its unsteady structure is three- dimensional, and, in supersonic jets, the first helical mode could dominate over the axisymmetric mode.

Three distinct scales can be identified. In the acoustic field, the disturbance scales with the acoustic wavelength. In the jet flow, the structure can be classified into two. One scales with the nozzle diameter and sometimes called the "jet column mode". The other sometimes called the shear-layer inability mode, scales with the initial momentum thickness of the shear-layer, which is about 3% of the nozzle diameter. The relevant frequencies can be as high as $St = 10$, where $St = fD/U_j$, f is the frequency in Hz, U_j is the jet exit velocity. In supersonic jets, besides mixing noise, shock-cell structure is formed and interacts with the unsteady disturbances producing broadband and shock associated noise and the screech tones. To resolve all these scales in such extended computational domain, one must have use a high-order scheme that requires minimum grid points per wavelength to accurately capture the disturbance field.

As the disturbance in the flow field propagate to the acoustic field, they become less than 10^{-4} of the mean flow. If both the mean flow and the disturbances are to be captured accurately, the numerical error must be much less than that. Furthermore, because of the long propagation length,

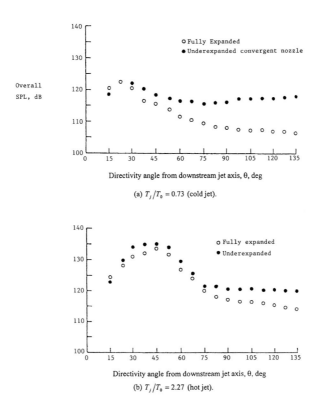

Fig. 18. Directivity of overall sound pressure level. $M_j = 1.372$.

numerical dispersion and dissipation errors accumulate. As such, dissipation and dispersion error should be minimized.

3.3 Wave-like sound source

Experimental observations [14, 15] indicate that turbulence in the noise-producing, initial region of shear layers, is dominated by large-scale organized structure. The fluid motion can, therefore, be split into three kinds of motion: a time-averaged motion, U; a periodic, organized, large-scale wavelike structure u'; and a fine-scale random turbulence, u'':

$$u(x, t) = U(x) + u'(x, t) + u''(x, t).$$

The pressure and density are similarly split. The integral energy approach to the study of this coherent structure seems to produce results consistent with observation (see for instance [17, 19, 21, 30, 31]. In this approach, as in

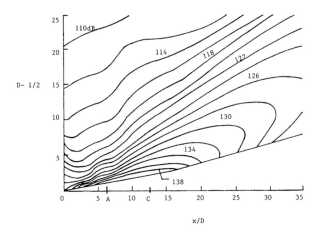

Fig. 19. Near field 1/3 octave band sound pressure level contours of Mach 1.5 supersonic jet at 10 kHz center frequency.

LES, the coherent component is calculated while accounting for the effect of the fine-grained turbulence on it. But rather than direct numerical computation, the following assumptions are made first based ideas for nonlinear instability waves. The coherent component composed of a finite number of wave-like frequency components, which can be expressed in the form:

Here ω is the frequency, n is the azimuthal wave number and $\Re$ denotes the real part. The assumption here is that the Fourier coefficient representing the coherent component can be separated into an amplitude that is a function of the downstream coordinate x, and a transversal shape function of the transversal coordinate (r or y) at a given location along the shear layer or jet. The amplitude, A, is determined from a nonlinear analysis. The transversal profile, $\hat{u}_{i,mn}(r)$, is taken as the eigenfunction given by the locally parallel linear stability theory. The phase angle, Ψ, is either given by the linear theory, as in numerous investigations, or governed by its own nonlinear evolution equation.

With these shape assumptions, the energy equations are derived and integrated in the transverse direction to yield the following system of simultaneous ordinary differential equations. The mean-flow momentum thickness $\delta_{\rm m}(x)$ is given by

$$\frac{1}{2}\frac{\mathrm{d}I_{\rm MA}}{\mathrm{d}\delta_{\rm m}}\frac{\mathrm{d}\delta_{\rm m}}{\mathrm{d}x} = I_{\rm MT}E_{\rm t} - \sum_{m,n} I_{{\rm MW},mn}E_{mn}.$$

(a) Fully expanded jet.

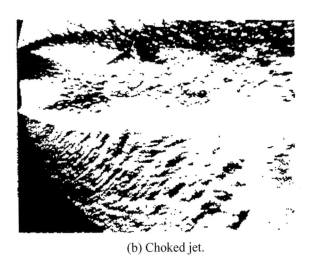

(b) Choked jet.

Fig. 20. Jet at supersonic speeds showing Mach waves outside jet boundary.

The turbulence kinetic energy $E_t(x)$ is given by

$$\frac{d}{dx}(I_{TA}E_t) = I_{MT}E_t + \sum_{m,n} I_{WT,mn}E_{mn}E_t - I_\epsilon E_t^{3/2}.$$

The energy of the mn wave $E_{mn}(x)$ is given by

$$\frac{d}{dx}(I_{WA,mn}E_{mn}) = I_{WM}E_{mn} - I_{WT,mn}E_{mn}E_t + W_{W_{mn}}.$$

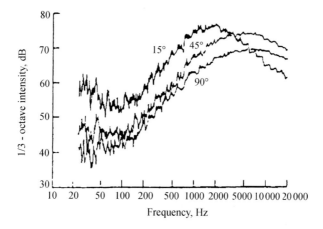

Fig. 21. Jet noise spectra at $\theta = 15°$, $45°$ and $90°$. $U_j = 195$ m/s.

The phase angle $\Psi_{mn}(x)$ is given by

$$E_{mn} I_{\mathrm{WA}} \frac{\mathrm{d}\Psi_{mn}}{\mathrm{d}x} = \pi St_{mn} E_{mn} + W_{\mathrm{W}_{mn}}.$$

Here $E_{mn} = |A_{mn}|^2$, and St is the Strouhal number. In this system of equations, I represents an integral that in general is a function of the momentum thickness δ_m, the Strouhal number St, and the azimuthal wavenumber n; and WW represents wave-wave interactions terms, which are given explicitly in [17].

In the integral energy approach, the nonlinear effects are accounted for by considering the interaction among various scales of motion (mean flow, turbulence, and various frequency components of the coherent structure). Since ordinary differential equations (rather than partial ones) are solved, the computational time is much reduced from that required for LES. Because of the assumptions involved, the integral-energy approach is restricted to the prediction of the sound source, and needs to be coupled with some other approach for sound propagation. The success of this approach depends mainly on the validity of the assumption regarding the profile for the coherent components. Using the integral energy approach, Mankbadi and Liu [22] compared the predicted growth of the excited coherent component to the experimental observations of Moore [24]. Satisfactory agreement was obtained as shown in Figure 17. The approach seems to be particularly successful in the initial region of developing shear layers, where most of the noise is generated.

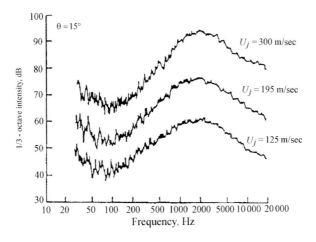

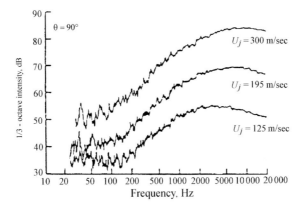

Fig. 22. Jet noise spectra at $\theta = 15°$, and $90°$ for various jet velocities.

3.4 Lighthill's theory

Lighthill [33, 34] considered sound radiation from a compact region of turbulent flow. The Navier–Stokes equations are re-arranged in the form:

$$\frac{\partial^2 \rho'}{\partial \tau^2} - c_0^2 \nabla^2 \rho' = \frac{\partial^2 T_{ij}}{\partial y_i \partial y_j} \tag{3.1}$$

where T_{ij} Lighthill's stress tensor,

$$T_{ij} = u_i u_j + \delta_{ij} \lfloor p - p_0 - c_0^2(\rho - \rho_0) \rfloor - e_{ij} \tag{3.2}$$

δ_{ij} is Kroneker delta, e_{ij} are the viscous stresses. Using Green's theorem, neglecting viscous stresses, the solution for the

$$p_s = \frac{1}{4\pi R_{0b} c_0^2} \iiint \left[\frac{\partial^2}{\partial x_i \partial x_j} \{\rho u_i u_j\} + \frac{1}{c_0^2} \frac{\partial^2}{\partial t^2} \{p - c_0^2 \rho\} \right] dV \qquad (3.3)$$

above equation can be given in terms of a volume integral.

Here p_s is the far-field sound pressure, u_i is the velocity vector, c_0 is the speed of sound in the medium surrounding the flow, and R_{0b} is the distance between the jet exit centerline and the observation point. The curly braces denote that the source term is calculated at the retarded time.

Upon making the assumption of a compact sound source, Lighthill's equation for the far-field sound can be approximated to

$$p_s = \frac{1}{4\pi R_{0b} a_0^2} \iiint \frac{\partial^2}{\partial t^2} \{\rho c_r^2\} r dr dx d\varphi, \qquad (3.4)$$

and the retarded time is given by

$$t_r = t - R_{0b}/c_0 + (x \cos\theta + r \sin\theta \cos\varphi)/c_0, \qquad (3.5)$$

where,

$$c_r = u \cos\theta + v \sin\theta \cos\varphi, \qquad (3.6)$$

θ is the emission angle, and φ is the azimuthal angle.

By writing the Fourier expansion in time of the sound source, the azimuthal integral can be performed and, for the axisymmetric case, Lighthill's equation reduces [23]

$$I = C \left| \iint F(x, r, St) e^{-2\pi i M_j x St \cos\theta} r dr dx \right|^2, \qquad (3.7)$$

where,

$$C = \frac{\pi^4}{(R_{0b}/D^2)} St^4 (U_e/C_0)^8, \qquad (3.8)$$

$$F(x, r, St) = A_0 J_0(\sigma) - i A_1 J_1(\sigma) - A_2 J_2(\sigma), \qquad (3.9)$$

$$A_0 = [\rho uu] \cos^2\theta + \frac{1}{2} [\rho vv] \sin^2\theta,$$

$$A_1 = [\rho uv] \sin 2\theta, \qquad A_2 = \frac{1}{2} [\rho vv] r \sin^2\theta, \qquad (3.10)$$

where square brackets denote Fourier-frequency component and J is the Bessel function of the argument

$$\sigma = 2\pi St M r \sin\theta. \qquad (3.11)$$

3.4.1 Application of Lighthill's theory

Early attempts for jet noise prediction relied heavily on Lighthill's theory. By adopting the view that the large-scale coherent structure is the dominant sound source in subsonic jets, several attempts have been directed toward calculating the corresponding sound field using Lighthill's theory. References [30,36,37] used analytical integrable, but modeled, coherent structure representation for the large-scale structure. Lighthill theory was then used to calculate the corresponding sound field.

To avoid encountering errors in the radiated field, due to the idealized source, Mankbadi and Liu [32] calculated the radiated field with coherent structures obtained from first-principles integral technique. Only the first-order noise was considered while the second-order terms were neglected. Their results have indicated that the radiation pattern of the axisymmetric modes resembles that of a longitudinal quadruple and that of the helical mode resembles that of a lateral quadruple (Figs. 23 and 24). The superposition gives a directional and spectral behavior that resembles some of the observed features. However, the directivity of the sound was weaker than the observed one. An attempt is made to account for these shortcomings by considering the self-noise contribution, and by considering sound radiation due to nonlinear interactions between various coherent structure modes Mankbadi [37]. The computed spectral pattern, directivity, and dependency on Mach number and Strouhal number, were found in close agreement with observations. But the level of noise attributed to the self-noise alone is much smaller than that of the shear noise or the measured one.

The application of Lighthill's theory in connection with large-eddy simulations is given in Mankbadi's *et al.*'s [23] wherein the sound source was obtained from LES. The directivity was found to be much weaker than what the observations indicate. This shortcoming might be attributed to the mean flow-acoustic interaction, which is not explicitly accounted for in Lighthill's theory. Furthermore, there are numerical difficulties associated with the use of Lighthill's theory as an extension technique coupled with LES simulation of the near field. To demonstrate this issue, Figure 25 shows the Lighthill's stress tensor $\rho u_i u_j$ which is used in Lighthill's integral as the sound source. This term is obtained from LES [45], and, as the figure shows, the source is wavy in nature. The source peaks in the radial direction around the shear layer. In the x-direction, it oscillates, grows and decays. The far field sound according to Lighthill's theory is the net cancellation upon performing the volume integral. Integration of an oscillatory integrand requires fine resolution and high accuracy. Another difficulty associated with using Lighthill's integral is that the source may not completely vanish within the computational domain, which is usually the case for supersonic jets. For such non-compact sources, Lighthill's volume integral may not converge

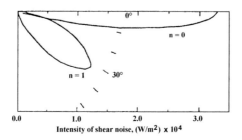

Fig. 23. Polar distribution of predicted shear noise intensity of coherent structure at $St = 0.8$ using Lighthill's Theory. $N = (a)$ 0, (b) 1.0.

within the computational domain. Thus, while for the subsonic results; for the supersonic case, Lighthill's results may be questionable.

The mathematical difficulty stems in part from the inability of the acoustic analogy to extract the sound "generation" problem from the interaction of acoustic waves with turbulence and the mean flow. It is possible to overcome some of these difficulties by moving certain terms from Lighthill's source term to the left-hand side of the equation [38, 39].

3.5 Kirckhoff's solution

Kirchhoff's formulation is extensively used in prediction of rotor-blades noise [40]. In this approach, the sound pressure is given in terms of a surface integral that includes information over a surface that enclosing all sound sources. The formulation for Kirchhoff's method for a sound source in a uniform stream U is given here [41]. Consider a surface S, that encloses all the sound sources. Outside this surface the convective wave equation applies:

$$\Delta^2 p' - \frac{1}{c_0^2} \left(U \frac{\partial}{\partial x} + \frac{\partial}{\partial t} \right)^2 p' = 0$$

where c_0 is the speed of sound, and U_0 is the free stream velocity, which is less than the speed of sound. The Prandtl-Glauert transformation

$$x_0 = x, \quad y_0 = By, \quad z_0 = Bz$$

is then used to convert the convective wave equation into the simple wave equation. After some algebra the pressure field can be expressed as:

$$p'(x, y, z, t) = -\frac{1}{4\pi} \int_S \frac{1}{R_0} \left[\frac{p'}{R_0} \frac{\partial R_0}{\partial N_0} + \frac{\partial p}{\partial N_0} + \frac{1}{c_0 B^2} \frac{\partial p}{\partial t} \frac{\partial}{\partial N_0} (R_0 - Mx_0) \right]_\tau \mathrm{d}S_0$$

$$(3.12)$$

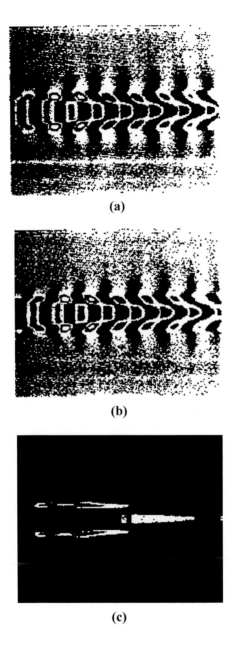

Fig. 24. Fourier transform of uu at the forcing frequency, a) real part b) imaginary part c) absolute values.

where the subscript $_0$ denotes the transformed variable, $N = (N_x, N_y, N_z)$ is the outward normal to the surface S, and the subscript τ indicates that all the values are calculated at the retarded time

$$t_r = t - \tau$$

R_0 is the distance between the observer and the surface point in Prandtl-Glauert coordinates:

$$R_0 = \left\{ (x - x_s)^2 + B^2 \left[(y - y_s)^2 + (z - z_s)^2 \right] \right\}^{\frac{1}{2}}$$

$$\tau = \frac{[R_0 + M(x - x_s)]}{c_0 B^2}$$

$$B = (1 - M^2)^{\frac{1}{2}}$$

and M is the free stream Mach number.

Equation (3.12) describes the sound pressure at a point (x, y, z) in terms o the information prescribed on the Kirchhoff's surface. The pressure and its normal and time derivative are required to be specified through numerical simulation, or some other means. This represents the difficulty with using Kirchhoff's method. The computational domain must be large enough so that the disturbances are already acoustic disturbances governed by the homogenous wave equation. Since the acoustic near field must already be given by some other means, Kirchhoff's method should be viewed as a means for extending an already calculated acoustic near field to the far field. The Kirchhoff method has been used by Lyrintzis and Mankbadi [42] for jet noise. The difficulty associated with in accruing of the method due to using numerically calculated pressure decivative (Fig. 9) has been added by Mankbadi et al. [43, 44].

4 Algorithms and boundary treatment

In CAA the objective is to capture wave-like disturbances. Finite difference schemes can suppress or distort these disturbances because of dispersion and dissipation errors. Furthermore, boundary treatment applied at a finite distance from the source can produce spurious modes that will render the computational results completely useless. As such, we discuss in this chapter discretization schemes and boundary treatment for CAA.

4.1 Algorithms for CAA

The importance of the dispersion and dissipation of a given scheme, when used in connection with CAA, was highlighted in [46, 47]. Consider a finite-difference representation of a wave-type variable

$$f = e^{i\omega t}.$$

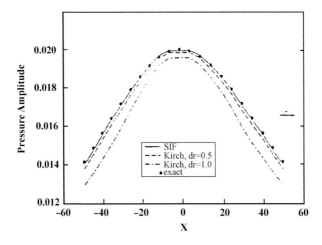

Fig. 25. Effect of grid spacing (dr) on the predicted pressure amplitude for a point source.

If a first-order, forward-difference scheme is used to discretize this function, then the discretized function f^- becomes

$$\frac{\mathrm{d}\breve{f}}{\mathrm{d}t} = \left[\left(\sin\frac{\omega}{2}\Delta t\right) / \left(\frac{\omega}{2}\Delta t\right)\right]\left[e^{i\frac{\omega}{2}\Delta t}\right]\frac{\mathrm{d}f}{\mathrm{d}t}.$$

The discrete approximation thus modifies the actual function by two factors. The first square- bracketed factor is a dissipative term that represents damping of the amplitude. The second square-bracketed factor is a dispersion term that represents a phase shift relative to the actual derivative of the function. Both effects are crucial in CAA, and can render the computed unsteady part of the solution completely unacceptable. As such, high-order accurate schemes are required for CAA.

Three workshops were recently held to address numerical issues pertaining to CAA. A number of benchmark problems were solved using various algorithms under controlled parameters. Several schemes were found to handle linear acoustic propagation quite well. Fewer schemes performed well for nonlinear problems. Problems involving shock-acoustic interactions or nonuniform curvilinear grids represent additional difficulties. Capturing shocks accurately usually requires dissipation, which could dampen the acoustic waves. Nonuniform curvilinear meshes could reduce the order of accuracy. Three discretization schemes, which have been successfully used for CAA of nonlinear problems, will be briefly given herein. Consider the one-dimensional equation:

$$\frac{\partial Q}{\partial t} + \frac{\partial F}{\partial x} = 0. \tag{4.1}$$

Where Q is the unknown vector and F is the flux in the x-direction.

4.1.1 The 2–4 scheme

The 2–4 scheme is fourth-order accurate in space and second-order accurate in time. This scheme is an extension of the classical second-order MacCormack [48]. As in the classical MacCormack scheme, a second-order accurate predictor-corrector method is used for the time marching, with one-sided spatial differences used in an alternating fashion as follows:

$$
\begin{aligned}
Q_i^{n+\frac{1}{2}} &= Q_i^n - \frac{\Delta t}{6\Delta x}(7F_i - 8F_{i-1} + F_{i-2})^n \\
Q_i^{n+1} &= \frac{1}{2}\left[Q_i^{n+\frac{1}{2}} + Q_i^n + \frac{\Delta t}{6\Delta x}(7F_i - 8F_{i+1} + F_{i+2})^{n+\frac{1}{2}}\right].
\end{aligned}
\tag{4.2}
$$

In a multidimensional application, the spatial differences can be applied as a series of one- dimensional operators. Because each operator is one-dimensional, the boundary treatment can be applied with relative ease.

4.1.2 The compact scheme

The compact scheme of Lele [49] is quite attractive, and can be written in the form:

$$\alpha D_{i-1} + (1 - 2\alpha)D_i + \alpha D_{i+1} = \frac{1}{\Delta x}(aF_{i-1} - aF_{i+1}) \tag{4.3}$$

where D is the operator and the coefficients "a" and α can be optimized to achieve the required degree of accuracy. The scheme has successfully been used to simulate several CAA problems. However, it is less friendly to boundary treatment, and often an exit zone or sponge layer needs to be imposed, such that spurious modes generated at the boundaries are gradually dampened. Hixon (2000) has attempted to overcome this difficulty by splitting the operator D into forward D^F and backward D^B ones, and writing the discretized equations in the form:
(add equation that $D = 1/2(Df + Db)$)

$$cD_{i-1}^F + [1 - (a+c)]D_i^F + aD_{i+1}^F = \frac{1}{\Delta x}[-(1-b)F_{i-1} - (2b-1)F_i + bF_{i+1}] \tag{4.4}$$

$$aD_{i-1}^B + [1 - (a+c)]D_i^B + cD_{i+1}^B = \frac{1}{\Delta x}[-bF_{i-1} + (2b-1)F_i + (1-b)F_{i+1}].$$

The stencil size for a 6th order compact is reduced from 5 points to 3 points due to the splitting, which helps the boundary stencil specification and stability.

4.1.3   The Dispersion-Relation-Preserving scheme

The Dispersion-Relation-Preserving (DRP) scheme of Tam and Webb [50] can be written as:

$$Q_{l,m}^{(n+1)} = Q_{l,m}^{(n)} + \Delta t \sum_{j=0}^{3} b_j K_{l,m}^{(n-j)}. \tag{4.5}$$

The Dispersion-Relation-Preservation (DRP) scheme of Tam and Webb (1993) is used with selective artificial damping in the form:

$$Q_{l,m}^{(n+1)} = Q_{l,m}^{(n)} + \Delta t \sum_{j=0}^{3} b_j K_{l,m}^{(n-j)} \tag{4.6}$$

where

$$
\begin{aligned}
K_{l,m}^{(n)} = {} & -\frac{1}{\Delta x} \sum_{j=-3}^{3} a_{l+j,m}^{(n)} - \frac{1}{\Delta r} \sum_{j=-3}^{3} a_j G_{l,m+j}^{(n)} \\
& -\frac{1}{r_m} G_{l,m}^{(n)} + S_{l,m}^{(n)} \\
& -\frac{\mu_a}{(\Delta r)^2} \sum_{j=-3}^{j=3} \mathrm{d}_j Q_{l+j,m}^{(n)} \\
& -\frac{\mu_a}{(\Delta r)^2} \sum_{j=-3}^{j=3} \mathrm{d}_j Q_{l,m+j}^{(n)}.
\end{aligned}
\tag{4.7}
$$

The coefficients appearing in the above two equations are given in Tam and Webb [50]. Here $K_{l,m}^{(n)}$ is a function of the flux. The coefficients of the scheme are chosen to represent more accurately the wave components over a wide range. For simple benchmark problems with DRP scheme requires fewer grid points per wavelength than in the fourth-order MacCormack case, but with more difficulty in boundary treatment. However, considerable progress has been made by Tam and his co-worker in developing boundary treatments suitable for the DRP scheme.

4.2   Boundary treatment for CAA

Unlike the physical problem, the computational domain is usually finite and numerical boundary treatments need to be applied at the boundaries

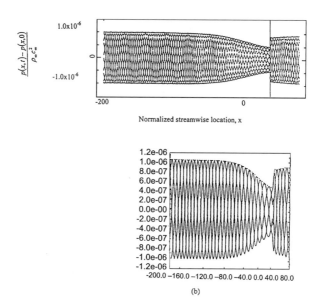

Fig. 26. Quasi -1D numerical simulation of acoustic-shock wave interaction in a convergent-divergent nozzle. a) Bui and Mankbadi's [51] unstructured scheme, b) Dong and Mankbadi [52] DRP scheme with selective damping.

to depict conditions at infinity. Errors in the boundary condition specification could generate spurious modes that propagate to the interior domain and render the computed solution entirely unacceptable. These modes arise from the boundary conditions imposed at the computational boundary, which mimics the physics only at infinity. It could also arise from the fact that the dispersion relation of the discretized equations differs from that of the differential one or from sudden changes in the differencing method or grid spacing. Unsteady boundary treatment represents a serious difficulty for CAA, but considerable progress has been made in this direction. There are several proposals for boundary treatments based on various ideas such as: characteristics-based boundary conditions, absorbing boundary treatment, sponge, buffer, or exit zone treatments. The Perfectly-Matched-Layers (PML) technique used in computational electromagnetics has been extended to computational aeroacoustics. For an illuminating description of boundary treatment requirements and a state-of-the-art review of available methods, the reader is referred to Tam [53]. We discuss herein some of the major ideas for boundary treatments.

4.2.1   Wall boundary conditions

On solid surfaces the usual no-slip, no-through flow boundary condition can be applied. But to maintain the accuracy of the scheme the minimum ghost points approach of Tam and Dong [54], Tam and Kurbatskii [55]. This has proven to be a successful treatment when coupled with biased spatial discretization.

4.2.2   Outflow boundary treatment

Asymptotic analysis of the Linearized Euler Equations (LEE) for large distances has been performed in several references [50,56,57]. For an outgoing-wave solution the boundary condition on the pressure can be stated as:

$$\frac{1}{\Omega}\frac{\partial p'}{\partial t} + \frac{\partial p'}{\partial R} + \frac{p'}{R} = 0$$

where

$$\Omega = c_0 \left[\frac{x}{R} M + \left(1 - \frac{r^2}{R^2} M^2 \right)^{1/2} \right]$$

and

$$R = \sqrt{x^2 + r^2}. \tag{4.8}$$

Here, p is the pressure, and prime denotes the disturbance. The speed of sound is c_0, and M is the local Mach number. However, the asymptotic analysis in Tam and Webb shows that the velocity and density disturbances are composed of both acoustic and flow disturbances. Thus, the governing equations, except the pressure (energy) equation, are not modified, but are one-sided differenced at the boundaries to account for the presence of flow disturbances. The pressure equation is replaced by its asymptotic approximation that ensures outgoing waves. The treatment produces acceptable solution as long as the mean flow at the outflow does not have steep velocity gradients.

4.2.3   Radiation boundary condition

At computational boundaries where acoustic radiation dominates, the conventional acoustic radiation condition based on the asymptotic analysis of the wave equation applies. Namely:

$$\{u',v',p'\}_t = -\Omega \left[\frac{\partial}{\partial R} + \frac{1}{R} \right] \{u',v',p'\} . \tag{4.9}$$

Fig. 27. Acoustic radiation from a point source in a uniform stream: effect of boundary treatment.

Here, p is the pressure, u and v are the velocities in the axial direction x, and the radial direction r, respectively and prime denotes the disturbance quantity. This treatment was found quite robust producing no reflection at radiation boundaries.

4.2.4 Inflow treatments

The effect of inflow disturbances is added to the computed flow at the inflow at each time step. For supersonic flow all characteristics are incoming, and supersonic flow treatment does not pose any difficulty. However, for subsonic inflow boundaries, reflections due to incorrect boundary specification could lead to an unstable solution. Unlike the cases of walls, radiation, and outflow, none of the existing treatments for subsonic inflow stand out as completely satisfactory. However, Ali and Mankbadi [48] and Hixon *et al.* [59–61] have made extensive comparisons of various inflow treatments. None were really perfect, but Engquist and Majda [62, 63], Giles [64] and Tam *et al.* [65] produce a stable solution marginally free from reflections. Each of these methods will work fine for some cases, but not for all. In the absence of a more robust inflow treatment, these three methods will be used, depending on their performance. These methods are summarized below.

4.2.4.1   Giles boundary treatment

Giles derived boundary conditions based on Fourier analysis of the two-dimensional linearized Euler equations with constant coefficients. To produce local boundary conditions the dispersion relation is expanded in a Taylor series around the one-dimensional solution. Boundary conditions of various degrees of approximation can thus be constructed. The outgoing characteristic is obtained from the interior numerical solution. The incoming characteristics are modified as

$$p'_t - C^2 \rho'_t = -V(p'_y - C^2 \rho'_y) \tag{4.10}$$

$$p'_t + \rho C u'_t = -\frac{\rho C}{2}(C - U)v'_y - V(p'_y + \rho C u'_y) \tag{4.11}$$

$$\rho C v'_t = -\rho C V v'_y - \frac{1}{2}(U + C)(p'_y + \rho C u'_y) \tag{4.12}$$

$$+\frac{1}{2}(U - C)p'_y - \rho C u'_y). \tag{4.13}$$

4.2.4.2   Engquist and Majda inflow treatment

Based on the asymptotic analysis of the Linearized Euler equations, Engquist and Majda [62] derived one-dimensional non-reflecting boundary conditions in the form:

$$\begin{aligned}
\rho'_t + (U - C)\rho'_x &= 0 \\
u'_t + (U - C)u'_x &= 0 \\
v'_t + (U - C)v'_x &= 0 \\
p'_t + (U - C)p'_x &= 0.
\end{aligned} \tag{4.14}$$

These relations are used to obtain the time update for the primitive variable during the axial sweep.

4.2.4.3 Tam *et al.* inflow treatment

Asymptotic analysis of LEE produces for outgoing waves

$$\frac{\partial F_{\text{out}}}{\partial t} = -\Omega \left[\frac{x}{R}\frac{\partial}{\partial x} + \frac{r}{R}\frac{\partial}{\partial r} + \frac{1}{R} \right] F_{\text{out}} \tag{4.15}$$

where

$$F_{\text{out}}^{(n)} = F_{l,m}^{(n)} - F \tag{4.16}$$

$$\Omega = C \left[\frac{x}{R}M + \left(1 - \left(\frac{r}{R}M\right)^2\right)^{1/2} \right] \tag{4.17}$$

and

$$R = \left[x^2 + r^2\right]^{1/2} \tag{4.18}$$

$F_{l,m}^{(n)}$ is the numerical solution of the discretized, while F is the incoming characteristics. In implementing this method, the derivatives appearing in the right-hand side of equation (4.16) are split into axial and radial ones. During the axial sweep, equation (4.16) is applied in the form:

$$\frac{\partial F_{l,m}^{(n)}}{\partial t} = -\Omega \left[\frac{x}{R}\frac{\partial}{\partial x}\right](F_{l,m}^{(n)} - F) \tag{4.19}$$

and the incoming waves are assumed in the form:

$$F = Re\left\{F(r)e^{\tilde{i}(\alpha x - \omega n \Delta t)}\right\}. \tag{4.20}$$

During the radial sweep, equation (4.16) is applied in the form

$$\frac{\partial F_{l,m}^{(n)}}{\partial t} = -\Omega \left[\frac{r}{R}\frac{\partial}{\partial r} + \frac{1}{R}\right](F_{l,m}^{(n)} - F). \tag{4.21}$$

After implementing the above equations, the axial and radial sweeps, the time derivative of the incoming disturbances is accounted for.

4.2.4.4. Thompson

In Thompson's analysis [66, 67], the nonlinear Euler equations for one-dimensional flow are decomposed into wave modes of definite velocity. The acoustic waves propagate at sound speed relative to the mean flow. The vorticity and the entropy waves are frozen patterns convicted downstream by the mean flow. The outward propagating waves are defined entirely by the state of the variables within the computational domain. The behavior of incoming waves is specified by data external to and on the boundary. For nonreflecting inflow, the amplitude of the inward propagating wave is set to zero. For extension to the two-dimensional case the transverse terms are taken as a passive source term. This method is acceptable if the incoming flow is nearly one- dimensional perpendicular or tangential to the inflow boundary.

5 Large-eddy simulations and linearized Euler

5.1 Large-eddy simulation

An alternative to DNS is large-eddy simulation (LES) in which the large-scale field is calculated directly from the solution of the filtered,

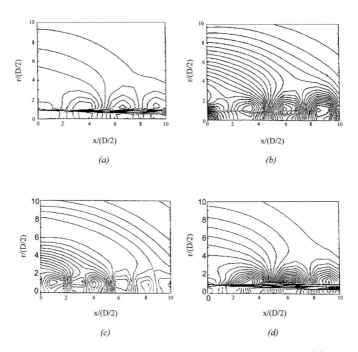

Fig. 28. Snapshot of the pressure disturbance field using: (a) Thompsom, (b) Giles, (c) Tam *et al.*, (d) Engquist and Majda.

time-dependent Navier–Stokes equations and the smaller, unresolved scales (subgrid scales, residual scales) are modeled. It is argued that the large scales are strongly dependent on the flow geometry and therefore LES should be capable of predicting them. The model for the subgrid scales then represents the effect of the small scales on the large- scale motion. However, it is believed that small scales are more universal in character for different flows than the large scales. Thus, it should be possible to use simple models with universal parameters for the subgrid scales. For further more details on large-eddy-simulation, the reader is referred to Lesieur [69–71], and Lumley [72].

Because in LES the smaller scales need not be resolved, fine resolution is not needed for LES. As pointed out by Reynolds (1989), a 32^3 LES is now done in a reasonable time on a 25 MHz 80386 machine. LES is thus useful in studying physics at high Reynolds numbers, in developing turbulence models, an in predicting flows of practical interest. As pointed out by Wyngard [73], the achievements of LES in atmospheric science, such as in the study of severe storms, are quite spectacular.

In very large eddy simulations (VLES) the grid size is coarser than in LES. The grid size is in the energy-containing eddies rather than in the inertial range. The term "LES" is usually used for the case where the filter width Δ_f corresponds to a wavenumber in the $k^{-5/3}$ inertial range. The term "VLES" is used when the residual field begins before the inertial range. VLES does not share the LES insensitivity to subgrid-scale model because its coarse grid can fail to resolve most of the flux. Thus, unlike LES, VLES requires more sophisticated turbulence models because of the coarser resolution.

5.1.1   Filtering

In LES the large scales are obtained by applying a filter to the full Navier–Stokes equations. The filter function has the property that the amplitude of the high-frequency Fourier components of any flow variable f are filtered out or substantially reduced. The filtered field, $-\overline{f}$ is given by

$$\overline{f}(x,t) = \int_V G(x - \xi)f(\xi, t)\mathrm{d}^3\xi \qquad (5.1)$$

where G is the filter function and the integration is over the flow domain. The filter function is usually normalized so that

$$\int_V G(x - \xi)\mathrm{d}^3\xi = 1$$

and G is usually taken to be a differentiable function; that is,

$$\frac{\overline{\partial f}}{\partial x_i} = \frac{\partial - f}{\partial x_i}.$$

A commonly used filter is the Gaussian filter

$$G(x - \xi, t) = \left(\frac{6}{\pi\Delta_f}\right)^{3/2} \exp\left[-6\frac{(x_i - \xi_i)^2}{\Delta_f^2}\right] \qquad (5.2)$$

where Δ_f is the filter width that sets the scale of the smallest resolved eddies. The coefficient is adjusted so that the filtered value of a constant is that constant, and the factor 6 makes the width representative of the scale of the smallest resolved eddies. The Gaussian filter is widely used for homogeneous turbulence and for inhomogeneous turbulence in directions of homogeneity, often with separate filter widths in the three directions.

On using finite difference to approximate the governing equations, the Gaussian filter filters out the scales that are smaller than the mesh spacing Δ_x. However, the larger scales are also filtered out to some extent. Thus the Gaussian filter is not sharply defined because it produces a filtered field that overlaps with the residual field in the wavenumber space.

In this case the subgrid-scale model may be sensitive to the numerical procedure adopted, and the mixed model should be considered.

In the sharp-cutoff filter [74] all Fourier modes having wavenumbers greater than a specified cutoff are put into the residual field, and all modes with smaller wavenumbers are retained in the resolved field. The filter for a given direction is given by

$$G(x_i - \xi_i) = \frac{2\sin[\pi(x_i - \xi_i)/\Delta_i]}{\pi(x_i - \xi_i)}. \tag{5.3}$$

This filter removes modes with wavenumbers

$$|k_i| > \pi/\Delta_i.$$

Moin *et al.* (1991) used a short-cutoff filter in the form

$$\begin{aligned} G(k_i) &= 1, \quad \text{if} \quad k_i \leq \pi/\delta_i \\ &0, \qquad\qquad \text{otherwise.} \end{aligned}$$

In a low-level finite difference representation a filter is implicit and represents a uniformly weighted averaging over a finite width in the y direction

$$\begin{aligned} G(y,\eta) &= 2/(\Delta_- + \Delta_+) \quad \text{for} \quad y - \Delta_-/2 < \eta < y + \Delta_+/2 \\ &0, \qquad\qquad\qquad\qquad \text{otherwise.} \end{aligned}$$

with Δ_- as the mesh spacing below and Δ_+ the spacing above.

Filtering in LES could be different from the conventional averaging in turbulence theory. For instance, in LES it could be that

$$\overline{f'}\_0, = f_- - f$$

because a second smoothing removes additional structure from the resolved field. However, these equalities do hold for sharp-cutoff filters.

5.1.2   Filtered equations

Starting from the full Navier–Stokes equations, the flow field is decomposed into filtered and residual fields by

$$f = \_f + f'$$

where an overbar denotes the resolved (filtered) field and a prime denotes the unresolved (subgrid) one. The mean of the filtered field is the mean of the total field. However, for the total turbulent stresses, for example, the equation is

$$\overline{u_i u_j} = \overline{\overline{u_i u_j}} + \overline{u_i' u_j'}.$$

Upon substituting the splitting in the full Navier–Stokes equation, the filtered compressible Navier–Stokes equation in cylindrical coordinates takes the form

$$\frac{\partial Q}{\partial t} + \frac{\partial F}{\partial x} + \frac{1}{r}\frac{\partial (rG)}{\partial r} + \frac{1}{r}\frac{\partial H}{\partial \phi} = S$$

$$Q = \begin{bmatrix} \overline{\rho} \\ \overline{\rho}\tilde{u} \\ \overline{\rho}\tilde{v} \\ \overline{\rho}\tilde{w} \\ \overline{\rho}\tilde{e} \end{bmatrix}$$

$$F = \begin{bmatrix} \overline{\rho}\tilde{u} \\ \overline{p} + \overline{\rho}\tilde{u}^2 - \tilde{\sigma}_{xx} - \tau_{xx} \\ \overline{\rho}\tilde{u}\tilde{v} - \tilde{\sigma}_{xr} - \tau_{xr} \\ \overline{\rho}\tilde{u}\tilde{v} - \tilde{\sigma}_{x\phi} - \tau_{x\phi} \\ \overline{\rho}\tilde{u}\tilde{I} - \tilde{u}\tilde{\sigma}_{xx} - \tilde{v}\tilde{\sigma}_{xr} - \tilde{w}\tilde{\sigma}_{r\phi} - \tilde{\kappa}\frac{\partial \tilde{T}}{\partial x} - c_{\mathrm{v}}q \end{bmatrix}$$

$$F = \begin{bmatrix} \overline{\rho}\tilde{v} \\ \overline{\rho}\tilde{u}\tilde{v} - \tilde{\sigma}_{xr} - \tau_{xr} \\ \overline{p} + \overline{\rho}\tilde{v}^2 - \tilde{\sigma}_{rr} - \tau_{rr} \\ \overline{\rho}\tilde{v}\tilde{w} - \tilde{\sigma}_{r\phi} - \tau_{r\phi} \\ \overline{\rho}\tilde{v}\tilde{I} - \tilde{u}\tilde{\sigma}_{xr} - \tilde{v}\tilde{\sigma}_{rr} - \tilde{w}\tilde{\sigma}_{r\phi} - \tilde{\kappa}\frac{\partial \tilde{T}}{\partial r} - c_{\mathrm{v}}q \end{bmatrix}$$

$$H = \begin{bmatrix} \overline{\rho}\tilde{w} \\ \overline{\rho}\tilde{u}\tilde{w} - \tilde{\sigma}_{x\phi} - \tau_{x\phi} \\ \overline{\rho}\tilde{v}\tilde{w} - \tilde{\sigma}_{r\phi} - \tau_{r\phi} \\ \overline{p} + \overline{\rho}\tilde{w}^2 - \tilde{\sigma}_{\phi\phi} - \tau_{\phi\phi} \\ \overline{\rho}\tilde{w}\tilde{I} - \tilde{u}\tilde{\sigma}_{x\phi} - \tilde{v}\tilde{\sigma}_{r\phi} - \tilde{w}\tilde{\sigma}_{\phi\phi} - \tilde{\kappa}\frac{\partial \tilde{T}}{\partial \phi} - c_{\mathrm{v}}q \end{bmatrix}$$

$$S = \frac{1}{r}\begin{bmatrix} 0 \\ 0 \\ \overline{p} + \overline{\rho}\tilde{w}^2 - \tilde{\sigma}_{\phi\phi} - \tau_{\phi\phi} \\ -\tilde{v}\tilde{w} - \tilde{\sigma}_{r\phi} - \tau_{r\phi} \\ 0 \end{bmatrix}. \tag{5.4}$$

Here Q is the unknown vector; F, G, and H are the fluxes in the x, r, and Ψ directions, respectively; S is the source term that arises in cylindrical polar coordinates; and κ is thermal conductivity. The enthalpy is I and the internal energy is e. Here, a tilde denotes Favre averaging,

$$-f = \overline{\rho f}/\overline{\rho} \tag{5.5}$$

and ρ is the viscous stresses

$$\sigma_{xx} = 2\mu\frac{\partial u}{\partial x} - \nabla \cdot u$$

$$\sigma_{rr} = 2\mu\frac{\partial v}{\partial r} - \nabla \cdot u$$

$$\sigma_{\phi\phi} = 2\mu\frac{\partial w}{r\partial\phi} - \nabla \cdot u$$

$$\sigma_{xr} = \mu\left(\frac{\partial u}{\partial r} + \frac{\partial v}{\partial x}\right)$$

$$\sigma_{x\phi} = \mu\left(\frac{\partial u}{r\partial\phi} + \frac{\partial w}{\partial x}\right)$$

$$\sigma_{r\phi} = \mu\left(\frac{\partial v}{r\partial\phi} + r\frac{\partial(w/r)}{\partial r}\right)$$

$$\nabla \cdot u = \frac{2}{3}\mu\left[\frac{\partial u}{\partial x} + \frac{1}{r}\frac{\partial(rv)}{\partial r} + \frac{\partial w}{r\partial\phi}\right]. \tag{5.6}$$

This system of equations is coupled with the equation of state

$$\bar{p} = \bar{\rho}R_{\mathrm{g}}\tilde{T}$$

Here,

$$e = \bar{\rho}c_{\mathrm{v}}\tilde{T}.$$

p is the thermodynamic pressure, T is the temperature, c_{v} is the specific heat at constant volume, which is usually assumed to be constant, and μ is the molecular viscosity, which is related to the temperature through Sutherland's formula (Schlichting 1979)

$$\frac{\mu}{\mu_{\mathrm{r}}} = \left(\frac{T}{T_{\mathrm{r}}}\right)^{3/2}\frac{T_{\mathrm{r}}s_1}{T + s_1} \tag{5.7}$$

where the constant s_1 is 100 K for air. The simpler power-law formula

$$\frac{\mu}{\mu_{\mathrm{r}}} = \left(\frac{T}{T_{\mathrm{r}}}\right)^{0.76}$$

is also sometimes used and the Prandtl number

$$Pr = c_{\mathrm{p}}\mu/k$$

is usually kept constant at 0.7.

Note that in filtering

$$\overline{\rho u_k} = \overline{\rho} \tilde{u}_k, \quad \overline{\rho u_k u_\ell} = \overline{\rho} u_k - u_\ell, \quad \overline{\rho u_k T} = \overline{\rho} u_k - T. \tag{5.8}$$

In equation (5.4) the unresolved stresses

$$\tau_{ij} = \overline{\rho}(\overline{u_i\_u_j} - \tilde{u}_i \tilde{u}_j)$$

and

$$q = \overline{\rho}(\_)$$

need to be modeled.

5.1.3   Modelling subgrid-scale turbulence

As indicated by the preceding equations, filtering the governing equation for a certain length scale results in the residual turbulent stresses σ_{ij} and in q, which need to be modeled.

5.1.3.1   LES with no models

It might be possible to perform such LES with no explicit residual turbulence model at all. In this case the "numerics" is the dissipative mechanism that sets the smallest scale, and there is no relation between the numerical viscosity and the fluid viscosity. Reynolds [75] warned that "this approach simply cannot predict important Reynolds number effects and consequently should be used with caution".

5.1.3.2   Smagorinsky's model

In Smagorinsky's [76] model the subgrid-turbulence stresses are modeled as

$$\tau_{ij} = k_{\mathrm{g}} \delta_{ij}/3 - 2\rho v_{\mathrm{R}} \left(\tilde{S}_{ij} - \frac{1}{3} \delta_{ij} \tilde{S}_{mm} \right)$$

where k_{g} is the kinetic energy of the residual turbulence. The strain rate of the resolved scale is given by

$$\tilde{S}_{ij} = \frac{1}{2} \left(\frac{\partial \tilde{u}_i}{\partial x_j} + \frac{\partial \tilde{u}_j}{\partial x_i} \right). \tag{5.9}$$

The summation S_{mm} is zero for incompressible flow, v_{R} is the effective viscosity of the residual field,

$$v_{\mathrm{R}} = (C_{\mathrm{S}} \Delta_{\mathrm{f}})^2 \sqrt{2 S_{mn} S_{mn}}$$

and Δ_f is the filter width given by

$$\Delta_f = (\Delta_x \Delta_r \Delta_\phi)^{1/3}.$$

For the heat equation Edison [77] proposed the eddy viscosity model

$$q = \overline{\rho} \frac{v_t}{Pr_t} \frac{\partial \tilde{T}}{\partial x_k}$$

where Pr_t is the subgrid-scale turbulent Prandtl number, which can be taken as 0.5.

Smagorinsky's constant C_S in equation (8.45) is given by Reynolds [75] as 0.23. In the presence of mean shear, however, this value was found to cause excessive damping or large-scale fluctuations. Piomelli et al. [78] found the optimum value of C_S to be 0.1.

The turbulent stresses are sometimes split as

$$\begin{aligned}
\tau &= L + C + S \\
L_{ij} &= \overline{\rho}(\tilde{u}_i \tilde{u}_{j-} - \tilde{u}_i \tilde{u}_j) \\
C_{ij} &= \overline{\rho}(\tilde{u}_{i-} u'_j + \tilde{u}_{j-} u'_i) \\
S_{ij} &= \overline{\rho} u'_{i-} u'_j
\end{aligned} \qquad (5.10)$$

which are the subgrid-scale Leonard, cross-correlation, and Reynolds stresses, respectively.

If needed, the Leonard stresses can be calculated directly and do not need to be modeled. Bardina et al. [79] suggested a mixed model in which the cross-correlation term is added to Smagorinsky's stresses and the former is modeled as

$$C_{ij} = \overline{\rho}(\tilde{u}_i \tilde{u}_j - u_i \approx u_j \approx)$$

which can be calculated directly. This mixed model is appropriate when a Gaussian filter is used but should not be used with the sharp-cutoff filter because the resolved and unresolved scales do not overlap.

Although Smagorinsky's model is one of the simplest models, it assumes a balance between the subgrid-scale energy production and its dissipation. This balance may not be true in several situations, such as in flows undergoing transition to turbulence, laminarization, or rapid change from one type to the other. Smagorinsky's model is too dissipative in the laminar-turbulent transition region and does not vanish in laminar flows. The model constants may need to be adjusted from one flow to the other and may need correction near solid surfaces. Furthermore, in Smagorinsky's model the subgrid turbulence energy is either neglected with respect to the thermodynamic pressure (as in Erlebacher et al. [80] or added to it and treated as a pressure head. Because of the strong anisotropy in a shear flow, however, there seems to be no systematic way to evaluate the subgrid-scale energy. Therefore, it is difficult to evaluate the pressure separately. Moreover, the model cannot account for energy transfer from the small scale to the larger scales (back-scattering). The following two models attempt to address some of these difficulties.

5.1.3.3 Dynamical subgrid Eddy viscosity model

Germano *et al.* [81] proposed a Smagorinsky type of model in which the model coefficient is computed dynamically as the calculations progress, rather than as an input a priori. Two filters are used: one is the grid filter, and the other is larger and is called the test filter. The test filter utilizes the spectral information on the energy content of the smallest resolved scales provided by LES calculation to dynamically adjust Smagorinsky's constant. This is done by assuming that the constant in Smagorinsky's model can be obtained by comparing the residual stresses obtained at the grid scale and at the test filter scale. An algebraic identify is obtained between subgrid-scale stresses at the two different filtered levels and the resolved turbulent stresses. Smagorinsky's constant is then given by

$$C(x,y,z,t) = -\frac{1}{2} \cdot \frac{\mathcal{L}_{ij}\overline{S}_{ij}}{\hat{\overline{\Delta}}^2 \left|\hat{\overline{S}}\right| \hat{\overline{S}}_{ij}\overline{S}_{ij} - \overline{\Delta}^2 \left|\hat{\overline{S}}\right| \overline{S}_{ij} \, \overline{S}_{ij}} \tag{5.11}$$

where
$$\mathcal{L}_{ij} = \overline{u_i \overline{u}_j} - \hat{\overline{u}}_i \, \hat{\overline{u}}_j$$

$\overline{\Delta}$ is the grid filter width, and $\hat{\Delta}$ is the test filter width. The ratio of the width of the two filters is usually taken as approximately 2. Germano *et al.* [81] pointed out that the subgrid-scale stresses obtained by using their model vanish in laminar flow and in solid boundaries. This model has been extended to the compressible case by Moin *et al.* [82].

5.1.3.4 One-equation models

Because of the shortcomings of Smagorinsky's model, Yoshizawa [84] and Yoshizawa and Horiuti [85] proposed a one-equation model for the subgrid turbulence energy k_{g}, which is governed by the equation

$$\frac{\partial k_{\mathrm{g}}}{\partial t} = -\overline{u}_i \frac{\partial k_{\mathrm{g}}}{\partial x_j} + -\overline{u_i' u_j'} \frac{\partial \overline{u}}{\partial x_i} -\cdot + E_{\mathrm{DG}} \tag{5.12}$$

where
$$E_{\mathrm{DG}} = -\frac{\partial}{\partial x_i}\left(\frac{1}{2}u_i'u_j'u_j' + p'u_i'\right)$$
$$v_{\mathrm{R}} = C_{\mathrm{v}}\Delta k_{\mathrm{g}}^{1/2}.$$

LES

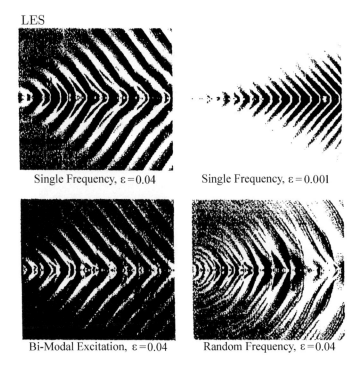

Single Frequency, ε = 0.04   Single Frequency, ε = 0.001

Bi-Modal Excitation, ε = 0.04   Random Frequency, ε = 0.04

Fig. 29.

The terms in the right-hand side of equation are, respectively, convection, production, dissipation, and diffusion. Yoshizawa and Horiuti [84] used the two-scale direct interaction theory to model the equation for k_g as

$$\frac{\partial k_g}{\partial t} = -\overline{u}_i \frac{\partial k_g}{\partial x_i} + \frac{1}{2} v_R \overline{e}_{ij}^2 - C_- \frac{k_g^{3/2}}{\Delta} + \frac{\partial}{\partial x_i} \left[\left(\frac{1}{R} + C_{kk} \Delta k_g^{1/2} \right) \frac{\partial k_g}{\partial x_i} \right].$$
(5.13)

They estimated the numerical constants for a flat-plate boundary layer as

$$C_v = 0.05, \qquad C_- = 0.1, \qquad C_{kk} = 0.1.$$

For channel flow their results have indicated a strong dependence on the model adopted, probably because of the dissipative nature of Smagorinsky's model. The diffusion term in the energy equation of the grid turbulence was found to play a dominant role in the budget of subgrid-scale turbulence energy.

The figures show the results for LES including the nonlinear flow regime and the acoustic fuel [85–89].

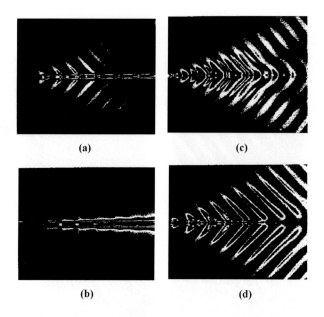

Fig. 30. Instantaneous distributions of a) density b) axial velocity c) pressure d) radial velocity.

5.2 Linearized Euler equations

In the linearized Euler equations (LEE) approach, the flow field is split into a time-averaged one and a time-dependent disturbance one. The latter is considered to be small with respect to the former. Starting from the full Navier–Stokes equations in conservative form, neglecting viscosity, and linearizing about a mean flow (U, V, W), the linearized Euler equations can be obtained in the form

$$\frac{\partial \hat{Q}}{\partial t} + \frac{\partial \hat{F}}{\partial x} + \frac{1}{r}\frac{\partial (r\hat{G})}{\partial r} + \frac{1}{r}\frac{\partial \hat{H}}{\partial \phi} = \hat{S}$$

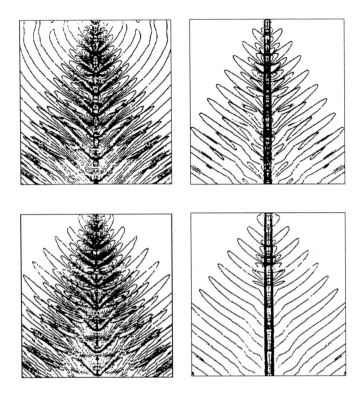

Fig. 31.

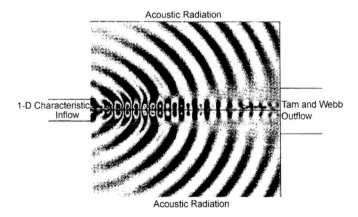

Fig. 32.

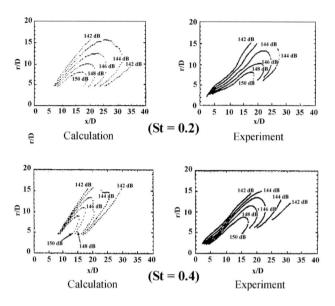

Fig. 33. Linear-Euler prediction of the SPL contours at $St = 0.2$ and 0.4 of a supersonic jet compared with Trout and MacLaughlin's data.

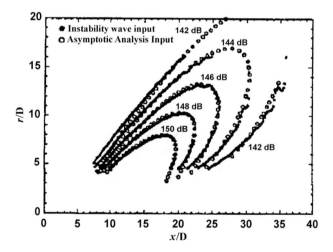

Fig. 34. Sound pressure-level contours of the present calculation compared with results of asymptotic analysis.

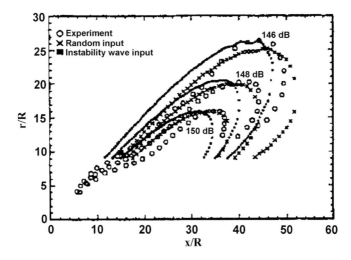

Fig. 35. Comparison of sound pressure-level contours for instability wave and random input with experiment.

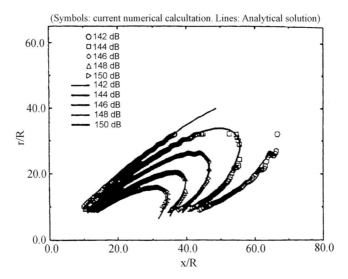

Fig. 36.

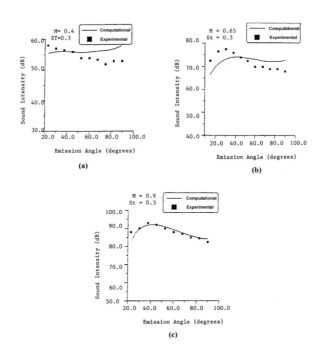

Fig. 37. Directivity of sound intensity at $St = 0.3$, a) $M = 0.4$, b) $M = 0.65$, c) $M = 0.9$.

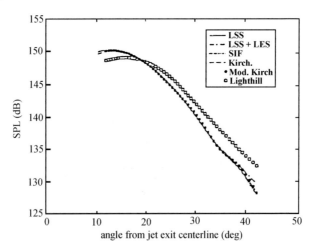

Fig. 38.

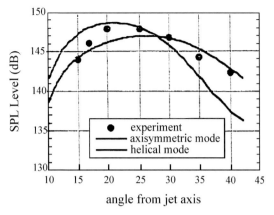

Fig. 39. Directivity of jet noise associated with the axisymmetric and the helical modes in a supersonic jet compared with Trout and McLaughlin's measurement.

where

$$\hat{Q} = [\hat{\rho}, \hat{u}, \hat{v}, \hat{w}, \hat{e}]^T$$

$$\hat{F} = \begin{bmatrix} \hat{u} \\ \hat{p} + 2\hat{u}U - \hat{\rho}U^2 \\ \hat{u}V + \hat{v}U - \hat{\rho}UV \\ \hat{u}W + \hat{w}U - \hat{\rho}UW \\ (\hat{p} + \hat{e})U + (\hat{u} - \hat{\rho}U)E \end{bmatrix}$$

$$\hat{G} = \begin{bmatrix} \hat{v} \\ \hat{u}V + \hat{v}U - \hat{\rho}UV \\ \hat{p} + 2\hat{v}V - \hat{\rho}V^2 \\ \hat{v}W + \tilde{w}V - \hat{\rho}VW \\ (\hat{p} + \hat{e})V + (\hat{v} - \hat{\rho}V)E \end{bmatrix}$$

$$\hat{H} = \begin{bmatrix} \hat{w} \\ \hat{u}W + \hat{w}U - \hat{\rho}UW \\ \hat{v}W + \hat{w}V - \hat{\rho}VW \\ \hat{p} + 2\hat{w}W - \hat{\rho}W^2 \\ (\hat{p} + \hat{e})W + (\hat{w} - \hat{\rho}W)E \end{bmatrix}$$

$$\hat{S} = \frac{1}{r} \begin{bmatrix} 0 \\ 0 \\ \hat{p} + 2\hat{w}W - \hat{\rho}W^2 \\ vW + wV - \rho VW \\ 0 \end{bmatrix} \tag{5.14}$$

$$\hat{p} = (\gamma - 1) \left[\hat{e} - (\hat{u}U + \hat{v}V + \hat{w}W) + \frac{1}{2}\hat{\rho}(U^2 + V^2 + W^2) \right]$$

$$P = (\gamma - 1)\rho \left[E - \frac{1}{2}(U^2 + V^2 + W^2) \right] \tag{5.15}$$

$$(\hat{u}, \hat{v}, \hat{w}, \hat{e}, \hat{\rho}, \hat{p}) = [(\rho u)', (\rho v)', (\rho w)', (\rho e)', \rho, p'].$$

Because of the linear nature of the equations, they are less computationally demanding than LES. Since no particular assumption is made regarding the form of the disturbance, LEE can be used for predicting both the flow and the acoustic disturbances. No boundary-layer type approximation is imposed on the LEE, and, therefore, it can be used for predicting the unsteady flow field of nonparallel mean flows. However, nonlinear effects are completely absent in LEE and the mean flow is assumed to be given by some other means. Figure numbers 32-35 shows the results using LEE for both the flow and acoustic fields [90, 91].

References

[1] K.J. Ahuja, K.C. Massey and A.C. Entrekin, *Contribution of Mixing Within an Ejector to the Far Field Noise Measurements*, AIAA Paper No. 96-0639.

[2] H.M. Atassi, S. Subramanian and J.R. Scott, *Acoustic Radiation from Lifting Airfoils in Compressible Subsonic Flow*, AIAA Paper No. 90-3911.

[3] J.F. Groeneweg, T.G. Sofrin, E.J. Rice and P.R. Glieb, Turbomachinery Noise, in *Aeroacoustics of Flight Vehicles: Theory and Practice*, Volume 1: Noise Sources, edited by H.H. Hubbard, NASA RP-1258.

[4] W.F. Ahtye, W.R. Miller and W.C. Meecham, *Wing and Flap Noise Measured by Near- and Far-Field Cross-Correlation Techniques*, AIAA Paper 79-0667 (1979).

[5] D.G. Crighton, Airframe Noise, in *Aeroacoustics of Flight Vehicles: Theory and Practice*, Volume 1: Noise Sources, edited by H.H. Hubbard, NASA RP 1258 (1991).

[6] M.R. Fink, *Airframe Noise Prediction Method*, FAA-RD-77-29 (1977).

[7] M. Fink and R.H. Schlinker, *J. Aircraft.* **17** (1980) 99-105.

[8] J.C. Hardin, *Airframe Self-Noise – Four Years of Research*, NASA TM X-73908 (1976).

[9] J.C. Hardin, *AIAA Journal* **18** (1980) 549-552.

[10] J.C. Hardin, D.J. Fratello, R.E. Hayden, Y. Kadman and S. Africk, *Prediction of Airframe Noise*, NASA TN D-7821 (1975).

[11] F.O. Thomas, R.C. Nelson and X. Liu, X., *An Experimental Investigation of Unsteady Effects in a High-Lift System*, AIAA 98-0704 (1988).

[12] M.E. Goldstein, *Aeroacoustics* (McGraw-Hill, New York, 1976).

[13] A. Hamed and R. Mankbadi, *Shock-Associated Noise in Jets and Nozzles*, Final Report, NASA Grant NAG3-1963 (1997).

[14] F.K. Browand and P.D. Weidman, *J. Fluid Mech.* **76** (1976) 127-144.

[15] S.C. Crow and F.H. Champagne, *J. Fluid Mech.* **48** (1971) 547-591.

[16] J.T.C. Liu, *J. Fluid Mech.* **62** (1974) 437-464.

[17] R.R. Mankbadi, *J. Fluid Mech.* **160** (1985) 385-419.

[18] R.R. Mankbadi, *AIAA J.* **24** (1986) 1941-1948.

[19] R.R. Mankbadi, *Phys. Fluids A* **3** (1991) 595-605; S.C. Crow, *Bull. Am. Phys. Soc.*, Paper IE.6 (1972).

[20] R.R. Mankbadi, *Transition, Turbulence, and Noise* (Kluwer Academic, Boston, 1994).

[21] R.R. Mankbadi, *Appl. Mech. Rev.* **45** (1992) 219-248.

[22] R.R. Mankbadi and J.T.C. Liu, *Philos. Trans. Roy. Soc. London A* **298** (1981) 541-602.

[23] R.R. Mankbadi, E. Hayder and L.A. Povinelli, *AIAA J.* **32** (1994).

[24] C.J. Moore, *J. Fluid Mech.* **80** (1977) 321-367.

[25] J.M. Seiner, T.R.S. Bhat and M.K. Ponton, *AIAA J.* **32** (1994) 234-235.

[26] J.M. Seiner, D.K. McLaughlin and C.H. Liu, *Supersonic Jet Noise Generated by Large-Scale Instabilities*, NASA TP-2072 (1982).

[27] J.M. Seiner, M.K. Ponton, B.J. Jansen and N.T. Langen, *The Effects of Temperature on Supersonic Jet Noise Emission*, AIAA Paper 92-02-046 (1992).

[28] C.K.W. Tam, *Annual Rev. Fluid Mech.* **27** (1995) 17-43.

[29] K.B.M.Q. Zaman, *J. Sound and Vibration* **106** (1986) 1-16.

[30] J.T.C. Liu, *Adv. Appl. Mech.* **26** (1988) 183-309.

[31] J.T.C. Liu, *Ann. Rev. Fluid Mech.* **21** (1989) 285-315.

[32] R.R. Mankbadi and J.T.C. Liu, *Philos. Trans. Roy. Soc. London A* **311** (1984) 183-217.

[33] M.J. Lighthill, *Proc. of the Roy. Soc. of London Ser. A* **211** (1952) 564-587.

[34] M.J. Lighthill, *A General Introduction to AeroAcoustics and Atmospheric Sound*, NASA CR 189717, ICASE Report No. 92-52 (1992).

[35] S.C. Crow, *Bull. Am. Phys. Soc.*, Paper IE.6 (1972).

[36] J.E. Ffowcs-Williams and A.J. Kempton, *J. Fluid Mech.* **84** (1978) 673-694.

[37] R.R. Mankbadi, *J. Appl. Mech.* **57** (1990) 241-246.

[38] M.E.Goldstein, *Aeroacoustics* (McGraw-Hill, New York, 1976).

[39] G.M. Lilley, *On the Noise from Jets. Noise Mechanisms*, Noise Mechanisms-AGARD Conference on Propagation and Reduction of Jet Noise, AGARD CP 131 (1974) pp. 13.1-13.12.

[40] A.S. Lyrintzis, The Use of Kirchhoff's Method in Computational Aeroacoustics, *Computational Aero-And Hydro-Acoustics*, edited by R.R. Mankbadi et al., FED-Vol. 147 (1993) pp. 35-40.

[41] L. Morino, *Appl. Mech. Rev.* **46** (1993) 445-466.

[42] A. Lyrintzis and R.R. Mankbadi, *AIAA J.* **34** (1996) 413-416.

[43] R.R. Mankbadi, S.H. Shih, D.R. Hixon, J.T. Stuart and L.A. Povinelli, *J. Comput. Acoust.* **6** (1997) 307-320, also AIAA 96-2651 (1998).

[44] S.H. Shih, D.R. Hixon and R.R. Mankbadi, *AIAA J. Aircraft* (1997) in Review, also, AIAA Paper, *Aerospace Sciences Meeting* (Reno, Nevada, 1997).

[45] D.R. Hixon, S. Shih and R.R. Mankbadi, *AIAA J. Propulsion & Power* **15** (1999), *Aerospace Sciences Meeting* (Reno, Nevada, 1995).

[46] J. Hardin and M.Y. Hussaini, *Computational Aeroacoustics* (Springer-Verlag, New York, 1993).

[47] J.C. Hardin, J.R. Ristorcelli and C.K.W. Tam, *ICAE/LaRC Workshop on Benchmark Problems In Computational Aeroacoustics*, NAS CP 3300 (1995).

[48] R.W. MacCormack, *The Effect of Viscosity in Hypervoloicity Impact Cratering*, AIA Paper, pp. 69-354.

[49] S.K. Lele, *J. Comput. Phys.* **103** (1992) 16-42.

[50] C.K.W. Tam and J.C. Webb, *J. Comput. Phys.* **107** (1993) 262-281.

[51] T. Bui and R.R. Mankbadi, *Int. J. Comput. Fluid Dynamics* **10** (1998) 281-290.

[52] T.Z. Dong and R.R. Mankbadi, *Direct Numerical Simulations of Engine Internal Noise Propagation and Radiation*, AIAA/CEAS, 16th Aeroacoustics Conference, 12-15 June 1995 (Munich, Germany).

[53] C.K.W. Tam, *J. Comput. Acoustics* (1998) (to appear), also, AIAA Paper 97-1774.

[54] C.K.W. Tam and T.Z. Dong, *Theoret. Comput. Fluid Dynamics* **6** (1994) 303-322.

[55] C.K.W. Tam and K.A. Kurbatskii, *AIAA J.* **35** (1997) 133-140.

[56] T. Hagstrom and S.I. Hariharan, *Comput. Acoust.* **2** (1990).

[57] T. Hagstrom and S.I. Hariharran, *Math. Comput.* **51** (1988) 581-597.

[58] A. Ali and R.R. Mankbadi, *J. Comput. Acoust.* **7** (1999).

[59] D.R. Hixon, S.H. Shih and R.R. Mankbadi, *On the Effect of Boundary Treatment on the Gust-Cascade Problem*, AIAA Paper, AIAA Theoretical Fluid Dynamics Meeting (1998).

[60] D.R. Hixon, S.H. Shih and R.R. Mankbadi, *AIAA J.* **33** (1995) 2006-2012.

[61] D.R. Hixon, S.H. Shih and R.R. Mankbadi, *Inflow Boundary Treatment for the Gust-Cascade Problem*, AIAA Theoretical Fluid Dynamics Conference (Albuquerque, New Mexico, 1998) pp. 14-17.

[62] B. Engquist and A. Majda, *Math. Comput.* **31** (1977) 629-651.

[63] B. Enqguist and A. Majda, *Comm. Pure Appl. Math.* **32** (1979) 313-357.

[64] M.B. Giles, *AIAA J.* **28** (1990) 2050-2058.

[65] C.K.W. Tam, J. Fang and K.A. Kurbatsill, *Inhomogeneous Radiation Conditions Simulating Incoming Acoustics Waves for Computational Aero-Acoustics*, ASME/Cairo University International Congress on Fluid Dynamics and Propulsion (1996).

[66] K.W. Thompson, *J. Comput. Phys.* **68** (1987) 1-24.

[67] K.W. Thompson, *J. Comput. Phys.* **89** (1990) 439-461.

[68] C.K.W. Tam, *AIAA J.* **33** (1995) 1788-1796.

[69] M. Lesieur, *Turbulence in Fluids* (Kluwer Academic Publishers, Dordrecht, 1990).

[70] M. Lesieur and R. Rogallo, *Phys. Fluids A* **1** (1989) 718-722.

[71] M. Lesieur, O. Métais, X. Normand and A. Silveira, *Spectral Large-Eddy Simulation of Turbulent Shear Flows*, International Conference: Large-Eddy Simulations, Where do We Stand? (St. Petersburg, Florida, Cambridge University Press, 1990).

[72] J.L. Lumley, *Whither Turbulence?:* Turbulence at the Crossroads, Proceedings of the Workshop (Springer-Verlag, New York, 1989) pp. 356-368.

[73] J. Wyngaard, *The Potential and Limitations of Direct and Large-Eddy Simulations*, Comment 3, in *Whither Turbulence?: Or Turbulence at the Crossroads*, Proceedings of the Workshop, edited by J.L. Lumley (Springer-Verlag, New York, 1989) pp. 369-373.

[74] A. Leonard, *Energy Cascade in Large-Eddy Simulation of Turbulent Flows*, Turbulent Diffusion in Environmental Pollution; Proceedings of the Second Symposium, Advances in Geophysics, edited by F.N. Frenkiel and R.E. Munn (Academic Press, New York, 1974) Vol. 18A, pp. 237-248.

[75] W.C. Reynolds, *The Potential and Limitations of Direct and Large Eddy Simulations, Whither Turbulence?:* Turbulence at the Crossroads, Proceedings of the Workshop, edited by J.L. Lumley (Springer-Verlag, New York, 1989) pp. 313-342.

[76] J. Smagorinsky, *Monthly Weather Rev.* **91** (1963) 99-164.

[77] T.M. Edison, *J. Fluid Mech.* **158** (1985) 245-268.

[78] U. Piomelli, P. Moin and J.H. Ferziger, *Phys. Fluids* **31** (1988) 1884-1891.

[79] J. Bardina, J.H. Ferziger and W.C. Reynolds, *Improved Turbulence Models Based on Large-Eddy Simulation of Homogeneous, Incompressible, Turbulent Flows*, NASA CR-166541 (1983).

[80] G. Erlebacher, M.Y. Hussaini, C.G. Speziale and T.A. Zang, *Toward the Large-Eddy Simulation of Compressible Turbulent Flows*, ICASE Report 90-76, Institute for Computer Applications in Science and Engineering (Hampton, VA, 1990).

[81] M.Germano, U. Piomelli, P. Moin and W.H. Cabot, *Phys. Fluids* **3** (1991) 1760-1765.

[82] P. Moin, K. Squires, W. Cabot and S. Lee, *Phys. Fluids A* **3** (1991) 2746-2757.

[83] A. Yoshizawa, *Phys. Fluids* **25** (1992) 1532-1538.

[84] A. Yoshizawa and K. Horiuti, *J. Phys. Soc. Jpn.* **54** (1985) 2834-2839.

[85] R.R. Mankbadi, D.R. Hixon, S.H. Shih and L.A. Povinelli, *Direct Prediction of Jet Noise Using Linearized Euler Equations*, AIAA Paper No. 95-0505, *Aerospace Sciences Meeting* (Reno, Nevada, 1995).

[86] R.R. Mankbadi, S.H. Shih and R. Hixon, *Direct Computation of Sound Radiation by Jet Flow Using Large-Scale Equations*, AIAA Paper No. 95-0680 (1995).

[87] S.H. Shih, D.R. Hixon and R.R. Mankbadi, *Three-Dimensional Structure in a Supersonic Jet: Behavior Near Centerline*, AIAA 95-0681 (Reno, Nevada, 1995).

[88] S.H. Shih, D.R. Hixon and R.R. Mankbadi, *AIAA J. Propulsion and Power* **13** 745-758.

[89] S.H. Shih, D.R. Hixon and R. Mankbadi, *Nonlinear, Viscous Computation of Large-Scale Structure in Shear Flows*, AIAA Paper, AIAA Fluid Dynamics Conference (1998).

[90] R.R. Mankbadi, D.R. Hixon, S.H. Shih and L.A. Povinelli, *AIAA J.* **36** (1998) 140-147.

[91] A. Ali and R.R. Mankbadi, *J. Comput. Acoust.* **7** (1999).

COURSE 6

THE TOPOLOGY OF TURBULENCE

H.K. MOFFATT

*Isaac Newton Institute
for Mathematical Sciences,
University of Cambridge,
20 Clarkson Road,
Cambridge CB3 0EH, U.K.*

Contents

THE TOPOLOGY OF TURBULENCE

H.K. Moffatt

1 Introduction

Topological considerations enter the study of turbulence through quantities, often expressed as integrals over the fluid domain, which in idealized (non-dissipative) circumstances, are constant in time, *i.e.* invariants of the flow. The simplest example occurs for two-dimensional incompressible inviscid flow for which the vorticity field $\omega(x, y, t)$ satisfies the equation

$$\frac{D\omega}{Dt} \equiv \frac{\partial \omega}{\partial t} + \mathbf{u} \cdot \nabla \omega = 0. \tag{1.1}$$

Here $\mathbf{u} = (u(x, y, t), v(x, y, t), 0)$ is the velocity field with $\omega = \partial v/\partial y - \partial u/\partial x$. Equation (1.1) tells us that the vorticity associated with any fluid element is constant. The "isovorticity lines" $\omega = $ cst. are "frozen" in the fluid, so that the area $A(\omega)$ inside any closed isovorticity line $\omega = $ cst. is conserved. This function, described (Moffatt 1986) as the "signature" of the vorticity field, may be thought of as a topological invariant under the continuous deformation of the vorticity field by its self-induced velocity field $\mathbf{u}$.

If $F(\omega)$ is an arbitrary function of ω, then

$$\frac{\mathrm{d}}{\mathrm{d}t} \int_{S_L} F(\omega)\mathrm{d}x\mathrm{d}y = \int_{S_L} F'(\omega)\frac{D\omega}{Dt}\mathrm{d}x\mathrm{d}y = 0, \tag{1.2}$$

where S_L is a "patch" of fluid bounded by a closed curve C_L that moves with the fluid. We use the suffix L here for "Lagrangian". This gives the family of invariants

$$I\{F\} = \int_{S_L} F(\omega)\mathrm{d}x\mathrm{d}y \tag{1.3}$$

(of which the best known, with $F(\omega) = \omega^2$ and S_L the whole fluid domain, is the enstrophy of the flow).

Two important questions arise from these elementary considerations: (i) how are these families of invariants to be generalized to three-dimensional flows? (ii) How are these "invariants" modified, and what role do they play, when weak dissipative (*e.g.* viscous) effects are taken into consideration?

In these lectures, we first provide some tentative approaches to answer these questions. Then we discuss some particular situations in which topological considerations play a central role.

2　The family of helicity invariants

Let $\mathbf{v}(\mathbf{x}, t)$ be an arbitrary three-dimensional incompressible flow and let $\mathbf{B}(\mathbf{x}, t)$ be an arbitrary solenoidal field satisfying the evolution equation

$$\frac{\partial \mathbf{B}}{\partial t} = \nabla \wedge (\mathbf{v} \wedge \mathbf{B}), \ \nabla \cdot \mathbf{B} = 0. \tag{2.1}$$

It is well-known that this equation describes "frozen-field" transport of $\mathbf{B}$, the flux of $\mathbf{B}$ across any material surface element being conserved. If $\mathbf{B}$ is interpreted as the magnetic field in a perfectly conducting fluid, then this is none other than Alfven's theorem in magnetohydrodynamics.

Now let $\mathbf{A}$ be a vector potential for $\mathbf{B}$ satisfying $\nabla \cdot \mathbf{A} = 0$; then "uncurling" (2.1) gives

$$\frac{\partial \mathbf{A}}{\partial t} = \mathbf{v} \wedge \mathbf{B} - \nabla \varphi \tag{2.2}$$

for some scalar field φ. Equations (2.1) and (2.2) can be written in equivalent Lagrangian form

$$\begin{aligned}
\frac{DB_j}{Dt} &= B_j \frac{\partial v_i}{\partial x_j} \\
\frac{DA_i}{Dt} &= v_j \frac{\partial}{\partial x_i} A_j - \frac{\partial \varphi}{\partial x_j}
\end{aligned} \tag{2.3}$$

where $D/Dt = \partial/\partial t + \mathbf{v} \cdot \nabla$. We define the helicity $\mathcal{H}$ of the field $\mathbf{B}$ within any (Lagrangian) magnetic surface S_L on which $\mathbf{n} \cdot \mathbf{B} = 0$ by

$$\mathcal{H} = \int_{V_L} \mathbf{A} \cdot \mathbf{B} dV. \tag{2.4}$$

It is easily shown that this integral is gauge invariant (*i.e.* invariant under replacement of $\mathbf{A}$ by $\mathbf{A} + \nabla \psi$). It is further easily shown that

$$\frac{d\mathcal{H}}{dt} = \int_{V_L} \frac{D}{Dt} (\mathbf{A} \cdot \mathbf{B}) dV = \int_{S_L} (\mathbf{n} \cdot \mathbf{B})(-\varphi + \mathbf{v} \cdot \mathbf{A}) dS \tag{2.5}$$

and this vanishes since $\mathbf{n} \cdot \mathbf{B} = 0$ on S_L. Hence $\mathcal{H}$ is invariant (Woltjer 1958).

The invariant has topological interpretation (Moffatt 1969; Arnol'd 1974); thus for example if the field $\mathbf{B}$ is identically zero except in two flux tubes carrying fluxes Φ_1, Φ_2, and linked with (Gauss) linking number n, then

$$\mathcal{H} = \pm 2n\Phi_1\Phi_2 \qquad (2.6)$$

(where V_L is taken to be a volume containing both tubes). This statement requires immediate qualification: it is true provided the $\mathbf{B}$-lines within either tube are closed curves with no linkage between any pair; in this case, each tube on its own makes no contribution to the helicity. It is just the linkage of the tubes that contributes, *via* the formula (2.6).

Suppose instead that we have a single flux tube with flux Φ and axis in the form of a circle, the $\mathbf{B}$-lines in the tube being all "parallel" circles. Then there is no linkage and the helicity is zero. Imagine now that we cut the tube, twist one cut end through an angle 2π and rejoin the ends. Now each pair of $\mathbf{B}$-lines is linked, with linking number 1; if we imagine this twisted tube as built up through the addition of incremental fluxes $d\varphi$, the total helicity is given by

$$\mathcal{H} = \pm 2 \int_0^\Phi \varphi \, d\varphi = \pm \Phi^2 \qquad (2.7)$$

the $+$ or $-$ being chosen according as the twist is right- or left-handed.

If the angle of twist is $2\pi h$, instead of 2π, then clearly the helicity is linear in h and is therefore given by

$$\mathcal{H} = \pm h\Phi^2. \qquad (2.8)$$

Returning now to the two linked flux tubes, we may allow for "twists" h_1 and h_2 in the two tubes, and the total helicity is then

$$\mathcal{H} = \pm h_1\Phi_1^2 \pm h_2\Phi_2^2 \pm 2n\Phi_1\Phi_2 \qquad (2.9)$$

where, for each term, the sign is chosen according as the twist (or linkage) is right- or left-handed.

2.1 Chaotic fields

In general, a field $\mathbf{B}$ in $\mathbb{R}^3$ may be expected to have chaotic field lines, *i.e.* field lines which do not lie on surfaces except in some sub-regions of "regularity" of the field. Examples of such fields may be found in Dombre *et al.* (1986) (the ABC-field) and in Bajer and Moffatt (1990) (the STF-field). Let V_L be a subdomain in which the **B**-lines are chaotic; then for such a subdomain, we have the single helicity invariant (2.4), and there appears to be no analogue of the infinite families of invariants $A(\omega)$ and $I\{F\}$ introduced in Section 1. It may be noted that it is for this situation that Arnol'd (1974) established the generalized interpretation of (2.5) as the "asymptotic Hopf invariant" (*i.e.* asymptotic linking number) of the field **B** in the chaotic region.

2.2 Simply degenerate fields

In many situations of interest however, the **B**-lines do lie on a family of surfaces. Suppose this family is the family $\psi(\mathbf{x}, t) = 0$; then evidently, since the **B**-lines move with the fluid, we must have

$$D\psi/Dt = 0 \quad \text{and} \quad \mathbf{B} \cdot \nabla\psi = 0. \tag{2.10}$$

We may clearly define an invariant "helicity function"

$$\hat{\mathcal{H}}(\psi) = \int_{V_L(\psi)} \mathbf{A} \cdot \mathbf{B}\mathrm{d}V \tag{2.11}$$

where $V_L(\psi)$ is the volume inside the Lagrangian surface $\psi = $ cst. Here we have an obvious analogue of the signature function $A(\omega)$ of Section 1. Equally however if we define

$$h(\psi) = \mathrm{d}\hat{\mathcal{H}}/\mathrm{d}\psi \tag{2.12}$$

so that $h(\psi)\mathrm{d}\psi$ represents the (invariant) helicity trapped between surfaces labeled ψ and $\psi + \mathrm{d}\psi$, then we may construct a family of invariants analogous to (1.3), namely

$$I\{F\} = \int_{V_L(\psi)} F(h(\psi))\mathrm{d}\psi \tag{2.13}$$

where F is an arbitrary function of h; for

$$\frac{\mathrm{d}I}{\mathrm{d}t} = \int F'(h)h'(\psi)\frac{D\psi}{Dt}\mathrm{d}\psi = 0 \tag{2.14}$$

by virtue of (2.10).

2.3 Doubly degenerate fields

In special circumstances, it may happen that every **B**-line is a closed curve; for example, the "twisted flux tube" described above, with angle of twist

$2\pi n$ (n an integer) is a field for which each **B**-line is an unknotted closed curve, each pair of **B**-lines having linking number n. More generally, we may consider a situation in which the **B**-lines are the intersections of surfaces $\psi = $ cst. and $\chi = $ cst. (a doubly-infinite family), with

$$D\psi/Dt = 0, \; D\chi/Dt = 0, \; \mathbf{B} = \nabla\psi \wedge \nabla\chi. \tag{2.15}$$

Then (*cf.* (2.11)), we may define

$$\hat{\mathcal{H}}(\Delta\psi, \Delta\chi) = \int_{V_L(\Delta\psi,\Delta\chi)} \mathbf{A} \cdot \mathbf{B}dV \tag{2.16}$$

where $\Delta\psi = \psi_1 - \psi_2$, $\Delta\chi = \chi_1 - \chi_2$, and V_L is the tube-like volume bounded by the surfaces labeled $\psi_1, \psi_2, \chi_1, \chi_2$. Defining $h(\psi, \chi)$ by

$$h(\psi, \chi)d\psi d\chi = \hat{\mathcal{H}}(d\psi, d\chi) \tag{2.17}$$

for increments $d\psi$, $d\chi$, we now have a family of invariants (*cf.* (2.13)) of the form

$$I\{F\} = \int_{V_L(\Delta\psi,\Delta\chi)} F(h(\psi,\chi))d\psi d\chi. \tag{2.18}$$

Note that the greater the degree of degeneracy of the field, the richer is the family of topological invariants that it exhibits. These invariants are all constructed from (2.4), in which the Lagrangian volume V_L is constrained only by the requirement that its surface S_L be a surface on which $\mathbf{n} \cdot \mathbf{B} = 0$; the greater the degeneracy of the field, the greater is the freedom in the choice of such volumes.

3   The special case of Euler dynamics

If, in the above theory, we choose $\mathbf{v} = \mathbf{u}(\mathbf{x}, t)$, a velocity field evolving under the incompressible Euler equations, and $\boldsymbol{\omega} = \nabla \wedge \mathbf{u}$, the corresponding vorticity field, then (2.1) is clearly satisfied since

$$\frac{\partial\boldsymbol{\omega}}{\partial t} = \nabla \wedge (\mathbf{u} \wedge \boldsymbol{\omega}), \; \nabla \cdot \boldsymbol{\omega} = 0. \tag{3.1}$$

The special feature here is that the field $\boldsymbol{\omega}$ is transported by its own "self-induced" velocity field $\mathbf{u}$. This coupling of $\mathbf{u}$ to $\boldsymbol{\omega}$ makes (3.1) nonlinear; but this does not invalidate the manipulation leading to (2.5) which, with

$$\mathcal{H} = \int_{V_L} \mathbf{u} \cdot \boldsymbol{\omega}dV, \tag{3.2}$$

now becomes

$$\frac{d\mathcal{H}}{dt} = \int_{S_L} (\mathbf{n} \cdot \boldsymbol{\omega}) \left(-p + \frac{1}{2} \mathbf{u}^2 \right) dV. \qquad (3.3)$$

Here p is the fluid pressure, and for simplicity of notation we have taken the fluid density to be $\rho = 1$. Thus again, $\mathcal{H} = $ cst. provided $\mathbf{n} \cdot \boldsymbol{\omega} = 0$ on S_L (a condition that persists under Euler evolution).

Thus the helicity $\mathcal{H}$ is an invariant of the nonlinear system (3.1). However, it should be interpreted within the broader canvass of Section 2, involving the system (2.1), which (for any given $\mathbf{v}$) is evidently linear in $\mathbf{B}$. From this point of view, it is evident that, under "artificial Euler dynamics" for which (3.1) is replaced by

$$\frac{\partial \boldsymbol{\omega}}{\partial t} = \nabla \wedge (\mathbf{v} \wedge \boldsymbol{\omega}), \ \nabla \cdot \boldsymbol{\omega} = 0 \qquad (3.4)$$

with $\boldsymbol{\omega} = \nabla \wedge \mathbf{u}$ and $\mathbf{v}$ an arbitrary field satisfying $\nabla \cdot \mathbf{v} = 0$, the helicity (3.2) is still invariant; this is because $\boldsymbol{\omega}$ is now transported by the $\mathbf{v}$-field and its topology is still conserved. In particular, if $\mathbf{v}$ is some functional of $\mathbf{u}$ (as for example in the artificial Euler dynamics of Vallis *et al.* 1989), helicity (and $\boldsymbol{\omega}$-topology) are still conserved. The more recent "averaged Euler equations" of Marsden *et al.* (2000) appear to have similar conservation properties.

4   Scalar field structure in 2D flows

The structure of a scalar field $s(\mathbf{x})$ may be described in terms of the critical points of the field where $\nabla s = 0$ and the structure of the iso-surfaces $s = $ cst. near these points (for details, see Moffatt 2001). Consider first the situation in a 2D periodic domain (as frequently adopted in numerical experiments on 2D turbulence). Here, the field s may be the streamfunction ψ of the flow, or the vorticity field $\omega = -\nabla^2 \psi$, or it may be some other scalar field of physical significance, such as the pressure field p. The critical points are either elliptic (extrema) or hyperbolic (saddle points). In a periodic domain, the number of extrema is equal to the number of saddle points (a consequence of Euler's index theorem). In an evolving situation, pairs of critical points (always one saddle and one extremum) can appear or disappear (through "saddle-node" bifurcation). If s is the vorticity field ω, then it is obvious that such bifurcation (with consequent change of topology of the ω-field) can occur only through the agency of viscosity.

In 2D turbulence, one may distinguish between the "eddies" that may be observed in instantaneous streamline plots $\psi = $ cst., and "vortices" that may be observed in instantaneous isovorticity plots $\omega = $ cst. Indeed, if we *define* an eddy as the region of closed streamlines circulating round an extremum of ψ, then the number of eddies N_ψ in the domain is precisely the number of extrema of ψ. We may define a vortex similarly in terms of the ω-field, so that the number of vortices N_ω is the number of extrema of ω (some positive, some negative). In a typical field of 2D turbulence at high Reynolds numbers, the vorticity field shows much more structure than the velocity field; this visual property suggests that N_ω is much greater than N_ψ. In fact, these numbers depend on the energy spectrum $E(k)$ of the turbulence. If there is an inertial range in which

$$E(k) \sim k^{-\lambda} \qquad (k_1 \ll k \ll k_2) \qquad (4.1)$$

with $1 < \lambda < 5$, then N_ψ and N_ω scale as

$$N_\psi \sim k_1^2 A, \; N_\omega \sim k_2^2 A \qquad (4.2)$$

(Moffatt 2001) where A is the area of the domain of periodicity of the flow, and the behavior $N_\omega \gg N_\psi$ is evident. The condition $1 < \lambda < 5$ is satisfied for all reasonable models of 2D turbulence (*e.g.* the enstrophy-cascade model of Kraichnan 1967 and Batchelor 1969 with $\lambda = 3$; or the spiral wind-up model of Gilbert 1988 for which $3 < \lambda < 4$).

5   Scalar field structure in 3D flows

The critical points of a scalar field $s(\mathbf{x})$ in 3D are again elliptic or hyperbolic, the elliptic points being either maxima or minima of the field. The hyperbolic (saddle)points are of two distinct types depending on whether s decreases in one principal direction away from the saddle (and increases in the other two directions), or *vice versa*. Bifurcations of such a field can create (or destroy) critical points in pairs: a minimum and a saddle point of the first kind; a maximum and a saddle point of the second kind; or a pair of saddle points, one of each kind. Each such topological transition is compatible with Euler's index theorem, which for the case of a space periodic field, takes the form

$$n_0 - n_1 + n_2 - n_3 = 0; \qquad (5.1)$$

here n_0 is the number of minima, n_3 the number of maxima and n_1, n_2 the number of saddle points of first and second kinds respectively. (n_α is the number of critical points, always assumed non-degenerate, with index α; α is the number of negative eigenvalues of the matrix $\partial^2 s / \partial x_i \partial x_j$ at the critical point.) It is worth noting that, since saddle points may be created in pairs, whereas extrema cannot be created without creating an equal number of saddle points, the number of saddle points will in general be greater (and possibly much greater) than the number of extrema.

A simple measure of the topological complexity of a scalar field $s(\mathbf{x})$ in 3D is given by the number N of critical points of the field per unit volume (irrespective of type). If $s(\mathbf{x})$ is a passive scalar field subjected to convection with negligible diffusion, then $Ds/Dt = 0$, and the number N is conserved (since topological transitions involving either increase or decrease of N) cannot occur.

In the presence of weak molecular diffusivity κ, the field s is governed by the advection-diffusion equation

$$Ds/Dt \equiv \partial s/\partial t + \mathbf{u} \cdot \nabla s = \kappa \nabla^2 s. \tag{5.2}$$

Following Batchelor (1959) and Batchelor et al. (1959), we may suppose that gradients of s are maintained at a large length-scale, and that fluctuations of s, relative to its mean, cascade to very small "diffusive" scales, at which they are eliminated. In these circumstances, the number N may be expected to attain a large mean value determined by the detailed process of molecular diffusion (which both creates and destroys critical point pairs in equal measure). When $\kappa \geq \nu$ (kinematic viscosity), the wave-number beyond which molecular diffusion becomes important is $k_c \sim (\epsilon/\kappa^3)^{1/4}$ (the "conduction cut-off") and the number N of critical points in a periodicity volume V may be estimated as

$$N \sim k_c^3 V = (\epsilon/\kappa^3)^{3/4} V. \tag{5.3}$$

The situation when $\kappa \ll \nu$ is physically less clear; all one can say in this case is that N/V must be at least of order k_v^3, where $k_v = (\epsilon/\nu^3)^{1/4}$ is the conventional "Kolmogorov cut-off".

Numerical simulations of the passive scalar field problem (with artificial "prescribed" $\mathbf{u}$-field having a Kolmogorov spectrum) can be carried out, and it should be a straightforward matter to locate and count the critical points of the scalar field, with a view to testing the above conclusions.

6 Vector field structure in 3D flows

If we pass from the passive scalar field problem to the passive vector field problem for which a field $\mathbf{B}(\mathbf{x}, t)$ (conventionally magnetic field) evolves according to the equation

$$\partial \mathbf{B}/\partial t = \nabla \wedge (\mathbf{u} \wedge \mathbf{B}) + \eta \nabla^2 \mathbf{B}, \quad \nabla \cdot \mathbf{B} = 0, \tag{6.1}$$

then we are faced with the extremely difficult problem of classifying possible topological structures for the field $\mathbf{B}$. We have already noted the possibility of a "degenerate" field whose $\mathbf{B}$-lines lie on surfaces $\psi = $ cst. However, for general $\mathbf{u}$, such a condition will persist only if $\eta = 0$. If $\eta \neq 0$, then the

B-lines cannot be expected to remain on surfaces (it would be interesting to investigate for what special class (if any) of flows **u**, the **B**-lines *do* remain on surfaces!). Generically, the **B**-lines must be expected to have a chaotic character. Two such fields (already mentioned in Sect. 2) have been studied in some detail. The first is the "ABC-field",

$$\mathbf{B} = (C\sin kz + B\cos ky,\ A\sin kx + C\cos kz,\ B\sin ky + A\cos kx) \quad (6.2)$$

which, for arbitrary constants A, B, C, satisfies the Beltrami condition $\nabla \wedge \mathbf{B} = k\mathbf{B}$. The chaotic character of the **B**-lines of this field (when $ABC \neq 0$) was conjectured by Arnold (1965), explored numerically by Hénon (1966), and analyzed further in detail by Dombre *et al.* (1986). The second is the "STF-field"

$$\mathbf{B} = (\alpha z - 8xy,\ 11x^2 + 3y^2 + z^2 + xy - 3,\ -\alpha x + 2yz - xy) \quad (6.3)$$

(Bajer and Moffatt 1990), which, as may be easily verified, satisfies $\nabla \cdot \mathbf{B} = 0$ and $\mathbf{n} \cdot \mathbf{B} = 0$ on $|\mathbf{x}| = 1$. (The field also satisfies the "Stokes condition" $\nabla^2(\nabla \wedge \mathbf{B}) = 0$.) When the parameter α equals zero, the field **B** is doubly degenerate, each **B**-line being a closed curve; this imposes a certain character on the chaotic wandering of **B**-lines for $\alpha \neq 0$. Note that the STF-field was originally constructed as a velocity field in a spherical ball $|\mathbf{x}| < 1$ having a combination of stretch, twist and fold ingredients – hence the STF-label.

Apart from these two fields, whose properties are now reasonably well documented, very little is known concerning the generic structure of solenoidal vector fields in 3D. The beginnings of a general structural theory may perhaps be seen in the work of Ghrist (1997) and Kuperberg (1999); but these are only the beginnings, and the general classification problem for such 3D fields is still wide open.

7   Helicity and the turbulent dynamo

The evolution of a magnetic field in a conducting fluid moving with velocity $\mathbf{u}(\mathbf{x}, t)$ is governed by equation (6.1). It has long been recognized that field intensification will result from the stretching of **B**-lines associated with the exponential separation of initially neighboring particles in turbulent flow. However, this intensification is achieved at the expense of a reduction of scale in the **B**-field, and a consequential enhancement of the effect of diffusion represented by the term $\eta\nabla^2\mathbf{B}$ in (6.1); this ultimately becomes important no matter how small the diffusivity parameter η may be. The question then arises whether sustained intensification of magnetic field (or "dynamo" action) can arise, and if so through what mechanism that somehow bypasses the "enhanced diffusivity" effect.

As pointed out earlier, if the velocity field $\mathbf{u}$ can be specified independently of $\mathbf{B}$, then (6.1) is linear in $\mathbf{B}$, which behaves as a "passive vector field". The only constraint that need be imposed on $\mathbf{u}$ is the kinematic constraint of incompressibility $\nabla \cdot \mathbf{u} = 0$; and the corresponding phase of field evolution is covered by the term "kinematic dynamo problem". If the field grows exponentially, then this phase cannot last forever, because the Lorentz force $\mathbf{j} \wedge \mathbf{B}$ (with $\mathbf{j} = \mu_0^{-1} \nabla \wedge \mathbf{B}$) obviously reacts back upon the dynamics of the flow, $i.e.$ $\mathbf{u}$ can no longer be specified independently of $\mathbf{B}$; we then enter upon the phase of the fully "magnetohydrodynamic dynamo problem", when it becomes necessary to specify also the nature of the forces agitating the fluid (or equivalently the nature of the source of kinetic energy of the turbulence).

7.1   The kinematic phase

A double-length-scale approach is usually adopted as a starting point. It is supposed that the dominant scale of the turbulence ($i.e.$ the scale of the "energy-containing eddies") is l_0, and one focuses on the evolution of the field $\bar{\mathbf{B}}(\mathbf{x}, t)$ on a scale L much greater than l_0. This is the "mean-field" approach pioneered by Steenbeck et $al.$ (1966) (see Moffatt 1978; Krause and Rädler 1980). If the turbulence is homogeneous with zero mean, then a mean electromotive force $\mathcal{E}(\mathbf{x}, t)$ is generated on the scale L, given by an expansion of the form

$$\mathcal{E}_i = \langle \mathbf{u} \wedge \mathbf{b} \rangle_i = \alpha_{ij} \bar{B}_j + \beta_{ijk} \partial \bar{B}_j / \partial x_k + \dots , \qquad (7.1)$$

where α_{ij}, β_{ijk}, ... are pseudo-tensors determined (in principle) by the statistical properties of the turbulence and the parameter η. The field $\mathbf{b}$ is the small-scale magnetic perturbation induced by the flow $\mathbf{u}$ across $\bar{\mathbf{B}}$. The important feature of (7.1) is the linearity between the fields $\mathcal{E}$ and $\bar{\mathbf{B}}$. This linearity results from the linearity of (6.1); no assumption need be made concerning the relative magnitudes of $|\mathbf{b}|$ and $|\bar{\mathbf{B}}|$ in order to deduce the structure (7.1). Conventional estimates suggest that

$$\mathbf{b} = O(R_{\mathrm{m}})\bar{\mathbf{B}}, \qquad (7.2)$$

where

$$R_{\mathrm{m}} = (\bar{u}^2)^{1/2} l_0 / \eta, \qquad (7.3)$$

the magnetic Reynolds number of the turbulence. In situations of interest in planetary physics and astrophysics, R_{m} is a least of order unity, and generally much greater than unity; in these circumstances, the fluctuating ingredient of the field $\mathbf{b}$ may be much greater than the mean $\bar{\mathbf{B}}$.

If the turbulence is isotropic as well as homogeneous, then the pseudo-tensors α_{ij}, β_{ijk}, ... must also be isotropic, $i.e.$

$$\alpha_{ij} = \alpha\delta_{ij}, \ \beta_{ijk} = \beta\epsilon_{ijk}, \ \ldots \tag{7.4}$$

where α is a pseudo-scalar and β a pure scalar (the "pseudo" property of β_{ijk} being taken up by the pseudo-tensor ϵ_{ijk}). The expansion (7.1) then takes the simpler form

$$\mathcal{E} = \alpha\bar{\mathbf{B}} - \beta\nabla\wedge\bar{\mathbf{B}} + \ldots, \tag{7.5}$$

so that (with α, β constants by virtue of homogeneity),

$$\nabla\wedge\mathcal{E} = \alpha\nabla\wedge\bar{\mathbf{B}} + \beta\nabla^2\bar{\mathbf{B}}. \tag{7.6}$$

The "mean-field" equation for $\bar{\mathbf{B}}$ then becomes

$$\partial\bar{\mathbf{B}}/\partial t = \alpha\nabla\wedge\bar{\mathbf{B}} + \eta_T\nabla^2\bar{\mathbf{B}}, \tag{7.7}$$

where $\eta_T = \eta+\beta$, a diffusivity augmented by the turbulent contribution of β; one would clearly expect β to be positive, although there is no guarantee of this, at this level of argument.

Equation (7.7) clearly admits exponentially growing modes of "force-free" structure satisfying

$$\nabla\wedge\bar{\mathbf{B}} = K\bar{\mathbf{B}}, \tag{7.8}$$

where K is a constant; for such modes satisfy

$$\partial\bar{\mathbf{B}}/\partial t = \alpha K\bar{\mathbf{B}} - \eta_T K^2\bar{\mathbf{B}}, \tag{7.9}$$

and hence $\bar{\mathbf{B}} \sim e^{pt}$ where

$$p = \alpha K - \eta_T K^2, \tag{7.10}$$

showing exponential growth provided

$$\alpha/K > \eta_T. \tag{7.11}$$

This condition is satisfied if K has the same sign as α and the scale $L \sim |K^{-1}|$ of $\bar{\mathbf{B}}$ is sufficiently large (consistent with the "two-scale" assumption $L \gg l_0$).

This dynamo mechanism is due to the "generation term" $\alpha\nabla\wedge\bar{\mathbf{B}}$ in (7.7), or equivalently to the term $\alpha\bar{\mathbf{B}}$ in (7.5). This is the famous "α-effect"; a growing understanding of this generic phenomenon has been one of the most dramatic developments of turbulence theory of the last half-century.

The fact that α is a pseudo-scalar implies that the effect can occur only in turbulence that "lacks reflexional symmetry". The mean helicity $\mathcal{H} = \langle \mathbf{u} \cdot \boldsymbol{\omega} \rangle$ of the turbulence (also a pseudo-scalar) is in general non-zero in such circumstances, and a link between α and $\mathcal{H}$ is to be expected.

To illustrate the essential mechanism by which an α-effect is produced, let us calculate α for the case of an "ABC"-type velocity field (with $A = B = C = u_0$)

$$\mathbf{u} = u_0(\sin(kz - \omega t) + \cos(ky - \omega t), \ \sin(kx - \omega t) + \cos(kz - \omega t),$$
$$\sin(ky - \omega t) + \cos(kx - \omega t)), \tag{7.12}$$

for which $\boldsymbol{\omega} = k\mathbf{u}$, and so

$$\mathcal{H} = \langle \mathbf{u} \cdot \boldsymbol{\omega} \rangle k\bar{\mathbf{u}}^2 = 3ku_0^2. \tag{7.13}$$

The helical character of (7.12) is evident in that the motion consists of a superposition of three circularly polarized waves propagating parallel to the axes Ox, Oy, Oz.

To calculate α, it is legitimate to assume a uniform mean field $\bar{\mathbf{B}}$, and the equation for the fluctuating field $\mathbf{b}$ (from 6.1) is then

$$\partial \mathbf{b}/\partial t = (\bar{\mathbf{B}} \cdot \nabla)\mathbf{u} + \nabla \wedge \mathbf{G} + \eta \nabla^2 \mathbf{b}, \tag{7.14}$$

where $\mathbf{G} = \mathbf{u} \wedge \mathbf{b} - \langle \mathbf{u} \wedge \mathbf{b} \rangle$. If the wave amplitude u_0 is sufficiently weak, then the nonlinear term $\nabla \wedge \mathbf{G}$ may be neglected (the "first-order smoothing" approximation); the field $\mathbf{b}$ may then be found by elementary methods, and $\mathcal{E} = \langle \mathbf{u} \wedge \mathbf{b} \rangle$ constructed. The result is indeed $\mathcal{E} = \alpha\bar{\mathbf{B}}$, where

$$\alpha = -\frac{1}{3}\left(\frac{\eta k^2}{\omega^2 + \eta^2 k^4}\right)\mathcal{H} \tag{7.15}$$

showing the expected dependence on mean helicity. Note also (i) the negative sign in (7.15), i.e. positive helicity generates a negative α-effect, and (ii) the fact that (for $\omega \neq 0$), $\alpha \to 0$ as $\eta \to 0$, i.e. an α-effect (at least in this first-order smoothing approximation) requires a non-zero molecular diffusivity η.

The helicity $\langle \mathbf{b} \cdot \nabla \wedge \mathbf{b} \rangle$ of the fluctuation field may be calculated under first-order smoothing (Moffatt 1978, Sect. 11.2). Not surprisingly, it has the same sign as the "driving" kinetic helicity $\langle \mathbf{u} \cdot \boldsymbol{\omega} \rangle$. Now look again at the mean-field equation (7.7) (with now slowly varying $\bar{\mathbf{B}}$); if (say) $\langle \mathbf{u} \cdot \boldsymbol{\omega} \rangle$ is positive, then α is negative and so negative helicity $\bar{\mathbf{B}} \cdot \nabla \wedge \bar{\mathbf{B}}$ is generated in the large-scale field. It seems that the positive magnetic helicity that is generated (and dissipated) on scales of order l_0 is at least partly compensated by negative magnetic helicity generated on the large scales of order $L \gg l_0$. We know that if $\eta = 0$, then total magnetic helicity is conserved. It would appear that dissipation of small-scale helicity is essential for dynamo action precisely because this allows the sustained increase of magnetic helicity (of opposite sign) in the large-scale field.

This manifestation of the effect of helicity in the turbulence was recognized in the seminal paper of Pouquet *et al.* (1976), who used the "eddy-damped quasi-normal Markovian" (EDQNM) closure scheme to analyze spectral evolution of fully magnetohydrodynamic turbulence. More recent direct numerical simulation of MHD turbulence (*e.g.* Brandenberg 1992), although limited to rather modest Reynolds and magnetic Reynolds numbers, shows similar trends.

7.2   The dynamic phase

Let us now consider briefly what happens when dynamo action occurs as a result of the α-effect, and a large-scale magnetic field grows exponentially until the stage at which the back-reaction of the Lorentz force on the flow becomes important. As previously indicated, it is necessary at this stage to specify the nature of the source of energy for the motion. Let us suppose for the sake of argument that there is a random body force $\mathbf{f}(\mathbf{x}, t)$ on the scale l_0, and that this force is independent of both $\mathbf{u}$ and $\mathbf{B}$.

The large-scale field $\bar{\mathbf{B}}$ evolves relatively slowly, and may, for the purpose of this analysis, be treated as constant. When $\bar{\mathbf{B}}$ is sufficiently strong, its effect is to severely control the amplitude of motion on scales of order l_0. It is reasonable then to linearize the equations for the velocity $\mathbf{u}$ and fluctuating field $\mathbf{b}$. These equations (in units such that $\mu_0 \rho = 1$) become

$$\left. \begin{array}{l} \partial \mathbf{u}/\partial t = -\nabla P + \bar{\mathbf{B}} \cdot \nabla \mathbf{b} + \nu \nabla^2 \mathbf{u} + \mathbf{f} \\ \partial \mathbf{b}/\partial t = \bar{\mathbf{B}} \cdot \nabla \mathbf{u} + \eta \nabla^2 \mathbf{b} \end{array} \right\} \tag{7.16}$$

where P is the sum of fluid and magnetic pressure. These are just the equations for forced Alfven waves traveling on the field $\bar{\mathbf{B}}$; indeed if we consider a single Fourier component $\hat{f} \exp i(\mathbf{k} \cdot \mathbf{x} - \omega t)$ of $\mathbf{f}$, the corresponding solution of (7.16) is

$$\hat{\mathbf{u}} = D^{-1}(-i\omega + \eta k^2)\hat{f}, \quad \hat{\mathbf{b}} = D^{-1}i(\bar{\mathbf{B}} \cdot k)\hat{f}, \tag{7.17}$$

where

$$D = \omega^2 + i\omega(\nu + \eta)k^2 - \nu\eta k^4 - (\bar{\mathbf{B}} \cdot \mathbf{k})^2. \tag{7.18}$$

Where ν and η are both small, there is a sharp resonance near $\omega = \pm \bar{\mathbf{B}} \cdot \mathbf{k}$, the frequency of freely propagating non-dissipative Alfven waves. When we construct $\langle \mathbf{u} \wedge \mathbf{b} \rangle$ for a spectrum of forcing, the result is dominated by the amplitude and width of these resonances in $(\omega, \mathbf{k})$ space.

The details, which are quite subtle, have been worked out with the additional complication of Coriolis effects in a rotating body of fluid (Moffatt 1972). The combined effects of Lorentz and Coriolis forces induce anisotropy in the turbulence; the "effective α" (e.g. $\alpha = \frac{1}{3}\alpha_{ii}$) can still however be calculated; this is now a decreasing function of mean field strength $|\bar{\mathbf{B}}|$, on account of the decreasing width of the resonant layers in $(\omega, \mathbf{k})$ space as $|\bar{\mathbf{B}}|$ increases.

This effect in which $\alpha(\bar{\mathbf{B}})$ is a decreasing function of $|\bar{\mathbf{B}}|$ (tending to zero as $|\bar{\mathbf{B}}| \to \infty$) has since been described as "α-quenching". Clearly, it leads to saturation in the growth of the mean field. For the simple model of Section 7.1, this saturation will occur for scale $L \sim K^{-1}$ when

$$\alpha(\bar{\mathbf{B}}) = \eta_{\mathrm{T}} K. \tag{7.19}$$

Note however that a field saturating at this level can still apparently grow on scales much larger than K^{-1}. Thus, under sustained forcing on some scale l_0 which, either through the intrinsic character of the forcing, or through a Coriolis effect, generates turbulence with non-zero helicity, we may expect the dynamo-generated magnetic field to saturate at successively larger length scales, the spectral shape of the resulting field being given, in qualitative terms, by solving (7.19) for $\bar{\mathbf{B}}$:

$$\bar{\mathbf{B}} = \alpha^{-1}(\eta_{\mathrm{T}} K) \tag{7.20}$$

and by interpreting this as providing the spectrum (in $\mathbf{K}$-space) of $\bar{\mathbf{B}}$.

Of course the parameter β (contributing to η_{T}) will also be subject to a measure of quenching, which could also be included in the above type of analysis.

8   Magnetic relaxation

A situation of great interest arises when we consider a different type of initial value problem: suppose that at time $t = 0$, a random magnetic field $\mathbf{B}_0(\mathbf{x})$ exists in a fluid which is at rest. The associated current is $\mathbf{j}_0 = \nabla \wedge \mathbf{B}_0$ and the Lorentz force $\mathbf{F}_0 = \mathbf{j}_0 \wedge \mathbf{B}_0$ is in general rotational (i.e. $\nabla \wedge \mathbf{F}_0 \neq 0$). This force cannot therefore be balanced by pressure gradients, and the fluid will move; energy is then dissipated by viscosity. At the same time, the field is transported by the flow, and the Lorentz force distribution $\mathbf{F}(\mathbf{x}, t)$ evolves in time.

The situation is of particular interest if the fluid is a perfect conductor (i.e. $\eta = 0$), since then the field topology is conserved during this "relaxation" process. Energy is however still dissipated by viscosity. The magnetic helicity is conserved, and this acts as a "topological barrier" that prevents the magnetic energy from decaying to zero. We are then faced with a variational problem with an unusual twist: to find the minimum energy state of a field whose topology is prescribed as that of the initial field $\mathbf{B}_0(\mathbf{x})$.

The equations of magnetohydrodynamics (with $\eta = 0$, $\nu \neq 0$) provide the natural dynamics that drives the system towards such a minimum energy state (Moffatt 1985). In the minimum energy state, the velocity is again zero (since otherwise it would continue to dissipate energy); the corresponding field, $\mathbf{B}^{\mathrm{E}}(\mathbf{x})$ say, is therefore magnetostatic, $i.e.$

$$\mathbf{j}^{\mathrm{E}} \wedge \mathbf{B}^{\mathrm{E}} = \nabla p^{\mathrm{E}} \qquad (8.1)$$

for some scalar (pressure) field p^{E}, and $\mathbf{j}^{\mathrm{E}} = \nabla \wedge \mathbf{B}^{\mathrm{E}}$. The field $\mathbf{B}^{\mathrm{E}}$ is "topologically accessible" from $\mathbf{B}_0(\mathbf{x})$, in the sense that it is obtained by continuous distortion by a velocity field $\mathbf{v}(\mathbf{x}, t)(0 < t < \infty)$ which dissipates a finite total amount of energy (the difference between the energies of the fields $\mathbf{B}_0$ and $\mathbf{B}^{\mathrm{E}}$).

Although the above relaxation process is simple to describe, and seems transparently clear, it should be noted that point-wise convergence of the field $\mathbf{B}(\mathbf{x}, t)$ to an equilibrium field $\mathbf{B}^{\mathrm{E}}(\mathbf{x})$ has not been proved, and remains an open problem.

8.1   The analogy with Euler flows

There is an exact analogy between (8.1) and the equation

$$\mathbf{u} \wedge \boldsymbol{\omega} = \nabla h \qquad (8.2)$$

describing steady Euler flows with $\boldsymbol{\omega} = \nabla \wedge \mathbf{u}$. The analogy is between the fields $\mathbf{B}^{\mathrm{E}}$ and $\mathbf{u}$, (or equivalently between $\mathbf{j}^{\mathrm{E}}$ and $\boldsymbol{\omega}$). Note that h in (8.2) must be regarded as the analogue of $-p^{\mathrm{E}}$. To any solution of (8.1), there corresponds via this analogy a corresponding solution of (8.2). It is thus apparent that (subject to pointwise convergence of the magnetic relaxation process) there exists a steady solution $\mathbf{u}(\mathbf{x})$ of the Euler equations having arbitrarily prescribed topology ($i.e.$ that of the initial field $\mathbf{B}_0(\mathbf{x})$ in the magnetic relaxation problem).

Note here that it is the topology of the velocity field (rather than that of the vorticity field) that can be prescribed. It would be more interesting if a relaxation procedure conserving the topology of $\boldsymbol{\omega}$ ($i.e.$ that of $\mathbf{j}$ in the magnetic relaxation problem!) could be devised, because that would be "more natural" for Euler dynamics. Only in 2D flows has this been found to be possible (Vallis et $al.$ 1989). In this context, an upper bound can be placed on the energy of a flow of prescribed enstrophy, and a "relaxation" procedure that $increases$ energy to a limiting value can be constructed.

In 3D, the steady Euler flows obtained *via* the magnetic relaxation technique, are all apparently unstable (Rouchon 1991). This is probably highly significant for turbulence! Nevertheless, it has been hypothesized (Moffatt 1990) that long-lived coherent vortices may be associated with "maximal helicity" regions where $\boldsymbol{\omega}$ is parallel to $\mathbf{u}$ (so that (8.2) is certainly satisfied). Recent analysis of 3D turbulent flows using wavelet transforms (Farge *et al.* 2001) lend some support to this hypothesis.

9   The blow-up problem

The question of smoothness of solutions of the Navier–Stokes and/or Euler equations for an incompressible fluid has remained open since first posed by Leray (1934). Numerical investigations (*e.g.* Kerr 1993; Pelz 1997) provide evidence for the blow-up of solutions of the Euler equations at finite time; but numerical codes have limited validity where singularities are concerned, and numerical results can be no more than suggestive in this context. On the analytical side, it is known (Beale *et al.* 1986) that if any breakdown of regularity of solutions of the Euler equations occurs at some finite time $t = t^*$, then the maximum value of the vorticity magnitude $|\boldsymbol{\omega}_{\max}(t)|$ must blow-up as $t \to t^*$ in such a way as to make the integral of this quantity (from t to t^*) diverge. The simplest possibility is that

$$|\omega_{\max}(t)| \sim \frac{1}{t^* - t} \qquad \text{as} \qquad t \uparrow t^*, \tag{9.1}$$

and this is indeed the sort of behavior that has been inferred from numerical experiments (Pumir and Siggia 1988; Pelz 1997).

Why, it may be asked, are we so interested in blow-up in the turbulence context? The reason is that if blow-up is a generic feature of any fully 3D time-dependent flow at very high Reynolds number, then the "spotty" or intermittent character of turbulent dissipation is immediately understandable in terms of the behavior near points where singularities of vorticity (and the related deformation tensor) occur.

Even more interesting is the question of how singularities (if they occur) must in practice be resolved. Singularities of the Euler equations may conceivably be resolved through inclusion of the effects of weak viscosity. But if it turns out that singular behavior can persist even for the Navier–Stokes equations (*i.e.* even when weak viscous effects are indicated), then we must look to other physical effects to resolve such behavior. The obvious effect that should then be considered is compressibility; for a singularity of vorticity will also imply a singularity of pressure (there being a large reduction of pressure in the core of an intense vortex). Compressibility in a gas results in the propagation of acoustic waves by the Lighthill mechanism

(Lighthill 1953), a mechanism that must surely prevent the formation of any pressure singularity in the fluid interior. In a liquid there is another mechanism: the liquid will cavitate wherever the pressure falls below the vapor pressure, and small bubbles will form, an effect that will be obviously dependent on the mean pressure applied to the system. This mechanism also mitigates against the formation of pressure singularities. the detailed mechanism of energy dissipation will be influenced by the presence of such cavitation bubbles; this deserves study!

9.1 Interaction of skewed vortices

All studies of the potential blow-up of vorticity have focused on the behavior of skewed vortex tubes, when for one reason or another these are driven into close proximity with one another. The simplest scenario (Moffatt 2000) is that in which two skewed vortex pairs propagate on a collision course towards each other. These interact strongly when the separation between the pairs becomes of the same order as the separation of the vortices within each pair. We may define an inner "interaction" zone whose scale $a(t)$ is a decreasing function of time; and an outer zone, where the vortex pairs continue to propagate with negligible interaction.

If all characteristic length-scales (*e.g.* vortex separations, vortex radii of curvature, vortex core radii, ...) decrease in proportion to $a(t)$ (and this is a big IF), then the behavior in the inner zone is self-similar and can be described by the Leray (1934) scaling:

$$\mathbf{u}(\mathbf{x},t) = \left(\frac{\Gamma}{t^* - t}\right)^{1/2} \mathbf{U}(\mathbf{X}), \quad \mathbf{X} = \frac{\mathbf{x}}{(\Gamma(t^* - t))^{1/2}} \cdot \quad (9.2)$$

where Γ is a constant having the dimensions of a circulation. The corresponding vorticity is then

$$\boldsymbol{\omega}(\mathbf{x},t) = \frac{1}{t^* - t}\boldsymbol{\Omega}(\mathbf{X}), \quad \boldsymbol{\Omega} = \nabla_{\mathbf{x}} \wedge \mathbf{U}, \quad (9.3)$$

and the vorticity equation transforms to

$$0 = \nabla \wedge \left[\left(\mathbf{U} + \frac{1}{2}\mathbf{X}\right) \wedge \boldsymbol{\Omega}\right] + \epsilon\nabla^2\boldsymbol{\Omega}, \quad (9.4)$$

where $\epsilon = \nu/\Gamma$. If any smooth solution of this equation, satisfying "acceptable" boundary conditions, can be found, then the corresponding solution of the native Navier–Stokes equation clearly has a singularity (with maximum vorticity behaving as in (9.1)) at $\mathbf{x} = 0$ as $t \to t^*$.

But what are the acceptable boundary conditions? Suppose that

$$|\boldsymbol{\Omega}(\mathbf{X})| \sim |\mathbf{X}|^{-\alpha} \quad \text{as} \quad |\mathbf{X}| \to \infty.$$

Then, for each fixed finite $\mathbf{x}$, from (9.2) and (9.3),

$$\boldsymbol{\omega}(\mathbf{x}, t) \sim \frac{1}{|\mathbf{x}|^*} \frac{\Gamma^{\alpha/2}}{(t^* - t)^{1-\alpha/2}} \; . \tag{9.5}$$

If $\alpha < 2$, this vorticity blows up for all finite $\mathbf{x}$ as $t \to t^*$, a behavior that is totally implausible. If $\alpha > 2$, then the vorticity goes identically to zero for all $\mathbf{x}$ as $t \to t^*$, a behavior that is equally implausible. The only realistic possibility therefore is that $\alpha = 2$, so that

$$|\boldsymbol{\Omega}(\mathbf{X})| \sim |\mathbf{X}|^{-2} \quad \text{as} \quad |\mathbf{X}| \to \infty \tag{9.6}$$

(and correspondingly $|\mathbf{U}| \sim |\mathbf{X}|^{-1}$) as would be realized, for example, by conical expansion of vortex tubes as they leave the interaction zone. Correspondingly we require that

$$|\boldsymbol{\omega}(\mathbf{x}, t)| \sim \Gamma/|\mathbf{x}|^2 \quad \text{as} \quad |\mathbf{x}| \to 0 \tag{9.7}$$

as the inner "boundary condition" for the outer region. Equations (9.6) and (9.7) indicate only radial behavior; angular dependence is unconstrained. Again, there is evidence of this behavior in the numerical work of Pelz (1997) who studied, by vortex filament techniques, the implosion of 6 vortex pairs towards the origin, the whole configuration having cubic symmetry.

Against this scenario, two theorems have been proved by functional analytic techniques. Nečas *et al.* (1996) have shown that, for $\epsilon > 0$, (9.4) has no nontrivial solution $\mathbf{U}(\mathbf{X})$ in $L^3(\mathbb{R}^3)$ (*i.e.* for which $|\mathbf{U}(\mathbf{X})|^3$ is integrable over the whole $\mathbf{X}$-space). This immediately rules out solutions for which $|\mathbf{U}| \sim |\mathbf{X}|^{-q}$ (and $|\boldsymbol{\Omega}| \sim |\mathbf{X}|^{-(q+1)}$) for $q > 1$; it does not rule out the behavior (9.6), which, perhaps significantly, lies just outside this function space. However, more seriously, Tsai (1998) has shown (again for $\epsilon > 0$) that the theorem of Nečas *et al.* (1996) can be extended to cover the nonexistence of nontrivial solutions $\mathbf{U}(\mathbf{x})$ of (9.4) in $L^q(\mathbb{R}^3)$ for all q in the range $3 < q < \infty$; this certainly does exclude the behavior (9.6). The full implications of Tsai's theorem are as yet unclear (to this writer!), but it does imply that singularities of the Navier–Stokes equations with $\nu > 0$ cannot in fact be described in terms of the Leray scaling (9.2).

The above remarks do not apply to the Euler limit ($\epsilon = 0$ in (9.4)); in this limit, we may integrate (9.4) to give

$$\left(\mathbf{U} + \frac{1}{2}\mathbf{X} \right) \wedge \boldsymbol{\Omega} = \nabla H \tag{9.8}$$

for some scalar $H(\mathbf{X})$ (the scaled Bernoulli function). Since $\boldsymbol{\Omega} \cdot \nabla H = 0$, vortex lines lie on surfaces $H = \text{cst.}$, which in effect define the vortex tubes of the flow in the inner (Leray) region. The self-induced velocity $\mathbf{U}(\mathbf{X})$ must satisfy

$$\left(\mathbf{U} + \frac{1}{2}\mathbf{X} \right) \cdot \nabla H = 0, \qquad (9.9)$$

i.e. there must be an inward flow across each vortex tube to compensate the outward transport represented by the term $\frac{1}{2}\mathbf{X} \cdot \nabla H$ (which can be traced to the space-scaling relating $\mathbf{X}$ and $\mathbf{x}$ in (9.2)). This inflow can in principle be compensated by outflow along the vortex tubes emanating from the interaction zone. But the \$ million question is whether there is any vortex configuration which induces a velocity field which in turn keep the configuration steady *via* the condition (9.8). It seems likely that this question will continue to present a profound challenge over the next few years, if not decades!

References

[1] V.I. Arnol'd, *C.R. Acad. Sci. Paris* **261** (1965) 17-20.

[2] V.I. Arnol'd, *Selecta Math. Sovet.* **5** (1986) 327-345 (1974).

[3] K. Bajer and H.K. Moffatt, *J. Fluid Mech.* **212** (1990) 337-363.

[4] G.K. Batchelor, *J. Fluid Mech.* **5** (1959) 113-133.

[5] G.K. Batchelor, *Phys. Fluids. Supp. II* (1969) 233-239.

[6] G.K. Batchelor, I. Howells and A.A. Townsend, *J. Fluid Mech.* **5** (1959) 134-139.

[7] J.T. Beale, T. Kato and A. Majda, *Comm. Math. Phys.* **94** (1984) 61-66.

[8] A. Brandenberg, *Phys. Rev. Lett.* **69** (1992) 605-608.

[9] T. Dombre, U. Frisch, J.M. Greene, M. Hénon, A. Mehr and A.M. Soward, *J. Fluid Mech.* **167** (1986) 353-391.

[10] M. Farge, G. Pellegrino and K. Schneider, *Coherent vortex extraction in 3D turbulent flows using orthogonal wavelets* (2001) (preprint).

[11] A.D. Gilbert, *J. Fluid Mech.* **195** (1988) 475-497.

[12] R.W. Ghrist, *Topology* **36** (1997) 423-448.

[13] M. Hénon, *C.R. Acad. Sci. Paris* **262** (1966) 312-314.

[14] R.M. Kerr, *Phys. Fluids A* **5** (1993) 1725-1746.

[15] R. Kraichnan, *Phys. Fluids* **10** (1967) 1417-1423.

[16] F. Krause and K.-H. Rädler, *Mean-field Magnetohydrodynamics and Dynamo Theory* (Pergamon Press, 1980).

[17] K.M. Kuperberg, *A C^∞ counterexample to the Seifert conjecture in dimension three* (AMS Bulletin, 1999).

[18] J. Leray, *Acta Math.* **63** (1934) 193-248.

[19] M.J. Lighthill, *Proc. Roy. Soc. A* **211** (1952) 564-587.

[20] J. Marsden, T. Ratiu and T. Shkoller, *Geom. Funct. Anal.* **10** (2000) 582-599.

[21] H.K. Moffatt, *J. Fluid Mech.* **35** (1969) 117-129.

[22] H.K. Moffatt, *J. Fluid Mech.* **53** (1972) 385-399.

[23] H.K. Moffatt, *Magnetic Field Generation in Electrically Conducting Fluids* (Cambridge University Press, 1978).

[24] H.K. Moffatt, *J. Fluid Mech.* **159** (1985) 359-378.

[25] H.K. Moffatt, *J. Fluid Mech.* **173** (1986) 289-302.

[26] H.K. Moffatt, Fixed points of turbulent dynamical systems and suppression of nonlinearity, in *Whither Turbulence? Turbulence at the Crossroads*, edited by J.L. Lumley, *Lecture Notes in Physics* **357** (Springer-Verlag, 1990) pp. 250-257.

[27] H.K. Moffatt, *J. Fluid Mech.* **409** (2000) 51-68.

[28] H.K. Moffatt, The topology of scalar fields in 2D and 3D turbulence, in *Geometry and Statistics of Turbulence*, edited by T. Kambe (Kluwer, 2001) (to appear).

[29] J. Nečas, M. Røužička and V. Šverák, *Acta Math.* **176** (1996) 283-294.

[30] R.B. Pelz, *Phys. Rev. E* **55** (1997) 1617-1626.

[31] A. Pouquet, U. Frisch and J. Léorat, *J. Fluid Mech.* **77** (1976) 321-354.

[32] A. Pumir and E. Siggia, *Phys. Fluids* **2** (1990) 220-241.

[33] P. Rouchon, *Eur. J. Mech., B/Fluids* **10** (1991) 651-661.

[34] M. Steenbeck, F. Krause and K.-H. Rädler, *Z. Naturforsch* **21a** (1966) 369-376.

[35] T.-P. Tsai, *Arch. Rational Mech. Anal.* **143** (1998) 29-51.

[36] G.K. Vallis, G.F. Carnevale and W.R. Young, *J. Fluid Mech.* **207** (1989) 133-152.

COURSE 7

"BURGULENCE"

U. FRISCH and J. BEC

Laboratoire G.D. Cassini, UMR 6529,
Observatoire de la Côte d'Azur,
BP. 4229, 06304 Nice Cedex 4, France

Contents

"BURGULENCE"

U. Frisch and J. Bec

1   Introduction

These lectures are about the d-dimensional Burgers equation

$$\partial_t \mathbf{v} + (\mathbf{v} \cdot \nabla)\, \mathbf{v} = \nu \nabla^2 \mathbf{v}, \qquad \mathbf{v} = -\nabla \psi. \qquad (1.1)$$

Note that the constraint that $\mathbf{v}$ be derived from a (velocity) potential ψ is trivially satisfied if $d = 1$. The word "burgulence", as we use it here, is a contraction of "Burgers" and "turbulence". It means "the study of random solutions to the Burgers equation". The randomness may arise because random initial conditions $\mathbf{v}_0 = -\nabla \psi_0$ are given or because a random driving force $\mathbf{f} = -\nabla F$ is added to the r.h.s. of (1.1), or both. When $\mathbf{f} = 0$ one speaks about "decaying burgulence".

In the thirties when the Dutch scientist J.M. Burgers introduced the equation in the one-dimensional case, he hoped to contribute to the study of turbulence with a simple model which, obviously, has a lot in common with the Navier–Stokes equation:

- same type of advective nonlinearity;

- presence of a diffusion term from which a Reynolds number may be defined;

- many invariance and conservation laws in common: invariance under translations in space and time, parity invariance, conservation of momentum and energy (only for $\nu = 0$ and $d = 1$).

Such hopes appeared to be shattered when, in the fifties, Hopf [1] and Cole [2] discovered – some say rediscovered – that the Burgers equation can actually be integrated explicitly (we shall return to this matter later). Indeed, an important property of the Navier–Stokes equation, not shared by the Burgers equation, is the sensitivity to small changes in the initial conditions in the presence of boundaries or driving forces and at sufficiently high Reynolds numbers. Hence, the Burgers equation is not a good model for one of the most important aspects of turbulence: the spontaneous arise of randomness by chaotic dynamics.

In spite of this there has been a strong renewal of interest in the Burgers equation, starting in the eighties, for a variety of reasons which we shall now explain. As a quantitative measure of the current interest, Table 1 shows some web-based statistical figures on the number of hits as of August 2000 (Google is an all-purpose search engine and "Los Alamos" stands for the nlin (ex-chao-dyn) preprint archive): the Burgers equation, which obviously describes a compressible flow (in one dimension there exist only trivial incompressible flows), has found many applications in *nonlinear acoustics* and other nonlinear wave problems. A review may be found in reference [3].

Table 1. Web-based statistical data.

	Navier–Stokes equation	Burgers equation
Google	15 000	4000
Los Alamos	100	75

1.1 The Burgers equation in cosmology

The Burgers equation has found interesting applications in cosmology, where it is known, in one instance, as the "Zel'dovich approximation" [4] and, in another instance, as the "adhesion model" [5]. Here, we shall give a brief introduction to how the Burgers equation arises in cosmology. More details may be found in [6–9]. Just after the baryon-photon decoupling in the early Universe, there may have been a rarefied medium formed by collisionless dustlike particles without pressure, interacting only *via* Newtonian gravity [8]. The gravitational potential is then determined from the fluctuations in mass density by a Poisson equation. Limiting ourselves to the case of a single type of matter, we can schematically write the acceleration of a fluid particle as follows:

$$
\begin{array}{ccccccc}
\text{acceleration} = & \text{pressure} & + & \text{viscous} & + & \text{expansion} & + & \text{gravit.} \\
& \text{term} & & \text{term} & & \text{term} & & \text{term} \\
& & & & & \propto \mathbf{v} \text{ in} & & \\
\partial_t \mathbf{v} + \mathbf{v} \cdot \nabla \mathbf{v} & \text{negligible} & & ? & & \text{comov. coord.} & &
\end{array}
$$

On the left hand side (l.h.s.) we recognize the familiar terms of the Burgers equation. The pressure is usually neglected because the matter is very cold. We shall come back to the viscous term later. The expansion term, proportional to the velocity, arises because the equation is written in a frame comoving with the expansion of the Universe.

It turns out that when the problem of self-gravitating gas in an expanding universe is examined in the linear approximation (small density fluctuations) an instability is obtained in which the dominant mode has the following properties [8,9]:

- it is potential ($\mathbf{v} = -\nabla\psi$);

- the expansion and gravitational terms cancel.

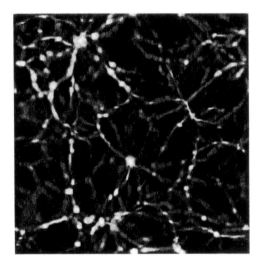

Fig. 1. N-body simulation by the Virgo Consortium (see http://star-www.dur.ac.uk/~frazerp/virgo/virgo.html). The simulation has 256^3 particles and was done on two large Cray T3D parallel supercomputers at the computing centers Garching (D) and Edinburgh (GB). The brightness is proportional to the log of the density of the particles.

In 1970, Zel'dovich [4] proposed to extend these properties into the nonlinear regime where density fluctuations become strong and mass condensation forming large-scale structures appear. Furthermore, this "Zel'dovich approximation" is exact in one dimension, irrespective of the strength of fluctuations. Clearly, in the Zel'dovich approximation each fluid particle is just moving in a straight line with constant velocity (after a suitable nonlinear change of variable of the time). Just like a family of straight light rays forms generally caustics along which the intensity is infinite, the material particle lines form singular objects along which the mass density is infinite.

Fig. 2. Mass density field from a 512^2 simulation of the Burgers equation with random scale-free initial data (from Ref. [7]).

Arnold *et al.* [10] studied the various kind of singularities which can form in this way, to see if they could account for known large-scale structures such as galaxies and clusters. Observations and numerical simulations have now revealed that large-scale structures are much simpler than the mathematical objects generated in a caustic-type theory.

Consider for example Figure 1 which shows a thin slice of a simulated universe using the so-called ΛCDM model (cold dark matter with cosmological constant). The line-like and node-like features on this slice are actually sheets and filaments embedded in the three-dimensional space. Sheets (walls), filaments and nodes (clusters) are the most common structures observed in such simulations. As was shown by Gurbatov and Saichev [5] these are precisely the structures obtained if one modifies the Zel'dovich approximation by requiring that particles should not cross but rather adhere. This *adhesion model* is just the three-dimensional Burgers equation (1.1), taken in the limit of vanishing viscosity. Numerical experiments indicate that the adhesion model reproduces quite well the early skeleton of large-scale structures in N-body numerical simulations (see, for example, Figs. 6a and 6b of Ref. [11]). Since dark matter is essentially collisionless, it is not clear at the moment what is the physics behind this agreement which seems to require some viscosity-generating mechanism to prevent – or dramatically slow down – particle crossing. Furthermore, the adhesion model cannot cope with many important aspects of gravitational dynamics. For example, in N-body simulations, one frequently observes the collapse of a filament into an isolated node (cluster). As we shall see, there is nothing of this sort in Burgers dynamics.

1.2   The Burgers equation in condensed matter and statistical physics

The Burgers equation arises in a number of condensed matter and statistical physics problems and even in non-physics problems such as vehicular traffic (for review see Ref. [12]) A frequently studied problem is the Kardar–Parisi–Zhang or KPZ equation [13] (see also Ref. [14])

$$\partial_t \psi = \frac{1}{2} \left| \nabla \psi \right|^2 + \nu \nabla^2 \psi + F, \tag{1.2}$$

which appears in studying the motion of an interface under deposition. Here, ψ is the vertical displacement of the interface as a function of $d - 1$ horizontal coordinates and of the time. It is immediately checked, by taking the horizontal gradient of (1.2), that one obtains the Burgers equation (1.1) with an additional forcing term $\mathbf{f} = -\nabla F$. Burgers equation also arises in studying directed polymers (see, *e.g.* [15, 16]), but with the time variable now interpreted as a space variable in the direction of main extension of the polymers. On all these problems there is considerable literature which it is not our purpose to review here.

1.3   The Burgers equation as testing ground for Navier–Stokes

The Burgers equation, because of its known solutions, is frequently used for testing numerical schemes, particularly those intended for compressible flow (many of the Google hits are of this kind). If one is mostly interested in turbulence, as is the case for participants of the present School, Burgers equation turns out to be quite useful for testing – and mostly discarding – certain types of theories of turbulence. Indeed, there have been many attempts to tackle the problem of the statistical theory of turbulence by adapting to it tools borrowed from field theory (for reviews, see [17–20]). Such methods had little impact on the field until recently when they have permitted a real breakthrough in understanding the mechanism for intermittency and anomalous scaling (see, *e.g.* the lectures by Falkovich *et al.* in the same volume). In the past such field-theoretic methods have frequently involved formal expansions in powers of the nonlinearity, with Feynman graphs used for the bookkeeping of all the terms generated after averaging over Gaussian initial conditions and/or random forces. Since the Burgers equation has the same type of nonlinearity as the Navier–Stokes equation such methods are typically also applicable to the Burgers equation. Hence it is possible to find what they predict for the latter and to compare the results with those obtained by more reliable methods. From this point of view, that is of using the Burgers equation as testing ground, it is desirable to know the answers to questions similar to those generally asked for Navier–Stokes turbulence. For example, what are the scaling properties of structure functions; what are the probability distribution functions (pdf) of velocity increments and velocity gradients? Such questions will be at the center of these lectures. Whenever possible we shall comment on the corresponding Navier–Stoke issues. The emphasis will be exclusively on what happens in the real space-time domain in the limit of vanishing viscosity, which is of course not the same as naively putting the viscosity equal to zero. A number of interesting questions, requiring a finite viscosity, such as the pole decomposition [21, 22] will thus be left out.

2   Basic tools

In this section we introduce various analytical, geometrical and numerical tools which are useful for constructing solutions to the decaying (unforced) Burgers equation (1.1). Mostly, we shall deal with the deterministic equation, while making occasional comments on consequences for burgulence.

2.1   The Hopf–Cole transformation and the maximum representation

If in (1.2) with $F = 0$ we set $\psi = 2\nu \ln \theta$ we obtain the d-dimensional heat equation [1, 2]

$$\partial_t \theta = \nu \nabla^2 \theta, \qquad (2.1)$$

which can be solved explicitly if there are no boundaries. One thus obtains

$$\psi(\mathbf{r}, t) = 2\nu \ln \left\{ \frac{1}{(4\pi\nu t)^{d/2}} \int_{\mathbb{R}^d} \exp\left[\frac{1}{2\nu} \left(\psi_0(\mathbf{a}) - \frac{|\mathbf{r} - \mathbf{a}|^2}{2t} \right) \right] d^d a \right\}, \quad (2.2)$$

where $\psi_0(\mathbf{a})$ is the initial potential. The limit of vanishing viscosity ($\nu \to 0$), obtained by steepest descent, has the following "maximum representation"

$$\psi(\mathbf{r}, t) = \max_{\mathbf{a}} \left(\psi_0(\mathbf{a}) - \frac{|\mathbf{r} - \mathbf{a}|^2}{2t} \right). \qquad (2.3)$$

Note that the operation of taking a maximum is global in nature, whereas the viscous Burgers equation is a local partial differential equation. If $\psi_0(\mathbf{a})$ is differentiable (*i.e.* the initial velocity $\mathbf{u}_0(\mathbf{a})$ exists as an ordinary function rather than a distribution), the maximum in (2.3) will be achieved at one or several points $\mathbf{a}$ where the gradient of the r.h.s. vanishes, that is, where

$$\mathbf{r} = \mathbf{a} + t\mathbf{v}_0(\mathbf{a}). \qquad (2.4)$$

In other words, $\mathbf{r}$ is the position at time t of the fluid particle starting at $\mathbf{a}$ and retaining its initial velocity $\mathbf{v}_0(\mathbf{a})$. Hence, we can interpret $\mathbf{a}$ and

r as being, respectively, Lagrangian and Eulerian coordinates. Along this Lagrangian trajectory, the velocity being conserved, we have

$$\mathbf{v}(\mathbf{r}, t) = \mathbf{v}_0(\mathbf{a}). \tag{2.5}$$

The map $\mathbf{a} \mapsto \mathbf{r}$ defined by (2.4) is called the naive Lagrangian map. It is not necessarily invertible: if there are several Lagrangian locations satisfying (2.4) for a given **r** the only acceptable one is that which maximizes the argument on the r.h.s. of (2.3). As long as the Jacobian of the naive Lagrangian map (2.4)

$$J(\mathbf{a}, t) = \det \left(\delta_{ij} - t \frac{\partial^2 \psi_0}{\partial a_i \partial a_j} \right) \tag{2.6}$$

does not vanish the map is guaranteed to be invertible and the solution of the Burgers equation cannot have a singularity. For sufficiently smooth initial data with bounded second derivatives of ψ_0 the first singularity appears at

$$t_\star = \frac{1}{\max_{\mathbf{a}} [\lambda(\mathbf{a})]}, \tag{2.7}$$

where $\lambda(\mathbf{a})$ is the largest eigenvalue of the Hessian matrix $\partial^2 \psi_0 / \partial a_i \partial a_j$.

In one dimension, we denote the velocity by u. Now, the time $t_\star$ is the inverse of the absolute value of the most negative initial velocity derivative $du_0(a)/da$. It is the first time at which the characteristics $x = a + tu_0(a)$ of the hyperbolic inviscid Burgers equation are crossing (Fig. 3). The first

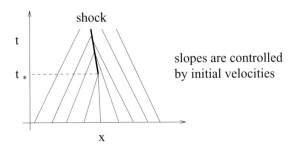

Fig. 3. Characteristics for the unforced one-dimensional Burgers equation in the (x, t) plane.

singularity in one or more dimension, is known as a "preshock" [23] and plays an important role in the theory of pdf for velocity gradients and densities (Sects. 6 and 7).

Note that, for Gaussian random initial conditions, $t_\star$ is itself random and can become arbitrarily small with very small but nonvanishing probability.

As a consequence, most averaged quantities (*e.g.* the two-point correlation function) will have an essential singularity at $t = 0$. Note also that the distribution of eigenvalues of the Hessian matrix extends to infinite values in any finite dimension d, but becomes compactly supported (on a semi-circle) when dividing the eigenvalues by $\sqrt{d}$ and letting $d \to \infty$. This is indeed a consequence of the properties of large random symmetric matrices, called Wigner matrices (see, *e.g.*, Ref. [25]).

2.2 Shocks in one dimension

After the time $t_\star$ the Lagrangian map (2.4) ceases to be invertible. More precisely, for a given Eulerian position $\mathbf{r}$ there is more than one Lagrangian position $\mathbf{a}$ satisfying (2.4). This implies the presence of shocks in the Eulerian velocity field. In this subsection we consider the one-dimensional case and give various geometrical constructions of the solution (including shocks).

First, let us define in the (x, ψ)-plane the Lagrangian manifold (a curve in one dimension)

$$x \quad \equiv \quad a + t u_0(a) \tag{2.8}$$

$$\psi \quad \equiv \quad \psi_0(a) - \frac{t}{2} u_0^2(a), \tag{2.9}$$

where the second line is just the r.h.s. of (2.3) without the maximum, evaluated at the (naive) Eulerian position $a + t u_0(a)$. Figure 4 (upper) shows this Lagrangian manifold after the time $t_\star$. Hence, above some Eulerian locations x there is more than one branch and cusps are present at Eulerian locations such that the number of branches changes. Clearly, the correct Eulerian potential is obtained by taking the maximum, *i.e.* always the highest branch. Note that this potential will have one or several points with discontinuous slope, the right derivative being always greater than the left one. Hence the velocity, which is the negative space derivative of the potential (shown in the lower part of Fig. 4) will have discontinuities at shock locations with $u_- > u_+$. It is also possible to directly construct the velocity starting from the Lagrangian manifold in the (x, u)-plane

$$x \quad \equiv \quad a + t u_0(a) \tag{2.10}$$

$$u \quad \equiv \quad u_0(a). \tag{2.11}$$

If there is a single shock present, it follows obviously that its position is determined by a Maxwell rule: the hashed loops shown in Figure 4 (lower part) right and left of the shock should have equal areas. A Maxwell rule construction can become very cumbersome if there are several shocks present.

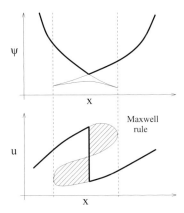

Fig. 4. Lagrangian manifolds for the potential in the (x, ψ)-plane (upper) and the velocity in the (x, u)-plane (lower). The heavy lines correspond to the correct Eulerian solutions. The vertical dashed lines delineate the multivalued region.

Another geometrical construction uses the *Lagrangian potential*

$$\varphi(a, t) \equiv t\psi_0(a) - \frac{a^2}{2}, \tag{2.12}$$

whose negative gradient is obviously the naive Lagrangian map. We can rewrite (2.3) as

$$t\psi(x, t) + \frac{x^2}{2} = \max_a \left[\varphi(a, t) + ax\right], \tag{2.13}$$

which represents the potential as, basically, a Legendre transform of the Lagrangian potential. (Note that the Legendre transformation is also used in the theory of multifractals.) The r.h.s. of (2.13) is equivalent to finding the largest algebraic vertical distance between the graph of the Lagrangian potential and the line of slope $-x$ through the origin. If the graph is convex (second derivative negative everywhere), the maximum is attained at the unique point where the derivative has the value $-x$. Otherwise, it suffices to replace the graph of φ by its convex hull φ_c, that is the intersection of all half-planes containing the graph. This is illustrated in Figure 5, which shows both regular points (Lagrangian points which have not fallen into a shock) and one shock interval, situated below the segment which is part of the convex hull. Again, it is possible to work directly with the (negative) derivative of the Lagrangian potential, namely, the naive Lagrangian map. The convex hull construction becomes then a Maxwell rule as shown in Figure 6. From this one can easily show that the speed of a shock is the half-sum of the velocities immediately to the right and to the left.

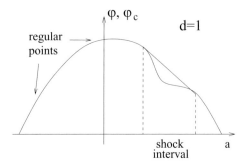

Fig. 5. Convex hull construction in terms of the Lagrangian potential.

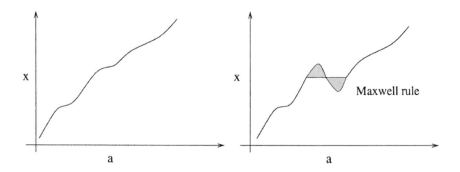

Fig. 6. Naive Lagrangian map before (left) and after (right) appearance of a shock. The correct Lagrangian map is obtained by a Maxwell rule.

Finally, the maximum formula (2.3) yields directly a "parabola construction", illustrated in Figure 7: a parabola with apex at x and radius of curvature proportional to t is moved down until it touches the initial potential $\psi_0(a)$ at the Lagrangian location associated to x (or at two such locations if there is a shock). Which of the five geometrical methods given is more convenient depends on the application considered. The parabola construction is best for understanding evolution in time (*cf.* Sect. 4). It may be used, for example, to show that the long-time Eulerian solution has a sawtooth structure with shocks separated by ramps of slope $1/t$ (see Fig. 14). The ramps are associated to high local maxima in the potential ψ_0.

With random and homogeneous initial conditions there will be shocks (discontinuities) at random Eulerian locations which do not cluster (unless we use non-smooth initial conditions as in Sect. 5). From this it is easily

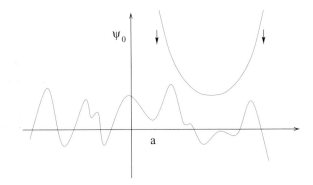

Fig. 7. Parabola construction of the solution.

inferred that, for $p > 0$, the structure functions

$$S_{\mathrm{p}}(\Delta x, t) \equiv \langle |u(x + \Delta x, t) - u(x, t)|^p \rangle \qquad (2.14)$$

behave, for small Δx, as

$$S_{\mathrm{p}}(\Delta x, t) \sim C_{\mathrm{p}}|\Delta x|^p + C'_{\mathrm{p}}|\Delta x|, \qquad (2.15)$$

where the first term comes from regular (smooth) parts of the Eulerian velocity, while the second comes from the $O(|\Delta x|)$ probability to have a shock somewhere in an interval of Eulerian length $|\Delta x|$. For $0 < p < 1$ the first term dominates as $\Delta x \to 0$, while, for $p > 1$, it is the second. Hence, $S_{\mathrm{p}} \sim |\Delta x|^{\zeta_{\mathrm{p}}}$, with the exponents ζ_{p} as shown in Figure 8. There are also higher-order corrections to the simple scaling law given in (2.15) which cannot be obtained by such simple arguments [30]. Note that a second-order structure functions with a behavior $\propto |\Delta x|$ at small distances implies an energy spectrum $E(k) \propto k^{-2}$ as $k \to \infty$.

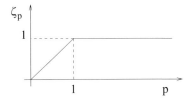

Fig. 8. Exponent of the structure function of order p. Note the "phase transition" at $p = 1$.

The "phase transition" at $p = 1$ seen in Figure 8 is due to the isolated character of the dissipative structures (the shocks), a feature not present in incompressible three-dimensional Navier–Stokes turbulence.

2.3   Convex hull construction in more than one dimension

Some of the methods used for the one-dimensional case are readily extended to dimensions $d > 1$, for example the construction from the Lagrangian manifold in the $(d+1)$ dimensional space $(\mathbf{x}, \psi)$. In Section 7 we shall use the multidimensional generalization of the convex hull construction, which we now briefly outline. We define the Lagrangian potential

$$\varphi(\mathbf{a}, t) \equiv -\frac{|\mathbf{a}|^2}{2} + t\psi_0(\mathbf{a}). \qquad (2.16)$$

and find, from (2.3), that

$$t\psi(\mathbf{r}, t) + \frac{|\mathbf{r}|^2}{2} = \max_{\mathbf{a}} \left[\varphi(\mathbf{a}, t) + \mathbf{r} \cdot \mathbf{a}\right]. \qquad (2.17)$$

As before, this involves a (multidimensional) Legendre transformation which leads us to the construction of the convex hull in a $(d+1)$ dimensional space of the graph of the Lagrangian potential. In more than one dimension, singularities of convex hulls are considerably more involved. As a consequence, the equivalent of shocks are discontinuities across $(d-1)$ manifolds, but there are many other singularities of higher codimension (the codimension is d minus the dimension of the object).

   In two dimensions the convex hull consists generically of four kinds of objects: (i) parts of the original graph, (ii) pieces of ruled surfaces, (iii) "kurtoparabolic points", to which we shall come back, and (iv) triangles (see Fig. 9). The associated Eulerian objects are, respectively, (i) regular points, (ii) shock lines, (iii) end points of shocks and (iv) shock nodes. Likewise, in three dimensions we have two-dimensional shock surfaces meeting in triples at shock lines, meeting in quadruples at shock nodes. (Nodes are always connected to shock lines and never isolated.) Note that the Eulerian part of Figure 9 looks just like a thermodynamic phase diagram, with the three shock lines playing the role of the liquid-gas, liquid-solid and solid-gas transition lines, the node playing the role of the triple point and the end point the role of the critical point. This is not accidental. In thermodynamics, equilibrium states are obtained by minimizing the Gibbs potential. This is equivalent to taking a Legendre transform of the internal energy in which the pressure and the temperature play the role of the Eulerian coordinates [26]. This analogy holds also in higher dimensions: the classification of "Legendrian singularities" can be used both for studying the Burgers equation [27] and for studying multi-variable phase transitions [28].

   A more complete description of singularities is obtained by considering the metamorphoses of singularities as time elapses. A complete classification in two and three dimensions may be found in the appendix (supplement 2) by Arnold *et al.* of reference [3].

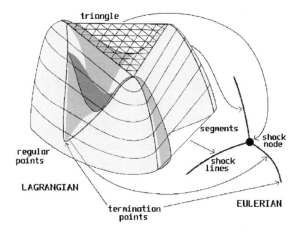

Fig. 9. Construction of the convex hull for a two-dimensional Lagrangian potential and associated Eulerian picture (figure adapted from Ref. [7]).

For random initial conditions the structure functions scale the same way as in one dimension. For example, the probability of having a $(d-1)$-dimensional shock intercepting an Eulerian segment of small length r is $O(r)$. The higher codimension structures give only subdominant corrections.

2.4 Remarks on numerical methods

Here, we give just some indications on how the (decaying) Burgers equation can be solved numerically. (For more details, see [7, 29, 30] or any textbook on numerical methods for nonlinear hyperbolic equations; for the case with forcing, see [15, 30] and references therein.) First, one can of course, solve the Burgers equation with viscosity. This should be avoided unless one is interested in what happens at dissipative scales (*e.g.*, inside shocks). For the inviscid limit and only in the decaying case it is possible to construct the solution at time t directly from the initial condition without recourse to any time marching. One way is to directly use the maximum representation (2.3), assuming that Lagrangian and Eulerian locations have been discretized on the same grid. Then, for a given Eulerian $\mathbf{r}$ one searches the Lagrangian $\mathbf{a}$ which maximizes the r.h.s. If there are N grid points, this seems to require $O(N^2)$ operations, but it can actually be done in $O(N \log_2 N)$ operations [7, 29]. Such a strategy must be combined with suitable interpolations to increase accuracy and avoid getting complete garbage for derivatives [30]. In one dimension one can also use Lagrangian strategies with particle and/or shock tracking. To be consistent with the inviscid limit, the particles must stick upon collisions. for sufficiently smooth initial data, it may be possible to construct the solution from the Lagrangian manifold (2.8, 2.9) or its multidimensional generalization, by just searching the maximum, for a given $\mathbf{x}$, of the finitely many branches present.

3 The Fourier–Lagrange representation and artefacts

In this section we show that formal manipulations of the inviscid Burgers equation with random initial conditions, even though they include apparently terms of all orders, can nevertheless lead to completely incorrect results, $e.g.$ for the energy spectrum. This section is entirely based on work by Fournier and Frisch [23]. The theory is given in one dimension but similar results can be established in higher dimensions.

In one dimension, it follows from (2.10, 2.11), that the Eulerian solution to the initial value problem for the decaying Burgers has the following implicit representation:

$$
\begin{aligned}
u(x,t) &= u_0(a) \\
x &= a + tu_0(a).
\end{aligned}
\tag{3.1}
$$

This becomes explicit if, instead of working with $u(x,t)$, we use its spatial Fourier transform (2π-periodicity is assumed for convenience)

$$
\hat{u}(k,t) \equiv \frac{1}{2\pi} \int_0^{2\pi} e^{-ikx} u(x,t) \, dx
\tag{3.2}
$$

and make the change of variables $x \mapsto a$, to obtain

$$
\hat{u}(k,t) = \frac{1}{2\pi} \int_0^{2\pi} e^{-ikx(a,t)} u_0(a) \frac{\partial x}{\partial a} \, da, \quad x(a,t) \equiv a + tu_0(a).
\tag{3.3}
$$

Equation (3.3) is called the *Fourier–Lagrangian* representation. A first integration by parts yields

$$
\hat{u}(k,t) = \frac{1}{2\pi} \frac{1}{ik} \int_0^{2\pi} e^{-ik(a+tu_0(a))} u_0'(a) \, da.
\tag{3.4}
$$

A second integration by parts leads then to

$$
\hat{u}(k,t) = \frac{1}{2\pi} \frac{1}{ikt} \int_0^{2\pi} e^{-ik(a+tu_0(a))} \, da, \quad k \neq 0.
\tag{3.5}
$$

If we now take random homogeneous Gaussian initial conditions, we can easily calculate moments of $\hat{u}(k,t)$ because they just involve averages of exponentials having the Gaussian initial velocity in their arguments. For example, the energy spectrum, related to the correlation function by

$$\langle \hat{u}(k,t)\hat{u}(k',t)\rangle = E(k,t)\delta_{k,k'}, \tag{3.6}$$

where $\delta_{k,k'}$ is a Kronecker delta, has the following expression

$$E(k,t) = \frac{1}{2\pi}\frac{1}{k^2t^2}\int_0^{2\pi} e^{-ikh}e^{-\frac{1}{2}k^2t^2 S_2(h,0)}\,dh, \tag{3.7}$$

where $S_2(h,0) \equiv \left\langle [u_0(h) - u_0(0)]^2\right\rangle$ is the second-order structure function of the initial velocity field. If the latter is smooth, as we shall assume, we have $S_2(h,0) \propto h^2$ for $h \to 0$. It then follows by a simple Laplace-type asymptotic expansion of (3.7) that

$$E(k,t) \propto k^{-3} \quad \text{when} \quad k \to \infty. \tag{3.8}$$

This is obviously the wrong answer: for Gaussian initial conditions there will be shocks with a non-vanishing probability for any $t > 0$. Their signature is a k^{-2} law in the energy spectrum at high wavenumbers, as shown in Section 2.2.

What went wrong? After the appearance of the first shock the Lagrangian map $a \mapsto x$ is not monotonic and the change of variable from (3.2) to (3.3) is valid only outside of the Lagrangian shock interval. Hence, in (3.3) we should excise this interval from the domain of integration. If we do not remove it, we are actually calculating the Fourier transform of a function obtained by superposing the three branches shown in Figure 10 with a plus sign for the two direct branches and a minus sign for the retrograde branch (the sign comes from the lack of an absolute value on the Jacobian $\partial x/\partial a$ in (3.3)). Obviously, this superposition has two square-root cusps as shown in Figure 10. This produces $k^{-3/2}$ tails in the Fourier transform and, hence, explains the spurious k^{-3} energy spectrum. Note also that this superposition of three branches is not a solution to the Burgers equation, the latter being nonlinear. This phenomenon is not related to the well known non-uniqueness of the solution to the Burgers equation with zero viscosity without proper additional conditions [24].

The problem is actually worse than suggested so far. It is easily shown that if the the initial velocity is deterministic and smooth, the function of the time defined by (3.3), for *fixed* wavenumber k, is entire, that is, its Taylor series around $t = 0$ has an infinite radius of convergence. There is no way to see the time $t_\star$ of the first preshock from this function. A preshock is indeed an "ultraviolet" singularity which is not seen in the temporal behavior of a single spatial Fourier component. This result has an important

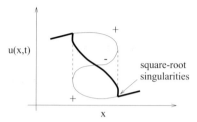

Fig. 10. Spurious solution of Burgers equation when three branches of a multi-valued solution are combined into one.

consequence for the case of random Gaussian initial conditions. Suppose we simply ignore the viscosity in the Burgers equation and expand the solution to all orders in a temporal Taylor series around $t = 0$ and then calculate various correlation functions and use Feynman graphs for bookkeeping of all the terms generated from averaging. We then find that the whole set can be resummed exactly and gives a spectrum with a k^{-3} tail. Of course, the origin of the "resummation miracle" is the Fourier–Lagrangian representation.

4   The law of energy decay

An important issue in burgulence and turbulence is that of the law of decay at long times when the viscosity is very small. Before turning to the Burgers equation let us recall a few things about the Navier–Stokes case. It is generally believed that high-Reynolds number turbulence has universal and non-trivial small-scale properties. In contrast, large scales, important for practical applications such as transport of heat or pollutants, are believed to be non-universal. This is however so only for the toy model of turbulence maintained by prescribed large-scale random forces. Very high-Reynolds number turbulence, decaying away from its production source, and far from boundaries can relax under its internal nonlinear dynamics to a (self-similarly evolving) state with universal and non-trivial statistical properties *at all scales*. Kármán and Howarth [31], investigating the decay of high-Reynolds number, homogeneous isotropic three-dimensional turbulence, proposed a self-preservation (self-similarity) ansatz for the spatial correlation function of the velocity: the correlation function keeps a fixed functional shape; the integral scale $L(t)$, characteristic of the energy-carrying eddies, grows in time and the mean kinetic energy $E(t) = u^2(t)$ decays, both following power laws; there are two exponents which can be related by the condition that the energy dissipation per unit mass $|\dot{E}(t)|$ should be proportional to u^3/L. But *an additional relation* is needed to actually determine the exponents. The invariance in time of the energy spectrum at low wavenumbers, known as the "permanence of large eddies" [19, 20, 34] can be used to derive the law of self-similar decay when the initial spectrum $E_0(k) \propto k^n$ at small wavenumbers k, with n below a critical value equal to 3 or 4, the actual value being disputed because of the "Gurbatov phenomenon" (see the end of this section). One then obtains a law of decay $E(t) \propto t^{-2(n+1)/(3+n)}$. (Kolmogorov [32] proposed a law of energy decay $u^2(t) \propto t^{-10/7}$, which corresponds to $n = 4$ and used in its derivation the so-called "Loitsyansky invariant", a quantity actually not conserved, as shown by Proudman and Reid [33].) When the initial energy spectrum at low wavenumbers goes to zero too quickly, the permanence of large eddies cannot be used, because the energy gets backscattered to low wavenumbers by nonlinear interactions. For Navier–Stokes turbulence the true law of decay is then known only within the framework of closure theories (see, *e.g.* [20]).

For one-dimensional burgulence, many of these questions are completely settled. First, we observe that the problem of decay is quite simple if a finite spatial periodicity is assumed. Indeed, eventually, all the shocks produced will merge into a single shock per period, as shown in Figure 11. The position of the shock is random and the two ramps have slope $1/t$, as is easily shown using the parabola construction of Section 2.2. Hence, the law of decay is simply $E(t) \propto t^{-2}$. Nontrivial laws of decay are obtained

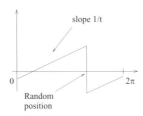

Fig. 11. Snapshot of solution of decaying burgulence at long times when spatial periodicity is assumed.

if the burgulence is homogeneous in an unbounded domain and has the "mixing" property (which means, roughly, that correlations are decreasing with separation). The number of shocks is then typically infinite but their density per unit length is finite and decreases in time because shocks are constantly merging. The $E(t) \propto t^{-2(n+1)/(3+n)}$ law mentioned above can be derived for burgulence from the permanence of large eddies when $n \leq 1$ [34]. For $n = 0$, this $t^{-2/3}$ law was actually derived by Burgers himself [35].

The hardest problem is again when permanence of large eddies does not determine the outcome, namely for $n > 1$. This problem was solved by Kida [36] (see also Refs. [3, 23, 34]).

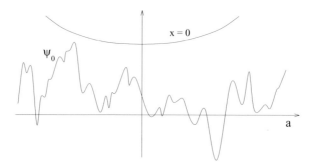

Fig. 12. An initial potential which is everywhere below the parabola $a^2/(2t) + \psi$. The probability of such events gives the cumulative probability to have a potential at time t less than ψ.

We now give some key ideas regarding the derivation of Kida's law of energy decay. We assume Gaussian, homogeneous smooth initial conditions, such that the potential is homogeneous. Since a homogeneous function is not, in general, the derivative of another homogeneous function, we assume that the initial energy spectrum

$$E_0(k) \propto k^n, \quad n > 1; \quad k \to 0. \tag{4.1}$$

This condition implies that the mean square initial potential $\int k^{-2} E_0(k)\,\mathrm{d}k$ has no infrared (small-k) divergence (the absence of an ultraviolet divergence is guaranteed by the assumed smoothness).

A very useful property of decaying burgulence, with no known counterpart for Navier–Stokes turbulence, is the relation

$$E(t) = \frac{\partial}{\partial t} \langle \psi \rangle, \tag{4.2}$$

which follows by taking the mean of (1.2) in the absence of a driving force. Hence, the law of energy decay can be obtained from the law for the mean potential. The latter can be derived from the cumulative probability of the potential which, by homogeneity, does not depend on the position. By (2.3), its expression at $x = 0$ is

$$\text{Prob}\left\{\text{Potential} < \psi\right\} = \text{Prob}\left\{\forall a, \ \psi_0(a) < \frac{a^2}{2t} + \psi\right\}. \tag{4.3}$$

Expressed in words, we want to find the probability that the initial potential does not cross the parabola $a^2/(2t) + \psi$ (see Fig. 12). Since, at long times t, the relevant ψ is going to be large, the problem becomes that of not crossing a parabola with small curvature and very high apex. Such crossings, more precisely the upcrossings, are spatially quite rare. As a consequence of the mixing property, for long t, they form a Poisson process [37] for which

$$\text{Prob}\{\text{no crossing}\} \simeq e^{-\langle N(t) \rangle}, \qquad (4.4)$$

where $\langle N(t) \rangle$ is the mean number of upcrossings. By the Rice formula (a consequence of the identity $\delta(\lambda x) = (1/|\lambda|)\delta(x)$,

$$\langle N(t) \rangle = \left\langle \int_{-\infty}^{+\infty} da\, \delta\left(m(a) - \psi\right) \frac{dm}{da}\, H\left(\frac{dm}{da}\right) \right\rangle, \qquad (4.5)$$

where H is the Heaviside function and

$$m(a) \equiv \psi_0(a) - \frac{a^2}{2t}. \qquad (4.6)$$

Since $\psi_0(a)$ is Gaussian, the r.h.s. of (4.5) can be easily expressed in terms of integrals over the probability densities of $\psi_0(a)$ and of $d\psi_0(a)/da$ (as a consequence of homogeneity these variables are uncorrelated and, hence, independent). The resulting integral can then be expanded by Laplace's method for large t, yielding

$$\langle N(t) \rangle \sim t^{1/2} \psi^{-1/2} e^{-\psi^2}, \quad t \to \infty. \qquad (4.7)$$

When this expression is used in (4.4) and the result is differentiated with respect to ψ to obtain the pdf of $p(\psi)$, the latter is found to be concentrated around $\psi_\star = (\ln t)^{1/2}$ (see Fig. 13). It then follows that, at large times, we

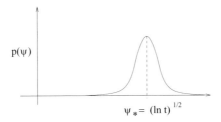

Fig. 13. A sketch of the pdf of the potential at long times.

have Kida's log-corrected $1/t$ law for the energy decay

$$\langle \psi \rangle \sim (\ln t)^{1/2}, \quad E(t) \sim \frac{1}{t(\ln t)^{1/2}}, \quad L(t) \sim \frac{t^{1/4}}{(\ln t)^{1/4}}. \qquad (4.8)$$

The Eulerian solution, at long times, has the ramp structure shown in Figure 14 with shocks of typical strength $u(t) = E^{1/2}(t)$, separated typically by a distance $L(t)$. The growth in time of $L(t)$ takes place because

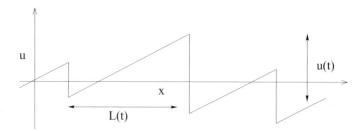

Fig. 14. The Eulerian solution at long times t. The ramps have slope $1/t$. In time-independent scales, the figure would be stretched horizontally and squeezed vertically by a factor proportional to t.

correlated particles, which initially cannot be much apart, may propagate to far-apart locations at long times.

The fact that Kida's law is valid for any $n > 1$, and not just for $n \geq 2$ as thought originally, gives rise to an interesting phenomenon now known as the "Gurbatov effect": if $1 < n < 2$ the long-time evolution of the energy spectrum cannot be globally self-similar. Indeed, the permanence of large eddies, which is valid for any $n < 2$ dictates that the spectrum should preserve exactly its initial $C_n k^n$ behavior at small wavenumbers k, with a constant-in-time C_n. Global self-similarity would then imply a $t^{-2(n+1)/(3+n)}$ law for the energy decay, which would contradict Kida's law. Actually, as shown in [34], for $1 < n < 2$ there are two characteristic wavenumbers with different time dependences, the integral wavenumber $k_L(t) \sim (L(t))^{-1}$ and a switching wavenumber $k_s(t) \ll k_L(t)$ below which holds the permanence of large eddies. It was shown that the same phenomenon is also present in the decay of a passive scalar [38]. Whether or not a similar phenomenon is present in three-dimensional Navier–Stokes incompressible turbulence or closure models thereto is a controversial matter [39, 40].

For decaying burgulence, if we leave aside the Gurbatov phenomenon which does not affect energy-carrying scales, the following may be shown. If we rescale distances by a factor $L(t)$ and velocities amplitudes by a factor $u(t) = E^{1/2}(t)$ and then let $t \to \infty$, the spatial (single-time) statistical properties of the whole random velocity field become time-independent. In other words, there is a self-similar evolution at long times. Hence, dimensionless ratios such as the velocity flatness

$$F(t) \equiv \frac{\langle u^4 \rangle (t)}{[\langle u^2 \rangle (t)]^2} \tag{4.9}$$

have a finite limit as $t \to \infty$. A similar property holds for the the decay of passive scalars [41]. We do not know if this property holds also for Navier–Stokes incompressible turbulence or if, say, the velocity flatness grows without bound at long times.

5 One-dimensional case with Brownian initial velocity

Burgers equation, when the initial velocity is Gaussian with a power-law spectrum $\propto k^{-n}$, is what cosmologists call *scale-free* initial conditions (see Refs. [8,9]). Here, we consider the one-dimensional case with Brownian motion (in the space variable) as initial velocity, corresponding to $n = 2$. The general case, including higher dimensions, is discussed in [7] (an example of a 2-D simulation with scale-free initial data is shown in Fig. 2).

Brownian motion is continuous but not differentiable (see Fig. 15); hence, shocks appear after arbitrarily short times and are actually *dense* (see Fig. 16). Numerically supported conjectures made in [6], have led to a proof by Sinai [42] of the following result: in Lagrangian coordinates, the regular points, that is fluid particles which have not yet fallen into shocks, form a fractal set of Hausdorff dimension $1/2$. This implies that there is a Devil's staircase of dimension $1/2$ in the Lagrangian map (see Fig. 18). Note that when the initial velocity is Brownian, the Lagrangian potential has a second space derivative which is delta-correlated in space; this can be approximately pictured as a situation where the Lagrangian potential has very strong oscillations in curvature. Hence, it is not surprising that very few points of its graph can belong to its convex hull (see Fig. 17).

We will now give some highlights of Sinai's proof of this result. For this problem, it turns out that the Hausdorff dimension of the regular points (determined in Ref. [42]) is also equal to its box-counting dimension, which is easier to determine. One obtains the latter by finding the probability that

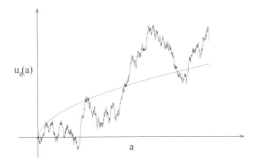

Fig. 15. A realization of the Brownian motion curve. The parabola shows the root-mean-square velocity $\propto a^{1/2}$.

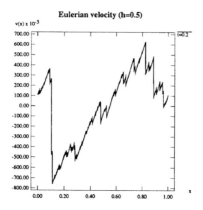

Fig. 16. Snapshot of the velocity resulting from Brownian initial data. Notice the dense proliferation of shocks (from Ref. [7]).

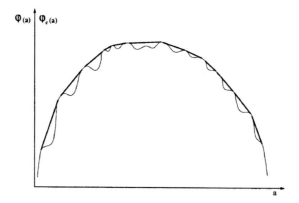

Fig. 17. Sketch of the Lagrangian potential together with its convex hull (from Ref. [7]).

a small Lagrangian interval of length ℓ contains at least one regular point which belongs simultaneously to the graph of the Lagrangian potential φ and to its convex hull. In other words, one looks for points, such as R, with the property that the graph of φ lies below its tangent at R (see Fig. 19). Sinai does this by the box construction with the following constraints on the graph:

Left: the graph of the potential should be below the half line Γ_-;

Right: the graph of the potential should be below the half line Γ_+;

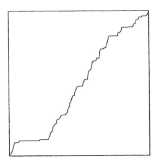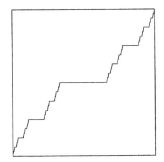

Fig. 18. Left: the Lagrangian map looks like a devil's staircase. Right: standard devil's staircase over the triadic Cantor set, which is constant almost everywhere, except on a fractal (from Ref. [7]).

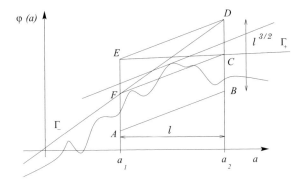

Fig. 19. The box construction used to find a regular point R (point of tangency with the graph entirely on one side of the tangent) within a Lagrangian interval of length ℓ (from Refs. [7, 42]).

$$Box:\begin{cases} 1:\ \text{enter } (AF) \text{ with a slope larger than that of } \Gamma_- \text{ by } O(\ell^{1/2}) \\ 2:\ \text{exit } (CB) \text{ with a slope less than that of } \Gamma_+ \text{ by } O(\ell^{1/2}) \\ 3:\ \text{cross } (FC) \text{ and stay below } (ED). \end{cases}$$

It is obvious that such conditions ensure the existence of at least one regular point. (Move (ED) down parallel to itself until it touches the graph.) Note that A and the slope of (AB) are prescribed. Hence, one is calculating conditional probabilities; but it may be shown that the conditioning is not affecting the scaling dependence on ℓ.

As the Brownian motion $u_0(a)$ is a *Markov process*, the constraints *Left*, *Box* and *Right* are independent and hence,

$$P^{\text{reg.}}(\ell) \equiv \text{Prob}\{\text{regular point in interval of length } \ell\}$$
$$= \text{Prob}\{Left\} \times \text{Prob}\{Box\} \times \text{Prob}\{Right\}. \qquad (5.1)$$

The scales of the box were chosen so that $\text{Prob}\{Box\}$ is independent of ℓ:

$$\text{Prob}\{Box\} \sim \ell^0. \qquad (5.2)$$

Indeed, Brownian motion and its integral have scaling exponent $1/2$ and $3/2$, respectively and the problem with $\ell \ll 1$ can be rescaled into that with $\ell = 1$ without changing probabilities.

It is clear by symmetry that $\text{Prob}\{Left\}$ and $\text{Prob}\{Right\}$ have the same scaling in ℓ. Let us concentrate on $\text{Prob}\{Right\}$. We can write the equation for the half line Γ_+ in the form

$$\Gamma_+ : a \mapsto \varphi(a_2) + \delta\ell^{3/2} + \left(\partial_a\varphi(a_2) + \gamma\ell^{1/2}\right)(a - a_2), \qquad (5.3)$$

where γ and δ are positive $O(1)$ quantities. Hence, introducing $\alpha \equiv a - a_2$, the condition *Right* can be written to the leading order as

$$\int_0^\alpha \left(u_0(a) + \gamma\ell^{1/2}\right) da + \delta\ell^{3/2} + \frac{\alpha^2}{2} > 0, \text{ for all } \alpha > 0. \qquad (5.4)$$

By the change of variable $\alpha = \beta\ell$ and use of the fact that the Brownian motion has scaling exponent $1/2$, one can write the condition *Right* as

$$\int_0^\beta (u_0(a) + \gamma) da > -\delta, \text{ for all } \beta \in [0, \ell^{-1}]. \qquad (5.5)$$

Without affecting the leading order, one can replace the Brownian motion by a stepwise constant random walk with jumps of ± 1 at integer a's. The integral in (5.5) has a geometric interpretation, as highlighted in Figure 20, which shows a random walk starting from the ordinate γ and the arches determined by successive zero-passings. The areas of these arches are denoted $S_\star, S_1, ...S_n, S_{\star\star}$. It is easily seen that

$$\text{Prob}\{Right\} \sim \text{Prob}\{S_1 > 0, S_1 + S_2 > 0, ..., S_1 + ... + S_n > 0\}, \qquad (5.6)$$

where $n = O(\ell^{-1/2})$ is the number of zero-passings of the random walk in the interval $[0, \ell^{-1}]$. The probability (5.6) can be evaluated by random walk methods (see, *e.g.* [43], Chap. 12, Sect. 7), yielding

$$\text{Prob}\{Right\} \sim \text{Prob}\{n \text{ first sums} > 0\} \propto n^{-1/2} \propto \ell^{1/4}. \qquad (5.7)$$

By (5.1, 5.2) and (5.7), the probability to have a regular point in a small interval of length ℓ behaves as $\ell^{1/2}$ when $\ell \to 0$. Thus, the regular points have a box-counting dimension $1/2$.

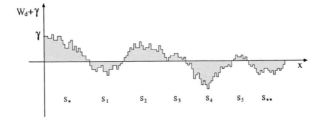

Fig. 20. The arches construction which uses the zero-passings of a random walk to estimate the integral of Brownian motion (from Refs. [7, 42]).

6 Preshocks and the pdf of velocity gradients in one dimension

In this section we shall determine the tail behavior of the probability density function (pdf) of the velocity gradient for one-dimensional decaying burgulence. To explain some of the motivations for this study, it is useful to make a digression concerning the forced one-dimensional Burgers equation:

$$\partial_t u + u \partial_x u = \nu \partial_{xx} u + f(x,t), \tag{6.1}$$

$$u(x, t_0) = u_0(x). \tag{6.2}$$

The latter displays much richer features than the unforced problem. The case where the force is random has often been studied as a prototype for a wide range of problems in non-equilibrium statistical physics (see Sect. 1.2).

Equation (6.1) can also be used in the same spirit as the forced Navier–Stokes equation, namely to investigate universality of various statistical properties with respect to the forcing. For Navier–Stokes turbulence, when the force is confined to large spatial scales and the Reynolds number is very high, small-scale (inertial range) statistical scaling properties are generally conjectured not to depend on the forcing, except through overall numerical factors. Similar conjectures have been made for burgulence with large-scale forcing. For example, there is little doubt that, because of the presence of shocks, structure functions of order $p > 1$ have universal exponents equal to unity, as in the decaying case (see, *e.g.* [16, 44]). More controversial is the tail behavior of the probability density function (pdf) of velocity gradients and velocity increments in the limit of zero viscosity when the force is a white-noise process in time. For increments, the problem was addressed for the first time by Chekhlov and Yakhot [45], who considered a force with a power-law spectrum, acting both at large and at small scales. Concerning the pdf $p(\xi)$ at large negative gradients ξ, it is generally believed that it follows a power law

$$p(\xi) \propto |\xi|^\alpha, \quad \text{for } \xi \to -\infty, \tag{6.3}$$

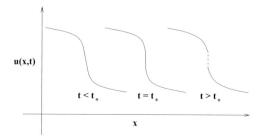

Fig. 21. Eulerian structure of the solution just before a preshock, at the time t_* of a preshock and just after.

but the conjectured values of α differ markedly. Polyakov [46] and Boldyrev [47], using a field-theoretical operator product expansion, predicted $\alpha = -5/2$; E *et al.* [44], using a semi-heuristic approach in which preshocks (nascent shocks) are key, predicted $\alpha = -7/2$; Gotoh and Kraichnan [48], using a Fokker–Planck equation approach, predicted $\alpha = -3$; more recent work by Kraichnan [49] favored $\alpha = -7/2$. E and Vanden Eijnden [50, 51] developed a probabilistic formalism adapted to solutions with shocks and giving insight into many aspects of the problem; they proved that $\alpha < -3$, and made a good case for $\alpha = -7/2$. The question of the correct law for the case of white-noise forcing remains however open (we shall come back to this later).

Actually, there is a situation much simpler than that originally considered in [44], for which the argument in favor of $\alpha = -7/2$ can be made rigorous, namely decaying burgulence. This closes our digression; in the remainder of this section we concentrate on the the unforced problem

$$\partial_t u + u\partial_x u = \nu \partial_x^2 u, \tag{6.4}$$

in the limit of vanishing viscosity $\nu \to 0$ and we follow references [52,53]. We assume a random initial velocity $u_0 = -(\mathrm{d}\psi_0/\mathrm{d}a)$, deriving from a smooth initial potential. Homogeneity is not required. The value $\alpha = -7/2$ for the exponent of the pdf at large negative gradients is easily understood in this case. It is just the signature of the preshocks, the cubic root singularities in Eulerian coordinates, which appear when new shocks are created (see Fig. 21). Preshocks constitute discrete events in space-time, contrary to shocks which persist in time (until they merge). These preshocks are the only structures giving large *finite* negative gradients: shocks give infinite negative gradients (unless a finite viscosity is introduced) and the gradients in the immediate spatial neighborhood of a mature shock are not particularly large. A simplified presentation is given hereafter for the case of a single preshock; the contributions of several preshocks to the pdf are just additive.

Let us suppose that the initial gradient du_0/da has a minimum at $a = 0$ (corresponding to an inflection point with negative derivative of the initial velocity), so that a shock will appear at time $t = t_\star = -1/((du_0/da)(0))$ and at $x = t_\star u_0(0)$. Without loss of generality, we assume $u_0(0) = 0$ (otherwise we perform a Galilean transformation to bring it to zero). As the initial velocity is supposed to be sufficiently smooth, we can perform a Taylor expansion of the initial potential in the neighborhood of $a = 0$. We then have, locally,

$$\psi_0(a) = c_1 a^2 - c_2 a^4 + \text{h.o.t.}, \tag{6.5}$$

where c_1 and c_2 are positive (random) constants and "h.o.t." stands for higher-order terms. The Lagrangian potential is locally

$$\varphi = -\frac{a^2}{2} + t\psi_0(a) = \frac{\tau}{2}a^2 - tc_2 a^4 + \text{h.o.t.}, \tag{6.6}$$

where $\tau = (t - t_*)/t_*$. The Lagrangian map outside the shock is thus

$$x(a, t) = -\partial_a \varphi(a, t) = -\tau a + 4tc_2 a^3 + \text{h.o.t.} \tag{6.7}$$

The Lagrangian potential, together with its convex hull, are shown in Figure 22. It is convex for $t \leq t_\star$. At $t = t_\star$, there is a degenerate maximum with quartic behavior, and, immediately after t_* (for $\tau > 0$), convexity is lost and a shock interval is born. Given the symmetry, resulting from our choice of coordinates, the convex hull contains a horizontal segment extending between the two maxima $a_\pm = \pm(\tau/(4c_2))^{1/2}$. The velocity gradient

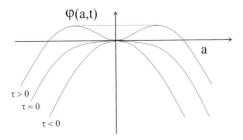

Fig. 22. Normal form of the Lagrangian potential in the neighborhood of a preshock in one dimension. At the time of the preshock ($\tau = (t - t_*)/t_* = 0$), the Lagrangian potential changes from a single extremum to three extrema and develops a non-trivial convex hull (shown as a dashed line).

can be written locally as

$$\partial_x u(x,t) = \frac{(du_0/da)(a)}{\partial_a x(a,t)} = \frac{2/t_\star}{-\tau + 12c_2 a^2}, \tag{6.8}$$

where a is the unique preimage of x by the naive Lagrangian map outside of the shock interval $]a_-, a_+[$. Since, by (6.7), the relation between x and a is cubic at $\tau = 0$, the velocity gradient $\partial_x u(x,t_\star) \propto |x|^{-2/3}$, which is unbounded. For any $t \neq t_\star$, the gradient remains bounded, except at the shock location. For $\tau < 0$, just before creation of the shock, the cubic relation between x and a still holds, except in a region of Lagrangian width of the order of $\tau^{1/2}$, and hence of Eulerian width $\sim \tau^{3/2}$, where the relation becomes linear to leading order.

The question is now: what is the fraction of Eulerian space-time where $\partial_x u < \xi$, with ξ a large negative number? Because of the cubic root structure, x must be in a small interval of width $\sim |\xi|^{-3/2}$. The time must be sufficiently close to $t_\star$ for this interval still to be in the region of validity of the cubic relation, that is, within $\sim |x|^{2/3} \sim |\xi|^{-1}$. Hence, the relevant space-time fraction or, in other words, the cumulative probability to have $\partial_x u < \xi$ is proportional to $|\xi|^{-5/2}$. This gives a pdf $p(\xi) \propto |\xi|^{-7/2}$ at large negative ξ's.

Actually, there is another contribution, also proportional to $|\xi|^{-7/2}$ stemming from a small time interval $\tau \sim |x|^{2/3} \sim |\xi|^{-1}$ just *after* $t_\star$ when small-amplitude shocks are present which have not yet completely destroyed the cubic root structure (see Ref. [52]). Similar arguments can be used to show that there are power-law ranges with exponent $-7/2$ and $+1$ in the pdf of velocity increments for decaying burgulence [52].

The preshock argument has first been introduced phenomenologically in [44] to predict pdf's of velocity gradients and increments for the case of white-noise in time forcing at large scales. In principle, in the presence of forcing, spatio-temporal accumulations of preshocks, invalidating the $-7/2$ law, cannot be ruled out. Nevertheless, numerical evidence in favor of the $-7/2$ law has been recently obtained by one of us (JB), using particle tracking simulations with a shot-noise approximation to white noise.

7 The pdf of density

In cosmological applications of the adhesion model/Burgers equation, it is of special interest to analyze the behavior of the density of matter, since the large-scale structures may also be characterized as mass condensations. In Eulerian coordinates, the mass density ρ satisfies the continuity equation

$$\partial_t \rho + \nabla \cdot (\rho \mathbf{v}) = 0. \tag{7.1}$$

The initial density is denoted by $\rho_0(\mathbf{a})$.

The question we intend to address here is the behavior, in the limit of vanishing viscosity and at large ρ's, of the pdf of mass density $p(\rho)$, when the initial velocity is random and smooth (and not necessarily homogeneous). This problem was studied in [53], where it was shown that density pdf's have universal power-law tails with exponent $-7/2$ in any dimension. This behavior stems from singularities, other than shocks, whose nature is quite different in one and several dimensions. (Similar results can in principle be obtained for velocity gradients and increments which are, however, not scalars in more than one dimension.)

In one dimension, the pdf of the mass density at large arguments is basically the same as the pdf of gradients at large negative arguments. Indeed, it is easy to show that, for any x not at a shock location,

$$\rho(x, t) = \rho_0(a) \left(1 - t\partial_x u(x, t) \right), \qquad (7.2)$$

where a is the preimage of x by the Lagrangian map [3]. If now ρ_0 is bounded from below and above (e.g., for uniform ρ_0), the result of the previous section implies that, for $\rho \to \infty$, the pdf $p(\rho)$ of the mass density satisfies a $\rho^{-7/2}$ law, which is again the signature of preshocks.

The key to studying this problem in more than one dimension is the geometric construction of the solution via the convex hull of the Lagrangian potential (see Sect. 2.3). Conservation of mass (7.1) implies that the density is given at regular points by

$$\rho(\mathbf{x}, t) = \frac{\rho_0(\mathbf{a})}{J(\mathbf{a}, t)}, \qquad (7.3)$$

where J is the Jacobian of the Lagrangian map. (The density is infinite within shocks.) Since the Jacobian is (up to a factor $(-1)^d$) equal to the Hessian of the Lagrangian potential (determinant of the matrix of second space derivatives), it follows that large densities are typically obtained only near parabolic points (where the Hessian vanishes). However, arbitrarily close to a parabolic point there are generically hyperbolic points where the surface defined by φ crosses its tangent (hyper)plane and which, therefore, do not belong to its convex hull. Yet, there exist in general exceptional "kurtoparabolic" points which are parabolic and belong to the boundary of the set of regular points (kurtos means convex in Greek). Near such points, arbitrarily large densities are obtained. In one dimension, the only kurtoparabolic points are the preshocks which are discrete space-time events in both Eulerian and Lagrangian coordinates. In two and more dimensions, kurtoparabolic points are also born at preshocks but live in general for a finite time; they reside on manifolds of spatial dimension $(d-1)$ (see

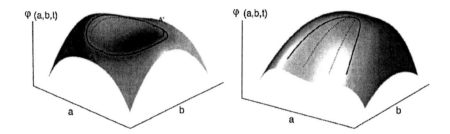

Fig. 23. Lagrangian potential in two dimensions with (a, b) coordinates, just after a preshock (left) and in the immediate neighborhood of a kurtoparabolic point (right). Continuous lines: separatrices between the regular part and the ruled surfaces of the convex hull; dotted-lines: vanishing of the Jacobian of the Lagrangian map. A and A' are a pair of kurtoparabolic points born with the shock.

Fig. 7). In Eulerian space, they are associated to boundaries of shocks (*e.g.* end points of shock lines for $d = 2$).

The determination of the large-ρ tail of the cumulative probability distribution of the density, $P^>(\rho)$, is equivalent to finding the fraction of Eulerian space-time where ρ exceeds a given value (see Ref. [53] for details). The latter is determined by changing from Eulerian to Lagrangian coordinates and Taylor-expanding to the relevant order the Lagrangian potential near a kurtoparabolic point, in a suitable coordinate frame:

$$\varphi(\mathbf{a}, t) \simeq \zeta a_1^4 + \sum_{j>1} \left[-\frac{\mu_j}{2} a_j^2 + \beta_j a_1^2 a_j \right]. \tag{7.4}$$

From (7.4), it is then easy to determine explicitly the line of vanishing Jacobian, the separatrix of the convex hull and the area where the density exceeds the value ρ (as illustrated in Fig. 24 for the 2-D case). When $\rho \to \infty$, the cumulative probability can be estimated as follows

$$P^>(\rho) \propto \underbrace{\rho^{-3/2}}_{\text{from } a_1} \times \underbrace{\rho^{-1}}_{\text{from } a_2} \times \underbrace{1 \times ... \times 1}_{\text{from } a_3 ... a_d} \times \underbrace{1}_{\text{from time}}. \tag{7.5}$$

Hence, the cumulative probability $P^>(\rho) \propto \rho^{-5/2}$ in any dimension; so that the pdf of the mass density has a universal power-law behavior with exponent $-7/2$. We have seen that the theory is rather different in one dimension and higher dimensions, because kurtoparabolic points are persistent only in the latter case. However, the scaling law for the resulting pdf is the same in all dimensions. Actually, two orthogonal spatial directions, a_1 and a_2 in (7.5), play the same role as space and time in one dimension.

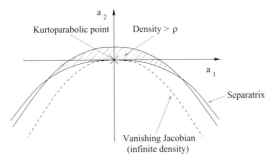

Fig. 24. Projection in the two-dimensional case of the neighborhood of a kurtoparabolic point.

It is now clear that, for burgulence, the algebraic tails of the pdf of velocity gradients or of the density stem from singularities. Turning briefly to *incompressible three-dimensional Navier–Stokes turbulence*, we note that measurements of pdf's for space or time derivatives of Eulerian velocities have not yet revealed power-law tails, but such tails may just have been, so far, "lost in the experimental noise". There has indeed been considerable speculations about singularities of the Navier–Stokes equations in the inviscid limit [19]. If singularities with divergent gradients are present, they will give power-law tails, at least as intermediate asymptotics when the viscosity is small (the converse is however not true, since statistical effects not related to singularities can also give power laws). The confirmed absence of power laws would probably rule out singularities.

8 Kicked burgulence

8.1 Forced Burgers equation and variational formulation

In the limit of vanishing viscosity and when no force is applied, the Burgers equation just means that fluid particles keep their initial velocity until they stick together in a shock. So, until merger, the position $X(t)$ of a given fluid particle will depend linearly on time:

$$X(t) = X(t_0) + (t - t_0)u_0(X(t_0)), \quad u(X(t), t) = u_0(X(t_0)). \tag{8.1}$$

When a force is applied, fluid particles trajectories, before merger with a shock, follow forcing-dependent continuous trajectories governed by

$$\frac{d^2}{dt^2} X(t) = f(X(t), t), \quad \frac{d}{dt} X(t) = u(X(t), t); \tag{8.2}$$

thus their dynamics can be rather complex (see Fig. 25).

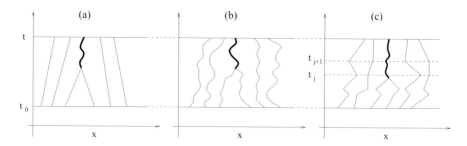

Fig. 25. Trajectories of particles in the decaying case (a), in the continuously forced case (b) and in the kicked case (c). The thick lines are shock trajectories.

Following reference [30], we shall be concerned here with the one-dimensional case where the force is a sum of impulses (or kicks), concentrated at discrete times t_j's:

$$f(x,t) = \sum_{j > j_0} f_j(x)\delta(t - t_j), \tag{8.3}$$

where $t_{j_0} = t_0$ is the initial time, and t_{j_0+1} is the time of the first kick. The kicking times t_j's and the kick $f_j(\cdot)$'s are prescribed. They can be either fixed or random. The meaning of such a forcing is that, between kicks, we let the solution evolve as a solution of the unforced problem. At each kicking time t_j, we discontinuously change the velocity field by the amount $f_j(x)$:

$$u(x, t_{j+}) = u(x, t_{j-}) + f_j(x). \tag{8.4}$$

This is an intermediate case between decay and time-continuous forcing. Such forcing implies a piecewise-linear time dependence of the position of a given fluid particle (see Fig. 25).

It is of interest to notice that this kind of discrete-in-time forcing can be applied also to the Navier–Stokes equations, with features of decaying turbulence still present to some extent. The original motivation for introducing such a forcing was to approximate white-noise-in-time forcing by discrete random noise, also called shot noise. But actually, the kicked case displays interesting features of its own. As will be shown later, the problem can be understood in terms of area-preserving mappings to which we can apply KAM theory (see Ref. [54] and references therein) and Aubry–Mather theory [55, 56].

We will focus on the space-periodic case. Namely, we assume that both the initial condition $u_0(\cdot)$ and the kicks $f_j(\cdot)$ are periodic with period 1, with respect to the space variable. For the moment, let us also assume that the initial velocity and the kicks both have zero spatial mean value over the space period $[0, 1[$. Since the mean velocity is conserved by Burgers dynamics, we have

$$\int_0^1 u(x,t)\mathrm{d}x = 0 \tag{8.5}$$

at all times. This constraint implies that the velocity potential $\psi(x,t)$, defined by $u(x,t) = -\partial_x \psi(x,t)$, is itself periodic in space. Let us define the kicking potentials $F_j(\cdot)$, so that

$$f_j(x) = -\frac{\mathrm{d}}{\mathrm{d}x}F_j(x). \tag{8.6}$$

It is then easy to write the potential at any time t, using between successive kicks the standard maximum representation (2.3) for decaying solution in the limit of vanishing viscosity (this is reexpressed here as a minimum in order to minimize a suitable action function), to obtain

$$\psi(x,t) = -\min_{y_J}\left[\frac{(x - y_J)^2}{2(t - t_J)} - \psi(y_J, t_{J-}) - F_J(y_J)\right], \tag{8.7}$$

where the index J is such that $t_J < t \le t_{J+1}$. Repeating this step as often as necessary to work our way back to the initial time, we obtain

$$\psi(x,t) = -\min_{\{y_j\}_{j_0 \le j \le J}}\left[A\left(x,t;\{y_j\}\right) - \psi_0(y_{j_0})\right], \tag{8.8}$$

where A is an action which has to be minimized,

$$A\left(x,t;\{y_j\}\right) = \frac{(x - y_J)^2}{2(t - t_J)} + \sum_{j=j_0}^{J-1}\left[\frac{(y_{j+1} - y_j)^2}{2(t_{j+1} - t_j)} - F_{j+1}(y_{j+1})\right]. \tag{8.9}$$

There is a similar representation for the case where the forcing $f(x,t) = -\partial_x F(x,t)$ is continuously applied, namely

$$\psi(x,t) = -\min_{y(\cdot)}\left[A\left(x,t;y(\cdot)\right) - \psi_0(y(t_0))\right]. \tag{8.10}$$

The minimum is now taken over continuous curves $y(\cdot)$ such that $y(t) = x$, the action being given by

$$A\left(x,t;y(\cdot)\right) = \int_{t_0}^t\left[\frac{1}{2}(\dot{y}(s))^2 - F(y(s),s)\right]\mathrm{d}s. \tag{8.11}$$

This representation goes back to work by Oleinik [57] on general conservation laws. It can be derived as the continuous limit of the discrete formulation when letting the time between kicks tend to zero. Many features of the forced Burgers equation were obtained by E *et al.* [58]. As we will see, the key notions introduced by E *et al.*, such as minimizers, global minimizer and main shock, are still valid in the case of discrete-in-time forcing.

First, we will introduce the notion of minimizing sequence (or minimizer). In terms of fluid particles trajectories, the minimum representation (8.8) just means that, to obtain the solution at time t and at some Eulerian location x, one has to look at all possible trajectories reaching x, and choose between them those which minimize the action. The sequence for which the minimum is achieved is, by definition, a minimizer. In general, there is only one minimizing trajectory arising at a given x. But for a countable set of x-values, there are several minimizing trajectories. These correspond to particles coalescing in a shock.

A minimizer can be explicitly characterized by requiring the vanishing of the derivatives, with respect to all the y_j's, of the argument of the minimum in (8.8). A minimizing sequence then has to verify the following Euler–Lagrange equations:

$$v_{j+1} \;=\; v_j + f_j(y_j), \tag{8.12}$$
$$y_{j+1} \;=\; y_j + (t_{j+1} - t_j)\,[v_j + f_j(y_j)], \tag{8.13}$$

where $v_j \equiv (y_j - y_{j-1})/(t_j - t_{j-1})$ is the velocity at the location y_j just before the kick. These equations have to be supplemented by the following initial and final conditions:

$$v_{j_0} \;=\; u_0(y_{j_0}), \tag{8.14}$$
$$x \;=\; y_J + (t - t_J)v_{J+1}. \tag{8.15}$$

Note that $v_{J+1} = u(x,t)$. The Euler–Lagrange map is area-preserving. It is also explicitly invertible, so that for a given (x, v), one can reconstruct the past history of a particle, except if a shock sits at x.

8.2   Periodic kicks

From now, we will focus on a particular case of forcing which displays globally the same features as random forcing but is much easier to handle. Namely, following reference [30], we consider the case of time-periodic kicks: the kicking potential is the same at each kick, $F_j(x) = G(x)$ for all j, and

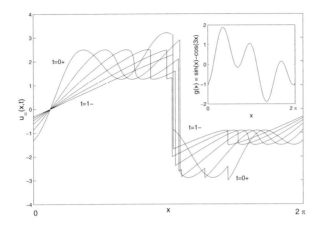

Fig. 26. Snapshots over one time period of the velocity for the limiting solution corresponding to the kicking force $g(x) = \sin x - \cos(3x)$ on the space period $[0, 2\pi[$ (see upper inset). The main shock is located around $x = \pi$; the global minimizer, here a fixed point, is the point of vanishing velocity common to all curves. Notice that during each period, two new shocks are born and two mergers occur.

the time interval is constant, $t_{j+1} - t_j = 1$, for convenience. The force can then be written

$$f(x,t) = \sum_{j > j0} g(x)\delta(t-j), \qquad (8.16)$$

where $g = -\mathrm{d}G/\mathrm{d}x$.

We now show, following reference [30], that the solution to the Burgers equation with this kind of forcing converges exponentially fast in time to a periodic solution $u_\infty(x,t)$. Snapshots of the time-periodic solution for one instance of kicking are shown in Figure 26; Figure 27 shows the exponential relaxation to $u_\infty(x,t)$.

Actually, the convergence to a unique solution at long times is related to properties near a fixed point of the two-dimensional dynamical system defined by the Euler–Lagrange map which reads here

$$v_{j+1} = v_j + g(y_j), \qquad (8.17)$$
$$y_{j+1} = y_j + v_j + g(y_j). \qquad (8.18)$$

A fixed point $(y_\star, v_\star)$, obviously, satisfies $v_\star = 0$ and $g(y_\star) = 0$. The latter expresses that the kicking potential achieves an extremum at $x = y_\star$. Let $P = (x_c, 0)$ be the particular fixed point of the map (8.17, 8.18), which corresponds to the location where the forcing potential achieves its maximum over the space period. This point is hyperbolic because the linearized

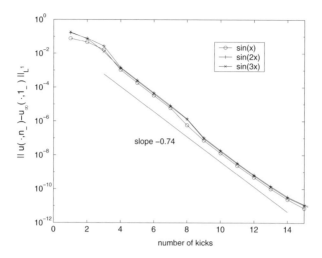

Fig. 27. Exponential relaxation to a time-periodic solution for the same forcing as in Figure 26, with three different initial conditions, as labeled. $\int_0^{2\pi} |u(x, n_-) - u_\infty(x, 1_-)| dx/(2\pi)$ is plotted vs the number of kicks.

system in its neighborhood has two real eigenvalues $\lambda > 1$ and $1/\lambda$, where

$$\lambda = 1 + c + \sqrt{c^2 + 2c}, \quad c = -\frac{1}{2}\frac{d^2}{dx^2}G(x_c). \qquad (8.19)$$

In the phase space (x, v), two globally invariant curves are associated to the corresponding eigendirections. These are (i) the stable manifold $\Gamma^{(s)}$, associated to $1/\lambda$, which is the set of points (x, v) converging to the fixed point under iteration of the map (because the eigenvalue is less than one), and (ii) the unstable manifold $\Gamma^{(s)}$, associated to λ, and generated by inverse iteration (see Fig. 28). An arbitrary continuous curve in the (x, v) plane which intersects the stable manifold will, under iteration, converge exponentially fast to the unstable manifold at the rate $1/\lambda$.

In the language of Burgers dynamics, the curve in the (x, v) plane defined by an initial condition $u_0(x)$ will be mapped after some kicks into a curve very close to the unstable manifold. To understand this mechanism of convergence, let us take an initial time t_0 tending to $-\infty$ and look at the behavior of the solution at time $t = 0$. The trajectory of the hyperbolic fixed point P corresponds to the so-called global minimizer. The global minimizer is the trajectory of a fluid particle never to be absorbed by a shock. Such a global minimizer is unique, and every minimizing trajectory converges exponentially fast to the global minimizer as $t \to -\infty$ [58]. This is illustrated in Figure 29a. By definition of the unstable manifold, each

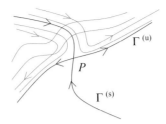

Fig. 28. Sketch of a hyperbolic fixed point P with its stable ($\Gamma^{(s)}$) and its unstable ($\Gamma^{(u)}$) manifolds. A curve, which intersects $\Gamma^{(s)}$, will eventually converge to $\Gamma^{(u)}$ under iteration of the map.

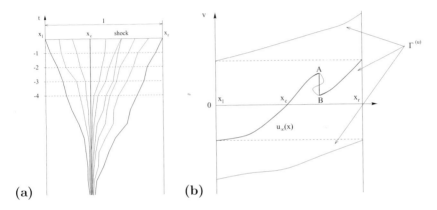

(a) **(b)**

Fig. 29. (a) Minimizers on the (x,t) cylinder; initial time t_0 is taken at $-\infty$. Shock locations are characterized by having two minimizers (an instance is at x_1); the fat line $x = x_c$ is the global minimizer. (b) Unstable manifold $\Gamma^{(u)}$ on the (x,v) cylinder which passes through the fixed point $P = (x_c, 0)$; the bold line is the graph of the limiting periodic solution. The main shock is located at $x_1 = x_r$, and another shock at x_1 corresponds to a local zig-zag of $\Gamma^{(u)}$ between A and B.

point (y_j, v_j) of a minimizer belongs thus to $\Gamma^{(u)}$ and every regular part of the graph of the limiting velocity belongs to the unstable manifold. Now we turn to the construction of the main shock. Since $\lambda > 0$, every minimizing trajectory starting from a point on the right (resp. left) of the global minimizer approaches it as $t \to -\infty$ from the right (resp. left). Hence, there exists x_r (resp. x_1), the rightmost (resp. leftmost) location from which a minimizer approaches the global minimizer from the right (resp. left). By periodicity in space and uniqueness of the global minimizer, these two points are actually the same: $x_r = x_1 \bmod 1$. If we shift the periodicity interval to $[x_1, x_r]$, we can draw $\Gamma^{(u)}$ on the (x,v)-cylinder. The regular parts of the limiting solution belong to this graph. By construction there is thus a shock at $x = x_l = x_r$ (see Fig. 29b). This is the *main shock*, the unique shock which exists for an infinite time. In Burgers dynamics, shocks are born and then they may merge. The main shock is a shock which has always existed when letting the initial time tend to $-\infty$. The other shocks are associated to the regions where $\Gamma^{(u)}$ is multi-valuated in x. Their locations are determined by requiring that the action be the same at points such as A and B in Figure 29b.

8.3   Connections with Aubry–Mather theory

So far, we have exclusively considered zero-mean-value initial conditions. Let us briefly consider the case where

$$\int_0^1 u(x,t)\mathrm{d}x = \int_0^1 u_0(x)\mathrm{d}x = a > 0. \tag{8.20}$$

The Burgers problem is then in exact correspondence with the description of equilibrium states of the Frenkel–Kontorova model [59]. In the latter, one has a one-dimensional chain of atoms connected by elastic springs in the presence of a space-periodic potential. The potential energy, which must be minimized to obtain the (classical) ground state, is

$$H(\{y_j\}) = \sum_j \frac{1}{2}\left(y_{j+1} - y_j - a\right)^2 - G(x), \tag{8.21}$$

where a is the unstretched distance between atoms. This problem was investigated by Aubry [55] and Mather [56]. The representation (8.21) matches the action minimizing representation for Burgers equation with a mean velocity a. The connection between the forced Burgers equation and Aubry–Mather theory was investigated by Jauslin *et al.* [60], E [61] and Sobolevski [62].

For $a = 0$, the global minimizer is a trivial ground state, associated to a fixed point, but for $a \neq 0$, it is much more complex. Within some intervals of the parameter a, the global minimizer lives on a periodic orbit associated to a rational rotation number ρ (asymptotic slope of the trajectory when $t \to -\infty$). The graph of ρ as a function of a is actually a Devil's staircase. The transitions between the intervals of the mean velocity corresponding to rational rotation numbers display interesting phenomena, such as accumulations of shocks (see Fig. 30).

We are grateful to V. Arnold, M. Blank, I.A. Bogaevski, G. Eyink, J.D. Fournier, W. E, K. Khanin, J. Lukovich, R. Mohayaee, Ya. Sinai, M.R. Rahimi Tabar, E. Vanden Eijnden, A. Sobolevski, M. Vergassola, B. Villone for useful comments. Part of this work was done

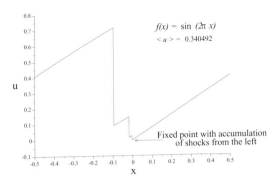

Fig. 30. Velocity profile at the first transition to a rational nonvanishing rotation number when increasing the mean velocity $a = \langle u \rangle$ from 0. Note the accumulation of shocks.

while JB and UF were visiting the Center of Nonlinear Studies (Los Alamos, U.S.A.) and the Isaac Newton Institute (Cambridge, U.K.); their support is gratefully acknowledged. This work was also supported by the European Union under contract HPRN-CT-2000-00162.

References

[1] E. Hopf, *Comm. Pure Appl. Math.* **3** (1950) 201-230.

[2] J.D. Cole, *Quart. Appl. Math.* **9** (1951) 225-236.

[3] S.N. Gurbatov, A.N. Malakhov and A.I. Saichev, *Non-linear Random Waves and Turbulence in Nondispersive Media: Waves, Rays, Particles* (Manchester University Press, Manchester, 1991).

[4] Y.B. Zel'dovich, *Astron. Astrophys.* **5** (1970) 84-89.

[5] S.N. Gurbatov and A.I. Saichev, *Radiophys. Quant. Electr.* **27** (1984) 303-313.

[6] Z. She, E. Aurell and U. Frisch, *Comm. Math. Phys.* **148** (1992) 623-641.

[7] M. Vergassola, B. Dubrulle, U. Frisch and A. Noullez, *Astron. Astrophys.* **289** (1994) 325-356.

[8] P.J.E. Peebles, *Principles of Physical Cosmology* (Princeton University Press, Princeton, 1993).

[9] P. Coles and F. Lucchin, *Cosmology: The Origin and Evolution of Cosmic Structures* (J. Wiley and Sons, Chichester, 1995).

[10] V.I. Arnold, S.F. Shandarin and Y.B. Zel'dovich, *Geophys. Astrophys. Fluid Dynam.* **20** (1982) 111-130.

[11] L. Kofman, E. Bertschinger, J. Gelb and A. Nusser, *Astrophys. J.* **420** (1994) 44-57.

[12] D. Chowdhury, L. Santen and A. Schadschneider, *Phys. Rep.* **329** (2000) 199-329.

[13] M. Kardar, G. Parisi and Y.-C. Zhang, *Phys. Rev. Lett.* **56** (1986) 889-892.

[14] A.L. Barabási and H.E. Stanley, *Fractal Concepts in Surface Growth* (Cambridge University Press, Cambridge, 1995).

[15] M. Kardar and Y.-C. Zhang, *Phys. Rev. Lett.* **58** (1987) 2087-2090.

[16] J.P. Bouchaud, M. Mézard and G. Parisi, *Phys. Rev. E* **52** (1995) 3656-3674.

[17] S.A. Orszag, *Statistical Theory of Turbulence*, in Fluid Dynamics, Les Houches 1973, edited by R. Balian and J.L. Peube (Gordon and Breach, New York, 1977) pp. 237-374.

[18] H.A. Rose and P.L. Sulem, *J. Phys. France* **39** (1978) 441-484.

[19] U. Frisch, *Turbulence: The Legacy of A.N. Kolmogorov* (Cambridge University Press, Cambridge, 1995).

[20] M. Lesieur, *Fluid Mechanics and Its Applications* **40** (Kluwer, 1997).

[21] F. Calogero, *Lett. Nuovo Cimento* **13** (1975) 411-416.

[22] U. Frisch and R. Morf, *Phys. Rev. A* **23** (1981) 2673-2705.

[23] J.D. Fournier and U. Frisch, *J. Méc. Théor. Appl.* **2** (1983) 699-750.

[24] P.-D. Lax, *Comm. Pure Appl. Math.* **10** (1957) 537-566.

[25] U. Frisch and R. Bourret, *J. Math. Phys.* **11** (1970) 364-390.

[26] L.D. Landau and E.M. Lifshitz, *Statistical Physics* (Butterworth-Heinemann, Oxford, 1980).

[27] I.A. Bogaevski, *Singularities of convex hulls as fronts of Legendre varieties*, in Geometry and Topology of Caustics – Caustics '98, Banach Center publications **50** (1999) 61-74, Institute of Mathematics, Polish Academy of Sciences (Warsaw).

[28] F. Aicardi, On the classification of generic phenomena in thermodynamic binary mixtures, *Physica D* (2001) (submitted).

[29] A. Noullez and M. Vergassola, *J. Sci. Comput.* **9** (1994) 259-281.

[30] J. Bec, U. Frisch and K. Khanin, *J. Fluid Mech.* **416** (2000) 239-267.

[31] T. von Kármán and L. Howarth, *Proc. R. Soc. Lond. A* **164** (1938) 192-215.

[32] A.N. Kolmogorov, *Dokl. Akad. Nauk SSSR* **31** (1941) 538-540.

[33] I. Proudman and W.H. Reid, *Phil. Trans. R. Soc. Lond. A* **247** (1954) 163-189.

[34] S.N. Gurbatov, S.I. Simdyankin, E. Aurell, U. Frisch and G. Toth, *J. Fluid Mech.* **344** (1997) 339-374.

[35] J.M. Burgers, *The Nonlinear Diffusion Equation* (D. Reidel, Dordrecht, 1974).

[36] S. Kida, *J. Fluid Mech.* **93** (1979) 337-377.

[37] S.A. Molchanov, D. Surgailis and W.A. Woyczynski, *Comm. Math. Phys.* **168** (1995) 209-226.

[38] G.L. Eyink and J. Xin, *J. Stat. Phys.* **100** (2000) 679-741.

[39] G.L. Eyink and D.J. Thomson, *Phys. Fluids* **12** (2000) 477-479.

[40] S. Ossia and M. Lesieur, *J. Turbulence* **1** (2000) 007.

[41] M. Chaves, G. Eyink, U. Frisch and M. Vergassola, *Phys. Rev. Lett.* **86** (2001) 2305-2308.

[42] Ya. Sinai, *Comm. Math. Phys.* **148** (1992) 601-622.

[43] W. Feller, *An Introduction to Probability Theory and its Applications*, Vol. 2 (J. Wiley and Sons, Chichester, 1995).

[44] W. E, K. Khanin, A. Mazel and Ya. Sinai, *Phys. Rev. Lett.* **78** (1997) 1904-1907.

[45] A. Chekhlov and V. Yakhot, *Phys. Rev. E* **52** (1995) 5681-5684.

[46] A.M. Polyakov, *Phys. Rev. E* **52** (1995) 6183-6188.

[47] S.A. Boldyrev, *Phys. Rev. E* **55** (1997) 6907-6910.

[48] T. Gotoh and R.H. Kraichnan, *Phys. Fluids* **10** (1998) 2859-2866.

[49] R.H. Kraichnan, *Phys. Fluids* **11** (1999) 3738-3742.

[50] W. E and E. Vanden Eijnden, *Phys. Rev. Lett.* **83** (1999) 2572-2575.

[51] W. E and E. Vanden Eijnden, *Comm. Pure Appl. Math.* **53** (2000) 852-901.

[52] J. Bec and U. Frisch, *Phys. Rev. E* **61** (2000) 1395-1402.

[53] U. Frisch, J. Bec and B. Villone, *Physica D* (2000) (in press), cond-mat/9912110.

[54] J.V. José and E.J. Saletan, *Classical Dynamics: A Contemporary Approach* (Cambridge University Press, Cambridge, 1998).

[55] S. Aubry, *Physica D* **7** (1983) 240-258.

[56] J.N. Mather, *Topology* **21** (1982) 457-467.

[57] O. Oleinik, *Uspekhi Mat. Nauk* **12** (1957) 3-73. (*Russ. Math. Survey, Amer. Math. Transl. Series 2* **26**, 95-172.)

[58] W. E, K. Khanin, A. Mazel and Ya. Sinai, *Ann. Math.* **151** (2000) 877-960.

[59] T. Kontorova and Y.I. Frenkel, *Zh. Eskp. & Teor. Fiz.* **8** (1938) 1340.

[60] H.R. Jauslin, H.O. Kreiss and J. Moser, *Proc. Symp. Pure Math.* **65** (1999) 133-153.

[61] W. E, *Comm. Pure Appl. Math.* **52** (1999) 811-828.

[62] A.N. Sobolevski, *Sbornik Math.* **190** (1999) 1487-1504.

COURSE 8

TWO-DIMENSIONAL TURBULENCE

J. SOMMERIA

LEGI/CORIOLIS, 21 avenue des Martyrs, 38000 Grenoble, France

Contents

TWO-DIMENSIONAL TURBULENCE

J. Sommeria

1 Introduction

Courses on turbulence generally begin with the delicate question of defining turbulence. As usual for important concepts, a clearcut definition is not possible, and the problem is still worse for two-dimensional (2D) turbulence. It can be said that turbulence is a flow which is disordered in time and space. The following properties more precisely characterize turbulence, see for instance Lesieur [64].

- Unpredictability of flow realization, in the sense of amplification of small errors (but the statistical properties are generally quite reproducible);

- Continuum flow phenomena, governed by the equations of fluid mechanics (which excludes for instance Brownian motion), and dominated by advective inertial effects (which excludes random wave phenomena, dominated by some restoring force);

- Interaction of a wide range of eddy scales (spatial complexity), which implies high Reynolds numbers and excludes chaos of low-dimensional dynamical systems;

- Increased mixing properties for transported quantities (e.g. chemicals, heat).

Then 2D turbulence is naturally defined as a turbulent flow depending only on two space coordinates x, y or alternatively as a flow confined to a surface (which does not need to be plane, for instance a sphere). In the former case, there is a possibly non-zero third velocity component u_z, along the z direction, but independent of z. Then the equation of motion states that this velocity component is passively transported (like the concentration of a chemical) by the flow u_x, u_y in the plane, so this third velocity component can be ignored in the theoretical description.

Two additional properties are often considered in defining turbulence (see *e.g.* Tennekes and Lumley [103]): the existence of strong vorticity fluctuations and strong energy dissipation. Although vorticity dynamics is also essential in 2D turbulence, there is no mechanism of vorticity amplification. We shall see in Section 2 that as a consequence energy dissipation is forbidden in the limit of small viscosity: this is the main dynamical signature of a 2D turbulence dynamics. In that sense, 2D turbulence is quite different from usual turbulence, but still the defining properties listed above can be satisfied in two dimensions.

Fig. 1. Grid turbulence in a soap film (from Rutgers, 1996 (http://www.physics.ohio-state.edu/~maarten). The fluid is moving from left to right, at velocity 2 m/s, across the comb with mesh 0.3 cm, while the total width is 4 cm. Visualization is provided by interference fringes due to small fluctuations of the film thickness (this is like the color patterns in usual soap bubbles). The increase of turbulent scale with distance to the grid is clearly visible, and it has been measured by laser Doppler velocimetry [71]. Note that this technique for producing 2D turbulence has been first developed by Couder [30].

The very existence of 2D turbulence has been questioned in the past. It has been considered as "a statistical extension of XIXth century fluid dynamics", limited to ideal 2D flow problems remote from the real physical world. Indeed the two cases of 2D turbulence considered above may seem at first sight equally unrealistic: z independent flows are (by definition) unstable when they become turbulent, and develop three-dimensional instabilities, while confinement to a thin layer requires external forces associated with severe friction effects or other perturbations. At the beginning of a classical review paper on 2D turbulence by Kraichnan and Montgomery [58], 20 years ago, it was said that "Two-dimensional turbulence has the special distinction that it is nowhere realized in nature or in the laboratory but only in computer simulations". Since then, laboratory realization have been obtained for an astounding variety of physical configurations: thin liquid films (Fig. 1), flows in rotating tanks, liquid metal flows or electron plasma in the presence of a uniform magnetic field. Observations of eddy fields in the ocean, or in Jupiter's atmosphere (Fig. 2) provide a strong motivation for studies of 2D turbulence. These systems are more closely approached by 2D turbulence concepts than the Earth atmosphere, for which the thermal forcing and the friction by 3D turbulence in the boundary layer have important influence (see [66] for a recent discussion of the relevance of 2D turbulence to the Earth atmosphere). The existence of coherent eddies like the Great Red Spot is a striking feature of 2D turbulence, as discussed in Sections 3.4 and 5. Of course 2D turbulence can be only an approximation of reality, but the theoretical concepts developed for this ideal case appear more and more useful in understanding the strange properties of turbulence in some real flows, and they can provide quantitative predictions.

Another difficulty for observing 2D turbulence, even for an ideally 2D flow dynamics, is that the onset of turbulence is not guaranteed: for instance simple shear flow, like Poiseuille flow, remain stable at any Reynolds number, which is never the case in usual turbulence: all flows become turbulent at sufficiently high Reynolds number, and reach a seemingly universal Kolmogorov regime. Therefore the influence of the forcing mechanism, or initial condition, is more important than in usual, 3D turbulence. It must be further remarked that 2D turbulence tends to eventually organize into steady coherent flows, loosing the unpredictability character. However this organization is itself the result of spatial complexity and mixing occurring in a transient stage.

Beside its practical relevance, a strong incentive to study 2D turbulence is its strange statistical properties, which were recognized in the early theoretical studies, in particular by Onsager [79] and Kraichnan [56]. The conservation of vorticity by fluid particles prevents the energy cascade towards small scales (see Sect. 4), resulting in the conservation of energy in the limit of small viscosity, in strong contrast with 3D turbulence. The flow organization into steady coherent structures is also a remarkable feature of 2D turbulence, as already mentioned. A reasonable understanding of 2D turbulence is a prerequisite before studying more complex turbulence problems in atmospheric or oceanic contexts.

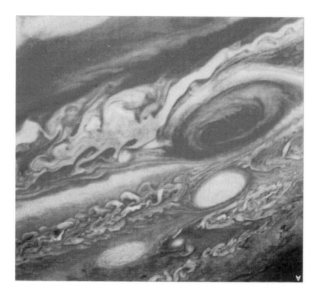

Fig. 2. The Great Red Spot (top) and White Oval (bottom) of Jupiter are large
vortices remaining remarkably coherent among turbulent eddies, as seen here by
the Voyager 1 spacecraft in February 1979. The length of the Great Red Spot
is 22 000 km. The mean zonal flow made of alternating jets is probably deeply
rooted in the fluid planetary interior, while the observed turbulence is limited
to a shallow active layer, dynamically separated from below by a stable density
stratification. This observation of a very active turbulence, made visible by cloud
motion, was a great surprise due to the weak available forcing. The high velocities
can only be explained by assuming that the dynamics is fundamentally 2D, with
negligible energy dissipation (although it is of course a layer-wise complex system),
as shown by the following arguments. The observed clouds are at a pressure level
3 bars, which must be equal to the column weight for the atmosphere above. Since
the gravity is 25 ms^{-2}, the corresponding mass is 1.2×10^4 kg/m^2, so the surface
density Σ of the active layer has at least this value (we do not know how deep is the
active layer below the observed cloud level). With typical velocities $U = 50$ m/s,
the corresponding energy density $E = \Sigma U^2/2$ is at least 15×10^6 J/m^2. The
free decay time, equal to the forcing time in a permanent regime, is E/P, where
P is the injected power. The heat flux coming from the planetary interior is
5 W/m^2, of the same order as the solar heat flux. The efficiency of conversion
to mechanical energy by convective effects is not more than the Carnot efficiency,
about 1% since the typical temperature differences involved are only a few K, for
a mean temperature 200 K. Therefore the forcing power is of $P \sim 0.05$ W/m^2, so
the decay time E/P is at least 30×10^7 s, ten earth years, much longer than the
eddy turnover time of a few days. For 3D turbulence, the decay time would be
by contrast of the same order as the turnover time.

2 Equations and conservation laws

2.1 Euler vs. Navier–Stokes equations

Turbulence is generally described as a complex solution of the Navier–Stokes equations, restricted here to an incompressible fluid. The status of viscosity in the description of turbulence is a often a subject of debate. The Euler equations (without viscosity) provide the most direct approach, historically also, but lead to many paradoxes, which are avoided by introducing viscosity. Viscous flows are well understood, and going progressively to turbulence by decreasing viscosity is a reassuring approach. The development of bifurcation theories and chaos comforted this view. However the use of viscosity probably skips the true insight into fluid turbulence, whose genuine properties are clearly controlled by inertial effects, which lead to a breakdown of spatial regularity. This is described by the Euler equations. Its strange behavior is repelling, but it may just correspond to the main difficulty of turbulence that cannot be avoided. In 3D turbulence, the introduction of viscosity is justified on physical grounds, as all real fluids are viscous (except superfluid). Furthermore, the regularity of the Euler equations in 3D is not known, so that it may not be a well posed problem (*i.e.* there is not a unique solution for a given initial condition).

These arguments are not valid in two dimensions. In many physical systems, the motion is not 2D down to the smallest scales. The dissipative mechanisms then depend on the system, for instance they are quite different in atmospheric flows, electron plasma or laboratory scale fluid experiments. Yet properties of 2D turbulence are expected to be common to these different systems. Furthermore, the inviscid equations are well posed: starting from any regular initial velocity field, the Euler equations have a unique regular solution for all time [2, 54, 108]. This property can be extended to any initial conditions with bounded measurable initial vorticity fields [28], for instance patches with uniform vorticity inside and vorticity discontinuity at the edge.

The inviscid dynamics develops increasing spatial complexity, with smaller and smaller scales of motion. This does not lead to mathematical singularities, as stated above, but actual computations are restricted to a finite resolution, and some smoothing is then necessary. For that purpose we introduce a smoothing operator $\mathcal{V}$, which can be for instance an ordinary Laplacian Δ, a higher order smoothing operator (hyper-viscosity) $(-1)^n \Delta^n$, or some spatial filtering of small scale oscillations. Such smoothing always alters the dynamics to some extent, as it will be discussed in Section 6.

Therefore we start with the Euler equations with a smoothing operator $\mathcal{V}$.

$$\partial_t \mathbf{u} + \mathbf{u}.\nabla \mathbf{u} = -\nabla p + \mathcal{V}\mathbf{u} \qquad (2.1)$$

$$\nabla.\mathbf{u} = 0 \tag{2.2}$$

$$\mathbf{u.n} = 0, \quad \text{on boundaries (\mathbf{n} normal).} \tag{2.3}$$

Note that the impermeability boundary condition (2.3) is sufficient for the Euler equations, while a smoothing operator requires an additional condition, whose choice is not obvious. The choice of a physical viscosity, with no-slip condition, may not be appropriate, as physical effects beyond the 2D model often occur in boundary layers, and resolving the thin boundary layer raises numerical difficulties at high Reynolds number. To avoid boundary problems, periodic boundary conditions (equivalent to a toric surface), or a spherical geometry, are often considered for fundamental studies.

2.2   Vorticity representation

It is often convenient to use a representation in terms of vorticity $\omega(t, \mathbf{r})$ and stream function $\psi(t, \mathbf{r})$,

$$\omega = (\nabla \times \mathbf{u})_z \tag{2.4}$$

$$\mathbf{u} = \nabla \times (\psi \mathbf{e_z}) \equiv -\mathbf{e_z} \times \nabla \psi \tag{2.5}$$

$$\partial_t \omega + \mathbf{u}.\nabla \omega = \nabla \times (\mathcal{V}\mathbf{u}). \tag{2.6}$$

We can identify the vorticity vector, along the z direction (with unit vector $\mathbf{e_z}$), with its z component, a scalar. The evolution equation (2.6), obtained by taking the curl of (2.1) just states that vorticity is advected and conserved by the flow (in the absence of the smoothing operator $\mathcal{V}$). The stream function ψ and resulting flow $\mathbf{u}$ are themselves determined from the vorticity field by solving the Poisson equation, obtained by combining (2.5) and (2.4),

$$-\Delta \psi = \omega, \quad \psi = 0 \text{ on boundaries,} \tag{2.7}$$

so that the whole flow evolution is determined by the scalar vorticity field only.

This Poisson equation (2.7) can be solved in terms of a Green function $G_\psi(\mathbf{r}, \mathbf{r}')$, representing the flow induced at point $\mathbf{r}$ by a singular point vortex (a Dirac vorticity distribution $\delta(\mathbf{r} - \mathbf{r}')$ located at position $\mathbf{r}'$),

$$\psi(t, \mathbf{r}) = \int G_\psi(\mathbf{r}, \mathbf{r}')\omega(t, \mathbf{r}')\mathrm{d}^2\mathbf{r}' \tag{2.8}$$

$$\text{with} \quad -\Delta G_\psi = \delta(\mathbf{r} - \mathbf{r}'), \quad G_\psi = 0 \text{ on boundaries.} \qquad (2.9)$$

The flows induced by all elementary vorticity elements at positions $\mathbf{r}'$ are summed in the integral (2.8).

Far from boundaries, G_ψ has the axisymmetric form,

$$G_\psi(\mathbf{r}, \mathbf{r}') = -(2\pi)^{-1} \ln(|\mathbf{r} - \mathbf{r}'|/L) \qquad (2.10)$$

where L is the typical domain size, which comes into play as an additive constant, due to the arbitrary choice $\psi = 0$ at the boundary. The corresponding azimuthal velocity, in $1/|\mathbf{r} - \mathbf{r}'|$, is analogous to the magnetic field produced by a line current representing the vortex line. Near boundaries, the effect of virtual mirror vortices outside the fluid domain must be added.

Such a vorticity representation is particularly useful in 2D. It can be used also in 3D, but the evolution of vorticity is more complex due to vortex stretching, possibly leading to singularities, and a vector potential must then replace the scalar stream function.

2.3 Conservation laws

• Casimirs

The incompressibility $\nabla.\mathbf{u} = 0$ ensures that any material area is conserved as its contour is transported by the fluid motion. Furthermore we have seen that the vorticity scalar ω is conserved for each fluid particle (in the absence of smoothing operator $\mathcal{V}$). This is a consequence of the more general Kelvin's theorem stating that $\omega/H = \text{const.}$ for a small vortex tube element with length H, which remains constant in the 2D case. Physically this is due to the conservation of angular momentum for a small fluid element, defined with respect to the center of gravity of the element. The conservation of ω for each fluid element implies that the corresponding value $f(\omega)$ is also conserved for any continuous function f, so that, since the surface element $d^2\mathbf{r}$ is also conserved, any functional of the form

$$C_f = \int f(\omega) d^2\mathbf{r} \qquad (2.11)$$

(called a Casimir integral) is conserved. This can be more straightforwardly demonstrated (for a differentiable function f), by multiplying (2.6) by the derivative $f'(\omega)$, which yields $\partial_t f(\omega) + \nabla.(f(\omega)\mathbf{u}) = f'(\omega)(\nabla \times \mathcal{V}\mathbf{u})$ (taking into account that $\nabla.\mathbf{u} = 0$). The domain integral of the second term transforms into a boundary integral which vanishes due to the impermeability condition $\mathbf{u}.\mathbf{n} = 0$, so that the time derivative

$$\dot{C}_f = \int f'(\omega)(\nabla \times \mathcal{V}\mathbf{u})\mathrm{d}^2\mathbf{r} \tag{2.12}$$

indeed vanishes in the absence of the smoothing operator $\mathcal{V}$.

In the case of a power function $f(\omega) \propto \omega^n$, we get for $n = 1$ and 2 respectively the conservation of the circulation Γ and enstrophy Γ_2,

$$\Gamma = \int \omega \mathrm{d}^2\mathbf{r}, \quad \Gamma_2 = \frac{1}{2}\int \omega^2 \mathrm{d}^2\mathbf{r}. \tag{2.13}$$

With a constant viscosity smoothing operator $\mathcal{V}\mathbf{u} = \nu\Delta\mathbf{u}$, the time evolution (2.12) of these quantities can be rewritten by replacing the integral of a divergence by boundary flux,

$$\dot{\Gamma} = \nu \oint \mathbf{n}.\nabla\omega, \quad \dot{\Gamma}_2 = -\nu \int (\nabla\omega)^2 \mathrm{d}^2\mathbf{r} + \nu \oint \mathbf{n}.\omega\nabla\omega \tag{2.14}$$

(an integration by parts has been used to express $\dot{\Gamma}_2$). Note that the circulation is still conserved in the presence of viscosity, except for possible boundary effects. These vanish for "super-slip" boundary conditions $\mathbf{n}.\nabla\omega = 0$ (but not for the more common "free slip" boundary condition $\omega = 0$). The enstrophy Γ_2 decays by viscous effects in the interior, and boundary effects vanish both for free slip ($\omega = 0$) and super-slip $\mathbf{n}.\nabla\omega = 0$ boundary conditions.

The extrema $\omega_{\min}$ and $\omega_{\max}$ of the vorticity field remains constant for the Euler equation (in the absence of forcing or friction effects): vorticity is just transported and cannot be amplified by the inertial flow evolution, unlike in three dimensions (these conservation laws can be also obtained from Casimirs with functions f dominated by the extremal values of ω, for instance $f(\omega) = \exp \pm n\omega$ with n very large). A uniform viscosity can only lower the maximum with time and raise the minimum, as by definition $\Delta\omega \geq 0$ at the vorticity minimum, and $\Delta\omega \leq 0$ at the maximum.

• Energy

The kinetic energy $\mathcal{E} = \frac{1}{2}\int \mathbf{u}^2 \mathrm{d}^2\mathbf{r}$ is conserved by the Euler equations. This is easily seen by taking the scalar product of (2.1) with $\mathbf{u}$ (with $\mathcal{V} = 0$). Rewriting the advective term with the classical identity $\mathbf{u}.\nabla\mathbf{u} = \omega \times \mathbf{u} + \nabla(\mathbf{u}^2/2)$, the first term is orthogonal to $\mathbf{u}$, while the second is incorporated in the pressure, and the $\mathbf{u}.\nabla p' = \nabla.(p'\mathbf{u})$, whose domain integral vanishes due to the impermeability condition. Note that this demonstration equally applies in 3D, but it requires differentiability of the velocity field, while we expect that in 3D, energy dissipation would occur after a finite time due to the formation of singularities (then the Euler equations themselves could be only defined in the sense of distributions, not ordinary fields). By contrast, in 2D, the flow remains regular for all times, so energy is truly conserved.

A more classical point of view is to consider the effect of a small viscosity. Then the energy varies as $\dot{\mathcal{E}} = \nu \int \mathbf{u}.\Delta\mathbf{u}\mathrm{d}^2\mathbf{r}$. Noting the identities $\Delta\mathbf{u} = -\nabla \times \omega$ (since $\nabla.\mathbf{u} = 0$), and $\mathbf{u}.(\nabla \times \omega) = \nabla.(\omega \times \mathbf{u}) + \omega^2$, we get

$$\dot{\mathcal{E}} = -2\nu\Gamma_2 - \nu \oint (\omega \times \mathbf{u}).\mathbf{n}. \tag{2.15}$$

The boundary term vanishes both for the no-slip ($\mathbf{u} = 0$) and free slip ($\omega = 0$) boundary conditions, and the interior term always makes the energy decay, as expected.

The enstrophy Γ_2 also decays by (2.14), in the absence of boundary effects. Then the rate of energy decay (2.15) is bounded by the initial enstrophy, and tends to 0 in the limit of small viscosity ν: energy is conserved in this limit[1]. By contrast in fully developed 3D turbulence the enstrophy increases as the viscosity is reduced (smaller and smaller scales are excited), such that the product of these two quantities, determining the energy dissipation, becomes independent of viscosity: it is controlled by the inertial cascade process.

Hyper-viscosity terms are often introduced in simulations of 2D turbulence to better approach the inviscid limit of zero energy dissipation. Then vorticity fluctuations are smoothed out like with viscosity, the enstrophy Γ_2 decays, but as a spurious effect the extrema ω_{min} and ω_{max} may be amplified (also higher order boundary conditions need to be introduced).

In 2D flows it is often convenient to rewrite the energy using an integration by parts, so that

$$\mathcal{E} = \frac{1}{2} \int \psi\omega\mathrm{d}^2\mathbf{r}. \tag{2.16}$$

Note that the integration by parts also yields a boundary term $\oint \psi\mathbf{u}.\mathrm{d}\mathbf{l}$, but it vanishes thanks to the boundary condition $\psi = 0$. Another choice $\psi =$ const. would introduce a boundary term in $\Gamma \times$ const. which is unimportant as it is constant in time. Note that when considering vortices interacting in a limited region of an infinite domain, the physical energy $\frac{1}{2} \int \mathbf{u}^2\mathrm{d}^2\mathbf{r}$ diverges for a non-zero circulation Γ (as the induced velocity only decays in Γ/r the energy integral logarithmically diverges), and only the form (2.16)

---

[1]This conclusion is unchanged with other boundary conditions, for instance no-slip: then a boundary layer of thickness $\delta = (\nu L/U)^{1/2}$ forms (U typical velocity), contributing to enstrophy as $\Gamma_2 \sim U^2L/\delta$. Then the energy dissipation $\nu\Gamma_2$ is in $\nu^{1/2}$, which also tends to zero in the inviscid limit.

can be used (see [4]). This kinetic energy has remarkably the same form as the electrostatic energy of a charge density field ω inducing a potential ψ, satisfying the Poisson equation (2.7).

• Momentum and angular momentum

In an infinite domain the momentum vector $\mathbf{P} = \int \omega \mathbf{r} \times \mathbf{e}_z \, \mathrm{d}^2\mathbf{r}$ is conserved, as well as the angular momentum with respect to any origin $L = \int \omega \mathbf{r}^2 \, \mathrm{d}^2\mathbf{r}$, see for instance [4] or [23]. These conservation laws are associated with symmetries of the system: invariance by translation for $\mathbf{P}$ and invariance by rotation for L and they are conserved as well in domains whose boundaries respect these symmetries: the x-wise momentum component is conserved in a channel along the x-direction (see *e.g.* [98]), and the angular momentum in a disk (taking the origin at the center). Note that this global angular momentum L has to be distinguished from the local angular momentum of a fluid particle, which is conserved for all geometries.

Finally the circulation $\oint \mathbf{u}.\mathrm{dl}$ along any boundary contour is conserved. For a simply connected domain, this is just the circulation Γ already considered, but the circulation along any obstacle is also conserved, and is not related to $\int \omega \mathrm{d}^2\mathbf{r}$, for instance along the inner wall of an annular domain. This conservation law is directly demonstrated from (2.1), rewriting the advective term with the identity $\mathbf{u}.\nabla \mathbf{u} = \omega \times \mathbf{u} + \nabla(\mathbf{u}^2/2)$, whose integral vanishes on the wall since $\omega \times \mathbf{u}$ is normal to the wall (as $\mathbf{u}$ is along the wall due to the impermeability condition) and a closed contour integral of a gradient vanishes.

• Other conservation laws

We have listed here all the explicit conservation laws for the 2D Euler equations: it can be shown [92] that there are no other conserved quantities with an explicit form $\int F(\mathbf{r}, \mathbf{u}(\mathbf{r}), \partial_i u_j(\mathbf{r})) \, \mathrm{d}^2\mathbf{r}$. Other conservation laws however exist, for instance "topological constraints": two initial uniform vorticity patches remain always distinct and they cannot fully merge in a single patch. However this constraint plays little role in practice, as the two patches can irreversibly deform and become more and more intertwined in the merging process discussed in next section.

2.4 Steady solutions of the Euler equations

It is often useful to discuss steady solutions of the 2D Euler equations, as they will appear as the result of turbulent mixing. For steady flows the particle trajectories are streamlines, so that ω, which is conserved along trajectories, will be constant along any streamline. This means that ω is a function of ψ only, at least in some sub-region: $\omega = F(\psi)$. In fact the same value of ψ can occur on several streamlines, so that different functions F can characterize different regions, as will be shown in the example of the dipole, next section. Reciprocally, it is clear that if $\omega = F(\psi)$, then the advective term $\mathbf{u}.\nabla\omega = -\nabla\psi \times \nabla\omega$ vanishes as $\nabla\omega = F'(\psi)\nabla\psi$ is parallel to $\nabla\psi$. Therefore the property of steady flow is indeed equivalent to the property $\omega = F(\psi)$ in subregions. The interface between these subregions must be a streamline with velocity continuous across it (but discontinuous vorticity in general).

It is also useful to consider steadily translating solutions, with a constant translation velocity vector $\mathbf{U}$, such that $\omega(t, \mathbf{r}) = \omega(\mathbf{r} - \mathbf{U}t)$. This is equivalent to a steady solution in a reference frame translating at velocity $\mathbf{U}$, with the same vorticity and stream function $\psi' = \psi - \mathbf{U}.\mathbf{r}$, so that $\omega = F(\psi - \mathbf{U}.\mathbf{r})$. Note that this is only possible in an infinite domain or a channel along the $\mathbf{U}$ direction.

Similarly we can consider purely rotating solutions $\omega(t, \mathbf{r}) = \omega(\mathbf{r} - (\Omega \times \mathbf{r})t)$, which is possible in an infinite domain or a circular geometry (disk or annulus). The general form of such flow patterns in solid body rotation is $\omega = F(\psi + \Omega\mathbf{r}^2/2)$. This can be shown directly on the Euler equations, or by using a rotating reference frame at angular velocity Ω^2.

3 Vortex dynamics

As turbulence is part of fluid dynamics, it is always useful to keep in mind elementary flow processes. This is particularly true for 2D turbulence, which displays "coherent structures" more clearly than 3D turbulence. Some discussion of inviscid vortex dynamics is therefore useful. Interesting results were already obtained in the XIXth century, and some of them "rediscovered" and extended recently in the context of 2D turbulence and ocean-atmosphere dynamics. The classical textbooks of Lamb [60] and Batchelor [4] provide good introductions to this field, and more advanced properties of discrete vortices are treated by Chorin [26, 27] and Aref [1], and vortex patches by Saffman [91]. The use of point vortices as a numerical discretization of continuous fluid motion is treated in a recent book by Cottet and Koumoutsakos [29]. The main motivation of the XIXth century

---

[2]While the invariance of the system by translation (Galilean change of reference frame) is warranted as a general physical principle, this is not so for a rotating reference frame, in which centrifugal and Coriolis force appear. However both forces are pure gradients in incompressible 2D flows, so they are exactly balanced by pressure gradients. Indeed the centrifugal force is proportional to $\nabla(\Omega\mathbf{r}^2/2)$, with a constant density factor, and the Coriolis force is proportional to $-2\Omega \times \mathbf{u} = \Omega\nabla\psi$. Note that the so-called geostrophic balance between Coriolis force and pressure gradient is only realized in 2D flows.

researchers was different: they were seeking mechanical models to build
theories of electromagnetism and atomic physics.

3.1 Systems of discrete vortices

Replacing the continuous vorticity field by a set of singular point vortices
(or vortex lines in the z direction) can be a good approach to many 2D flow
phenomena. Relation (2.8) then reduces to the discrete sum of the flows
induced by each point vortex at position $\mathbf{r}_j(t)$ and circulation γ_j,

$$\psi(t, \mathbf{r}) = \sum_j \gamma_j G_\psi(\mathbf{r}, \mathbf{r}_j(t)). \tag{3.1}$$

Each vortex is transported by the flow induced by all the other vortices.
The self interaction of the vortex (leading to a diverging ψ) can be ignored,
as seen by defining a point vortex as the limit of small vorticity patches with
vorticity $a_i \to \infty$ and infinitely small area γ_i/a_i, such that the circulation
γ_i remains constant. Then self-interaction just produces a local rotation of
the patch with no influence on the limiting point vortex. Furthermore the
circulation γ_i of each vortex is conserved in the flow evolution, since both the
small vortex patch area and vorticity a_i are conserved. Thus each vortex i
is transported by the velocity derived from the stream function (3.1), with
the sum restricted to $i \neq j$ which yields (the sum is made on indices $i < j$
to avoid double counting of the same term),

$$\dot{x}_i = \partial E_{\text{int}}/\partial y_i, \quad \dot{y}_i = -\partial E_{\text{int}}/\partial x_i \tag{3.2}$$

$$\text{with } E_{\text{int}} = \sum_{i<j} \gamma_i \gamma_j G_\psi(\mathbf{r_i}, \mathbf{r_j}). \tag{3.3}$$

The N point vortices therefore move like N interacting particles. The study
of point vortices was initiated by Helmholtz in 1858, and this general dynamical equation first derived by Kirchhoff.

The dynamical equation is first order in time, unlike the usual second
order Newton equation. However it has quite remarkably a Hamiltonian
structure, but the conjugate variables are the space coordinates x_i and y_i
instead of the positions and momenta of the particles. The Hamiltonian
E_{int} is conserved with flow evolution, and it corresponds to an interaction
energy of vortices. Note that the true physical energy is infinite due to
the self-energy associated with each vortex (the velocity tends to infinity
in $1/r$ around each vortex core but this has no influence in the vortex
interaction). The expression (2.10) of the interaction energy is like the
electrostatic interaction energy for long charged rods (notice however that it

corresponds physically to a kinetic energy of the flow, and the analogy with electrostatics is not complete, due to the different dynamical equation). The pair interaction decays only slowly with distance, so that vortex interactions are highly non-local, and we expect collective effects to be important, rather than binary collisions.

3.2 Vortex pairs

• Case of point vortices

Suppose two point vortices of like sign and circulation γ are separated by a distance d, far from boundaries. Then (3.2) just states that the two vortices rotate at velocity $\gamma/(2\pi d)$, keeping a constant distance d (Fig. 3a). Two vortices with unequal strength rotate around their "center of mass". If the two vortices have equal circulation with opposite sign, the center of mass does not exist, and both vortices translate with constant spacing d (Fig. 3b).

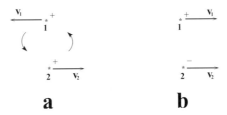

Fig. 3. Sketch of the motion of a vortex pair. (a) Rotation with like signs and (b) translation with opposite signs.

Note that in superfluids, vortices of opposite sign tend to eventually attract and annihilate each other. This is due to interactions with additional degrees of freedom, which can extracts energy from the fluid system. An external force on a vortex, for instance due to pinning on a solid substrate, can result in drift of the vortex core with respect to the local flow, and the occurrence of a Magnus force perpendicular to this drift, resulting in reduction of the vortex distance. Such effects are absent in the ideal flow problems considered here, and the distance d does remain constant.

In the translating case it is interesting to note that a region of the flow is transported and follows the translating motion. Therefore this flow contains momentum, representing the translating motion of some fluid area, see Figure 4 left (this is the 2D analog of a vortex ring).

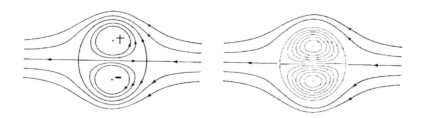

Fig. 4. Flow (left) in a point vortex pair and (right) in a Chaplygin's continuous pair. The streamlines are represented in a reference frame moving with the structure (so the flow is steady). Note that the fluid area inside the closed streamlines is entrained by the pair in its motion.

• **Translating vortex pairs with continuous vorticity**

The translating motion of vortex pairs with opposite sign is a remarkably robust feature, obtained when flow momentum is injected, even when vortices are far from pointwise. Figure 5 shows how an initial jet organizes into a vortex pair in a stratified fluid [46]. Similar results had been previously obtained from a wake in a soap film [31] or in an electromagnetically driven flow [77]. We see that, remarkably, the turbulence "self-organizes" into simple coherent structures, here vortex pairs, and this will be the subject of Section 5.

A vortex pair solution proposed by Chaplygin [15] (see also the review paper [72]) provides a more realistic description of such features, using continuous vorticity fields. We have seen in Section 2.4 that a general steadily translating solution is obtained with a vorticity of the form $\omega = -\Delta\psi = F(\psi - \mathbf{U.r})$ in some region. A natural idea is to choose a linear function F, then solving a Helmholtz equation for ψ in the vortex domain, matching an irrotational flow outside with continuous velocity. It turns out that a good matching is then obtained only with a circular domain, and one obtains Chaplygin's solution

$$\psi = -\frac{2U}{k}\frac{J_1(kr)}{J_0(ka)}\sin\theta \ \text{ for } r \leq a, \ \ \psi = U\left(r - \frac{a^2}{r}\right)\sin\theta \ \text{ for } r \geq a \quad (3.4)$$

in terms of the polar coordinates r and θ, where J_0 and J_1 are Bessel functions and ka the first zero of J_1.

Note that this solution, represented in Figure 4 right, emerges in many experiments and numerical simulations by spontaneous organization after complex flow evolution. Similar asymmetric dipoles, with a rotating motion

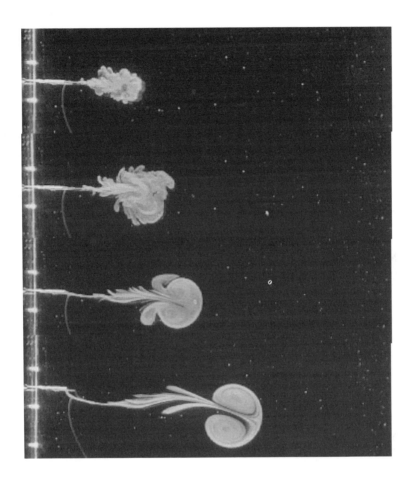

Fig. 5. Organization of a short jet injected in a density stratified flow (from Ref. [46]). We see in the successive views of the same jet the initial 3D stage, followed by a collapse to a quasi-2D state, which organizes into a dipole.

are also obtained. For any initial momentum, angular momentum, and energy, one can determine a corresponding dipole or axisymmetric monopole solution [23], which is expected to be obtained after complex vorticity stirring in some region of space. However non-linear relationships between vorticity and streamfunction can be obtained as well [77].

• **Vortex merging**

Two vorticity patches with the same sign rotate around each other like point vortices when their distance is sufficiently large in comparison with their size. Each patch is just slightly deformed by vortex interactions: this is like tidal effects between two gravitating bodies. This is however no longer true below some critical distance: the two vortices irreversibly deform leading to a single vortex, as shown in Figure 6. This figure is obtained from an experiment with an electron plasma trapped in a magnetic field, which remarkably follows the 2D Euler equations. The flow eventually tends to an axisymmetric configuration when small scale vorticity oscillations are smoothed out. The threshold of distance leading to irreversible deformation has been studied in detail using contour dynamics for vortex patches [81]. The resulting final merging can be understood on general grounds as a process of entropy maximization (see Sect. 5): the most probable state of the system is axisymmetric.

The initial vortex deformation leading to merging is due to the effect of the strain induced by the other vortex. This is a motivation for studying the influence of a uniform pure strain on a single vortex. A weak vortex is clearly passively deformed by the strain, while it resists deformation when its vorticity reaches a value of the same order as the strain rate.

The same merging mechanism is observed for vortices of unequal size and strength (but equal sign). This has been checked with electron plasma experiments [75] as well as various numerical simulations[3].

• Interaction of more than two point vortices

Interaction of three point vortices yields a variety of motion. The problem has some similarity [78] with a triad in Fourier space, which shall be discussed in Section 4. An interesting curiosity is the possibility, for particular initial conditions and vortex circulations, that the three vortices spiral inward to a singular point [1]. Singularities are however forbidden in the case of more realistic finite core vortices. For four vortices and more, chaotic motion is possible, as well as stable configurations. Tripoles made of a central vortex and two satellite vortices of opposite sign have been observed both in laboratory [48] and oceanic flows. With point vortices, stable patterns are obtained for more than three vortices, but none of them seem robust for extended vortices: two vortices of the same sign tend to merge.

---

[3]Note however that contour dynamics simulations indicate a variety of other possibilities for unequal size vortices [34]: in some cases merging is only partial, and small satellite vortices are produced. Such processes could be relevant in controlling the population of vortices of different size in 2D turbulence as discussed in Section 4, so this problem would require more careful examination, comparing different numerical methods.

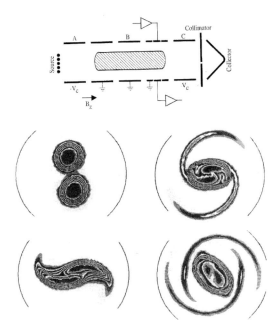

Fig. 6. Vortex merger in an electron plasma experiment (from [75]). Maps of the vorticity field are represented at four successive times (0, 16, 41 and 76 μs). The initial distance of the vortex centers is 1.2 times the vortex diameter. The experimental device is sketched at the top of the figure. The electrons are extracted from a source in the high vacuum cylindrical cell. Then they are transversally confined by the uniform axial magnetic field $\mathbf{B}_z$ and they drift with the velocity $\mathbf{u} = c(\mathbf{E} \times \mathbf{B}_z)/B_z$ perpendicular to the electric field $\mathbf{E}$ induced by the electron space charge. This drift flow remarkably satisfies the Euler equations: the incompressibility condition $\nabla.\mathbf{u} = 0$ is a consequence of $\nabla \times \mathbf{E} = 0$ while the vorticity $\nabla \times \mathbf{u} \propto \nabla.\mathbf{E} = n/\epsilon_0$ is proportional to the charge density n, conserved by the flow. This charge density field is visualized and measured by suddenly accelerating the electrons on the collector (destroying the system).

3.3 Instability of shear flows and vortex lattices

• Parallel flows

The classical stability criterion of Rayleigh applies to 2D inviscid flows. Thus classical flows with vorticity extrema, shear layers, jets and wakes develop 2D turbulence. The turbulent region grows linearly with time or stream-wise coordinate, in a similar way whether or not 3D perturbations are allowed to develop [65].

By contrast Poiseuille flows and boundary layers behave quite differently. These flows are linearly stable according to the Rayleigh criterion (they have no vorticity extrema), and they indeed remain stable in purely 2D flows, whatever the Reynolds number. The instabilities occurring in these flows are genuinely three-dimensional and are suppressed by the constraint of two-dimensionality.

• Vortex lattices

Vortex lattices can initiate 2D turbulence when they are unstable. The square lattice of alternating sign vortices is highly unstable and initiates the inverse energy cascade of 2D turbulence. In contrast, triangular lattices of equal sign vortices and hexagonal lattices with alternating sign vortices turn out to be stable [104]. Such stability properties can be more easily determined if one restricts the analysis to perturbations at large scales with respect to the lattice mesh, so that asymptotic expansions can be used. It is then shown that a minimal degree of anisotropy is needed to get instability [43].

In conclusion, while 3D flows at very high Reynolds systematically develop turbulence, with quasi-universal behavior, this is not true in 2D. There is much more influence of the generating mechanism, and turbulence may not be produced at all in some cases.

3.4 Statistical mechanics of point vortices

• The statistical mechanics approach

Statistical mechanics, as developed by Maxwell, Boltzmann, Gibbs and their followers, has been remarkably successful in predicting the behavior of systems with many degrees of freedom. Its main success has been in predicting the statistical (thermodynamic) equilibrium for a conservative (Hamiltonian) systems, for instance a gas made of many molecules. 3D turbulence is, in contrast, a strongly dissipative system. However, 2D turbulence conserves energy, which raises hope for equilibrium statistical mechanics approaches. This is not an obvious matter however, since 2D turbulence undergoes irreversible transformations with dissipation of vorticity fluctuations (enstrophy), instead of energy. The point vortex model is precisely a Hamiltonian system, as we have seen in Section 3.1, so that the standard methods of equilibrium statistical mechanics readily apply. The relevance to actual continuous flows will be discussed later.

The statistical mechanics of point vortices was first discussed by Onsager [79]. First of all, it is an interesting exercise in statistical mechanics, since "negative temperature" states are obtained. In such states vortices of like sign tend to clump together, forming large coherent vortices. Onsager stressed the importance of such coherent structures with remarkable foresight, and he pointed out the fundamental difference with the energy cascade of 3D turbulence, which had been recently formalized by Kolmogorov in 1941. Although short, his paper contains far-reaching remarks on both 2D and 3D turbulence, and its reading is highly recommended.

The general principles of equilibrium statistical mechanics are explained in many textbooks of physics, but this is always a subtle subject. Since applications to 2D turbulence are unusual, they require a good understanding (and re-discussion) of the basic principles, and it may be useful to recall them in the context of vortex dynamics.

The starting point of equilibrium statistical mechanics is to list the *conserved quantities* of the system, which are clear constraints to the dynamics. For a set of many point vortices, the only known conserved quantity is energy, as is the case for usual thermodynamic systems (but we shall see additional conserved quantities with alternative models of 2D turbulence). Then it is assumed that the system evenly explores all its possible states (the "*microscopic states*") allowed by the given value of its conserved quantities (here just energy). This assumption (the *ergodic hypothesis*) has been rigorously demonstrated only for a system of hard spheres in elastic collisions, but is believed to be true in many cases.

The output of the theory is the probability distribution of "*macroscopic*" states of interest, for instance the vortex density field. The *entropy S* of the macroscopic state is defined as the logarithm of the "number" of possible microscopic states corresponding to this macroscopic state. Then it follows from the ergodic hypothesis that the probability of this macroscopic state is just proportional to the exponential of the entropy. The most probable macroscopic state is therefore that which maximizes the entropy. In the limit of a very large number of particles, this maximum tends to be very sharp: an overwhelming majority of microscopic states tends to concentrate near the macroscopic state of maximum entropy. Therefore *deterministic* predictions result from the statistics of many particles, for instance the density of vortices will fluctuate less and less as the vortex density increases.

A microscopic state is defined by the coordinates of each of the N vortices. To count the possible states, we need first to discretize the coordinates into elementary cells. Let us take a mesh h in each coordinate[4] (and consider a fluid domain with surface unity for simplicity), so that the total number of possible states for a single vortex is just the cell number $1/h^2$. For N vortices it is $1/h^{2N}$ (we assume that several vortices can occupy the same cell without restriction, which is true for ideal point vortices). Among these states we must select and count the ones which have a given energy E_{int}, relation (3.3).

This is a tremendous task in general, but let us first neglect this interaction. Then we expect a uniform vortex density. To show this, consider the density field $n(\mathbf{r})$ as the macroscopic state, and let us count the corresponding number of configurations. We make a partition of the fluid domain in p sub-domains, with area $A = 1/p$ each, and consider the vortex numbers $n_1, \dots n_p$ in each sub-domain as the macroscopic state. We must first distribute the vortices in packets with $n_1, \dots n_p$ vortices respectively. The number of possibilities is $\frac{N!}{n_1! \dots n_p!}$ (this is the total number of permutations divided by the number of permutations within each packet, which does not change the distribution). Then for sub-domain 1 the number of possible vortex configurations (positions) is $(ph^2)^{-n_1}$ and we have to multiply by the similar formula with the other sub-domains. The number of configurations with $n_1, \dots n_p$ vortices is therefore $\frac{N!}{n_1! \dots n_p!}(ph^2)^{-N}$. The entropy is the logarithm of this quantity. For large vortex numbers, we can use the Sterling formula, $n! \simeq n \ln n$, so that the entropy is

$$S = -\sum n_i \ln n_i \rightarrow -\int n \ln n\, d^2\mathbf{r} \qquad (3.5)$$

in the continuous limit (up to an unimportant constant, depending on the discretization mesh).

Maximizing this entropy with the constraint of a given total vortex number $N = \int n\, d^2\mathbf{r}$ gives a uniform density. To check that, we introduce a Lagrange parameter α associated with the constraint N, and write the condition for the first variations $\delta S - \alpha \delta N = 0$. Differentiating the expression

---

[4]The uniform discretization used for the counting seems here a natural choice, but it may be wrong with other coordinates. For instance a uniform discretization in the polar coordinates r, θ would give very small cells $dr d\theta$ near the pole, resulting in excessive statistical weight. The justification lies in the Hamiltonian form (3.2) of the dynamical equations, from which the *Liouville theorem* is readily demonstrated: considering the evolution of many identical systems, this theorem states that the volume element $dx_1 \dots dx_N dy_1 \dots dy_N$ in phase space is conserved with time. Indeed the divergence of the "velocity vector" $\dot{x}_1, \dots, \dot{x}_N, \dot{y}_1, \dots \dot{y}_N$ is clearly 0, due to the Hamiltonian form (it is the analog for the phase space flow of a stream function for a usual 2D flow). Then the uniform sampling in the coordinates $x_1, \dots, x_N, y_1, \dots, y_N$ will *remain uniform* with the time evolution of the system. This is only true for the so-called canonical coordinates for which this usual canonical form (3.3) of the Hamiltonian system can be written. It is not the case, for instance, with polar coordinates.

of the entropy gives $\delta S = -\int(\ln n + 1)\delta n d^2\mathbf{r}$, so that the condition on first variations becomes

$$\int(\ln n + 1 + \alpha)\delta n d^2\mathbf{r}. \qquad (3.6)$$

This has to be satisfied for any variation δn (function of position) around the optimum state, which is only possible if the term in parenthesis is uniform, so that the density n is uniform: non-interacting particles uniformly mix due to entropy maximization.

• **The mean field approximation**

Coming back to the interacting particles, a great simplification is provided by the *mean field* approximation, as developed by Joyce and Montgomery [52]. The idea is that, due to the long range interactions, each vortex feels the influence of the mean field ψ due to many others, so that we can write the interaction energy with the continuous field expression as

$$E = \frac{1}{2}\int \psi n \gamma d^2\mathbf{r} \qquad (3.7)$$

replacing the vorticity ω in (2.16) by the local density $n\gamma$. We suppose first that all the vortices have the same circulation γ, but generalization to several vortex species is straightforward by just adding their contributions to ω. The field ψ is itself given by the Poisson equation (2.7), which becomes $-\Delta\psi = \gamma n$.

The condition on energy brings the new constraint (3.7) for entropy maximization, and a corresponding Lagrange parameter β must be introduced accordingly. Then the condition on first order variations becomes

$$\delta S - \alpha \delta N - \beta \delta E = 0. \qquad (3.8)$$

We calculate $\delta E = (\gamma/2)\int(\psi\delta n + n\delta\psi)d^2\mathbf{r}$. In fact the second term is equal to the first, as checked by using the Poisson equation and an integration by parts. The condition on first variations then becomes $\int(\ln n + 1 + \alpha + \beta\gamma\psi)\delta n d^2\mathbf{r}$, which implies that

$$n = n_0 \exp(-\beta\gamma\psi) \qquad (3.9)$$

(with $n_0 \equiv e^{-\alpha}$). Combining (3.9) and the Poisson equation, we get the self-consistent mean field equation

$$-\Delta\psi = \gamma n_0 \exp(-\beta\gamma\psi), \quad \psi = 0 \text{ on boundaries.} \qquad (3.10)$$

Since the locally averaged vorticity $-\Delta\psi$ is a function of ψ, this remarkably represents a steady solution of the Euler equation. A general justification of self-organization into large scale steady flows is thus provided: this is the most probable outcome for the wandering of many small vortices.

The two constants n_0 and β are indirectly given by the constraints on energy and total vortex number. In fact what is given is the product γn_0 (related to the total circulation of the system), while n_0 tends to infinity, and γ tends to 0. The validity of the mean field approximation has been rigorously demonstrated in this limit [36]. Similarly the important parameter is $\beta\gamma$ instead of β, and we can rewrite (3.10) with the non-dimensional variable $\phi = (\gamma n_0)^{-1}\psi$,

$$-\Delta\phi = \exp(-B\phi), \quad \phi = 0 \text{ on boundaries} \tag{3.11}$$

depending on the single parameter $B = \beta\gamma^2 n_0$.

Note that the expression (3.9) can be obtained in general for a particle in a field with potential energy $\psi\gamma$ in contact with a "thermal" bath with temperature $1/\beta$. This is the so-called canonical approach, in contrast with the micro-canonical approach used her, dealing with an isolated system. These two approaches are generally equivalent, but it is not always so for systems with long range interactions. Note also that some textbooks consider statistical mechanics as the limit of large systems, making the volume goes to infinity. What is important is the limit of a large number of particles, which is here considered in a given domain of finite size. The system is not extensive, on the contrary its spatial confinement is essential.

• Discussion of results

Supposing for instance $\gamma > 0$, it is clear from (3.10) that ψ is a convex function of the coordinates, which is everywhere positive. For positive "temperature" $\beta > 0$, the vorticity tends to be depleted where ψ is maximum, and maximum near the domain boundary, where ψ is set to zero. By contrast for $\beta < 0$, the vorticity tends to be maximum at the vortex center, leading to sharper and sharper maximum of ψ as β is more negative.

It is interesting to represent the entropy of the statistical equilibrium *versus* its energy, which has always a bell shape, as shown in Figure 7. Due to (3.8), the derivative dS/dE (for N fixed) is just β the inverse of the temperature. Therefore the positive temperature is obtained for small energy and negative temperature for large energy. The maximum corresponds to the uniform vortex density: the energy has just the right value to allow for uniform density, which is the state of maximum entropy in the absence of energy constraint, as shown above. Higher energy requires the vortices to remain closely packed, while low energy requires them to remain near the boundaries. The existence of negative temperature states is forbidden with ordinary particles whose Hamiltonian has a quadratic term in the momentum (the usual kinetic energy): then the entropy always increases with

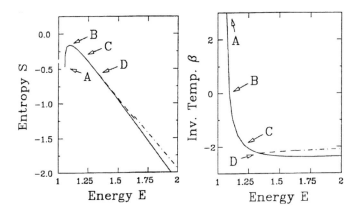

Fig. 7. Entropy (left) and inverse "temperature" $\beta = \mathrm{d}E/\mathrm{d}S$ (right) *versus* energy E for the statistical equilibria of a set of N identical point vortices in a disk (from [95]). For small energies (point A) vortices remain near the disk periphery, and the "temperature is positive", while for large energies they remain clumped in a global vortex, more and more concentrated as energy is larger. Then the "temperature" is negative. The entropy maximum (point B) corresponds to a uniform vorticity in the disk, with $\beta = 0$. The solid curves represents axisymmetric equilibria, while the point-dash curve represents a "bifurcated state", in which the vortex spontaneously forms out of the disk axis. This bifurcated state is more likely than the axisymmetric one as it has a higher entropy. It has been indeed observed in a laboratory experiment using a mercury flow in a magnetic field [33]. Note however that, far from boundaries, the axisymmetric equilibrium state is always predicted (the bifurcation is due to the confinement by the disk periphery).

energy. However in the vortex system, such a quadratic term is absent: negative temperature states do occur, and they correspond to the interesting case of clumping of like sign vortices into large coherent vortices.

For several vortex species, with circulation γ_i, the density of each species satisfies a mean field relation (3.9). Then the relationship (3.10) between vorticity and stream function is replaced by a sum of exponentials (or an integral for a continuous distribution of elementary vortex circulation γ,

$$-\Delta\psi = \sum_i \gamma_i n_i \exp(-\beta\gamma_i\psi), \quad \psi = 0 \text{ on boundaries.} \tag{3.12}$$

For a symmetric set of positive and negative vortices with circulation $\pm\gamma$, we get a sinh function. Note however that symmetry breaking is also possible, so that $n_+ \neq n_-$ even for an equal number of positive and negative vortices. This has been first shown by Pointin and Lundgren [84] in a square domain.

In Figure 8 we show striking examples of symmetry breaking. The statistical equilibrium reached by two initial vortex lines of opposite sign, forming a jet, is considered. The confinement in a channel with periodic boundary conditions is necessary to get an equilibrium. However when the wall is far from the initial jet width d, and the allowed period L sufficiently long, we get an organization with the topology of the Karman vortex street (state DD in the figure). This may explain the observed trend of plane wakes to form such a structure, even in the presence of a strong turbulence (although wakes freely expand with time so they never reach equilibrium). For other parameters shown in Figure 8, an additional symmetry breaking occurs: vortices on one sign clump together while vortices of opposite sign are dispersed (which favors entropy, while the coherent vortex is necessary to satisfy the energy constraint). The solitary vortex state (SV) qualitatively explains the organization of a turbulent jet observed in an annular channel [100], modeling many dynamical aspects of the Great Red Spot of Jupiter.

• Limitations

The point vortex statistical mechanics explains self-organization. However it does not provide a consistent and quantitative prediction for the Euler system with continuous velocity fields. Of course it is always possible to approximate the continuous velocity fields by a set of many point vortices with a small circulation γ and spacing h, such as $h^2\gamma = \omega$. The limit of small spacing h provides a consistent, stable and convergent approximation [44]. Vortex methods can be used indeed in practice for numerical simulations of the Euler equations [26, 27, 29]. However any approximation to a dynamical system is valid for a finite time, and it may break down for sufficiently long time (which increases with the spatial resolution). The system of N vortices will eventually behave differently from the continuous system, and the equilibrium statistical mechanics, dealing with the limit of very long times, is different (in more mathematical terms, taking the results of the two limits $h \to 0$ and $t \to \infty$ depends on the order in which they are taken).

There is first the possibility that the maximum vorticity of the statistical equilibrium exceeds the maximum value of the initial condition, which is inconsistent with the Euler equations. Secondly there are several ways to model the same continuous initial condition: we can for instance use either constant vortices with a variable spacing h, or a constant spacing h and different vortex circulations, proportional to the local vorticity. The point vortex statistical equilibrium will be different in the two cases: in the first case we shall get the result (3.10), while in the second case we shall get the result (3.12), with a sum of exponential terms. These difficulties will be solved in Section 5 using a different approach.

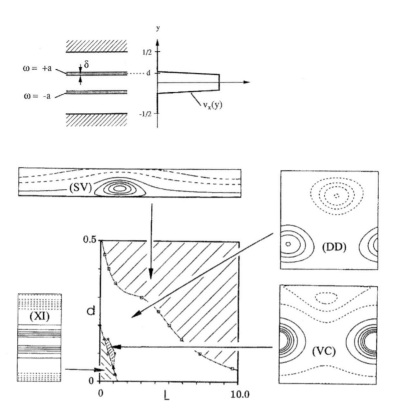

Fig. 8. (From [53].) Statistical equilibria resulting from the mixing of two vortex lines of opposite sign, representing an initial jet in a x-periodic channel, as sketched at the top of the figure. The equilibrium states are represented as a function of the domain length L and the half jet width d (while the channel width is set to unity). The different flow configurations are represented by vorticity isolines for the parameters indicated by the arrows. The boundaries between these configurations are drawn on the diagram. In addition to the x-independent states (XI), we get states breaking the translation symmetry, like the "sinuous mode" (DD). This mode is preferred for small initial jet width d and sufficiently large L, for which the effect of boundary confinement is minimum. For the other states (VC or SV), the additional symmetry breaking between positive and negative vortices is obtained.

4 Spectral properties, energy and enstrophy cascade

Since Kolmogorov, the use of spectral representations is at the heart of the study of turbulence. This approach quantifies the energy transfers among the different scales of motion. Kolmogorov's ideas have been applied to 2D turbulence by Kraichnan in 1967 [56], see [58] and [64] for good reviews. The most interesting result is the prediction of an inverse energy cascade toward large scales, whose existence is now firmly established both from numerical simulations and laboratory experiments. The existence of a direct enstrophy cascade toward small scales has been also predicted, but its relevance remains controversial. The emergence of isolated vortices plays an important role, at least in some cases.

4.1 Spectrally truncated equilibrium states

Instead of using point vortices, a quite different approach is to expand the velocity (and vorticity) field in the eigenmodes ϕ_n of the Laplacian for the fluid domain. These are the Fourier modes for the usual periodic conditions. We therefore expand the vorticity as

$$\omega(t, \mathbf{r}) = \sum a_n(t)\phi_n(\mathbf{r}), \quad \text{with } \Delta\phi_n = -k_n^2\phi_n \tag{4.1}$$

and the streamfunction is $\psi = \sum a_n k_n^{-2}\phi_n$. Other basis, like wavelets, allow to make scale analysis depending on position, which may be more sensible in the presence of coherent structures, see [39] in this book. However the classical Fourier have the advantage of nice dynamical properties, in addition to their simplicity. In particular each mode ϕ_n is a steady solution of the Euler equations, since its vorticity $-\Delta\phi_n$ is a function of ϕ_n (see Sect. 2.4). The energy E and enstrophy Γ_2 are readily expressed, due to the orthonormality of the eigenmodes,

$$E = \frac{1}{2}\sum a_n^2 k_n^{-2}, \quad \Gamma_2 = \frac{1}{2}\sum a_n^2. \tag{4.2}$$

Since the Euler equations contain only quadratic terms in velocity, it can be written in the general form

$$\dot{a}_n = \sum_{r,s} A_{nrs}a_r a_s \tag{4.3}$$

with fixed interaction coefficients A_{nrs} for the triad interactions. These coefficients satisfy the "detailed conservation of energy" among each triad

$$k_n^{-2}A_{nrs} + k_r^{-2}A_{rsn} + k_s^{-2}A_{snr} = 0 \tag{4.4}$$

as well as the detailed conservation of enstrophy

$$A_{nrs} + A_{rsn} + A_{snr} = 0. \tag{4.5}$$

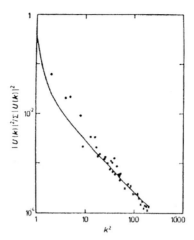

Fig. 9. Normalized modal energies (dots), averaged over 100 time steps, from the numerical solutions of the evolution equation (4.3) for the truncated spectral model, compared to the statistical equilibrium prediction (4.8) in solid line. This run (from [93]) corresponds to a negative β regime with 70% of the energy condensed in the fundamental mode.

Starting with some initial condition with energy limited to a few modes, higher and higher modes will be excited: this is the cascade process of turbulence. It can be understood as a tendency of the system to explore all the available modes, a general effect of entropy increase. However a statistical equilibrium, maximizing entropy, is really reached only if we artificially set a bound to the accessible modes, keeping a finite number N of modes. This stops the fundamental irreversibility of turbulence but gives indications on the general trends of the system.

The spectrally truncated system (4.3) then becomes a closed dynamical system which conserves energy, and the general methods of equilibrium statistical mechanics can be used. Although conservative, it does not have an Hamiltonian form, unlike the vortex system (3.2). Still a Liouville theorem exists

$$\partial \dot{a}_n / \partial a_n = 0 \tag{4.6}$$

which is the required condition for applying statistical mechanics (the volume in phase space is then conserved as mentioned in the footnote of Sect. 3.4). This condition is easily derived: the terms in (4.3) with n, r, s all different give no contribution to $\partial \dot{a}_n / \partial a_n$. The coefficients A_{nss} vanish by the stationary property of the eigenmode ϕ_s. A_{nnn} vanishes in particular, and by (4.4), $A_{nsn} = A_{nns} = 0$. This exhausts the possibilities and we have (4.6).

In the presence of a "thermal bath" with inverse temperature β, the probability of a microscopic state with energy E is in $\exp(-\beta E)$, like we have seen for point vortices in Section 3.4. Since there is a second conserved quantity Γ_2, with associated Lagrange multiplyer α, the probability of a state $a_1, \ldots a_N$ is proportional to $\exp(-\beta E - \alpha \Gamma_2)$. Due to the quadratic expressions (4.2), we have just independent Gaussian probabilities for each mode

$$p(a_n) \propto \exp[-(\beta k_n^{-2} + \alpha)a_n^2/2] \tag{4.7}$$

so the mean energy of a mode is

$$k_n^{-2}\langle a_n^2 \rangle = k_n^{-2} \int p(a_n)a_n^2 \, \mathrm{d}a_n / \int p(a_n)\mathrm{d}a_n = \frac{1}{2}(\beta + \alpha k_n^2)^{-1}. \tag{4.8}$$

For large wavenumbers, it is convenient to replace the mode number n by the wavevector $\mathbf{k}$, and the discrete mode amplitudes a_n by the Fourier transform $\hat{\omega}(\mathbf{k})$ of the vorticity field. Furthermore the energy spectrum $E(k)$, or the enstrophy spectrum $k^2 E(k)$ is used. It is defined by integrating $\hat{\omega}^2(\mathbf{k})$ over a circle with $|\mathbf{k}| = k$, so that for an isotropic field, $k^2 E(k) = 2\pi k \hat{\omega}^2(\mathbf{k})$. The total energy is the integral of the energy spectrum, $E = \int_0^\infty E(k)\mathrm{d}k$, and similarly for the enstrophy, $\Gamma_2 = \int_0^\infty k^2 E(k)\mathrm{d}k$. The equilibrium energy spectrum is therefore in $k/(\beta + \alpha k^2)$.

For 3D turbulence, we have only the conservation of energy, so that $\alpha = 0$, and the energy is equally spread over the modes. This is not in agreement with actual turbulence, but it justifies the tendency of energy to spread over all the accessible modes, and therefore to undergo an energy cascade toward the small scales where a majority of modes is located. In contrast for 2D turbulence we can have a variety of states, depending on β and α, which can be indirectly related to the mean energy and enstrophy of the system. A particularly interesting case corresponds to $\beta \rightarrow -\alpha k_0^2$ for which the energy can accumulate in the lowest available mode k_0, as represented in Figure 9. This figure also indicates that the statistical mechanics prediction is well verified by numerical computations of the dynamical system (4.3), artificially truncated by keeping a finite number of modes without dissipation.

This helps to understand the self-organization into the lowest mode, with a model different from that of point vortices. The artificial truncation of high modes have then little influence. Note however that the truncation suppresses conservation laws (2.11) other than enstrophy. Moreover the use of a canonical distribution is then questionable, as energy is concentrated in the single lowest mode, with large fluctuations. For instance the most probable state corresponds to the center of the Gaussian (4.7), with zero energy. The use of a micro-canonical approach, keeping the energy constant, would lead to a different and more realistic result, although the trend for concentration in the lowest mode should be the same.

4.2 The enstrophy and inverse energy cascades of forced turbulence

• The double cascade of Kraichnan (1967)

In the absence of confinement at small wavenumbers (*e.g.* boundaries) and artificial cutoff at high wavenumbers the cascades can freely develop. To study stationary regimes, it is convenient to consider a statistically permanent forcing concentrated at a given wavenumber k_I. The dimensional analysis leading to the Kolmogorov cascade can be carried out in 2D as well as in 3D and it yields the same inertial range,

$$E(k) = C\epsilon^{2/3}k^{-5/3}. \tag{4.9}$$

The direction of the energy flux is not given by dimensional analysis. However a direct cascade toward high wave numbers is forbidden by the absence of energy dissipation (see Sect. 2.3). Therefore Kraichnan suggested a cascade toward large scales (small k). He further justifies this cascade direction as a trend of the system to go toward the statistical equilibrium described above (although it never reaches it in the absence of spectral truncation). There is no need for energy dissipation at large scales, at least in the ideal case of an infinite fluid domain, since the integral $\int k^{-5/3}\mathrm{d}k$ diverges at 0: energy progressively accumulates toward lower and lower k. More physically, we shall see that a friction force proportional to the velocity (Rayleigh friction) can progressively pump out the energy along the inverse cascade.

Now what happens toward the small scales? Enstrophy must be injected by the forcing, at a rate $\eta = k_I^2\epsilon$, where k_I is the injection wavenumber. Then an enstrophy cascade is expected, which can be predicted by dimensional analysis in the same way as the energy cascade, just replacing energy by enstrophy, and ϵ by η. This yields an energy spectrum

$$E(k) = C'\eta^{2/3}k^{-3}. \tag{4.10}$$

We shall see that these two cascades are mutually exclusive (for infinitely extended cascades): the rate of enstrophy transfer vanishes in the energy cascade and the rate of energy transfer vanishes in the enstrophy cascade. Therefore since energy goes to the larger scales, the enstrophy cascade must be toward small scales. Enstrophy can be dissipated at small scales by viscosity unlike energy. Even without viscosity the enstrophy cascade can theoretically extends with time to higher and higher wavenumbers. Figure 10 summarizes this theoretical double cascading spectrum.

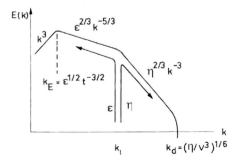

Fig. 10. Sketch of the classical double cascade of 2D turbulence (from Lesieur [64]).

For a more precise discussion of these cascade processes, let us express the forcing as a vorticity source $f(x, y, t)$ added on the right hand side of (2.6), and introduce some energy dissipation. Then the mode enstrophies $\langle a_n^2 \rangle$ satisfy the equations

$$\frac{1}{2} \frac{\mathrm{d}\langle a_n^2 \rangle}{\mathrm{d}t} = \langle a_n \dot{a}_n \rangle = \sum_{r,s} A_{nrs} \langle a_r a_s a_n \rangle + \langle f_n a_n \rangle - \text{diss} \qquad (4.11)$$

where f_n is the amplitude of f in the mode n. Note that the enstrophy production is the correlation $\langle f_n a_n \rangle$ between the vorticity and its source f, and the corresponding energy production $k_n^{-2} \langle f_n a_n \rangle$. The sums of these quantities over the modes are the total enstrophy production η and energy production ϵ respectively.

For the considered homogeneous isotropic turbulence, the discrete mode amplitudes a_n is replaced by the Fourier transform $\hat{\omega}(\mathbf{k})$, and the energy equation (4.11) takes the form of a conservation equation for the energy spectrum $E(k)$,

$$\frac{\partial E}{\partial t} = -\frac{\partial \Pi}{\partial k} + \text{forcing} - \text{diss} \qquad (4.12)$$

Π can be viewed as the energy flux due to nonlinear interactions and $\partial \Pi / \partial k$ its divergence. The same equation can be also written in a form displaying enstrophy conservation

$$k^2 \frac{\partial E}{\partial t} = -\frac{\partial Z}{\partial k} + \text{forcing} - \text{diss} \qquad (4.13)$$

with an enstrophy flux Z. These fluxes are explicitly obtained [56, 64] by Fourier transform of the Euler equations and integration over wavevector directions. They are expressed as integral of the rates of triad energy transfer $T(k, p, q)$, depending on the triple correlations $\langle \hat{\omega}(\mathbf{k})\hat{\omega}(\mathbf{p})\hat{\omega}(\mathbf{q})\rangle$, where the vectors $(\mathbf{k}, \mathbf{p}, \mathbf{q})$ form a triangle $(\mathbf{k} + \mathbf{p} + \mathbf{q} = 0)$ with sides k, p, q,

$$\Pi(k) = \frac{1}{2}\int_k^\infty dk' \int_0^k T(k', p, q)dpdq - \frac{1}{2}\int_0^k dk' \int_k^\infty T(k', p, q)dpdq$$
$$(4.14)$$

$$Z(k) = \frac{1}{2}\int_k^\infty k'^2 dk' \int_0^k T(k', p, q)dpdq - \frac{1}{2}\int_0^k k'^2 dk' \int_k^\infty T(k', p, q)dpdq.$$
$$(4.15)$$

At this stage a possible approach is to introduce a closure hypothesis to express the triple correlations and obtain dynamical equations for the energy spectrum. This has been done by assuming that the fluctuations of the mode amplitudes have statistics close to a Gaussian. After a failed attempt by Millionshtchikov (the quasi-normal theory), this approach has been widely developed by Kraichnan. His models have the property of relaxing the system toward the spectrally truncated statistical equilibrium in the absence of forcing and dissipation. With forcing at a given wavenumber k_I, these closure models lead to the double cascade sketched in Figure 10 [3, 49], see [64] for a review.

However Kraichnan did not use closure hypothesis in his original paper of 1967. He assumes instead an infinite cascade, with forcing and dissipation replaced by constant flux in wavenumber space. He further assumes that the transfer rate $T(k, p, q)$ scales in power law: $T(k, p, q) = k^{-m}T(1, p/k, q/k)$, where $T(1, p/k, q/k)$ depends only on the angles in the triad. Then the energy flux (4.14) is expressed as k^{3-m} (due to the triple integrals) multiplied by angular integrals over triad directions p/k and q/k. The only possibility for a flux $\Pi(k)$ independent of k is therefore $m = 3$. By dimensional analysis, $T(k, p, q) \sim u^3/k^2$, where u is the typical velocity at scale $1/k$, and $E(k) \sim u^2/k$. Therefore, $T \sim k^{-3}$ implies $u \sim k^{-1/3}$, corresponding to an energy spectrum in $k^{-5/3}$. This dimensional analysis is however valid only if the triple correlations scale in u^3 while the double correlations scale in u^2, i.e. in the absence of intermittency. The cascade is more fundamentally defined by the scaling of the transfer rate than the scaling of the energy spectrum (double correlations).

Such constant energy flux cascades can be as well obtained in 3D, but in 2D one can furthermore imagine cascades with a constant enstrophy flux $Z(k)$, corresponding to the k^{-3} energy spectrum. Then the energy flux $\Pi(k)$ should scale in k^{-2}. However by using the detail conservation laws (4.4) and (4.5), which translate into similar relations for $T(k, p, q)$, Kraichnan shows that the angular integration in (4.14) exactly cancels: the energy flux is zero. Similarly the enstrophy flux cancels in the energy cascade. As

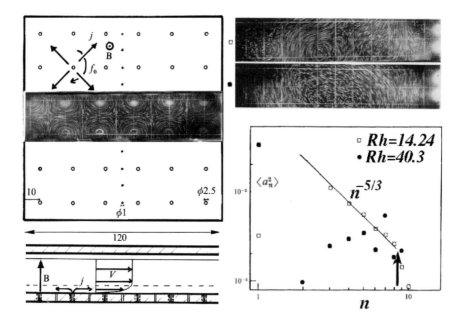

Fig. 11. Laboratory observation of the inverse energy cascade, from [97]. The experimental cell is sketched on the left as a top view and side view below. The flow is maintained 2D in a horizontal mercury layer by a vertical magnetic field which prevents the built up of 3D perturbations. The electric current, steadily injected with alternating sign in an electrode lattice, generates a square vortex lattice by interaction with the magnetic field. The flow is visualized from above by the streaks of particles floating on the mercury free surface (the field of view is limited to a central band due to the constraint of the electromagnet). The dynamics is controlled by a friction parameter Rh representing the ratio of the inertial effect to the friction on the bottom surface. For $Rh < 1.78$, the flow is steady as shown on the left, over the apparatus sketch. For higher Rh instabilities develop, exciting larger scales of motion, as shown on the two photos on the right top. The corresponding energy spectra reveal the built up of an inverse energy cascade for $Rh = 14.24$, and the condensation in the fundamental mode $n = 1$ for $Rh = 40.3$. This mode corresponds to a global rotation of the flow, spontaneously breaking the symmetry between positive and negative vorticity. The spectra are obtained from the spatial Fourier transform of the electric potential measured along a line of small electrodes ($\emptyset$ 1): the induced electric potential is proportional to the transverse velocity.

stressed by Kraichnan, a cascade cannot be viewed just as a carrying belt in wavenumber space, transporting together the energy $E(k)$ and its related

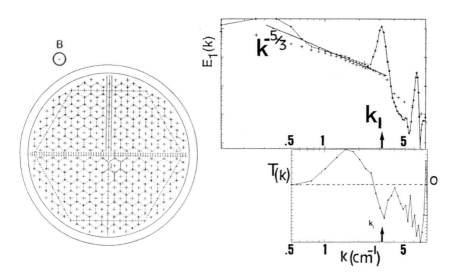

Fig. 12. Forcing 2D turbulence with the same technique as in Figure 11, but with an hexagonal lattice (from [76]). The energy spectra indicate a persistent peak at the forcing wavenumber, which is removed if the mean velocity is subtracted from the data, keeping only the turbulent part with a hint of a $k^{-5/3}$ range. The spectral energy transfer $T(k) = -\mathrm{d}\Pi/\mathrm{d}k$ have been also obtained from the triple velocity correlations, assuming isotropy. One checks that the energy transfer indeed feeds the small wavenumbers while it removes energy from the forcing range.

enstrophy $k^2 E(k)$. The transport is rather the result of overlapping triads, which makes possible a flux of energy or enstrophy alone. Eyink [35] has obtained mathematical results supporting this double cascade theory. He has shown that the flux of higher vorticity moments, *e.g.* ω^4 is related to the enstrophy cascade and is also toward high wavenumbers.

This work of Kraichnan provides constraints on the nature of putative cascades, but it does not guarantee that such states should be approached. As mentioned above, closure models provide a first support of these ideas. The enstrophy cascade has received a more precise theoretical justification by a different approach [37], presented at the same school [38]. The idea is derived from the study of the passive scalar stirring in a random large scale strain. Batchelor in 1959 first predicted k^{-1} spectra for the scalar variance, and this has been confirmed by rigorous approaches. Vorticity in 2D turbulence is transported like a scalar, and we notice that the k^{-3} energy spectrum corresponds indeed to a k^{-1} enstrophy spectrum [57]. Of course the strain is not limited to large scales: the strain produced by flow structures at scale k^{-1} can be estimated as $uk \sim (k^3 E(k))^{1/2}$. This strain is independent of k for a k^{-3} energy spectrum so the contribution of all scales is the same, and the "nonlocal interactions" with the large scale strain is dominant (unlike in the $k^{-5/3}$ spectrum), but only marginally. The precise analysis [37] gives a k^{-3} energy spectrum corrected by a logarithmic factor. Kraichnan also proposed [57] such a logarithmic correction to avoid some divergence in the calculations.

• Observations of the cascades

It is difficult to simultaneously observe the two cascades due to the required spatial resolution and high Reynolds number: observing two decades for each cascade in a numerical computation would require at least 10^4 grid points in each direction. Simulations or laboratory experiments must be optimized to study one of the cascades.

The existence of the inverse energy cascade is now well established both from direct numerical simulations [42, 96] and from laboratory experiments [76, 83, 97]. Note that the experiments of Sommeria [97] have been performed with a steady forcing (in a square vortex lattice), and the inverse cascade is therefore spontaneously generated, see Figure 11. Spontaneous generation of an inverse energy cascade has been also observed with an hexagonal lattice, see Figure 12, although a significant steady flow component coexists with the turbulence, unlike with the square lattice, more efficient at generating 2D turbulence (see also Sect. 3.3). In contrast, both the experiments of Paret and Tabeling [83] and the numerical simulations have been performed with some random forcing. The Kolmogorov constant found is about 7, which means that this cascade is less "efficient" than in 3D (for which $C = 1.5$): for a given value of the spectrum $E(k)$ the transfer rate ϵ is smaller than in 3D. Notice that this value of the coefficient fits well with the prediction of Kraichnan using the test field closure model [57].

Remarkably, intermittency seems absent [83, 96], or at least very weak: the successive moments of the two-point velocity difference $\langle (\delta \mathbf{u})^n \rangle$ scale with point separation r in $r^{n/3}$. The ideas of Kolmogorov (1941) turns out to be more appropriate for 2D than for 3D turbulence! Furthermore the probability distribution for $\delta \mathbf{u}$ is close to a Gaussian at all scales. It cannot be exactly a Gaussian as the energy transfer is associated with a non zero third order moment, as seen above. However the cascade is less efficient than in 3D (the Kolmogorov constant is larger) so we can understand that the system is closer to an equilibrium with Gaussian statistics.

To get a steady cascade in a finite size domain, some energy dissipation acting at large scale is necessary. In laboratory experiments the friction on the support of the fluid layer, proportional to velocity, plays this role. In numerical simulations, a large scale dissipation consistent with the inverse energy cascade must be chosen [101]. When the inverse cascade is limited

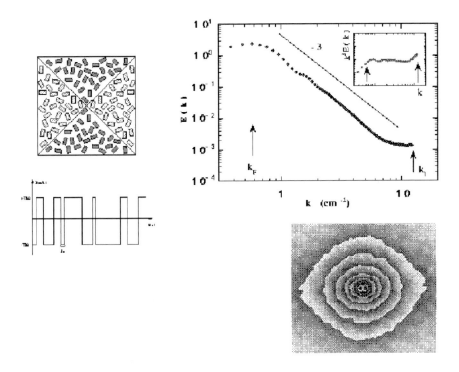

Fig. 13. Laboratory observation of the direct enstrophy cascade in a thin water layer, stratified in density by salinity to restrict 3D recirculation (from [82]). The flow is generated by a set of magnets (sketch on the upper left) interacting with a uniform horizontal electric current randomly switched in time (curve below). The flow is thus generated at large scale allowing to focus the study on the enstrophy cascade. Spectra in k^{-3} are indeed obtained (upper right), while the spectral isocontours below illustrate the isotropy of the cascade, although the forcing at small k is not isotropic, as revealed by the two peaks near the center.

by the domain size instead, the cascade breaks down and condensation of energy in the lowest mode is obtained as predicted above from truncated spectral equilibria. This condensation has been well observed both in laboratory experiments [83,97] (see Fig. 11) and numerical computations [96].

The existence of the enstrophy cascade is also confirmed by laboratory experiments [82], represented in Figure 13. Note that measurements of the steep k^{-3} energy spectra over large wavenumber ranges is difficult as it requires a very high precision to distinguish the small scale fluctuations from the much stronger large scales (obtaining two decades requires a precision on velocity better than 10^{-3}). Numerical simulations with high resolution [5], up to 4096^2 provide more extended spectral ranges, with a constant enstrophy flux over 2 decades. The k^{-3} energy spectrum has logarithmic corrections, while a nearly perfect k^{-3} spectrum has been obtained with a slightly different forcing [67]. Note that earlier simulations [63] point out that the presence of coherent vortices modifies the cascade, leading to steeper energy spectra. The permanent production of random vorticity tends however to prevent the build up of the coherent vortices [5], unlike in freely decaying flows discussed next.

4.3   The enstrophy cascade of freely evolving turbulence

Unlike in 3D, the behavior of freely evolving 2D turbulence is quite different than that of forced turbulence. This difference can be understood by considering that the dynamical time for eddies with scale k^{-1} is in $(k^3 E(k))^{-1/2}$, so that in a $k^{-5/3}$ energy spectrum the small scales have a much shorter dynamical time than the large scales (it is in $k^{-2/3}$). Therefore in a direct energy cascade the effect of large scales is reasonably similar to a permanent forcing for the smaller scales. This is not true for the effect of small scales on the large scales in the inverse energy cascade.

A classical prediction for a freely evolving turbulence have been given by Batchelor in 1969, assuming a self-similar spectral evolution, with a (single) time scale and length scale evolving in time. An enstrophy cascade is still predicted, with a spectrum in $t^{-2}k^{-3}$. The inverse energy cascade is then replaced by a growth of the integral scale, such that the total enstrophy is predicted to decay in t^{-2}.

Energy spectra close to k^{-3} are indeed obtained in laboratory experiments or numerical simulations of freely decaying 2D turbulence. Experimentally, it has been first measured in grid turbulence in liquid metal duct flows submitted to a transverse magnetic field [55]. This was the first experimental observation of 2D turbulence ever reported. Recent measurements in soap films yield similar results [71]. Numerical simulations starting with some random field at wavenumber k_{I} first show the onset of a k^{-3} energy spectral range, when the enstrophy dissipation is maximum [8,9]. However the spectra tend to become progressively steeper as isolated vortices form, see below.

Numerical simulations [16], as well as laboratory experiments [102], all indicate that enstrophy decays much less rapidly than the Batchelor's prediction in t^{-2} (the exponent found in various cases is in the range $0.3 - 1$ instead of 2). This global decay is related to the spectral behavior in the limit of small wavenumbers. Recent numerical simulations [80] allowing a good resolution at small wavenumber confirms a behavior in k^3 predicted with closure models [3]. This probably depends however on assumptions on the initial condition [32]. In laboratory experiments such large scales are clearly less universal, and more difficult to control than the small scale behavior.

4.4   The emergence and evolution of isolated vortices

The emergence of isolated vortices is a fascinating aspect of 2D turbulence. This phenomenon has been first documented by Mc Williams [107], and then obtained in many similar numerical computations. Most vortices are monopoles but some dipoles, and even tripoles, can be temporarily formed. Such vortex formation has been observed as well in laboratory experiments with thin water layer [102] and more spectacularly in electron plasma experiments [40], where the vorticity field is directly visualized as electron density. This organization can be explained in terms of statistical mechanics, see Section 5, as a local equilibrium around an initial vorticity maximum.

Once formed these vortices interact and merge, so they become fewer and fewer, while their maximum vorticity only weakly decays. Therefore they dominate more and more the background vorticity, which decays by enstrophy cascade. Note that some contour dynamics computations [34] indicate the possibility of formation of new secondary vortices of various sizes by the reorganization of vorticity filaments resulting from vortex interactions. Persistence of non-axisymmetric vortices is also observed in such computations. These results are however not supported by electron plasma experiments [75], as well as more classical numerical computations with increasing resolution [8]: significant vortex formation only occurs during the initial stage of organization from the random initial condition. Statistical mechanics also indicates a general trend for merging and reorganization into axisymmetric vortices.

The formation of organized vortices has a clear consequence in terms of global statistics. In particular the Kurtosis $Ku = \langle \omega^4 \rangle / \langle \omega^2 \rangle^2$ considerably increases with time: it is just the inverse of the area proportion occupied by the vortices. Starting from a value $Ku = 3$ characterizing the Gaussian statistics of the initial condition, a typical value of $Ku = 50$ can be reached, which characterizes a highly intermittent field: the distribution of vorticity levels has big "tails" corresponding to a significant probability of finding vorticity values much larger than the vorticity root mean square. Steepening of the energy spectrum beyond the k^{-3} prediction is also observed to be associated with vortex predominance, while k^{-3} spectra are observed in regions outside the vortices [39].

The condition of formation of such organized vortices is still a debated question. They are best observed for an initial condition spectrum with a dominant peak, while a wideband range of scales (like a $k^{-5/3}$ inertial range) tends to prevent their formation. The conditions at large scales seem to have a significant influence. The effect of the dissipation operator at small scale is also important: the classically used hyperviscous operators can spuriously enhance the vortex influence (the peak vorticity can increase), while the usual viscosity makes then wider. "Contour surgery" methods lead to different results as mentioned above. Well tested high resolution numerical studies are still needed to make safe claims about the asymptotic inertial limit, in particular about the statistical distribution of vortex strength and size.

The evolution of the vortex population is an interesting theoretical issue. At moderate Reynolds numbers, the vortices have all a similar size and strength. At higher Reynolds numbers, vortices with different sizes coexist, but their probability distribution seems to reach a steady shape [8], so that further evolution is controlled in all cases by an increase of the typical vortex radius r_a and a decrease of the vortex number N. Power law evolutions $N \propto t^{-\xi}$ and $r_a \propto t^{\xi/4}$ are observed for these quantities, while the typical peak vorticity remains constant. The relation between the exponent for vortex radius and vortex number is justified by the conservation of energy. This indeed implies the conservation of the typical flow velocity, which is induced by the vortices with a scaling in γ/l, where $\gamma \propto r_a^2$ is the vortex circulation and $l \propto N^{-1/2}$ is the typical vortex separation. Therefore a constant typical velocity indeed implies that $N r_a^4 \propto$ const. The total vortex area $N r_a^2 \propto t^{-\xi/2}$, then decreases with time, so that some vorticity is mixed away in the background during merging. Since enstrophy becomes dominated by the vortices, this total vortex area is just proportional to the enstrophy.

An exponent $\xi \simeq 0.7$ is obtained in direct numerical simulations [8] and laboratory experiments [102], and it is reproduced by a model of "punctuated dynamics" [14]: a set of N point vortices interact according to the Kirchhoff equations (3.2, 3.3), and a merging rule is used when two vortices get closer than their radius. Such a model has been recently improved [94] using a procedure of "numerical renormalization" to reach much longer evolution times: after each merging, the domain of computation is increased, introducing randomly a new vortex to keep the same vortex density. This allows to keep a constant vortex number in the computations, while previous methods required a very large initial vortex number to get good statistics at later times. These calculations agree with the previous ones at early times, but a progressive increase of the exponent ξ is observed, with a final value $\xi = 1$.

This final value can be understood by an elementary kinetic model, considering vortices like atoms in ballistic motion. The rate of binary collisions is then proportional to the square $N^2 \propto t^{-2\xi}$ of the vortex density, to the vortex size $r_a \propto t^{\xi/4}$ (its cross-section of interaction) and to the constant vortex velocity. This would imply $dN/dt \propto t^{-(7/4)\xi}$ so that $\xi = 4/3$, clearly inconsistent with the results. However Sire and

Chavanis [94] show that three-body collisions are necessary to get merging. Typically a vortex dipole (two opposite sign vortices) with separation $\propto r_a$ collides with a monopole. In that case the pair translates at velocity $\gamma/r_a \propto r_a$. Assuming uncorrelated random pair formation, the density of dipoles is $\propto N \times N r_a^2$ (the vortex density multiplied by the probability of finding another vortex at a distance $\propto r_a$). The collision probability therefore becomes $dN/dt \propto N \times N^2 r_a^2 \times r_a \times r_a$ (the successive product of the vortex density, the dipole density, the collision cross section r_a, and the dipole velocity). This yields the exponent $\xi = 1$.

This work therefore clarifies the problem of punctuated vortex dynamics. Its practical relevance for 2D turbulence is however remote, as it is limited to very large time scales and domain sizes, with accordingly extremely high Reynolds numbers. Furthermore the question of energy spectra is open. Assuming random uncorrelated point vortices completely determines in principle the statistics of the velocity field [20]. In particular the energy spectrum is [78] in k^{-1} for $k > 1/l$, where l is the typical distance between vortices (this can be understood by remarking that the Fourier transform of a Dirac function is constant, leading to an enstrophy spectrum in k). Therefore some correlations between vortex positions must occur to explain the steeper spectra numerically observed. Statistical mechanics of point vortices, beyond the mean field approximation discussed in Section 3.4, should be relevant there.

5 Equilibrium statistical mechanics and self-organization

5.1 Statistical mechanics of non-singular vorticity fields

We have seen in Section 3.4 that the statistical mechanics of point vortices explains self-organization of 2D turbulence into large steady coherent structures. However we have noted that the modeling of continuous flows by point vortices leads to some difficulties. A solution to this problem has been proposed by Kuz'min (1982), rediscovered and justified by Robert [85], Robert and Sommeria [88], and independently by Miller [74]. This equilibrium statistical theory is performed directly on the Euler equations[5]. Then, the standard procedure for Hamiltonian systems of particles is not available, but the method is still justified (on a weaker basis) by a set of rigorous properties [90]. The result is again a steady solution of the Euler equations, on which fine scale vorticity fluctuations are superimposed. The relationship between vorticity and streamfunction is different that of the point vortex

---

[5]A similar statistical mechanics had been previously proposed [68] for the Vlasov equation used to describe the organization of galaxies with stellar dynamics. The analogies with the Euler equations have been put forward only recently [18, 25].

model, and it is now quite consistent with the properties of the Euler equations with nonsingular vorticity.

• The macroscopic description

The Euler equations are known to develop very complex vorticity filaments, at finer and finer scales, and a deterministic ("microscopic") description of the flow would require a rapidly increasing amount of information as time goes on. We are rather interested in some local vorticity average $\bar{\omega}$. However to keep track of the conservation laws, we need to introduce a more precise "macroscopic" description, as the probability $\rho(\mathbf{r}, \sigma)$ of finding the vorticity level σ in a small neighborhood of the position $\mathbf{r}$ (this is a Young's measure in mathematical terms). The locally averaged vorticity field is then expressed in terms of this probability density as:

$$\bar{\omega}(\mathbf{r}) = \int \rho(\mathbf{r}, \sigma)\sigma d\sigma. \tag{5.1}$$

This probability can be viewed as the local area proportion occupied by each vorticity level σ, and it must satisfy at each point the normalization condition:

$$\int \rho(\mathbf{r}, \sigma)d\sigma = 1 \tag{5.2}$$

and the associated (macroscopic) stream function satisfies in the fluid domain ($\mathcal{D}$):

$$\bar{\omega} = -\Delta\psi \quad \text{with} \quad \psi = 0 \quad \text{on} \quad (\partial\mathcal{D}). \tag{5.3}$$

Note that since the streamfunction is expressed by space integrals of vorticity, it smoothes out the local vorticity fluctuations, supposed at very fine scale, so ψ has negligible fluctuations.

It is then possible to express the conserved quantities as integrals of the macroscopic fields. A first set of conserved quantities is the global probability distribution of vorticity $\gamma(\sigma)$ (i.e. the total area of each vorticity level):

$$\gamma(\sigma) = \int \rho(\mathbf{r}, \sigma)d^2\mathbf{r}. \tag{5.4}$$

As a consequence the integral of any function of the vorticity is conserved (the vorticity elements are just rearranged within the bounded fluid domain as time goes on).

The energy (2.16) is also conserved. As discussed above the streamfunction can be considered as smooth, so we can express the energy in terms of the locally averaged vorticity:

$$E = \frac{1}{2} \int \psi\bar{\omega}d^2\mathbf{r}. \tag{5.5}$$

In a domain with rotational or translational symmetries, additional quantities are conserved like the angular momentum in the disk, as discussed in Section 2.3.

• Entropy maximization

As in usual statistical mechanics, for instance in Section 3.4, we need to determine the entropy ("counting" the associated microscopic states) of a given macroscopic state. The macroscopic state which maximizes the entropy, with the constraint of the conserved quantities, will be the most likely to result from complex stirring. The expression of the entropy is the usual mixing entropy,

$$S = - \int \rho(\mathbf{r}, \sigma) \ln \rho(\mathbf{r}, \sigma) \mathrm{d}^2 \mathbf{r} \mathrm{d}\sigma. \tag{5.6}$$

The difference with point vortices lies in the local normalization condition (5.2): we count the possible rearrangements of small vorticity parcels which exclude each other on a given area unlike point vortices[6].

We therefore maximize the entropy with the constraints (5.4, 5.5) and (5.2) due to the conserved quantities and to the normalization. This variational problem is treated by introducing the corresponding Lagrange multipliers β, $\alpha(\sigma)$, $\zeta(\sigma)$ so that the first variations satisfy:

$$\delta S - \beta \delta E - \int \alpha(\sigma)\delta\gamma(\sigma)\mathrm{d}\sigma - \int \zeta(\mathbf{r})\delta\left(\int \rho(\mathbf{r}, \sigma)\mathrm{d}\sigma\right)\mathrm{d}^2\mathbf{r} = 0. \tag{5.7}$$

By analogy with usual thermodynamics, β can be viewed as the inverse temperature and $\alpha(\sigma)$ the chemical potential of species σ. Introducing the expressions (5.6) and (5.5) of entropy and energy, (5.7) becomes $\int [\ln \rho + 1 + \alpha(\sigma) + \zeta(\mathbf{r}) + \beta\sigma\psi] \, \delta\rho \, \mathrm{d}^2\mathbf{r}\mathrm{d}\sigma = 0$. This has to be satisfied for any variation $\delta\rho$, implying that the integrand vanishes. The resulting optimal probability density $\rho(\mathbf{r}, \sigma)$ is therefore related to the equilibrium streamfunction ψ by the relationship:

$$\rho(\mathbf{r}, \sigma) = \frac{1}{Z} g(\sigma) e^{-\beta\sigma\psi} \tag{5.8}$$

where $g(\sigma) \equiv e^{-\alpha(\sigma)}$ and $Z \equiv e^{\zeta(\mathbf{r})}$.

This is like in the case of point vortices, the vorticity level σ replacing the elementary vortex circulation γ. However the additional normalization

---

[6]This entropy can be further justified by considering the system as the limit of a series of spectrally truncated approximations of increasing resolution [90].

constraint (5.2) has to be satisfied at each point, which leads to:

$$Z(\psi) = \int g(\sigma) e^{-\beta\sigma\psi} d\sigma \qquad (5.9)$$

so that Z is a function of ψ, which we call the partition function by analogy with usual statistical mechanics. The locally averaged vorticity (5.1) is then expressed as a function of the streamfunction:

$$\overline{\omega} = \frac{\int g(\sigma)\sigma e^{-\beta\sigma\psi} d\sigma}{\int g(\sigma) e^{-\beta\sigma\psi} d\sigma} = -\frac{1}{\beta Z}\frac{\partial \ln Z}{\partial \psi} \equiv f_{\beta,g}(\psi) \qquad (5.10)$$

and the resulting flow can be calculated by solving the corresponding partial differential equation:

$$-\Delta\psi = f_{\beta,g}(\psi) \quad \text{with} \quad \psi = 0 \quad \text{on} \quad (\partial\mathcal{D}). \qquad (5.11)$$

Like in the point vortex case, random mixing yields a steady solution of the Euler equations once the local vorticity fluctuations have been averaged.

The parameters β and $g(\sigma)$ are indirectly determined by the conservation laws, and we call the resulting solutions of (5.11) the Gibbs states. This is only a necessary condition for a true statistical equilibrium: in addition the second variation of the entropy must be negative. A good way to select such maxima is to use a relaxation algorithm which increases the entropy while preserving the conserved quantities. The relaxation equations of Section 6 fulfil this goal, provided an appropriate discretization is implemented. A relaxation algorithm in discrete steps has been also implemented [106].

5.2   The Gibbs states

• Case of vortex patches

In the case of an initial condition made of patches with vorticity 0 or a, we can write ρ in terms of the Dirac distribution δ as $\rho(\sigma, \mathbf{r}) = p(\mathbf{r})\delta_{(\sigma-a)} + (1 - p(\mathbf{r}))\delta_{(\sigma)}$ involving the local area proportion $p(\mathbf{r})$ of the level a and the complementary $1 - p$ for the level 0. Then the result (5.11) reduces to

$$-\Delta\psi = pa = a\frac{e^{-\beta a\psi}}{g_0 + e^{-\beta a\psi}} \quad \text{with} \quad \psi = 0 \quad \text{on} \quad (\partial\mathcal{D}). \qquad (5.12)$$

Making a formal analogy with quantum gas statistics, this can be called a Fermi–Dirac distribution by contrast with the Boltzmann relation (3.10) for point vortex statistics. Similarly the local exclusion of the vorticity patches results in a saturation of the vorticity at the unmixed value a, when $\exp(-\beta a\psi) \gg g_0$. In the opposite limit, the relation (5.12) reduces to the point vortex result (3.10). We call it the dilute limit [88] as it corresponds to a small initial vorticity area which has been diluted among the dominant irrotational fluid.

For practical calculations in the more general case, the initial vorticity field has to be discretized in vorticity levels, so that the function $f_{\beta,g}$ can be expressed by sums of exponential terms at both numerator and denominator. The result generally converges already well when just a few levels are used.

An example of statistical equilibria is represented in Figure 14. The geometry is a channel, with periodic boundary conditions along x (which can be viewed as a simplified representation of an annulus). The mixing of a single level vorticity patch with a given initial area A is considered (here $A = 1/10$ of the total surface). The accessible energy is then restricted between a lower and an upper bound. At the lower bound, the vorticity is pushed to the walls, without any possibility of mixing, so the entropy remains equal to zero. At the upper bound, mixing is also forbidden, and the vorticity forms a central patch. This state breaks the translational symmetry. The branch of x-independent states has lower entropy and is not a maximum beyond the bifurcation (it is numerically obtained by suppressing all x-dependent perturbation). The entropy *versus* energy has a bell shape curve, whose slope is the inverse temperature β, equal to $+\infty$ at the low energy bound and to $-\infty$ at the high energy bound. Between these two bounds, the entropy reaches a maximum with $\beta = 0$, corresponding to a complete mixing, with a uniform coarse grained vorticity.

The point vortex mean field equilibrium is obtained from the present result by taking the limit of a small area A (for a fixed energy). An explicit family of solutions is then available in this channel geometry: the Stuart vortices [98]. This point vortex statistics leads to a similar bell shaped curve (like in Fig. 7), but without energy bound: the vortices can concentrate without limit, in contradiction with the conservation of the maximum vorticity.

Note that for negative temperature states, the equilibrium structure is self-confined along the transverse direction by energy conservation. The lateral walls have no influence (unlike in the jet case with zero global circulation represented in Fig. 8). In contrast the x wise periodicity sets the scale of the bifurcated vortex state. The Gibbs state equation (5.12) has also solutions with smaller x-wise periods, but they are not entropy maxima: the largest scale is always preferred, which justifies the tendency for vortex merging and growth of the free shear layer.

• General properties of the Gibbs states

For any global distribution $\gamma(\sigma)$ of vorticity levels, the accessible energy is restricted between a lower and an upper bound [13]. At the upper bound,

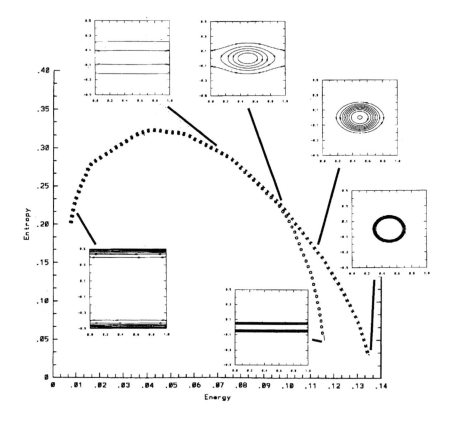

Fig. 14. Statistical equilibrium for an initial single level vorticity patch in a channel with periodic conditions along x (from [106]). The entropy is plotted *versus* energy, and representative vorticity fields $\overline{\omega}$ are given (isovorticity contours). For sufficiently high energy, vorticity clumps in a large vortex, breaking the translational symmetry.

any vorticity mixing is forbidden, and the entropy remains equal to zero, with an inverse temperature $\beta = -\infty$. The behavior at the lower bound depends on the total circulation Γ. For a non-zero circulation, the entropy *versus* energy has a bell shape curve, as in Figure 14. At its maximum, $\beta = 0$, the coarse grained vorticity is uniform. With $\Gamma = 0$, a state of zero energy can be reached, with complete mixing, $\overline{\omega} = 0$, so the entropy is maximum for $E = 0$. It decreases with increasing energy, so the inverse temperature β is always negative. This is the case for instance in a doubly periodic domain, for which $\Gamma = 0$ by construction.

General properties of the function $f_{\beta,g}$ can be shown [88]. First of all it is always bounded between the minimum and maximum values of the initial vorticity, which is expected since the vorticity $\overline{\omega}$ results from a *mixing* of the vorticity levels. Secondly $f_{\beta,g}$ is always a monotonic function of ψ. This is most easily shown by differentiating the expression (5.10) of $f_{\beta,g}$, which yields an expression for the local enstrophy[7]

$$\omega_2 \equiv \overline{\omega^2} - \overline{\omega}^2 = \frac{\int g(\sigma)\sigma^2 e^{-\beta\sigma\psi}d\sigma}{\int g(\sigma)e^{-\beta\sigma\psi}d\sigma} - \left(\frac{\int g(\sigma)\sigma e^{-\beta\sigma\psi}d\sigma}{\int g(\sigma)e^{-\beta\sigma\psi}d\sigma} \right)^2 = -\frac{1}{\beta}f'_{\beta\,g}(\psi).$$

$$(5.13)$$

Since $\omega_2 \geq 0$, the derivative $f'_{\beta,g}$ is of the sign opposite to β: $f_{\beta,g}$ is an increasing function for negative temperature and a decreasing function for positive temperature.

It can be shown [88] that the Gibbs state equation (5.11) has a unique solution for β larger than a negative bound, so there is no bifurcation, in particular for $\beta > 0$. The Gibbs state is then nonlinearly stable in the sense of Arnold. In practice stability is observed in all cases for the maximum entropy states, but there is no available demonstration in the presence of bifurcation. The statistical equilibria are also stable in all cases with respect to further mixing [88]: if we smooth out the fluctuations, taking $\overline{\omega}$ as the new initial state, the final state remains unmixed.

*Linearized cases and minimum enstrophy

Near the maximum of the entropy *versus* energy, $\beta \simeq 0$, so one can linearize the function $f_{\beta,g}$, and (5.11) becomes a linear (Helmholtz) equation. This "limit of strong mixing" provides nice possibilities for analytical results and classification of the bifurcations [21, 23]. Furthermore, the Gibbs state then depends only on the normalized energy E/Γ_2 and circulation $\Gamma/\Gamma_2^{1/2}$. Expansion of $f_{\beta\,g}$ in powers of ψ can be performed, and each successive term depends on successive higher moments of the vorticity. Therefore the statistical equilibrium for strong mixing does not depend on the detail distribution of the vorticity levels (only on the normalized energy and circulation), but it becomes more and more dependent as the mixing is prevented by energy conservation.

There is also a different possibility for obtaining a linear relationship $f_{\beta\,g}$ between vorticity and streamfunction. It corresponds to a Gaussian function $g(\sigma)$, as easily checked by substitution in (5.10). It corresponds

---

[7]This can be viewed as a relation between fluctuations and polarizability, like in magnetism [24]. Similarly the successive moments are related to the successive higher derivatives of $f_{\beta,g}$ [12].

to a particular distribution of vorticity levels $\gamma(\sigma)$, which depends on the energy.

A linear relationship between vorticity and streamfunction is also obtained by a principle of "minimum enstrophy" [62] or "selective decay" [45]. The rational is that the enstrophy decays in the limit of small viscosity, while energy, and possibly other robust integrals like the angular momentum, remain constant. Then a natural idea is that the system evolves until it minimizes its enstrophy for a given energy (and possibly other constraints). This yields a linear relationship between vorticity and stream function. This prediction is good in some cases, but not of general validity. For instance in the case of electron plasma, with vorticity always positive (proportional to the electron density), this can yield spurious negative vorticity, in the absence of an additional constraint [11]. The point of view of the statistical theory is that part of the initial enstrophy Γ_2 is irreversibly transferred into fine grained (microscopic) vorticity fluctuations, so that the final coarse grained enstrophy $\Gamma_2^{c.g} = (1/2) \int \overline{\omega}^2 d^2\mathbf{r}$ is always smaller than Γ_2,

$$\Gamma_2^{c.g} \equiv \frac{1}{2} \int \overline{\omega}^2 d^2\mathbf{r} = \frac{1}{2} \int \overline{\omega^2} d^2\mathbf{r} - \frac{1}{2} \int (\overline{\omega^2} - \overline{\omega}^2) d^2\mathbf{r} < \Gamma_2. \qquad (5.14)$$

However $\Gamma_2^{c.g}$ is truly minimized only in the linearized cases. In conclusion, a minimum enstrophy principle appears as a particular limit of entropy maximization, either in the limit of strong mixing either in the Gaussian case (see [21] for details).

5.3 Tests and discussion

A first test of the statistical mechanics predictions is shown in Figure 15 by comparison with numerical simulations of the Navier–Stokes equations at low viscosity. The shear flow in a channel with periodic boundary conditions develops vortices which self-organize in a steady flow after complex evolution. When plotted on a scatter-plot of the vorticity *versus* streamfunction, the points of the field collapse on a curve, confirming that the flow approaches a steady solution of the Euler equations (although a slow decay persists due the small viscosity). The global flow structure indeed corresponds to what is predicted by statistical mechanics, as shown in Figure 14. Moreover, a linear relationship is obtained in the vortex core between $\ln[\omega/(a - \omega)]$ and ψ, where a is the vorticity of the initial vorticity strip. This linearity is equivalent to (5.12). We observe however that the agreement is limited to the region of active stirring and that little mixing occurs outside. As a consequence, the maximum vorticity remains a little larger than predicted.

Similar results have been obtained for the usual vortex merging [33]. For a jet in a channel, the states (DD) and (VC) predicted in Figure 8 have

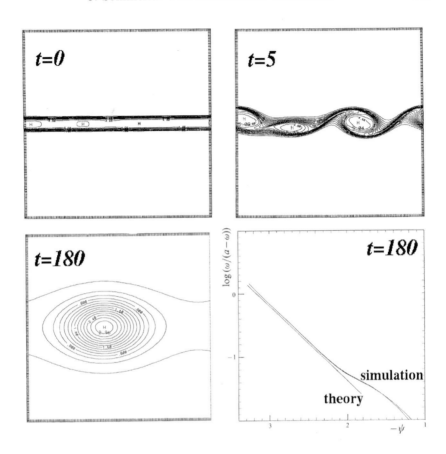

Fig. 15. Test of the statistical mechanics prediction (5.12) from direct numerical simulations of the Navier–Stokes equations (from [98]).

been remarkably checked by numerical simulations [53]. In a laboratory experiment, the organization into a single vortex by merging of a few initial vortices has been correctly predicted, while discrepancy progressively arises after successive merging events [70]. Similar conclusions arise from comparisons with simulations of turbulence in a periodic domain [10].

An explanation of this discrepancy is that viscosity modifies the distribution of vorticity levels in zones of strong strain. Under the effect of a strain s, we can estimate that the scale η of a vorticity structure exponentially decreases $\eta = \eta_0 \exp(-st)$, until smoothing occurs by viscous diffusion. This happens when the diffusion time η^2/ν equals the straining time s^{-1}, so that $\eta_0^2 \exp(-2st) = \nu/s$. Estimating $s \sim U/L$ and $\eta_0 \sim L$ from the large scale L and typical velocity U, we find that viscosity is influent after a time $t = (2s)^{-1} \ln Re$, which increases only logarithmically with the Reynolds number Re. In contrast the time for reaching the statistical equilibrium is controlled by vorticity, so that the prediction will be good only in zones where vorticity dominates strain, in the vortex core. The range of validity of the statistical theory is expected to improve with increasing Re, but only logarithmically.

Capel and Pasmanter [12] have recently proposed a model to take into account the modification of vorticity levels by a weak viscosity. In a different attempt, Turkington [105] has proposed to keep only the vorticity extrema as conserved quantities in the statistical theory, instead of the whole distribution of vorticity levels. Finally, Chavanis and Sommeria [23] have introduced the concept of "restricted statistical equilibrium", maximizing entropy in a "bag" with free boundaries. It is assumed isolated from the outside irrotational flow by kinetic restriction to mixing, due to vanishing fluctuations (see next section). The organization into steady dipoles, like observed in Figure 5, can be explained by this method. Entropy maximization would indeed make the dipole spread to infinity in the absence of restriction (while the monopoles are self-confined by energy conservation).

In laboratory experiments, boundary layer detachment can bring vorticity in the fluid interior and consequently modify the statistical equilibrium. This effect is striking in spin-up experiments in a rectangular container [47].

The formation of vortex lattices in an electron plasma disk [40] seems also in contradiction with the statistical theory: calculations with a single non-zero vorticity level predict merging in a single vortex. However, if the vortices are taken as given objects, their interaction with the background vorticity is remarkably described by statistical mechanics: both the lattice geometry and the density field in the background are quantitatively accounted [51]. These vortex lattice are probably meta-stable equilibrium states (local entropy maxima) with three vorticity levels: 0, an intermediate level in the background and the strong vortex cores. A slight change in the initial condition makes them organize in the main equilibrium state with a single vortex.

In conclusion, although various restrictions can prevent reaching the true statistical equilibrium, complex stirring clearly tends to increase the entropy, whose expression (5.6) is supported by many arguments. This will be used in next section to model the evolution of 2D turbulence.

6 Eddy diffusivity and sub-grid scale modeling

6.1 Thermodynamic approach

Diffusion processes are classically considered as a relaxation toward statistical equilibrium. The linear non-equilibrium thermodynamics states that the fluxes driving the system toward equilibrium are proportional to the gradient of the thermodynamic "potentials", for instance species concentration. We use a similar idea [89], expressed by means of a "Maximum Entropy Production" (MEP) principle, to drive eddy fluxes for 2D turbulence. The goal is two-fold: to compute the statistical equilibrium corresponding to a given initial condition, and to propose a subgrid-scale modelling for computations of the actual flow evolution with coarse resolution (LES).

Like for equilibrium states, the flow is described in terms of a local probability $\rho(\mathbf{r}, \sigma)$ of vorticity levels σ at position $\mathbf{r}$. However this probability is now assumed to evolve with time. The conservation of vorticity levels is written in terms of a transport equation for ρ, by both the explicit velocity $\mathbf{u}$ and an eddy flux $\mathcal{J}$ due to the subgrid-scales:

$$\partial_t \rho + \mathbf{u}.\nabla\rho = -\nabla.\mathcal{J}, \quad \text{with } \mathcal{J}.\mathbf{n} = 0, \quad \text{on } \partial\mathcal{D}. \tag{6.1}$$

At the wall the normal eddy flux $\mathcal{J}.\mathbf{n}$ must vanish due to impermeability. The explicit velocity $\mathbf{u}$ derives from the streamfunction ψ by (2.5), and its curl is equal to $\overline{\omega}$, related to the field ρ by

$$-\Delta\psi = \overline{\omega} = \int \sigma\rho \, \mathrm{d}\sigma, \quad \text{with } \psi = 0 \text{ on } \partial\mathcal{D}. \tag{6.2}$$

We can deduce from (6.1) an equation for the locally averaged vorticity $\overline{\omega}$ by integration over the vorticity levels σ,

$$\partial_t\overline{\omega} + \mathbf{u}.\nabla\overline{\omega} = -\nabla.\mathcal{J}_\omega \tag{6.3}$$

where we have introduced the vorticity flux

$$\mathcal{J}_\omega = \int \sigma\mathcal{J}\mathrm{d}\sigma. \tag{6.4}$$

We are mostly interested in the field $\overline{\omega}$, as the local fluctuations are in practice sensitive to viscous effects, but we cannot directly close (6.3) and, like for equilibrium states, we need to work with the probabilities ρ, solving the equations (6.1).

Of course we first need to determine the flux $\mathcal{J}$. We first express the rate of entropy increase, by time differentiating (5.6), expressing $\partial_t\rho$ by (6.1), and noting that $\rho\ln\rho$ is conserved by the advective term,

$$\dot{S} = -\int \mathcal{J}.\nabla(\ln\rho) \, \mathrm{d}^2\mathbf{r}\mathrm{d}\sigma. \tag{6.5}$$

In order to relax toward statistical equilibrium, the entropy must clearly increase with time.

In fact we determine $\mathcal{J}$ such that, for a given field ρ at each time t, $\mathcal{J}$ *maximizes* the entropy production $\dot{S}$, with the appropriate dynamical constrains, which are:

- the conservation of the local normalization (5.2), implying

$$\int \mathcal{J}\mathrm{d}\sigma = 0; \qquad (6.6)$$

- the energy conservation expressed from (5.5) and (6.3) as

$$\dot{E} = \int \mathcal{J}.\nabla\psi\mathrm{d}^2\mathbf{r} = 0; \qquad (6.7)$$

- a limitation on the eddy flux $\mathcal{J}$, characterized by a bound $C(\mathbf{r})$, which exists but is not specified.

$$\int \frac{\mathcal{J}^2}{2\rho}\mathrm{d}\sigma \le C(\mathbf{r}). \qquad (6.8)$$

A justification of this choice is that the quantity $\mathcal{J}/\rho$ can be considered as the velocity producing the flux $\mathcal{J}$, so the integral $\int \rho(\mathcal{J}/\rho)^2\mathrm{d}\sigma$ is the total energy of this diffusion velocity, a natural quantity to bound. Another justification is that it yields results consistent with the classical approach of linear non-equilibrium thermodynamics [89].

This variational problem is treated by introducing (at each time t) Lagrange multipliers, denoted $\zeta(\mathbf{r}), \beta, 1/A_E$ for the three respective constraints. It can be shown by a convexity argument that reaching the bound (6.8) is always favorable for increasing $\dot{S}$, so that this constraint can be replaced by an equality. Therefore the condition

$$\delta\dot{S} - \int \zeta(\mathbf{r})\delta\mathcal{J}\,\mathrm{d}^2\mathbf{r}\mathrm{d}\sigma - \beta\delta\dot{E} + \int \frac{1}{A_E(\mathbf{r})}\frac{\delta\mathcal{J}^2}{2\rho}\,\mathrm{d}^2\mathbf{r}\mathrm{d}\sigma = 0 \qquad (6.9)$$

must be satisfied for any variations $\delta\mathcal{J}(\sigma\,\mathbf{r})$, which yields

$$\mathcal{J} = -A_E(\mathbf{r}, t)\left[\nabla\rho + \beta\rho(\sigma - \overline{\omega})\nabla\psi\right]. \qquad (6.10)$$

The Lagrange multiplier $\zeta(\mathbf{r})$ has been eliminated, using the condition (6.6) of local normalization conservation.

The first term in the eddy flux (6.10) represents a usual diffusion: the flux of the quantity ρ is proportional to its gradient. The second term states that vorticity diffusion is constrained by the energy conservation of the induced flow: vorticity is not a passive quantity. Remembering the analogy of ψ with an interaction potential, this second term can be called a drift term, with a flux proportional to the "force" $\nabla\psi$, like sedimentation in a gravitational field.

At equilibrium, the flux must vanish, so the drift term balances diffusion. One can check that this yields again the Gibbs state (5.8), with β the corresponding inverse "temperature". During flow evolution this quantity varies and is determined by the condition of energy conservation. Introducing (6.10) in the condition (6.7) of energy conservation, we indeed obtain,

$$\beta = -\frac{\int A_E(\nabla\overline{\omega}).(\nabla\psi)d^2\mathbf{r}}{\int A_E(\nabla\psi)^2\omega_2 d^2\mathbf{r}} \tag{6.11}$$

where $\omega_2 \equiv \overline{\omega^2} - \overline{\omega}^2 \equiv \int \sigma^2\rho d\sigma - (\int \sigma\rho d\sigma)^2$ is the local enstrophy.

We have thus obtained a complete set of dynamical equations (6.1, 6.2, 6.10, 6.11), which exactly conserves the distribution of vorticity levels and energy. This system relaxes to statistical equilibrium at an optimum rate. We can express the entropy production (6.5) as $\dot{S} = \int \mathcal{J}^2(A_E\rho)^{-1}d^2\mathbf{r}d\sigma$ so the eddy diffusivity A_E must be *positive* to satisfy the condition of entropy increase. Except for its sign, the diffusivity is not determined by this thermodynamic approach: it is related to the unknown bound (6.8) on the flux.

These relaxations equations are suitable to calculate the statistical equilibrium resulting from any initial condition. Once numerically implemented, it provides a convenient way to solve the Gibbs state equation (5.8) with the appropriate constraints. Furthermore it selects an entropy *maximum* among these solutions, since it is obtained by an entropy increase[8].

For practical implementation, the simplest case is the evolution of vorticity patches with only one non-zero vorticity level $\sigma = a$. Then we have only one equation (6.1), and the vorticity is proportional to this density, $\overline{\omega} = \rho a$. Figure 16 shows an example of the evolution of a vortex ring in an annulus, compared with a high resolution numerical simulation of the same process. Although the relaxation equation smooths out the local vorticity fluctuations, it correctly handles the large scale dynamics. Moreover the final vortex remains in its statistical equilibrium, without any further diffusion. Comparisons with direct numerical simulations (DNS) in various cases, with both positive and negative vorticity patches (formation of dipoles and tripoles) show good agreement [86].

---

[8]It can be proved that stable steady solutions of the relaxation equations are indeed entropy maxima (Chavanis, in preparation).

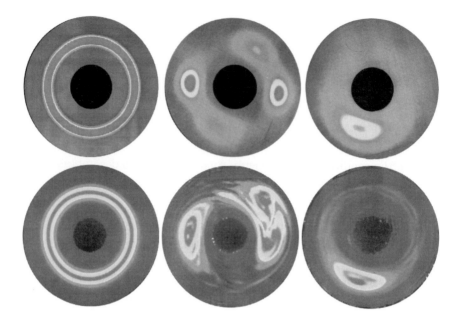

Fig. 16. Instability and final organization of a vorticity strip in an annular channel (at three successive times from left to right). The relaxation model (top), with low resolution (4000 grid points), is from Dumont, Laboratoire d'Analyse Numérique, Univ. Lyon. It is compared with a direct numerical simulation [69] (bottom).

For general initial conditions, one has to discretize the vorticity levels σ (in practice good convergence is already obtained with a few levels). The vorticity flux can be directly calculated by integration of (6.10) over the vorticity levels, $\mathcal{J}_\omega = -A_E(\mathbf{r}, t) [\nabla\overline{\omega} + \beta\omega_2\nabla\psi]$, but the local enstrophy ω_2 itself depends on the transport of the probability distribution ρ. An eddy flux of momentum, in $A_E(\mathbf{r}, t) [\nabla^2\mathbf{u} - \beta\omega_2\mathbf{u}]$, is also associated [24].

These relaxation equations have the advantage of being fully consistent with the properties of the Euler equations, and the comparison with DNS is quite good. While their justification by the MEP principle is somewhat philosophical, a more precise justification has been obtained by kinetic approaches, discussed in next section. These approaches provide estimates of the diffusion coefficient A_E, and justify the presence of the drift term, but they also stress some limitations.

Other limitations can be seen from the structure of the equations themselves. A first difficulty is that the constraint of energy conservation is only global. This is reasonable, due to the long range vortex interactions, but in a very large domain we expect that two sub-systems will evolve independently. The integral condition (6.11) should be replaced by a more local condition (the temperature should not remain uniform but diffuse with time toward an equilibrium). Furthermore the equations are not invariant by a change of reference frame, which is again problematic in a large domain. The MEP has been extended [22] to solve all these difficulties, but the model is more complex and has not been numerically implemented. Furthermore it involves additional unknown diffusivity coefficients for energy and momentum.

6.2   Kinetic models

The relaxation equations were justified in previous section by thermodynamical arguments, without discussing the mechanisms. Further insight has been recently provided by kinetic models inspired from the analogies with stellar systems and plasma physics [18].

A first approach is provided by the point vortex model, which has the advantage of fitting in the standard framework for N-body statistical mechanics. The vortices are assumed randomly positioned with the density close to the mean field statistical equilibrium. Each vortex diffuses by the random close encounters with the other vortices. Moreover each vortex induces a velocity field (added to the dominant mean field) which systematically displaces all the other vortices in its neighborhood. Chavanis [17] finds that the induced displacement (like a polarization effect in electrostatics) reacts back on the considered vortex. It results in a drift, in $-A_\mathrm{E}\beta\sigma\nabla\psi$, explaining the second term in the flux (6.10) (in the diluted case $\sigma \gg \overline{\omega}$ for which the point vortex approximation applies). At negative temperatures, the drift tends to attract together like-sign vortices, and its effect exactly balances diffusion at equilibrium. Both effects are proportional to the same coefficient A_E which can be explicitly calculated. Such a correspondence between the drift coefficient (friction) and diffusion is quite general, it is similar to the Einstein formula for the Brownian motion[9].

Coming back to non-singular vorticity fields, we can classically make a decomposition of the Euler vorticity equation (2.6) into explicit and implicit parts, $\omega = \overline{\omega} + \widetilde{\omega}$. We assume that $\overline{\omega}$ is an ensemble average, which therefore commutes with the spatial derivatives. In particular a fluctuating velocity $\widetilde{\mathbf{u}}(\mathbf{r}) = \int \widetilde{\omega}(\mathbf{r}')\mathbf{K}(\mathbf{r} - \mathbf{r}')\mathrm{d}^2\mathbf{r}'$ is induced by these fluctuations, where the Kernel $\mathbf{K}(\mathbf{r} - \mathbf{r}')$ expresses the velocity induced at point $\mathbf{r}$ by a unit singular vortex at position $\mathbf{r}'$, $\mathbf{K}(\mathbf{r} - \mathbf{r}') = (1/2\pi)\mathbf{e_z} \times (\mathbf{r} - \mathbf{r}')/|\mathbf{r} - \mathbf{r}'|^2$ plus the effect

---

[9]In this diluted case $\sigma \gg \overline{\omega}$ (or point vortex statistics), the relaxation equations are equivalent to Fokker–Planck equations describing diffusion with drift in a potential ψ.

of the image vortices near the boundary). Assuming fluctuations with a
short (Lagrangian) correlation time τ_c, the diffusion coefficient is classically
given as $A_E = (1/4)\tau_c\overline{\tilde{\mathbf{u}}^2}$ (Taylor 1921). From the previous expression of $\tilde{\mathbf{u}}$,
we get

$$\overline{\tilde{\mathbf{u}}^2} = \int \overline{\tilde{\omega}(\mathbf{r}')\tilde{\omega}(\mathbf{r}'')}\mathbf{K}(\mathbf{r} - \mathbf{r}').\mathbf{K}(\mathbf{r} - \mathbf{r}')\mathrm{d}^2\mathbf{r}'. \qquad (6.12)$$

Assuming a short correlation length ϵ for the vorticity, $\overline{\tilde{\omega}(\mathbf{r}')\tilde{\omega}(\mathbf{r}'')} = \epsilon^2\overline{\tilde{\omega}(\mathbf{r}')^2}\delta(\mathbf{r}' - \mathbf{r}'')$, we can write a local approximation of (6.12), assuming
$\overline{\tilde{\omega}(\mathbf{r}')^2} \simeq \overline{\tilde{\omega}(\mathbf{r})^2}$, $\overline{\tilde{\mathbf{u}}(\mathbf{r})^2} = \epsilon^2\overline{\tilde{\omega}(\mathbf{r})^2}\int_\epsilon^L (2\pi r')^{-1}\mathrm{d}r'$. We find that the vorticity
fluctuations at all scales contribute equally to the local velocity fluctuations,
and we have arbitrarily cut the integral between ϵ and the domain scale L.
Nevertheless the result, in $\ln(L/\epsilon)$, depends only weakly on these bounds,
so the local approximation yields a reasonable estimation,

$$A_E = \frac{\tau_c\epsilon^2}{8\pi}\ln\left(\frac{L}{\epsilon}\right)\overline{\tilde{\omega}^2}. \qquad (6.13)$$

The diffusivity is proportional to the local enstrophy $\omega_2 \equiv \overline{\tilde{\omega}^2}$ and increases
with the scale ϵ, typically the cutoff scale. The correlation time τ_c can be
estimated as $\omega_2^{-1/2}$, the time scale of the local fluctuations, or as the inverse of the strain rate of the explicit flow. Both times are of the same order
as the global time scales, so that the short correlation is only marginally
justified, like the local approximation made above. In any case, the diffusivity vanishes for weak local enstrophy fluctuations ω_2. This explains to
some extent the existence of the "restricted statistical equilibria" mentioned
above: relaxation toward equilibrium is efficient inside the active region but
weak mixing occurs at the periphery, due to the vanishing of the vorticity
fluctuations.

Chavanis [19] has proposed a more precise analysis, which furthermore
provides the drift term in addition to diffusion. He starts from the equations

$$\partial_t\overline{\omega} + \mathcal{L}\overline{\omega} = -\nabla.\overline{\tilde{\mathbf{u}}\tilde{\omega}}, \quad \text{with } \mathcal{L} \equiv \overline{\mathbf{u}}.\nabla \qquad (6.14)$$

$$\partial_t\tilde{\omega} + \mathcal{L}\tilde{\omega} = -\tilde{\mathbf{u}}\nabla\overline{\omega} \qquad (6.15)$$

similar to the quasi-linear approximation made in plasma physics. The first
one is the classical Reynolds averaged Euler equation (2.6), with an eddy
flux $\mathcal{J} = \overline{\tilde{\mathbf{u}}\tilde{\omega}}$. The second one is the equation for the fluctuations $\tilde{\omega} = \omega - \overline{\omega}$,
neglecting two terms, $-\tilde{\mathbf{u}}\nabla\tilde{\omega}$ and $\overline{\tilde{\mathbf{u}}\nabla\tilde{\omega}}$, assumed of weak influence on $\overline{\omega}$. In
contrast $\tilde{\mathbf{u}}\nabla\overline{\omega}$ builds up the eddy flux with a term $\partial_t\overline{\tilde{\mathbf{u}}\tilde{\omega}} = \overline{\tilde{\mathbf{u}}^2}\nabla\overline{\omega}$ producing
diffusion in (6.14).

Assuming again a delta-correlated vorticity field, Chavanis [19] finds an
eddy flux and a drift term as integrals over the Lagrangian trajectories of

Fig. 17. Wake behind a knife blade in a soap film, the flow comes from the left, (from Rutgers (1996) (`http://www.physics.ohio-state.edu/~maarten`)).

the explicit flow $\overline{\omega}$. Quite remarkably, the resulting kinetic equation satisfies a H-theorem: the entropy (5.6) always increases with time. This result provides a new, independent, support of the statistical equilibrium approach. The expressions of the flux are in general non-local in time and space. They reduce to the local flux (6.10) only close to equilibrium. Therefore, the relaxation equations obtained by the thermodynamic approach provide the right trends, but are probably of limited accuracy to describe the flow evolution.

The initial hypothesis of random fluctuations is not easily justified, in contrast with 3D turbulence. Some simulations suggest that the implicit scales are mostly strained by the large scales, which suggests to defined $\overline{\omega}$ as a filtered field rather than an ensemble average. Then new terms appear [61] in (6.14) and (6.15). Nevertheless, on longer time scales, chaos should develop and the probabilist approach may be more appropriate to capture the long time trends of the system.

7   Conclusions

Our knowledge of 2D turbulence has made great progress in the last ten years, due in particular to the availability of high resolution computations. Simulations at resolution 4096^2 are now currently available, and simulations at resolution 512^2, which were at the forefront of research ten years ago, are accessible on a good home computer. In spite of this great progress, with a growing consensus on some aspects, many results are still under debate. There is a need for careful computations at still higher resolution, comparing various numerical methods, initial and forcing conditions. One possible reason for the observed discrepancy may be that truly asymptotic results, forgetting the details of the initial conditions, require a very long time and high Reynolds numbers. The results on punctuated vortex dynamics discussed in Section 4.4 support this point of view.

Nevertheless some robust results are obtained in laboratory experiments. Many results of 2D turbulence seem to be remarkably insensitive to weak 3D perturbations, which are always present to some extent. The availability of various laboratory experiments has been a strong motivation for the recent rise of interest and progress in 2D turbulence.

The course has been focused on two fundamental aspects, the spectral cascades and the self-organization into coherent structures. The double-cascade theory of Kraichnan has received a remarkable confirmation by experiments and simulations. A new approach [38], adapting field theoretical methods developed for the passive scalar, brings a rigorous support to the enstrophy cascade forced by random large scale motion. Concerning the inverse cascade, the quasi-Gaussian statistics raises also hope for a deeper theoretical insight. In contrast with the forced case, the freely decaying turbulence progressively deviates from the classical enstrophy cascade law, as coherent vortices emerge and become more and more isolated, increasing the intermittency. The statistics and kinetics of this vortex system represents challenging problems of current interest.

The self-organization into coherent structures is well explained by statistical mechanics of vorticity. This theory provides good predictions in cases of rapid stirring, checked in both numerical simulations and laboratory experiments. Although various limitations can prevent the system for reaching true equilibrium, it still reveals the trends of the evolution. This provides new ideas for developing LES adapted to the peculiar properties of 2D turbulence. It supports the use of an ordinary (positive) eddy viscosity, whose energy dissipation is compensated by a drift term, acting at large scale. At statistical equilibrium these two terms balance each other. These results have been recently further justified in terms of kinetic models, but the diffusion flux and drift appear in general as integrals over space and previous times, rather than local effects.

Many questions about 2D turbulence have been left aside. The problem of predictability is for instance of great interest for applications to the atmosphere, see *e.g.* [64] for a discussion using closure models. Lorenz first addressed this question with severely truncated spectral models, the only models accessible to simulation at that time, and found his famous "butterfly" effect: the exponential growth of errors associated with chaos in low dimensional systems. However when all the degrees of freedom are recovered, a new regularity occurs, of statistical nature, as illustrated by the formation of organized structures at statistical equilibrium. Then the butterfly effect disappears [87].

The transport of a scalar is another important subject not discussed in this course, but with active recent theoretical progress both in 3D and 2D (see [38]). In fact most results and questions in usual turbulence have a counterpart in 2D, for instance in the classical shear flow problems. Channel or plane boundary layer flows are stable in the 2D case. By contrast the boundary layer detachment, with Kelvin–Helmholtz instability and vortex roll-up is basically a 2D process. Its analogy with decaying homogeneous 2D turbulence has been stressed [65]. Jets or wakes tend to organize in dipoles [31], and Figure 17 shows the fascinating structure of the wake behind a knife in a soap film. Understanding the role of these coherent structures in the global statistics of these 2D flows may be helpful for the 3D cases.

Many results in 2D turbulence can be relevant to a rotating and density stratified medium, like occurring in atmosphere, oceans, and even proto-planetary disks [7]. The statistical mechanics approaches of Sections 5 and 6 have been extended to the quasi-geostrophic model [50,73,99], with application to the Great Red Spot of Jupiter [6]. Extension to the more general shallow water system [24] can be readily applied to multilayer (isopycnal) models used in oceanography.

One should finally note the fascinating analogies with different physical systems. We have seen above that an electron plasma in a magnetic field satisfy the 2D Euler equations. A neutral plasma can be described by more complex 2D models, in analogy with some geostrophic planetary flow problems. The Vlasov equation, for plasma or stellar systems, has also formal analogies [18,25] with the Euler equation, but in the 6D phase space rather than in 2D. These analogies are further motivations to better understand 2D turbulence.

References

[1] H. Aref, *Ann. Rev. Fuid Mech.* **15** (1983) 345-89.

[2] C. Bardos, *J. Math. Anal. Appl.* **40** (1972) 769-790.

[3] C. Basdevant, M. Lesieur and R. Sadourny, *J. Atmos. Sci.* **35** (1978) 1028-1042.

[4] G.K. Batchelor, *An introduction to fluid dynamics* (Cambridge University Press, 1967).

[5] V. Borue, *Phys. Rev. Lett.* **71** (1993).

[6] F. Bouchet and J. Sommeria, Emergence of intense jets and Jupiter Great Red Spot as maximum entropy structures, *J. Fluid Mech.* (submitted) [e-print physics/0003079].

[7] A. Bracco, P.H. Chavanis, A. Provenzale and E. Spiegel, *Phys. Fluids* **11** (1999) 2280-2286.

[8] A. Bracco, J.C. McWilliams, G. Murante, A. Provenzale and J.B. Weiss, *Phys. Fluids* **12** (2000) 2931-2942.

[9] M. Brachet, M. Meneguzzi and S. Sulem, *J. Fluid Mech.* **194** (1988) 333-349.

[10] H. Brands, J. Stulemeyer, R.A. Pasmanter and T.J. Schep, *Phys. Fluids* **9** (1997) 2815; comments by W.J. Matthaeus and D. Montgomery and authors' answer, *Phys. Fluids* **10** (1998) 1237-1238.

[11] H. Brands, P.H. Chavanis, R. Pasmanter and J. Sommeria, *Phys. Fluids* **11** (1999) 3465-3477.

[12] H.W. Capel and R.A. Pasmanter, *Evolution of the vorticity-area density during the formation of coherent structures in two-dimensional flows* [e-print chao-dyn/9908010].

[13] G.F. Carnevale and G.K. Vallis, *J. Fluid Mech.* **213** (1990) 549-571.

[14] G.F. Carnevale, J.C. Mc Williams, Y. Pomeau, J.B. Weiss and W.R. Young, *Phys. Rev. Lett.* **66** (1991) 2735-2738.

[15] S.A. Chaplygin, *Proc. Phys. Sec. Natural Philos. Soc.* **11** (1902) 114.

[16] J.R. Chasnov, *Phys. Fluids* **9** (1997).

[17] P.H. Chavanis, *Phys. Rev. E* **58** (1998) R1199-R1202.

[18] P.H. Chavanis, *Ann. New York Acad. Sci.* **867** (1998) 120-141.

[19] P.H. Chavanis, *Phys. Rev. Lett.* **84** (2000) 5512-5515.

[20] P.H. Chavanis and C. Sire, The spatial correlations in the velocities arising from a random distribution of point vortices, *Phys. Fluids* (to appear) [arXiv:cond-mat/0004410].

[21] P.H. Chavanis and J. Sommeria, *J. Fluid Mech.* **314** (1996) 267-297.

[22] P.H. Chavanis and J. Sommeria, *Phys. Rev. Lett.* **78** (1997) 3302-3305.

[23] P.H. Chavanis and J. Sommeria, *J. Fluid Mech.* **356** (1998) 259-296.

[24] P.H. Chavanis and J. Sommeria, Statistical mechanics of the shallow water system, *Phys. Fluids* (submitted) [e-print physics/0004056].

[25] P.H. Chavanis, J. Sommeria and R. Robert, *ApJ* **471** (1996) 385-399.

[26] A.J. Chorin, Vortex methods, in *Les Houches LVIX, Computational fluid dynamics* (North-Holland, 1993) pp. 65-109.

[27] A.J. Chorin, *Vorticity and turbulence* (Springer, 1994).

[28] G.H. Cottet, *Analyse numérique des méthodes particulaires pour certains problèmes non-linéaires*, Thèse Univ. Paris VI (1987).

[29] G.H. Cottet and Koumoutsakos, *Vortex methods* (Cambridge University Press, 2000).

[30] Y. Couder, *J. Phys. Lett. Paris* **45** (1984) 353-360.

[31] Y. Couder and C. Basdevant, *J. Fluid Mech.* **173** (1986) 225-251.

[32] P. Davidson, The role of angular momentum in isotropic turbulence: Part 2, Two-dimensional turbulence, *J. Fluid Mech.* (submitted).

[33] M.A. Denoix, J. Sommeria and A. Thess, Two-dimensional turbulence: The prediction of coherent structures by statistical mechanics, in *Progress in Turbulence Research*, edited by H. Branover and Y. Unger, (1994) pp. 88-107.

[34] D.G. Dritschel, *J. Fluid Mech.* **293** (1995) 269-303.

[35] G.L. Eyink, *Physica D* **91** (1996) 97-142.

[36] G.L . Eyink and H. Spohn, *J. Stat. Phys.* **70** (1993) 833-886.

[37] G. Falkovitch and V. Lebedev, *Phys. Rev. E* **50** (1994).

[38] G. Falkovitch, this book.

[39] M. Farge, this book.

[40] K.S. Fine, A.C. Cass, W.G. Flynn and C.F. Driscoll, *Phys. Rev. Lett.* **75** (1995) 3277-3280.

[41] U. Frisch, *Turbulence: The legacy of A.N. Kolmogorov* (Cambridge University Press, 1995).

[42] U. Frisch and P.L. Sulem, *Phys. Fluids* **27** (1984) 1921-1923.

[43] S. Gama, M. Vergassola and U. Frisch, *J. Fluid Mech.* **260** (1994) 95-126.

[44] J. Goodman, T.Y. Hou and J. Lowengrub, *Comm. Pure Appl. Math.* **XLIII** (1990) 415-430.

[45] A. Hasegawa, *Adv. in Phys.* **34** (1985) 1-42.

[46] G.J.F. van Heijst and J.B. Flor, *Nature* **340** (1989) 212-215.

[47] G.J.F. van Heijst, P.A. Davies and R.G. Davis, *Phys. Fluids A* **2** (1990) 150.

[48] G.J.F. van Heijst and Kloosterziel, *Lett. to Nature* **338** (1989) 569-571.

[49] J. Herring and J. McWilliams, *J. Fluid Mech.* **153** (1985) 229-242.

[50] G. Holloway, *J. Phys. Ocean.* **22** (1992) 1033.

[51] D.Z. Jin and D.H.E. Dubin, *Phys. Rev. Lett.* **80** (1998) 4434-4437.

[52] G. Joyce and D. Montgomery, *J. Plasma Phys.* **10** (1973) 107-121.

[53] B. Juttner, A. Thess and J. Sommeria, *Phys. Fluids* **7** (1995) 2108-2110.

[54] T. Kato, *Arch. Rat. Mech. Anal.* **25** (1967) 302-324.

[55] Yu.B. Kolesnikov and A.B. Tsinober, *Isv. Akad. Nauk SSSR Mech. Zhid. i Gaza* **4** (1974) 146.

[56] R.H. Kraichnan, *Phys. Fluids* **10** (1967) 1417-1423.

[57] R.H. Kraichnan, *J. Fluid Mech.* **67** (1975) 155-175.

[58] R.H. Kraichnan and D. Montgomery, *Rep. Prog. Phys.* **43** (1980) 547-617.

[59] G.A. Kuz'min, Statistical mechanics of the organization into two-dimensional coherent structures, in *Structural Turbulence*, Acad. Naouk CCCP Novosibirsk, Institute of Thermophysics, edited by M.A. Goldshtik (1982) pp. 103-114.

[60] H. Lamb, *Hydrodynamics* (Dover, 1932).

[61] J.P. Laval, B. Dubrulle and S. Nazarenko, Nonlocality of interaction of scales in the dynamics of 2D incompressible fluids, *Phys. Fluids* (submitted).

[62] C.E. Leith, *Phys. Fluids* **27** (1984) 1388-95.

[63] B. Legras, P. Santangelo and R. Benzi, *Europhys. Lett.* **5** (1988) 37-42.

[64] M. Lesieur, *Turbulence in fluids* (Kluwer Academic, 1990).

[65] M. Lesieur, C. Staquet, P. Le Roy and P. Comte, *J. Fluid Mech.* **192** (1988) 511-534.

[66] E. Lindborg, *J. Fluid Mech.* **388** (1999) 259-288.

[67] E. Lindborg and K. Alvelius, *Phys. Fluids* **12** (2000).

[68] D. Lynden-Bell, *Monthly Notes of Roy. Astronom. Soc.* **136** (1967) pp. 101-121.

[69] P.S. Marcus, *Vortex dynamics in a shearing zonal flow*, Vol. 215 (1990) 393-430.

[70] D. Marteau, O. Cardoso and P. Tabeling, *Phys. Rev. E* **51** (1995) 5124.

[71] B.K. Martin, X.L. Wu, W.I. Goldburg and M.A. Rutgers, *Phys. Rev. Lett.* **80** (1998) 3964.

[72] V.V. Meleshko and G.J.F. van Heijst, *J. Fluid Mech.* **272** (1994) 157-182.

[73] J. Michel and R. Robert, *J. Stat. Phys.* **77** 645-666.

[74] J. Miller, *Phys. Rev. Lett.* **65** (1990) 2137-2140.

[75] T.B. Mitchell and C.F. Dritscoll, *Phys. Fluids* **8** (1996) 1828-1841.

[76] J.M. Nguyen Duc, P. Capéran and J. Sommeria, Experimental investigation of the two-dimensional inverse energy cascade, in *Current Trends in Turbulence Research*, Vol. 112 of Progress in Astronautics and Aeronautics, A.I.A.A (1988) pp. 78-86.

[77] J.M. Nguyen Duc and J. Sommeria, *J. Fluid Mech.* **192** (1988) 175-192.

[78] E.A. Novikov, *Sov. Phys. JETP* **41** (1976) 937-943.

[79] L. Onsager, *Nuovo Cimento Suppl.* **6** (1949) 279-287.

[80] S. Ossia, *Simulation numérique des échelles infrarouges en turbulence isotrope incompressible*, Thèse INPG-Grenoble, France (2000).

[81] E.A. Overman and N.J. Zabusky, *Phys. Fluids* **25** (1982) 1297.

[82] J. Paret, M.C. Jullien and P. Tabeling, *Phys. Rev. Lett.* **83** (1999) 3418-3421.

[83] J. Paret and P. Tabeling, *Phys. Fluids* **10** (1998) 3126-3136.

[84] Y.B. Pointin and T.S. Lundgren, *Phys. Fluids* **19** (1976) 1459-1470.

[85] R. Robert, *C.R. Acad. Sci. Paris Ser. I* **311** (1990) 575-578.

[86] R. Robert and C. Rosier, *J. Stat. Phys.* **86** (1997) 481.

[87] R. Robert and C. Rosier, Long range predictability of atmospheric flows, *Nonlinear Proc. Geophysics* (in press).

[88] R. Robert and J. Sommeria, *J. Fluid Mech.* **229** (1991) 291-310.

[89] R. Robert and J. Sommeria, *Phys. Rev. Lett.* **69** (1992) 2776-2779.

[90] R. Robert, *Comm. Math. Phys.* **212** (2000) 245-256.

[91] P. G. Saffman, *Vortex Dynamics* (Cambridge University Press, 1995).

[92] D. Serre, *Physica D* **13** (1984) 105-136.

[93] C.E. Seyler, Y. Salu, D. Montgomery and G. Knorr, *Phys. Fluids* **18** (1975) 803-813.

[94] C. Sire and P.H. Chavanis, *Phys. Rev. E* (2000) 6644-6653.

[95] R.A. Smith and T.M. O'Neil, *Phys. Fluids B* **2**, 2961-2975.

[96] L.M. Smith and V. Yakhot, *J. Fluid Mech.* **274** (1994) 115.

[97] J. Sommeria, *J. Fluid Mech.* **170** (1986) 139-168.

[98] J. Sommeria, C. Staquet and R. Robert, *J. Fluid Mech.* **233** (1991) 661-689.

[99] J. Sommeria, Statistical mechanics of potential vorticity for parameterizing meso-scale eddies, in *Ocean modeling and parameterization*, edited by P. Chassignet and J. Verron, NATO Science Series C, Vol. 516 (1998) pp. 303-326.

[100] J. Sommeria, S.D. Meyers and H.L. Swinney, *Nature* **331** (1988) 1.

[101] S. Sukoriansky, B. Galperin and A. Chekhlov, *Phys. Fluids* **11** (1999) 3043-3053.

[102] P. Tabeling, S. Burkhart, O. Cardoso and H. Willaime, *Phys. Rev. Lett.* **6** (1991) 3772.

[103] H. Tennekes and J.L. Lumley, *A first course in turbulence* (MIT Press, 1974).

[104] A. Thess and J. Sommeria, Two-dimensional turbulence: transition, in *Progress in Turbulence Research*, edited by H. Branover and Y. Unger, A.I.A.A. (1994) pp. 80-87.

[105] B. Turkington, *Comm. Pure Appl. Math.* **52** (1999) 781.

[106] B. Turkington and N. Whitaker, *SIAM J. Sci. Comput.* **17** (1996) 1414.

[107] J. Mc Williams, *J. Fluid Mech.* **146** (1984) 21-43.

[108] V.I. Youdovitch, *Zh. Vych. Mat.* **3** (1963) 1032-1066.

COURSE 9

ANALYSING AND COMPUTING
TURBULENT FLOWS
USING WAVELETS

M. FARGE   K. SCHNEIDER

LMD-CNRS,   *CMI,*
École Normale Supérieure,   *Université de Provence,*
24 rue Lhomond,   *39 rue Joliot–Curie,*
75231 Paris Cedex 05, France   *13453 Marseille Cedex 13, France*

Contents

IV Conclusion   500

ANALYSING AND COMPUTING
TURBULENT FLOWS
USING WAVELETS

M. Farge[1] and K. Schneider[2]

Abstract

These lecture notes are a review on wavelet techniques for analyzing and computing fully-developed turbulent flows, which correspond to the regime where nonlinear instabilities are dominant. The wavelet-based techniques we have been developing during the last decade are explained and the main results are presented. After introducing the continuous and discrete wavelet transforms we present classical and wavelet-based statistical diagnostics to study turbulent flows. We then present wavelet methods for extracting coherent vortices in two- and three-dimensional turbulent flows. Afterwards we present an adaptive wavelet solver for the two-dimensional Navier–Stokes equations and apply it to compute a time-developing turbulent mixing layer. Finally we draw some conclusions and present some perspectives for turbulence modelling.

1 Introduction

This chapter will focus on fully-developed turbulence in incompressible flows. By fully-developed turbulence we mean the limit for which the nonlinear advective term of Navier–Stokes equations is larger by several orders of magnitude than the linear dissipative term. The ratio between both terms is defined as the Reynolds number Re, which is related to the ratio of the large excited scales and the small scales where dissipation damps any instabilities. In practically relevant applications (*e.g.* aeronautics, meteorology, combustion...) Re varies in between 10^6 and 10^{12}. For Direct

---

[1]LMD-CNRS, École Normale Supérieure, 24 rue Lhomond, 75231 Paris Cedex 05, France.

[2]CMI, Université de Provence, 39 rue Joliot–Curie, 13453 Marseille Cedex 13, France.

Numerical Simulation (DNS), where all scales are resolved, the number of degrees of freedom to be computed scales as Re for two-dimensional flows and as $Re^{9/4}$ for three-dimensional flows. Consequently one cannot integrate Navier–Stokes equations in the fully-developed turbulence regime with present computers without using some *ad hoc* turbulence model. Its role consists in reducing the dimension of the system of equations to be computed. Typically the degrees of freedom are split into two subsets: the active modes to be computed and the passive modes to be modelled. The number of active modes should be as small as possible while the number of passive modes should be as large as possible.

A classical approach to compute fully-developed turbulent flows is Large Eddy Simulation (LES) [37] where the separation is done by means of linear filtering between large-scale modes, assumed to be active, and small-scale modes, assumed to be passive. This means that the flow evolution is calculated deterministically up to the cutoff scale, whereas the influence of the subgrid scales onto the resolved scales is statistically modelled, *e.g.* using Smagorinsky's parametrization. As consequence vortices in strong nonlinear interaction are smoothed and instabilities which may develop at subgrid scales are ignored. Indeed LES models have problems to deal with backscatter, *i.e.* transfers from subgrid scales towards resolved scales due to nonlinear instabilities. The dynamical LES model [32] takes into account backscatter, but only in a locally averaged way. A further step in the hierarchy of turbulence models are the Reynolds Averaged Navier–Stokes (RANS) equations where the time-averaged mean flow is computed while fluctuations are modelled, in which case only steady state solutions are predicted. This leads to turbulence models such as k-ϵ or Reynolds stress models, extensively used in industry. It should be stressed that such low-order turbulence models are lacking universality, in the sense that one should adjust the parameters of the model from laboratory measurements for each flow configuration, and sometimes different parameters are even needed for different regions of the flow.

Turbulent flows are characterized by their unpredictability, namely each flow realization is different although the statistics are reproductible as long as the flow configuration and parameters are the same. One observes in each flow realization the formation of localized coherent vortices whose motions are chaotic, resulting from their mutual interactions. The statistical theory of homogeneous and isotropic turbulence [2, 35, 36, 43] is based on L^2-norm ensemble averages and therefore is unsensitive to the presence of coherent vortices which contribute too weakly to the L^2-norm. In opposition to this approach one can consider that coherent vortices are the fundamental components of turbulent flows [49] and therefore both numerical and statistical models should take them into account. In the present lecture notes we propose a new semi-deterministic approach which reconciles both, the statistical and the deterministic points of view. This technique is based on the space and scale decomposition of the flow using the wavelet representation.

Wavelet methods have been introduced during the last decade to analyze, model and compute fully-developed turbulent flows [6,14,17,28,29,51]. For recent overview of wavelets and turbulence, we refer the reader to [20,22,53]. The main result is that the wavelet representation is able to disentangle coherent vortices from incoherent background flow in two-dimensional turbulent flows. Both components are multiscale but present different statistics with different correlations. The coherent vortex components present non-Gaussian distribution and long-range correlation, while the incoherent background flow components are characterized by Gaussian statistics and short-range correlation [16,19,21]. This leads to propose a new way to split turbulent dynamics into: active coherent vortex modes, to be computed in a wavelet basis dynamically adapted to follow their motion, and passive incoherent modes, to be statistically modelled as a Gaussian random process. This new approach, called Coherent Vortex Simulation (CVS) [21], differs significantly from LES. LES is based on linear filtering (defined either in physical space or in Fourier space) between large and small scales, but without a clearcut separation between Gaussian and non-Gaussian behaviours. CVS uses nonlinear filtering (defined in wavelet space) between Gaussian and non-Gaussian modes having different scaling laws, but without any clearcut scale separation. The advantage of CVS method compared to LES is to reduce the number of computed active modes for a given Reynolds number [16] and to control the Gaussianity of the passive degrees of freedom to be statistically modelled [21].

These lecture notes are organized in three parts: wavelet transforms, turbulence analysis and turbulence computation. In the first part we present both the continuous and the orthogonal wavelet transforms in one and several dimensions, together with the corresponding algorithms. In the second part, after a brief review of classical statistical tools for analysing turbulence, we present wavelet-based statistical tools, such as local and global wavelet spectra and discuss their relation with the Fourier spectrum. We also introduced several wavelet-based measures to characterize and quantify the intermittency, property which is generic for turbulence. In the last part, we present a new approach, called Coherent Vortex Simulation (CVS), which computes turbulent flows for regimes where their dynamics is dominated by coherent vortices. We first show how to extract coherent vortices out of turbulent flows, and we illustrate this method for a 3D turbulent mixing layer. Then, after a brief review of classical methods for computing turbulent flows, we expose the principle of CVS. Finally we describe the algorithm for computing 2D Navier–Stokes equations in an adaptive wavelet basis and we apply it to simulate a 2D turbulent mixing layer.

Part I

Wavelet Transforms

2   History

Wavelets have been developed in the beginning of eighties in France [31]. This recent mathematical technique is based on group theory and square integrable representations, which allows the decomposition of a signal, or a field, into both space and scale, and possibly directions [14]. To motivate the use of wavelets in relation with turbulence, we briefly recall some fundamental ideas and mention several fields of applications. For details we refer the reader to textbooks [8, 39, 42]. From an abstract point of view wavelets constitute new "atoms" and "molecules", *i.e.* basic building blocks of various function spaces out of which some can be used to contruct orthogonal bases. The starting point is a function $\psi(x)$, called mother wavelet. This function is assumed to be well-localized, *i.e.* ψ exhibits a fast decay for $|x|$ tending to infinity, is oscillating, *i.e.* ψ has at least a vanishing integral (the mean value is zero) or better the first m moments of ψ vanish, and is smooth, *i.e.* $\hat{\psi}$ the Fourier transform of ψ exhibits fast decay in wavenumber space $|k|$.

The mother wavelet then generates a family of wavelets $\psi_{l,x}(x')$ by dilatation (or contraction) by the parameter $l > 0$ and translation by the parameter $x \in \mathbb{R}$, *i.e.* $\psi_{l,x}(x') = l^{-1/2}\psi(\frac{x'-x}{l})$, where all wavelets being normalized in L^2-norm. An example of such a family (for discrete l and x) is depicted in Figure 1. The wavelet transform of a function f is then defined as a convolution of the analysing wavelet with the signal f, $\tilde{f}(l,x) = \int f(x')\psi_{l,x}(x')\mathrm{d}x'$. The wavelet coefficients $\tilde{f}(l,x)$ measure the fluctuations of f around the point x and scale (frequency) l. The function f can be reconstructed as a linear combination of wavelets $\psi_{l,x}(x')$ with coefficients $\tilde{f}(l,x)$ [31], *e.g.* $f(x) = 1/C_\psi \int \int \tilde{f}(l,x)\psi_{l,x}(x')l^{-2}\mathrm{d}l\mathrm{d}x$, C_ψ being a constant which depends on the wavelet ψ. Let us mention that due to the localization of wavelets in physical space the behaviour of the signal at infinity does not play any role. Therefore the wavelet analysis and synthesis can be performed locally as opposed to the Fourier transform where a global behaviour is intrinsically implicated through the nonlocal nature of the trigonometric functions.

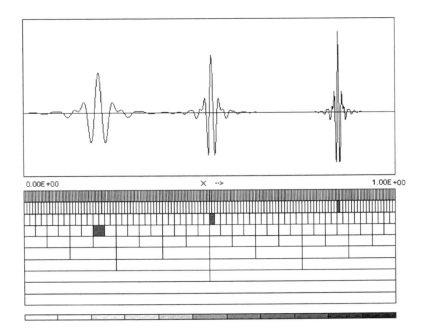

Fig. 1. Example: discrete wavelets $\psi_{j,k}(x) = l_0^{j/2}\psi(l_0^j x' - kx_0)$ with $l_0 = 2$ and $x_0 = 1$ and $l = l_0^j$, $x = kx_0 l_0^j$ for the case of orthogonal quintic spline wavelets $\psi_{5,6}(x')$, $\psi_{6,32}(x')$ $\psi_{7,108}(x')$.

The origin of wavelets is interdisciplinary. Wavelets come from different fields such as engineering (subband coding, quadrature mirror filters, time-frequency analysis), theoretical physics (coherent states of affine groups in quantum mechanics) and mathematics (Calderon–Zygmund operators, characterization of function spaces). Meanwhile a large spectrum of applications has grown and is still developing, ranging from signal or image analysis and processing, to numerical analysis.

3   The continuous wavelet transform

3.1   One dimension

3.1.1   Analyzing wavelet

Starting point for the wavelet transform is a real or complex valued function $\psi(x)$, called wavelet, which has to fulfill the admissibility condition,

$$C_\psi = \int_0^\infty \left|\widehat{\psi}(k)\right|^2 \frac{\mathrm{d}k}{|k|} < \infty \tag{3.1}$$

where

$$\widehat{\psi}(k) = \frac{1}{2\pi} \int_{-\infty}^{\infty} \psi(x) e^{-i2\pi kx} \mathrm{d}x \tag{3.2}$$

denotes the Fourier transform. If ψ is integrable this implies that ψ has zero mean,

$$\int_{-\infty}^{\infty} \psi(x)\mathrm{d}x = 0 \quad \text{or} \quad \widehat{\psi}(k=0) = 0. \tag{3.3}$$

In practice however one also requires that the wavelet ψ should be well-localized in both physical and Fourier spaces, which implies smoothness. We also require that higher order moments of ψ vanish, *i.e.*

$$\int_{-\infty}^{\infty} x^m \psi(x)\mathrm{d}x = 0 \quad \text{for} \quad m = 0, M \tag{3.4}$$

which means that monomials up to degree M are exactly reproduced. In Fourier space this property is equivalent to

$$\frac{\mathrm{d}^m}{\mathrm{d}k^m}\widehat{\psi}(k)|_{k=0} = 0 \quad \text{for} \quad m = 0, M \tag{3.5}$$

so that the Fourier transform of ψ decays smoothly at $k = 0$.

3.1.2   Wavelet analysis

From this function ψ, the so-called mother wavelet, we generate a family of continuously translated and dilated wavelets, normalized in L^2–norm

$$\psi_{l,x}(x') = l^{-1/2}\psi\left(\frac{x'-x}{l}\right) \quad \text{for} \quad l > 0 \quad \text{and} \quad x \in \mathbb{R} \tag{3.6}$$

where l denotes the scale dilation parameter, corresponding to the width of the wavelet and x the translation parameter, corresponding to the position of the wavelet.

In Fourier space this reads

$$\widehat{\psi}_{l,x}(k) = \sqrt{l}\,\widehat{\psi}(lk)e^{-i2\pi kx} \tag{3.7}$$

where the contraction with $1/l$ is reflected in a dilation with l and the translation with x implies a rotation in the complex plane.

The continuous wavelet transform of a signal f is then defined as a convolution of f with the wavelet family $\psi_{l,x}$

$$\tilde{f}(l,x) = \int_{-\infty}^{\infty} f(x')\overline{\psi}_{l,x}(x')\mathrm{d}x' \tag{3.8}$$

where $\overline{\psi}_{l,x}$ denotes in the case of complex valued wavelets the complex conjugate.

Using Parseval's identity we also get

$$\tilde{f}(l,x) = \int_{-\infty}^{\infty} \hat{f}(k)\overline{\hat{\psi}_{l,x}}(k)\mathrm{d}k \tag{3.9}$$

so that the wavelet transform may be interpreted as a frequency decomposition using band pass filters $\hat{\psi}_{l,x}$ centered at frequency $k = \frac{k_0}{l}$, where k_0 denotes the center of the wavelet in Fourier space, and having variable width $\frac{\Delta k}{k}$, so for increasing scales the bandwidth is getting wider.

3.1.3   Wavelet synthesis

The admissibility condition (3.1) of ψ implies the existence of a finite energy reproducing kernel, see *e.g.* [8], which is a necessary condition for being able to reconstruct a function from its wavelet coefficients. The signal can thus be reconstructed entirely from its wavelet coefficients,

$$f(x) = \frac{1}{C_\psi} \int_0^\infty \int_{-\infty}^\infty \tilde{f}(l,x)\psi_{l,x}(x')\frac{\mathrm{d}l\mathrm{d}x}{l^2} \tag{3.10}$$

which is the inverse wavelet transform.

3.1.4   Energy conservation

There also holds an energy conservation like for Fourier transforms, *i.e.* a Plancherel identity, which means that the total energy of a signal can be either calculated in physical space or in wavelet coefficient space,

$$\int_{-\infty}^\infty |f(x)|^2 \mathrm{d}x = \frac{1}{C_\psi} \int_0^\infty \int_{-\infty}^\infty |\tilde{f}(l,x)|^2 \frac{\mathrm{d}l\mathrm{d}x}{l^2}. \tag{3.11}$$

This formula is also the starting point for the definition of wavelet spectra.

3.2   Higher dimensions

The theory of the continuous wavelet transform can be generalized in several dimensions [44] using rotation in addition to dilatation and translation. For a two-dimensional function $f \in L^2(\mathbb{R}^2)$ (constructions for higher dimensions are analogous) we get

$$\tilde{f}(l,\vec{x},\theta) = \int\!\!\int_{\mathbb{R}^2} f(\vec{x}')\overline{\psi}_{l,\vec{x},\theta}(\vec{x}')\mathrm{d}^2\vec{x}' \quad \text{for} \quad l > 0, \theta \in [0, 2\pi[, \vec{x} \in \mathbb{R}^2 \tag{3.12}$$

where the family of functions $\psi_{l,\vec{x},\theta}$ is obtained from a single one ψ by dilatation with l^{-1}, by translation by $\vec{x}$ and by rotation of angle θ,

$$\psi_{l,\vec{x},\theta}(\vec{x}') = \frac{1}{l}\psi\left(R_\theta\left(\frac{\vec{x}'-\vec{x}}{l}\right)\right) \tag{3.13}$$

where R_θ denotes a rotation matrix.

Analoguously to the one-dimensional case this transformation is invertible and isometric, provided that ψ fulfills the admissibility condition

$$C_\psi = \int\int_{\mathbb{R}^2}\left|\widehat{\psi}(\vec{k})\right|^2\frac{\mathrm{d}^2\vec{k}}{|\vec{k}|^2} < \infty. \tag{3.14}$$

In the following we restrict ourselves to isotropic real-valued wavelets, so there is no more a dependence on θ. The wavelet coefficients can then be calculated using the formula

$$\tilde{f}(l,\vec{x}) = \int\int_{\mathbb{R}^2}\widehat{f}(\vec{k})l\overline{\widehat{\psi}(l\vec{k})}e^{i2\pi\vec{k}\cdot\vec{x}}\mathrm{d}^2\vec{k} \tag{3.15}$$

where the Fourier transform of the wavelet ψ_l is essentially supported on an annulus a radius l^{-1}.

The energy conservation reads

$$\int\int_{\mathbb{R}^2}|f(\vec{x})|^2\mathrm{d}\vec{x} = \frac{1}{C_\psi}\int_0^\infty\int\int_{\mathbb{R}^2}|\tilde{f}(l,\vec{x})|^2\frac{\mathrm{d}l\mathrm{d}^2\vec{x}}{l^3} \tag{3.16}$$

and the function f can be recontructed entirely from its wavelet coefficients by the relation

$$f(\vec{x}') = \frac{1}{C_\psi}\int_0^\infty\int\int_{\mathbb{R}^2}\tilde{f}(l,\vec{x})\psi_l(\vec{x}'-\vec{x})\frac{\mathrm{d}^2\vec{x}\mathrm{d}l}{l^3}. \tag{3.17}$$

3.3 Algorithm

To illustrate the practical implementation of the continuous wavelet transform we consider a one-dimensional signal $f(x)$ sampled on a regular grid with $N = 2^J$ points, i.e. $f(i/2^J)$ for $i = 0,...,2^J-1$ are given. We assume the signal to be periodic and compute the wavelet coefficients by means of the Fast Fourier Transform (FFT). The large scale corresponds to the domaine size which is by construction equal to 1. The smaller scales are discretized logarithmically,

$$l_j = l_0^{-j} \qquad j \geq 0. \tag{3.18}$$

The choise of l_0 is determined by the reconstruction formula to ensure a given precision for the reconstruction of the signal by a discretized version of formula (3.10). For wavelets being derivatives of Gaussians the choice of $l_0 = 2^{1/4}$ is sufficiently precise as discussed in [8]. The smallest scale of the discretization corresponds to $\frac{2}{N} = 2^{1-J}$.

To compute $\tilde{f}(l, x)$ we discretize the formula (3.9). First we compute the disrete Fourier transform of the signal samples. Then we multiply it with the wavelet filter in Fourier space and we obtain subsequently at each scale l_j for $j = 0$ to $J - 1$ simultaneously at all positions $x_n = \frac{n}{N}, n = 0, N - 1$ the wavelet coefficients by executing an inverse FFT

$$\tilde{f}(l_j, x_n) = \sum_{k=-N/2}^{N/2-1} \widehat{f_k} \sqrt{l_j} \overline{\widehat{\psi}(l_j k)} e^{+i2\pi kn/N} \tag{3.19}$$

where $\widehat{f_k}$ denotes the discrete Fourier transform of the samples $f\left(\frac{n}{N}\right)$,

$$\widehat{f_k} = \frac{1}{N} \sum_{n=0}^{N-1} f\left(\frac{n}{N}\right) e^{-i2\pi nk/N}. \tag{3.20}$$

Due to periodicity of the signal f no boundary effects are introduced by using the FFT. The complexity of the algorithm is of order $N \log N$ at each scale due to the use of the FFT.

The above algorithm can be applied analoguously in the two-dimensional case using tensor product discretization together with 2D FFTs.

4   The orthogonal wavelet transform

4.1   One dimension

In this section we recall some essential features of the discrete wavelet approximation on the real line, *i.e.* in $L^2(\mathbb{R})$ that will be important in the sequel. For an exhaustive treatment we refer the reader to [8, 39, 42].

4.1.1   1D Multi-Resolution Analysis

The discrete wavelet transform relies on the concept of Multi-Resolution Analysis (MRA) which is a sequence of imbedded subspaces V_j verifying

$$V_j \subset V_{j+1} \qquad \forall j \in \mathbb{Z} \tag{4.1}$$

$$\overline{\bigcup_{j \in \mathbb{Z}} V_j} = L^2(\mathbb{R}) \tag{4.2}$$

$$\bigcap_{j \in \mathbb{Z}} V_j = \{0\} \tag{4.3}$$

$$f(x) \in V_j \Leftrightarrow f(2x) \in V_{j+1}. \tag{4.4}$$

A scaling function $\phi(x)$ is required to exist. Their translates generate a basis in each V_j, i.e.

$$V_j = \overline{\text{span}}\{\phi_{ji}\}_{i \in \mathbb{Z}} \tag{4.5}$$

where

$$\phi_{ji}(x) = 2^{j/2}\phi(2^j x - i) \qquad j, i \in \mathbb{Z}. \tag{4.6}$$

In the classical case this basis is orthonormal, so that

$$\langle \phi_{ji}, \phi_{jk} \rangle_{\mathbb{R}} = \delta_{ik} \tag{4.7}$$

with $\langle f, g \rangle_{\mathbb{R}} = \int\limits_{-\infty}^{+\infty} f(x)\overline{g}(x)\mathrm{d}x$ being the inner product in $L^2(\mathbb{R})$. The main issue of the wavelet approach now is to work with the orthogonal complement spaces W_j defined by

$$V_{j+1} = V_j \oplus W_j. \tag{4.8}$$

Based on the function $\phi(x)$ one can find a function $\psi(x)$, the so-called mother wavelet. Their translates and dilates constitute a orthonormal bases of the spaces W_j,

$$W_j = \overline{\text{span}}\{\psi_{ji}\}_{i \in \mathbb{Z}} \tag{4.9}$$

where

$$\psi_{ji}(x) = 2^{j/2}\psi(2^j x - i) \qquad j, i \in \mathbb{Z}. \tag{4.10}$$

Each function $f \in L^2(\mathbb{R})$ can now be expressed as

$$f(x) = \sum_{i \in \mathbb{Z}} \overline{f}_{j_0 i}\phi_{j_0 i}(x) + \sum_{j=j_0}^{\infty} \sum_{i \in \mathbb{Z}} \tilde{f}_{ji}\psi_{ji}(x) \tag{4.11}$$

where

$$\overline{f}_{ji} = \langle f, \phi_{ji} \rangle_{\mathbb{R}} \qquad \tilde{f}_{ji} = \langle f, \psi_{ji} \rangle_{\mathbb{R}}. \tag{4.12}$$

In numerical applications the sums in (4.11) have to be truncated which corresponds to the projection of f onto a subspace of $V_J \subset L^2(\mathbb{R})$. The decomposition (4.11) is orthogonal, as, by construction,

$$\langle \psi_{ji}, \psi_{lk} \rangle_{\mathbb{R}} = \delta_{jl}\,\delta_{ik} \tag{4.13}$$

$$\langle \psi_{ji}, \phi_{lk} \rangle_{\mathbb{R}} = 0 \qquad j \geq l \tag{4.14}$$

in addition to (4.7).

4.1.2 Regularity and local decay of wavelet coefficients

The relation between the local or global regularity of a function and the decay of its wavelet coefficients is well known. The global regularity directly determines the error being made when the wavelet sum is truncated at some scale. Depending on the type of norm and whether global or local characterization is concerned, various relations of this kind have been developed, see e.g. [8, 39, 42] for an overview. As an example we consider the case of an α-Lipschitz function, with $\alpha \geq 1$ [33]. Suppose $f \in L^2(\mathbb{R})$, then for $[a, b] \subset \mathbb{R}$ the function f is α-Lipschitz for any $x_0 \in [a, b]$, i.e. $|f(x_0 + h) - f(x_0)| \leq C|h|^\alpha$, if and only if there exists a constant A such that $|\langle f, \psi_{ji} \rangle| \leq A2^{-j\alpha - \frac{1}{2}}$ for any (j, i) with $\frac{i}{2^j} \in]a, b[$. This shows the relation between the local regularity of a function and the decay of its wavelet coefficients in scale. The adaptive discretization discussed in the present lecture notes is precisely based on taking into account spatially varying regularity of the solution through a variable cut off in scale of its wavelet series.

4.2 Higher dimensions

This section consists of an extension of the previously presented one-dimensional construction to higher dimensions. For simplicity, we will consider only the two-dimensional case, since higher dimensions can be treated analogously [8, 39, 42]. We start with a brief description of the construction principle and then turn in more detail to the two-dimensional case with periodicity, which is relevant for the subsequent applications.

4.2.1 Tensor product construction

Having developed a one-dimensional orthonormal basis ψ_{ji} of $L^2(\mathbb{R})$ one would like to use these functions as building blocks in higher dimensions. One way of doing so is to take the tensor product of two one-dimensional bases [8] and to define

$$\psi_{j_x,j_y,i_x,i_y}(x, y) = \psi_{j_x,i_x}(x)\psi_{j_y,i_y}(y). \tag{4.15}$$

The resulting functions constitute an orthonormal wavelet basis of $L^2(\mathbb{R}^2)$. Each function $f \in L^2(\mathbb{R}^2)$ can then be developed into

$$f(x, y) = \sum_{j_x,i_x} \sum_{j_y,i_y} \tilde{f}_{j_x,j_y,i_x,i_y} \psi_{j_x,j_y,i_x,i_y}(x, y) \tag{4.16}$$

with $d_{j_x,j_y,i_x,i_y} = \langle f, \psi_{j_x,j_y,i_x,i_y} \rangle$. However in this basis the two variables x and y are dilated separately and therefore no longer form a multiresolution analysis. This means that the functions ψ_{j_x,j_y} involve two scales, 2^{j_x}

and 2^{j_y}, and each of the functions is essentially supported on a rectangle with these side lengths. Hence the decomposition is often called rectangular wavelet decomposition. This is closely related to the standard form of operators using the nomenclature of Beylkin [3]. From the algorithmic viewpoint this is equivalent to apply the one-dimensional wavelet transform to the rows and the columns of a matrix representing an operator or a two-dimensional function. For some applications such a basis is advantageous, for others not. For exemple in turbulence the notion of a scale has an important meaning and one would like to have a unique scale assigned to each basis function.

4.2.2 2D Multi-Resolution Analysis

A suitable concept which fulfills the above requirement of having a unique scale is the construction of a truly two-dimensional MRA of $L^2(\mathbb{R}^2)$. It can be obtained through the tensor product of two one-dimensional MRA's of $L^2(\mathbb{R})$ [42]. More precisely one defines the spaces $\mathbf{V}_j, j \in \mathbb{Z}$ by

$$\mathbf{V}_j = V_j \otimes V_j \tag{4.17}$$

and $\mathbf{V}_j = \overline{\mathrm{span}}\{\phi_{j,i_x,i_y}(x,y) = \phi_{j,i_x}(x)\phi_{j,i_y}(y), i_x, i_y \in \mathbb{Z}\}$ fulfilling analogous properties as in the one-dimensional case (3.1–3.4).

Likewise, we define the complement space $\mathbf{W}_j$ to be the orthogonal complement of $\mathbf{V}_j$ in $\mathbf{V}_{j+1}$, *i.e.*

$$
\begin{aligned}
\mathbf{V}_{j+1} &= V_{j+1} \otimes V_{j+1} = (V_j \oplus W_j) \otimes (V_j \oplus W_j) & (4.18) \\
&= V_j \otimes V_j \oplus ((W_j \otimes V_j) \oplus (V_j \otimes W_j) \oplus (W_j \otimes W_j)) & (4.19) \\
&= \mathbf{V}_j \oplus \mathbf{W}_j. & (4.20)
\end{aligned}
$$

It follows that the orthogonal complement $\mathbf{W}_j = \mathbf{V}_{j+1} \ominus \mathbf{V}_j$ consists of three different types of functions and is generated by three different wavelets

$$\psi^\kappa_{j,i_x,i_y}(x,y) = \begin{cases} \psi_{j,i_x}(x)\phi_{j,i_y}(y); & \kappa = 1 \\ \phi_{j,i_x}(x)\psi_{j,i_y}(y); & \kappa = 2 \\ \psi_{j,i_x}(x)\psi_{j,i_y}(y); & \kappa = 3. \end{cases} \tag{4.21}$$

Observe that here the scale parameter j simultaneously controlls the dilatation in x and in y. We recall that in d dimensions this construction yields $2^d - 1$ types of wavelets spanning $\mathbf{W}_j$.

Using (4.21) each function $f \in L^2(\mathbb{R}^2)$ can be developed into an MRA basis as

$$f(x,y) = \sum_j \sum_{i_x,i_y} \sum_{\kappa=1,2,3} \tilde{f}^\kappa_{j_x,j_y,i_x,i_y} \psi^\kappa_{j,i_x,i_y}(x,y) \tag{4.22}$$

with $d^\kappa_{j_x,j_y,i_x,i_y} = \langle f, \psi^\kappa_{j,i_x,i_y}\rangle$. The wavelets ψ^κ_{j,i_x,i_y} are the basis functions of the so-called square wavelet decomposition. The algorithmic structure of the one-dimensional transforms carries over to the two-dimensional case by simple tensorisation, *i.e.* applying the filters at each decomposition step to rows and columns. Applying this kind of transform to matrices representing operators (differential, integral, integro-differential) leads to the non-standard form in the terminology of Beylkin [3].

Remark: The described two-dimensional wavelets and scaling functions are separable. This advantage facilitates the generation of a multidimensional MRA from several one-dimensional MRA's. However the main drawback of this construction is that three wavelets are needed to span the orthogonal complement space $\mathbf{W}_j$ in two dimensions and seven in three dimensions. Another property should be mentioned. By construction the wavelets are anisotropic, *i.e.* horizontal, diagonal and vertical directions are preferred. This could be an advantage in digital signal processing to recognize corners and edges.

4.2.3   Periodic 2D Multi-Resolution Analysis

Using the tensor product construction of two-dimensional wavelets on the real line and the periodization technique, see *e.g.* [46], we now recall the essential features of periodic two-dimensional wavelets of $L^2(\mathbb{T}^2)$. For notational ease we drop from now on the tilde introduced to distinguish the periodic wavelets from those on the real line. In the latter applications the periodic basis is used throughout unless otherwise explicitly stated.

A two-dimensional MRA of $L^2(\mathbb{T}^2)$ is a sequence of embedded subspaces $\mathbf{V}_j \subset \mathbf{V}_{j+1}$, $j \in \mathbb{N}_0$. It can be obtained through the tensor product of two one-dimensional MRA's of $L^2(\mathbb{T})$ [42]. This induces a decomposition of $L^2(\mathbb{T}^2)$ into mutually orthogonal hierarchical subspaces

$$L^2(\mathbb{T}^2) = \mathbf{V}_0 \oplus_{j\geq 0} \mathbf{W}_j. \tag{4.23}$$

The space $\mathbf{V}_j$ is generated by the bivariate scaling functions

$$\mathbf{V}_j = \overline{\mathrm{span}}\{\phi_{j,i_x,i_y}(x,y) = \phi_{j,i_x}(x)\phi_{j,i_y}(y)\}_{i_x,i_y=0,\dots,2^j-1} \tag{4.24}$$

and the orthogonal complement space $\mathbf{W}_j = \mathbf{V}_{j+1} \ominus \mathbf{V}_j, j \geq 0$ by three different wavelets

$$\mathbf{W}_j = \overline{\mathrm{span}}\{\psi^\kappa_{j,i_x,i_y}(x,y)\}_{i_x,i_y=0,\dots,2^j-1,\kappa=1,2,3} \tag{4.25}$$

with ψ^κ_{j,i_x,i_y} defined as in (4.21) using the periodic analogons.

Correspondingly, any function $f \in L^2(\mathbb{T}^2)$ which is at least continuous can be projected onto $\mathbf{V}_J$ by collocation

$$f_J(x,y) = \sum_{i_x=0}^{2^J-1} \sum_{i_y=0}^{2^J-1} f\left(\frac{i_x}{2^J}, \frac{i_y}{2^J}\right) S_{J,i_x,i_y}(x,y) \qquad (4.26)$$

using the two-dimensional cardinal Lagrange function

$$S_{j,i_x,i_y}(x,y) = S_{j,i_x}(x)S_{j,i_y}(y). \qquad (4.27)$$

It can then be expressed as

$$f_J(x,y) = \overline{f}_{0,0,0}\phi_{0,0,0}(x,y) + \sum_{j=0}^{J-1} \sum_{i_x=0}^{2^j-1} \sum_{i_y=0}^{2^j-1} \sum_{\kappa=1}^{3} \tilde{f}_{j,i_x,i_y}^{\kappa} \psi_{j,i_x,i_y}^{\kappa}(x,y) \qquad (4.28)$$

with coefficients

$$\tilde{f}_{j,i_x,i_y}^{\kappa} = \langle f, \psi_{j,i_x,i_y}^{\kappa}\rangle, \qquad \overline{f}_{0,0,0} = \int_{\mathbb{T}^2} f(x,y)\mathrm{d}x\mathrm{d}y \qquad (4.29)$$

using that $\phi_{0,0,0} = 1$.

Representing a function in terms of wavelet coefficients has the following advantages. Smooth functions yield rapid decay of the coefficients in scale (depending on the number of vanishing moments of ψ_{ji}). At locations where u develops a singularity or an "almost singularity" only local coefficients have to be retained (depending on the decay of ψ_{ji} in space). Second, all employed basis functions are mutually orthogonal, a property which is the keystone of the algorithms.

4.3 Algorithm

In case of a regular sampling and periodic functions, computations can be done in physical space by periodizing the required filters (defined below) using Mallat's algorithm [39]. For long filters it is more economical, however, to use fast convolution in Fourier space employing FFT. Following [46] we describe such a transformation together with a computational trick for its acceleration [26]. The algorithm is based on the application of three discrete filters. The scaling coefficients $c_{J,k}$ are computed by application of the interpolation filter

$$I_n^J = \langle S_{J,n}, \phi_{J,0}\rangle, \quad \widehat{I^J} = 2^{-3J/2}\widehat{I}\left(\frac{k}{2^j}\right), \quad \widehat{I}(k) = \widehat{S}(k)/\widehat{\phi}(k) \qquad (4.30)$$

to the sampled values $\{f(\frac{i}{2^J})\}_i$.

The filters

$$G_n^j = \langle \phi_{j,n}, \psi_{j-1,0} \rangle, \qquad H_n^j = \langle \phi_{j,n}, \phi_{j-1,0} \rangle \qquad (4.31)$$

are classically used for computing the wavelet transform. They can be obtained in physical space for compactly supported bases and in Fourier space through

$$\widehat{H}(k) = \widehat{\phi}(2k)/\widehat{\phi}(k), \qquad \widehat{G}(k) = \widehat{\psi}(2k)/\widehat{\phi}(k). \qquad (4.32)$$

The algorithm then reads

Step 0. FFT of the values $\{f_i\}_{i=0,\ldots,2^J-1}$ at the points $\{x_i = \frac{i}{2^J}\}_{i=0,\ldots,2^J-1}$ to the Fourier coefficients $\{\widehat{f}_k\}_{k=0,\ldots,2^J-1}$.

Step 1. Interpolation using the Lagrange function $S_J(x)$ of the space V_J by computation in Fourier space: application of $\widehat{I_J}$ gives $(\widehat{f}_J)_k$, $k = 0,\ldots,2^J-1$.

Step 2. Application of Filters G and H in Fourier space (* indicating double length sequences)

$$\left(\widehat{\overline{f}^*_{J-1}}\right)_k = \overline{\widehat{H}_k}\left(\widehat{\overline{f}_J}\right)_k \qquad k = 0,\ldots,2^J-1 \qquad (4.33)$$

$$\left(\widehat{\tilde{f}^*_{J-1}}\right)_k = \overline{\widehat{G}_k}\left(\widehat{\overline{f}_J}\right)_k \qquad k = 0,\ldots,2^J-1. \qquad (4.34)$$

Step 3. Instead of setting

$$\overline{f}_{J-1,i} = \overline{f}^*_{J-1,2i} \qquad \tilde{f}_{J-1,i} = \tilde{f}^*_{J-1,2i} \qquad i = 0,\ldots,2^{J-1}-1 \quad (4.35)$$

in physical space, downsampling can be done directly in Fourier space through

$$\left(\widehat{\overline{f}_{J-1}}\right)_k = \left(\widehat{\overline{f}^*_{J-1}}\right)_k + \left(\widehat{\overline{f}^*_{J-1}}\right)_{k+2^{J-1}} \qquad k = 0,\ldots,2^{J-1} \qquad (4.36)$$

$$\left(\widehat{\tilde{f}_{J-1}}\right)_k = \left(\widehat{\tilde{f}^*_{J-1}}\right)_k + \left(\widehat{\tilde{f}^*_{J-1}}\right)_{k+2^{J-1}} \qquad k = 0,\ldots,2^{J-1}. \qquad (4.37)$$

Step 4. Inverse FFT of lenght 2^{J-1} to get $\{\tilde{f}_{J-1,i}\}_{i=0,\ldots,2^{J-1}-1}$.

Iterate steps 2 to 4 replacing J by $j = J-1,\ldots,0$. Observe that in the last step 2^{-1} is replaced by 0 and $(\widehat{f}_0)_0 = \overline{f}_{0,0}$.

The use of (4.36, 4.37) instead of (4.35) leads to a speed up by a factor 6 with respect to extracting the coefficients in physical space. The inverse transform is obtained by executing the above steps in reversed order omitting the conjugate complex in (4.33, 4.34) and replacing Step 3 with upsampling in Fourier space.

Part II
Statistical Analysis

5 Classical tools

5.1 Methodology

Turbulence research is based on either laboratory or numerical experiments. Typical quantities measured to characterize turbulent flows are scalar fields (temperature, concentration, pressure, etc.), vector fields (velocity, vorticity, etc.), tensor fields (stress, strain, etc.).

5.1.1 Laboratory experiments

Laboratory experiments are done (*e.g.* in wind tunnels or water tank) using flow visualisations and time measurements performed in few points of the flow (*e.g.* hot-wire anemometry, laser velocimetry). Flow visualizations give mostly qualitative information. Time measurements give quantitative information by accumulating well-sampled and well-converged time statistics, although only at very few spatial locations. By checking Taylor's hypothesis, namely that the time fluctuations are small compared to the mean flow velocity, one assumes that the time statistics can be identified with the space statistics. This allows to compare laboratory measurements with the predictions of the statistical theory of turbulence, and also with the statistics obtained from numerical experiments.

5.1.2 Numerical experiments

Numerical experiments are based on the accepted assumption that Navier–Stokes is the fundamental equation of fluid dynamics whatever the flow regime. The turbulent regime is reached when the nonlinear advective term dominates the linear dissipative term (limit $\nu \to 0$ or $Re \propto 1/\nu \to \infty$). In this highly nonlinear regime, Navier–Stokes solutions can only be computed by numerical approximation. The computation predicts the time evolution of one flow realization only. Statistical analyses are performed afterwards in three different ways:

- by computing spatial statistics of instanteneous turbulent fields, which is valid only if the computational domain is much larger than the integral scale where turbulence is produced;

- by computing time statistics of long flow history, which is valid only if the time evolution is much larger than the eddy turn over time characteristic of turbulent flow instabilities;

- by running a large number of numerical simulations for the same parameter and flow configuration, but with different initial conditions. The ensemble averages are computed afterwards. This procedure requires a number of independant realisations sufficient to ensure the stationarity of the probability distribution.

5.2 Averaging procedure

Turbulent flows are characterized by their unpredictability, namely each flow realization is different, although the statistics are reproducible as long as the flow configuration and parameters are the same. This is the reason why turbulence models predict only statistical quantities.

Another essential characteristic of turbulent flows is their intermittency, *i.e.* the fact that we observe in each flow realization well localized strong events (bursts). This intermittent behaviour is not very pronounced in the velocity field, but becomes dominant when one considers the velocity gradients or the vorticity fields. They are characterized by non-Gaussian probability distribution functions, whose tails correspond to the intermittent bursts. We think that the flow intermittency comes from the nonlinear dynamics of turbulent flows which, for incompressible fluids, tend to form well localized coherent vortices (vortex spots in two dimensions and vortex tubes in three dimensions) which move around in a chaotic way resulting from their mutual interactions.

A crucial difficulty in turbulence modelling is to define averages able to take into account intermittency. Actually the L^2-norm averages, *i.e.* $(\int |f(x)|^2 dx)^{1/2})$, classically used in turbulence (*e.g.* two-point correlations, second order structure functions, spectra) are "blind" to intermittency, because the well localized strong events responsible for intermittency are too rare to affect the L^2-norm, their weight remaining negligible in the integral.

To define averages able to take into account intermittency there are two possible strategies:

- either to consider L^p-norms, *i.e.* $((\int |f(x)|^p dx)^{1/p})$ with p large enough to have the values of the PDF tails (Probability Distribution Function) contributing significantly to the integral;

- or to extract the rare (intermittent) events, responsible for the heavy tails of the PDF, from the dense (non-intermittent) events, which contribute only to the center of the PDF, and perform classical L^2-norm averages for the dense events only.

The second approach corresponds to conditional averages and requires a criterium to separate the rare events from the dense events. As we have assumed that the intermittency of turbulent flows is due to the presence of coherent vortices, responsible for the rare events, we first need to identify them in order to then be able to extract them.

As we have shown [15, 17] the coherent vortices can be characterized by the fact that they correpond to the strongest wavelet coefficients of the vorticity field. Based on this property we have defined a procedure to extract them [19, 21], which consists in retaining only those vorticity wavelet coefficients $\tilde{\omega}$ which are larger than a threshold value $\tilde{\omega}_T = (2Z \log_{10} N)^{1/2}$, with Z the total enstrophy and N the resolution (*i.e.* the number of grid points or wavelet coefficients). We then verify that the PDF of the discarded coefficients, *i.e.* those smaller than the treshold $\tilde{\omega}_T$ which correspond to the dense non-intermittent events, is actually Gaussian.

5.3 Statistical diagnostics

5.3.1 Probability Distribution Function (PDF)

To motivate the introduction of a probability space $(\Xi, \mathcal{F}, \mathcal{P})$ we consider:

- the set of all possible configurations of the flow, *i.e.* the phase space of the Navier–Stokes equations, denoted by Ξ;

- the collection $\mathcal{F}$ of all experiments with a definite outcome which have been performed, called the flow realisations;

- a probability measure $\mathcal{P}$ assigned to $\mathcal{F}$, such that $\mathcal{P}(\emptyset) = 0$ and $\mathcal{P}(\Xi) = 1$, which assigns a probability to each experiment which has been performed.

We consider a stationary, homogeneous and isotropic random field $f(\xi, \vec{x}, t) \in \mathcal{F}$, with $\xi \in \Xi$, $\vec{x} \in \mathbb{R}^n$ and $t \in \mathbb{R}_0^+$. For fixed ξ the function $f(\vec{x}, t)$ is called a realization of the random field or a sample, *e.g.* one component of the velocity field.

Definition of the PDF: Using the probability measure $\mathcal{P}$ we define the distribution function $F(g) = \mathcal{P}(-\infty < f \leq g, f \in \Xi)$ which measures the probability of f having a value less or equal to g.

5.3.2 Radon–Nikodyn's theorem

If the probability measure $\mathcal{P}$ is absolutely continuous, there exists a probability density p of $\mathcal{P}$ such that $p(f) = \frac{\mathrm{d}\mathcal{P}}{\mathrm{d}f}$, which corresponds to the derivative $\frac{\mathrm{d}F}{\mathrm{d}g}$ of the distribution F, i.e. $p(f)\mathrm{d}f = \mathcal{P}(f < g \leq f + \mathrm{d}f, g \in \Xi)$.

The probability density is normalized such that $\int_{\mathbb{R}} p(x)\mathrm{d}x = 1$.

5.3.3 Definition of the joint probability

Let f and g be two random fields, one can define the joint probability

$$F(f, g) = \mathcal{P}(-\infty < f' \leq f, \ -\infty < g' \leq g, [f', g'] \in \Xi).$$

The corresponding joint probability density function is given by $p(f, g) = p(f)p(g) - p(f \cap g)$ (Bayes' theorem). If f and g are independent and identically distributed (i.i.d.), then $p(f, g) = p(f)p(g)$.

5.3.4 Statistical moments

The q-th order moments of the random field f are defined as

$$M_q(f) = \langle f^q \rangle = \int f^q p(f)\mathrm{d}f. \tag{5.1}$$

If $f \in \Xi$ is ergodic the moments can also be expressed as space averages

$$M_q(f) = \int (f(\vec{x}))^q \mathrm{d}^n \vec{x}, \tag{5.2}$$

where the integral is defined as

$$\int = \lim_{L \to \infty} \frac{1}{L^3} \int_0^L \int_0^L \int_0^L, \tag{5.3}$$

L being the size of the domain.

Ratios of moments are defined, such as

$$Q_{p,q}(f) = \frac{M_p(f)}{(M_q(f))^{p/q}}. \tag{5.4}$$

Classically one chooses $q = 2$, which leads to define statistical quantitities such as:

- skewness $S = Q_{3,2}(f)$;
- flatness $F = Q_{4,2}(f)$;
- hyperskewness $S_h = Q_{5,2}(f)$;
- hyperflatness $F_h = Q_{6,2}(f)$.

5.3.5 Structure functions

The p-th order structure function of a random scalar field f is defined as

$$S_{p,f}(\vec{l}) = \int (f(\vec{x} + \vec{l}) - f(\vec{x}))^p \mathrm{d}^n \vec{x}. \qquad (5.5)$$

5.3.6 Autocorrelation function

The autocorrelation function of the random scalar field f is defined as

$$R(\vec{l}) = \int f(\vec{x}) f(\vec{x} + \vec{l}) \mathrm{d}^n \vec{x} \qquad (5.6)$$

and for vector fields $\vec{f}$ we get the two-point correlation tensor

$$R_{ij}(\vec{l}) = \int f_i(\vec{x}) f_j(\vec{x} + \vec{l}) \mathrm{d}^n \vec{x}. \qquad (5.7)$$

Note that in turbulence the above quantity computed for the velocity field is called Reynolds stress tensor, and plays a key role in turbulence modelling.

5.3.7 Fourier spectrum

Definition of the spectrum: The spectrum of the random scalar field f is the Fourier transform of its autocorrelation function:

$$\Phi(\vec{k}) = \frac{1}{(2\pi)^n} \int R(\vec{l}) e^{-i\vec{k}\cdot\vec{l}} \mathrm{d}^n \vec{l}. \qquad (5.8)$$

For vector fields we obtain analogously

$$\Phi_{ij}(\vec{k}) = \frac{1}{(2\pi)^n} \int R_{ij}(\vec{l}) e^{-i\vec{k}\cdot\vec{l}} \mathrm{d}^n \vec{l}. \qquad (5.9)$$

One can integrate $\Phi(\vec{k})$ on shells of radius $k = |\vec{k}|$ which gives the one-dimensional spectrum

$$E(k) = \int \Phi(\vec{k}) k^{n-1} \mathrm{d}\theta. \qquad (5.10)$$

5.3.8   Wiener–Khinchin's theorem

For a function $R(l) \in L^1(\mathbb{R})$, to be the correlation function of a homogeneous field $f(x)$ which satisfies the condition $S_2(l) \to 0$ for $l \to 0$, it is necessary and sufficient that it has a representation of the form $R(l) = \int_{\mathbb{R}} E(k)e^{ikl}\mathrm{d}k$ where $E(k) \geq 0$ is the spectral density of the random variable $f(x)$.

Remark: If $R(l)$ is not in $L^1(\mathbb{R})$, then Wiener–Khinchin holds in a distributional sense only, *e.g.* for a Gaussian white noise $R(l) = \delta_{l,0}$ therefore it is not in L^1 and $E(k) = 1$.

The spectrum $E(k)$, the second order structure function $S_{2,f}(l)$ and the autocorrelation function $R(l)$ fulfill the following relations:

$$R(l) = \int f(x+l)f(x)\mathrm{d}x = 2\int_0^\infty \cos(2\pi kl)E(k)\mathrm{d}k \qquad (5.11)$$

and hence we get

$$
\begin{aligned}
S_{2,f}(l) &= \langle(f(x+l) - f(x))^2 = 2R(0) - 2R(l)\rangle \\
&= 2\int_0^\infty (1 - \cos(2\pi kl))E(k)\mathrm{d}k. \qquad (5.12)
\end{aligned}
$$

Remark: The above relation illustrates that the structure function corresponds to a high pass filtered spectrum although the corresponding filter is not very selective. We will propose wavelet tools to improve the filter selectivity.

6   Statistical tools based on the continuous wavelet transform

6.1   *Local and global wavelet spectra*

When analyzing velocity signals of turbulent flows one should calculate ensemble averages of the energy spectra from many realizations. In practice, to avoid performing ensemble averages, one assumes ergodicity of the turbulent motions and averages only one flow realization split into many pieces whose lengths are larger than the integral scale (which is the largest correlated scale in a turbulent signal). In statistical theory of homogeneous turbulence, only the modulus of the Fourier transform is used (*e.g.* the energy spectrum) and thus the phase information is lost. This is probably a major weakness of the traditional way of analyzing turbulence since it neglects any spatial organization of the turbulent fields, which happens in each flow realization although the averages are homogeneous. For statistically inhomogeneous flows the standard statistical tools, *e.g.* the energy spectrum which is the Fourier transform of the two-point correlation of the velocity increments, are too limited to analyze and model turbulence.

The wavelet transform extends the concept of energy spectrum so that one can define a local energy spectrum $\tilde{E}(x, k)$ using the wavelet transform (which, as we have seen conserves, the L^2–norm of a function), such that

$$\tilde{E}(k, x) = \frac{1}{2C_\psi k_\psi} \left| \tilde{f}\left(\frac{k_\psi}{k}, x\right) \right|^2 \quad \text{for} \quad k \geq 0 \qquad (6.1)$$

where k_ψ is the peak wave number of the analyzing wavelet ψ and C_ψ as defined in (3.1). By measuring $\tilde{E}(k, x)$ at different places in a turbulent flow one might estimate what parts of the flow contribute most to the overall Fourier energy spectrum and how the energy spectrum depends on local flow conditions. For example, one can determine the type of energy spectrum contributed by coherent structures, such as isolated vortices, and the type of energy spectrum contributed by the unorganized part of the flow.

Although the wavelet transform analyses the flow into wavelets rather than complex exponentials one shows [47] that the mean wavelet energy spectrum converges to the Fourier energy spectrum provided the analyzing wavelets have enough cancellations. More precisely the mean wavelet spectrum $\tilde{E}(k)$

$$\tilde{E}(k) = \int_0^{+\infty} \tilde{E}(k, x)\mathrm{d}x \qquad (6.2)$$

gives the correct Fourier exponent for a power-law Fourier energy spectrum $E(k) \propto k^{-\beta}$ if the analysing wavelet has at least $n > (\beta - 1)/2$ vanishing moments. This condition is the same as that for detecting singularities derived in the previous section since $\beta = 1 + 2\alpha$ for isolated cusps. Thus, the steeper the energy spectrum the more vanishing moments of the wavelet we need. The inertial range in turbulence has a power-law form. The ability to correctly characterize power-law energy spectra is therefore a very important property of the wavelet transform (which is related to its ability to detect and characterize singularities).

6.2   Relation with Fourier spectrum

The mean wavelet energy spectrum $\tilde{E}(k)$ is a smoothed version of the Fourier energy spectrum $E(k)$. This can be seen from the following relation between the two spectra

$$\tilde{E}(k) = \frac{1}{2C_\psi k_0} \int_0^{+\infty} E(k') \left| \hat{\psi}\left(\frac{k_0 k'}{k}\right) \right|^2 \mathrm{d}k' \qquad (6.3)$$

which shows that the mean wavelet spectrum is an average of the Fourier spectrum weighted by the square of the Fourier transform of the analysing wavelet shifted at wavenumber k. Note that the larger k is, the larger the averaging interval, because wavelets are bandpass filters at $\frac{\Delta k}{k}$ constant. This property of the mean wavelet energy spectrum is particularly useful for turbulent flows. Indeed, the Fourier energy spectrum of a single realization of a turbulent flow is too spiky to be able to clearly detects a slope, but it is no more the case for the mean wavelet energy spectrum which is much smoother.

The Mexican hat wavelet

$$\hat{\psi}(k) = k^2 \exp(-k^2/2) \tag{6.4}$$

has only two vanishing moments and thus can correctly measure energy spectrum exponents up to $\beta < 5$. Only the zeroth order moment of the Morlet wavelet

$$\hat{\psi}(k) = \frac{1}{2\pi} \exp(-(k - k_\psi)^2/2) \quad \text{for} \quad k > 0$$
$$\hat{\psi}(k) = 0 \quad \text{for} \quad k \leq 0 \tag{6.5}$$

is zero, but the higher n-th order moments are very small ($\propto k_\psi^n \exp(-k_\psi^2/2)$) provided that k_ψ is sufficiently large. Therefore the Morlet wavelet transform gives accurate estimates of the power-law exponent of the energy spectrum at least for approximately $\beta < 7$ (if $k_\psi = 6$).

There is also a family of wavelets [47] with an infinite number of cancellations

$$\hat{\psi}_n(k) = \alpha_n \exp\left(-\frac{1}{2}\left(k^2 + \frac{1}{k^{2n}}\right)\right), \quad n \geq 1, \tag{6.6}$$

where α_n is chosen for normalization. The wavelets defined in (6.6) can therefore correctly measure any power-law energy spectrum. Furthermore, these wavelets can detect the difference between a power-law energy spectrum and a Gaussian energy spectrum ($E(k) \propto \exp(-(k/k_0)^2)$). It is important to be able to determine at what wavenumber the power-law energy spectrum becomes exponential since this wavenumber defines the end of the inertial range of turbulence and the beginning of the dissipative range.

6.3   Application to turbulence

The first measurements of local energy spectra in turbulence were reported in [18] and [40]. Farge *et al.* [18] used a Morlet wavelet to obtain the local and global energy spectra for a 3D mixing layer computed by DNS. They showed that the deviation from the mean energy spectrum was very large due to intermittency and increased with the scales. Therefore they conjectured that the intermittency of the flow increases for increasing Re. Meneveau [40] used the discrete wavelet transform to measure local energy spectra in experimental and Direct Numerical Simulation (DNS) flows. He found that the standard deviation of the local energy (a measure of the spatial fluctuation of energy) was approximately 100% throughout the inertial range. He also calculated the spatial fluctuation of $T(k)$ which measures the transfer of energy from all wavenumbers to wavenumber k. On average $T(k)$ is negative for the large scales and positive for the small scales, indicating that in three-dimensional turbulence energy is, in average, transferred from the large scales to the small scales where it is dissipated. However, he found that at many locations in the flow the energy cascade actually operates in the opposite direction, from small to large scales, indicating a local inverse energy cascade (called back-scattering) which concerns a very important part of the transferred energy. This local spectral information, which links the physical and Fourier representations of turbulence, can be obtained using the wavelet transform but not with the Fourier transform.

7   Statistical tools based on the orthogonal wavelet transform

7.1   Local and global wavelet spectra

In this section we describe some statistical tools based on the orthogonal wavelet transform. We present them considering, as example, a two-dimensional scalar field $f(\vec{x})$ which has vanishing mean and is periodic (the extention to higher dimensions and vector fields is straightforward [54]). Hence we employ a periodic two-dimensional Multi–Resolution Analysis (MRA) [14, 42] and develop the field f, sampled on $N^2 = 2^{2J}$ points, as an orthonormal wavelet series from the largest scale $l_{\max} = 2^0$ to the smallest scale $l_{\min} = 2^{-J}$:

$$f(x,y) = \sum_{j=0}^{J-1} \sum_{i_x=0}^{2^j-1} \sum_{i_y=0}^{2^j-1} \sum_{\kappa=1}^{3} \tilde{f}_{j,i_x,i_y}^{\kappa} \psi_{j,i_x,i_y}^{\kappa}(x,y), \qquad (7.1)$$

with

$$\psi_{j,i_x,i_y}^{\kappa}(x,y) = \begin{cases} \psi_{j,i_x}(x)\phi_{j,i_y}(y); & \kappa = 1, \\ \phi_{j,i_x}(x)\psi_{j,i_y}(y); & \kappa = 2, \\ \psi_{j,i_x}(x)\psi_{j,i_y}(y); & \kappa = 3, \end{cases} \qquad (7.2)$$

where $\phi_{j,i}$ and $\psi_{j,i}$ are the 2π-periodic one-dimensional scaling function and the corresponding wavelet, respectively. The wavelets $\psi_{j,i_x,i_y}^{\kappa}$ correspond to horizontal, vertical and diagonal directions, for $\kappa = 1, 2, 3$, respectively. Due to orthogonality the coefficients are given by $\tilde{f}_{j,i_x,i_y}^{\kappa} = \langle f, \psi_{j,i_x,i_y}^{\kappa} \rangle$ where $\langle \cdot, \cdot \rangle$ denotes the L^2 inner product.

We can define the scale distribution of energy, also called scalogram, as

$$E_j = \sum_{i_x=0}^{2^j-1} \sum_{i_y=0}^{2^j-1} \sum_{\kappa=1}^{3} |\tilde{f}_{j,i_x,i_y}^{\kappa}|^2. \qquad (7.3)$$

Introducing the discrete mean square wavelet coefficient at scale 2^{-j} and at position $x_{i_x,i_y} = 2^{-j}(i_x + 1/2, i_y + 1/2)$ as

$$\overline{\overline{f}}(2^{-j}, 2^{-j}(i_x, i_y)) = \frac{1}{2}((\tilde{f}^1_{j,i_x,i_y})^2 + (\tilde{f}^1_{j,i_x+1,i_y})^2) + \frac{1}{2}((\tilde{f}^2_{j,i_x,i_y})^2$$
$$+ (\tilde{f}^2_{j,i_x,i_y+1})^2) + (\tilde{f}^3_{j,i_x,i_y})^2 \qquad (7.4)$$

we define a discrete local wavelet spectrum [10] by

$$\widetilde{E}(k_j, x_{i_x,i_y}) = \overline{\overline{f}}(2^{-j}, 2^{-j}(i_x, i_y)) \frac{2^{2j}}{\Delta k_j}. \qquad (7.5)$$

This quantitiy allows to study the space dependent spectral behaviour of f. By construction we have

$$\widetilde{E}(k_j) = \sum_{i_x=0}^{2^j-1} \sum_{i_y=0}^{2^j-1} \widetilde{E}(k_j, x_{i_x,i_y}). \qquad (7.6)$$

7.2 Relation with Fourier spectrum

Owing to the orthogonality of the wavelet decomposition, the total energy is preserved and we have $E = \sum_j E_j$. To be able to relate the scale distribution to the Fourier spectrum, we introduce the mean wavenumber k_0 of the wavelet ψ, defined by

$$k_0 = \frac{\int_0^\infty k|\widehat{\psi}(k)|\mathrm{d}k}{\int_0^\infty |\widehat{\psi}(k)|\mathrm{d}k}. \qquad (7.7)$$

Therewith each scale 2^{-j} of the wavelet ψ_j is related to the mean wavenumber $k_j = k_0 2^j$. With $\Delta k_j = \sqrt{k_j k_{j+1}} - \sqrt{k_j k_{j-1}}$, describing the mean radial wavenumber of the three two-dimensional wavelets ψ^κ_{j,i_x,i_y}, we define the global wavelet spectrum as

$$\widetilde{E}(k_j) = E_j/\Delta k_j \qquad (7.8)$$

which is related with the Fourier energy spectrum by

$$\widetilde{E}(k) = \frac{1}{k} \int_0^\infty E(k')|\widehat{\psi}(k_0 k'/k)|^2 \mathrm{d}k'. \qquad (7.9)$$

The wavelet spectrum is a smoothed Fourier spectrum weighted with the modulus of the Fourier transform of the analyzing wavelet [47]. Note that for increasing wavenumbers the averaging interval becomes larger [14]. A sufficient condition, guaranteeing the global wavelet spectrum to be able to detect the same power-law behaviour $k^{-\alpha}$ as the Fourier spectrum, is that ψ has enough vanishing moments [47], *i.e.*

$$\int_{-\infty}^{+\infty} x^n \psi(x) \mathrm{d}x = 0 \quad \text{for} \quad 0 \le n \le \frac{\alpha - 1}{2}. \tag{7.10}$$

If this condition is not fulfilled the global wavelet spectrum saturates at the critical cancellation order n and shows a power-law behaviour with a slope not steeper than $-2(n+1)$.

7.3   Intermittency measures

Useful diagnostics to quantify the intermittency of a field are the moments of its wavelet coefficients at different scales j [53, 54],

$$M_{p,j}(f) = \frac{1}{3 \cdot 2^{2j}} \sum_{i_x=0}^{2^j-1} \sum_{i_y=0}^{2^j-1} \sum_{\kappa=1}^{3} |\tilde{f}_{j,i_x,i_y}^{\kappa}|^p. \tag{7.11}$$

The sparsity of the wavelet coefficients at each scale can be measured and the intermittency of the field f can be quantified using ratios of the moments at different scales,

$$Q_{p,q,j}(f) = \frac{M_{p,j}(f)}{(M_{q,j}(f))^{p/q}}, \tag{7.12}$$

which may be interpreted as quotient norms between different L^p- and L^q-spaces. Classically, one chooses $q = 2$ to define typical statistical quantities as a function of scale. Recall that for $p = 4$ we obtain the scale dependent flatness $F_j = Q_{4,2,j}$ which is equal to 3 for a Gaussian white noise at all scales j, which proves that this signal is not intermittent. The scale dependent skewness, hyperflatness and hyperskewness are obtained for $p = 3, 5$ and 6, respectively. For intermittent signals $Q_{p,q,j}$ increases with j.

Part III

Computation

8   Coherent vortex extraction

8.1   CVS filtering

In [21, 23] we have proposed a wavelet-based method, called Coherent Vortex Simulation (CVS), to compute turbulent flows for regimes where the coherent vortices dominate the nonlinear dynamics. We first present the CVS filtering, which extracts coherent vortices out of each flow realization. We then describe the CVS computation, which calculates the time evolution of turbulent flows by deterministically computing the dynamics of coherent vortices, in an adaptive wavelet basis, and statistically modelling the effect of the coherent velocity field onto the incoherent background flow.

8.1.1 Vorticity decomposition

We describe the wavelet algorithm to extract coherent vortices out of turbulent flows and consider as example the 3D case (for the 2D case refer to [21]). We consider the vorticity field $\vec{\omega}(\vec{x}) = \nabla \times \vec{v}$, computed at resolution $N = 2^{3J}$, N being the number of grid points and J the number of octaves. Each component is developed into an orthogonal wavelet series from the largest scale $l_{\max} = 2^0$ to the smallest scale $l_{\min} = 2^{J-1}$ using a 3D multi-resolution analysis (MRA) [8, 14]:

$$
\begin{aligned}
\omega(\vec{x}) &= \bar{\omega}_{0,0,0}\phi_{0,0,0}(\vec{x}) \\
&+ \sum_{j=0}^{J-1}\sum_{i_x=0}^{2^j-1}\sum_{i_y=0}^{2^j-1}\sum_{i_z=0}^{2^j-1}\sum_{\mu=1}^{2^n-1} \tilde{\omega}^{\mu}_{j,i_x,i_y,i_z}\,\psi^{\mu}_{j,i_x,i_y,i_z}(\vec{x}),
\end{aligned} \tag{8.1}
$$

with $\phi_{j,i_x,i_yi,i_z}(\vec{x}) = \phi_{j,i_x}(x)\phi_{j,i_y}(y)\phi_{j,i_z}(z)$, and

$$
\psi^{\mu}_{j,i_x,i_y,i_z}(\vec{x}) =
\begin{cases}
\psi_{j,i_x}(x)\phi_{j,i_y}(y)\phi_{j,i_z}(z); & \mu = 1, \\
\phi_{j,i_x}(x)\psi_{j,i_y}(y)\phi_{j,i_z}(z); & \mu = 2, \\
\phi_{j,i_x}(x)\phi_{j,i_y}(y)\psi_{j,i_z}(z); & \mu = 3, \\
\psi_{j,i_x}(x)\phi_{j,i_y}(y)\psi_{j,i_z}(z); & \mu = 4, \\
\psi_{j,i_x}(x)\psi_{j,i_y}(y)\phi_{j,i_z}(z); & \mu = 5, \\
\phi_{j,i_x}(x)\psi_{j,i_y}(y)\psi_{j,i_z}(z); & \mu = 6, \\
\psi_{j,i_x}(x)\psi_{j,i_y}(y)\psi_{j,i_z}(z); & \mu = 7,
\end{cases} \tag{8.2}
$$

where $\phi_{j,i}$ and $\psi_{j,i}$ are the one-dimensional scaling function and the corresponding wavelet, respectively. Due to orthogonality, the scaling coefficients are given by $\bar{\omega}_{0,0,0} = \langle \omega, \ \phi_{0,0,0}\rangle$ and the wavelet coefficients are given by $\tilde{\omega}^{\mu}_{j,i_x,i_y,i_z} = \langle \omega, \ \psi^{\mu}_{j,i_x,i_y,i_z}\rangle$, where $\langle \cdot, \cdot \rangle$ denotes the L^2-inner product.

8.1.2 Nonlinear thresholding

We then split the vorticity field into $\vec{\omega}_{\mathrm{C}}(\vec{x})$ and $\vec{\omega}_{\mathrm{I}}(\vec{x})$ by applying a nonlinear thresholding to the wavelet coefficients. The threshold is defined as $\epsilon = (4/3Z \log N)^{1/2}$ and it only depends on the total enstrophy Z and on the number of grid points N without any adjustable parameters. The choice of this threshold is based on theorems [11,12] proving optimality of the wavelet representation to denoise signals in presence of Gaussian white noise, since

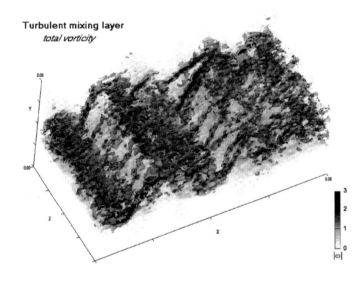

Fig. 2. Total vorticity of the turbulent mixing layer.

this wavelet-based estimator minimizes the maximal L^2-error for functions with inhomogeneous regularity.

8.1.3   Vorticity and velocity reconstruction

The coherent vorticity field $\vec{\omega}_C$ is reconstructed from the wavelet coefficients whose modulus is larger than ϵ and the incoherent vorticity field $\vec{\omega}_I$ from the wavelet coefficients whose modulus is smaller or equal to ϵ. The two fields thus obtained, $\vec{\omega}_C$ and $\vec{\omega}_I$, are orthogonal, which ensures a separation of the total enstrophy into $Z = Z_C + Z_I$ because the interaction term $\langle \vec{\omega}_C, \vec{\omega}_I \rangle$ vanishes. We then use Biot–Savart's relation $\vec{v} = -\nabla \times (\nabla^{-2} \vec{\omega})$ to reconstruct the coherent velocity $\vec{v}_C$ and the incoherent velocity $\vec{v}_I$ for the coherent and incoherent vorticities respectively.

8.2   Application to a 3D turbulent mixing layer

In the present lecture notes we apply the above algorithm to a high resolution DNS ($N = 512 \times 256 \times 128$) of a forced turbulent mixing layer [48] to check the potential for the CVS method in 3D shear flows. Figure 2 show the modulus of vorticity for the total flow. We observe longitudinal vortex tubes, resulting from 3D instability and called ribs, which are wrapped onto four transversal rollers, produced by the 2D Kelvin–Helmholtz instability.

The coherent part (see Fig. 3 top), which represents 3% of the total number of coefficients, captures most of the turbulent kinetic energy and 80% of the enstrophy, even at high wavenumbers, and the PDF of its vorticity is similar to that of the total flow (see Fig. 4). The incoherent part (see Fig. 3 bottom), which represents 97% of the total number of coefficients, contains little of the turbulent kinetic energy and 20% of the enstrophy. It is nearly homogenenous with a very low amplitude and contains no structure.

The corresponding 1D energy spectra in the streamwise direction shows that the coherent part presents, all along the inertial range, the same correlation as the total flow, while the incoherent part contains very little energy and is well decorrelated.

8.3 Comparison between CVS and LES filtering

The CVS method is in the spirit of the Large Eddy Simulation (LES) method [24, 37]. But, in contrast to LES, it uses a nonlinear filter that depends on each flow relisation. The CVS filter corresponds to an orthogonal projection, implying $(\vec{\omega}_I)_C = 0$, and is hence idempotent, *i.e.* $(\vec{\omega}_C)_C = \vec{\omega}_C$, which is not the case for all LES filters (*e.g.* the Gaussian filter).

In Figure 4 we compare the CVS and the LES filterings for the same number of retained coefficients ($3\%N$). Since the LES filtering, chosen here to be a Fourier low-pass filter, retains only the low wavenumbers (Fig. 4 top, right), the coherent vortices are smoothed and as a result the variability of vorticity is strongly reduced (see PDF in Fig. 4 bottom, right). In contrast the CVS filtering retains the organized features, whatever their scales are, and as a results the shape of the vorticity PDF is preserved, even for large values of $|\vec{\omega}|$ (Fig. 4 bottom, left).

Concerning turbulence parametrization, *i.e.* the statistical modelling of the effect of the discarded modes onto the retained modes, for CVS and LES method we can draw the following conclusion:

- the LES filtering has the drawback that the high wavenumber modes are not decorrelated and high amplitudes of vorticity are present. This may lead to nonlinear instabilities which trigger backscatter;

- the CVS filtering allows to disentangle the organized and random components of turbulent flows. As a results the discarded incoherent modes have very weak amplitude, are almost homogeneous in space and are well decorrelated.

In conclusion we conjecture that the derivation of a turbulence model is easier with a CVS filtering than with a LES filtering. However, the CVS

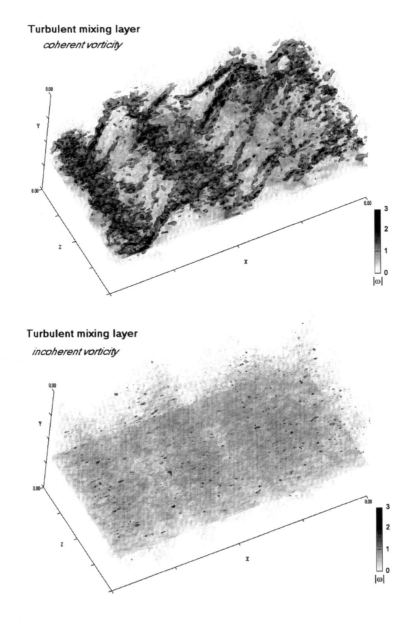

Fig. 3. Top: coherent vorticity of the turbulent mixing layer reconstructed from 3% of the wavelet coefficients and containing 80% of the total enstrophy. Bottom: incoherent vorticity reconstructed from 97% of the wavelet coefficients and containing 20% of the total enstrophy.

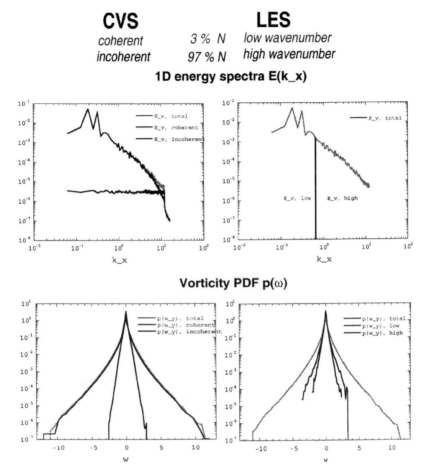

Fig. 4. Comparison of CVS (left) with LES (right) filtering. Energy spectra (top) and PDF of vorticity (bottom) of total, coherent and incoherent flow using CVS filtering and of low wavenumber and high wavenumber components using LES filtering.

filtering requires a dynamically adaptive mesh refinement for solving Navier–Stokes. In the next section we present such a method based on the wavelet representation that we have developed for the 2D Navier–Stokes equations.

9 Computation of turbulent flows

9.1 Navier–Stokes equations

9.1.1 Velocity–pressure formulation

The Navier–Stokes equation written in primitive variable formulation (velocity and pressure) describes the dynamics of a Newtonian (deformation proportional to velocity gradients) fluid

$$\partial_t \vec{v} + (\vec{v} \cdot \nabla)\vec{v} - \nu\nabla^2\vec{v} + \frac{1}{\rho}\nabla p = \vec{F} \qquad (9.1)$$

and, if we suppose that the fluid is incompressible (constant density of the fluid elements), it is complemented by the continuity equation

$$\nabla \cdot \vec{v} \;=\; 0 \qquad (9.2)$$

where $\vec{v} = (v_1(\vec{x}, t), v_2(\vec{x}, t), v_3(\vec{x}, t))$ and $p(\vec{x}, t)$ denote the fluid velocity and the pressure respectively, at point $\vec{x} = (x_1, x_2, x_3)$ and time t. $\vec{F}$ is the field of external forces per unit mass, ρ the density and ν the constant kinematic viscosity. This system of coupled PDE's must be supplemented by appropriate initial and boundary conditions.

9.1.2 Vorticity–velocity formulation

Taking the curl of (9.1), the pressure term can be eliminated and we get a dynamical equation for the vorticity

$$\partial_t \vec{\omega} + (\vec{v} \cdot \nabla)\vec{\omega} - \vec{\omega} \cdot \nabla\vec{v} - \nu\nabla^2\vec{\omega} = \nabla \times \vec{F}. \qquad (9.3)$$

This is an advection-diffusion equation for the vorticity with an additional term $\vec{\omega} \cdot \nabla\vec{v}$, which is responsible for the vortex stretching mechanism, *i.e.* vortex tubes can be stretched by velocity gradients, which leads to vorticity production. In two dimensions this term vanishes, because the vorticity is a pseudo-scalar $\vec{\omega} = (0, 0, \omega_3)$ perpendicular to the velocity gradients.

Taking the divergence of (9.1) and using the incompressibility of $\vec{v}$, we get a Poisson equation for the pressure which is used in many numerical schemes, such as projection methods or fractional step schemes [5, 56]

$$\frac{1}{\rho}\nabla^2 p = -\nabla \cdot ((\vec{v} \cdot \nabla)\vec{v}). \qquad (9.4)$$

The relation $\vec{\omega} = \nabla \times \vec{v}$ can be inverted for a star shaped domain $\mathcal{D}$ using Poincaré's lemma which leads to:

$$\vec{v} = -\nabla \times \nabla^{-2}\vec{\omega}. \qquad (9.5)$$

The above relation can be expressed as a convolution product of the vorticity with an operator K, called the Biot–Savart kernel, *i.e.* $\vec{v} = K \star \vec{\omega}$. As K decays slowly in physical space, *i.e.* as $|\vec{x}|^{-2}$ in three dimensions and as $|\vec{x}|^{-1}$ in two dimensions, the velocity is less localized than the vorticity.

9.2   Classical numerical methods

In the last 30 years the progress in numerical methods and the availability of supercomputers have had a significant impact on turbulence research. For example, the importance and role of coherent vortices in three-dimensional turbulence has been established largely by high resolution numerical simulations [1,57,58]. Von Neumann's vision, who had suggested in 1949 [45] that turbulence could be simulated numerically, has become a reality. In contrast to the statistical theory of turbulence and to most laboratory experiments, which deal with L^2-norm averaged quantities, numerical experiments deal with non-averaged instantaneous quantities. Numerical experiments deterministically compute the evolution of one flow realization at a time, and perform the desired averages afterwards. There are two ways of computing turbulent flows: either by Direct Numerical Simulation (DNS), or by Modelled Numerical Simulation (MNS).

9.2.1   Direct Numerical Simulation (DNS)

In DNS one computes all degrees of freedom of the flow and both the nonlinear dynamics and the linear dissipation are fully resolved by computing the time evolution of all these degrees of freedom. The DNS schemes currently in use may be classified into three categories:

- spectral and pseudo-spectral schemes;

- finite-difference, -volume, and -element methods;

- Lagrangian methods, *e.g.* vortex methods and contour dynamics.

The various methods are characterized by differences in their numerical complexity, accuracy and flexibility. Using DNS the evolution of all scales of turbulent flows can only be calculated for moderate Reynolds numbers with present supercomputers. The severe limit of DNS is that the number of degrees of freedom N for a regular discretization depends on the Reynolds number Re, such that $N \sim Re$ for two-dimensional flows and $N \sim Re^{9/4}$ for the three-dimensional case. At present only moderate Reynolds number flows ($Re \sim 10^3$ in three dimensions) can be simulated using DNS. Although most flows of engineering interest have higher Reynolds numbers ($Re \sim 10^8$), some physical insight can be gained from studying DNS of only moderate

Re flows. However, laboratory experiments have shown that new behaviour appears in the range $Re \sim 10^4 - 10^5$ [9]. As the number of degrees of freedom scales with Reynolds number, the simulation of such high Reynolds number flows in two or three dimensions requires schemes employing some sort of adaptive discretization.

To our knowledge no current non-wavelet DNS methods use a spatial discretization that adapts to the dynamics and structure of the flow. In [30] we have proposed an adaptive wavelet scheme for nonlinear PDE's and we have extended it to the two-dimensional Navier–Stokes equations [28, 51]. Since the wavelet basis functions are localized in both physical and spectral spaces this approach is a compromise between grid-point methods and spectral methods. The adaptive wavelet method is well suited for turbulence simulations because the characteristic structures encountered in turbulent flows are localized coherent vortices evolving under a multiscale nonlinear dynamics. Thus the space- and scale-adaptivity of the wavelet basis should be very efficient at representing turbulence structures and their dynamics. The fact that the basis is adapted to the solution and follows the time evolution of coherent vortices corresponds to a combination of both Eulerian and Lagrangian approaches.

9.2.2   Modelled Numerical Simulation (MNS)

In MNS, *e.g.* Unsteady Reynolds Averaged Navier–Stokes (URANS), LES and nonlinear Galerkin methods, one supposes that many modes can be discarded, provided that some term(s) or some new equation(s) are added to model the effect of the discarded modes onto the retained modes.

The time evolution of the resolved modes is deterministically computed using the same numerical methods as for DNS. Concerning the discarded modes, one supposes that they are slaved to the retained modes and passively follow their motion. Consequently the dynamics of the unresolved modes cannot become unstable and grow in such a way that they would deterministically affect the evolution of the resolved modes. To ensure this one should check that the unresolved modes have reached a statistical equilibrium state and are sufficiently decorrelated. In this case it is no longer necessary to compute the evolution of the unresolved modes in detail because, if they are in statistical equilibrium, their effect onto the retained modes can be entirely characterized by their averages. The model describing the effect of the unresolved modes onto the resolved modes can be specified once the averaged quantities of the unresolved modes can be parametrized as a function of the resolved modes.

Remark: Ideally, in order to reduce the computational cost as much as possible, the number of resolved modes should be much smaller than the number of discarded modes and should increase more slowly with Re than the total number of modes does.

9.3   Coherent Vortex Simulation (CVS)

9.3.1   Principle of CVS

Coherent Vortex Simulation (CVS) deterministically computes the evolution of the coherent vorticity $\vec{\omega}_C$ and statistically models the effect of the incoherent vorticity $\vec{\omega}_I$ and velocity $\vec{v}_I$. In the following we apply the CVS method to compute 2D turbulent flows. We filter the two-dimensional Navier–Stokes equations using CVS filtering and obtain the evolution equation for the coherent vorticity ω_C:

$$\partial_t \omega_C + \nabla \cdot (\omega \vec{v})_C - \nu \nabla^2 \omega_C \;=\; \nabla \times \vec{F}_C \qquad (9.6)$$
$$\nabla \cdot \vec{v}_C \;=\; 0.$$

To model the effect of the discarded coefficients, which corresponds to the incoherent stress, we propose (as in LES) to use a Boussinesq ansatz. For the nonlinear term we use Leonard's triple decomposition, because the nonlinear term is computed with the same adapted grid as the linear term (*i.e.* without dealiasing). We decompose the nonlinear term of (9.6) into

$$(\omega \vec{v})_C = \omega_C \vec{v}_C + L + C + R, \qquad (9.7)$$

where

$$L \;=\; (\omega_C \vec{v}_C)_C - \omega_C \vec{v}_C,$$
$$C \;=\; (\omega_I \vec{v}_C)_C + (\omega_C \vec{v}_I)_C,$$

and

$$R = (\omega_I \vec{v}_I)_C,$$

denoting the Leonard stress L, the cross stress C and the Reynolds stress R, respectively. The sum of these unknown terms corresponds to the incoherent stress:

$$\tau = (\omega \vec{v})_C - \omega_C \vec{v}_C = L + C + R, \qquad (9.8)$$

which describes the effect of the discarded incoherent terms on the resolved coherent terms. Note that, due to the localization property of the wavelet representation, the Leonard stress L is actually negligible because $(\omega_C \vec{v}_C)_C \simeq \omega_C \vec{v}_C$ [53].

The filtered Navier–Stokes equations (9.6) can be rewritten as:

$$\partial_t \omega_C + \nabla \cdot (\omega_C \vec{v}_C) - \nu \nabla^2 \omega_C = \nabla \times \vec{F}_C - \nabla \cdot \tau \qquad (9.9)$$
$$\nabla \cdot \vec{v}_C = 0.$$

9.3.2   CVS without turbulence model

If we consider a very small threshold, there is no longer any need to model the effect of the incoherent part because the incoherent stress is then negligible, and in this case CVS becomes DNS. Note that even when the wavelet threshold tends to zero, the number of discarded incoherent modes may still be large, due to the excellent compression properties of the wavelet representation for turbulent flows. This is reflected by the fact that many wavelet coefficients are essentially zero and can therefore be discarded without loosing a significant amount of enstrophy.

To obtain the coherent variables ω_C and $\vec{v}_C$ we deterministically integrate (9.6) since the variables are non-Gaussian and correspond to a dynamical system out of statistical equilibrium. We solve these equations in an adaptive wavelet basis [30, 51, 53] (*cf.* the next section). The separation into coherent and incoherent components is performed at each time step. The adaptive wavelet basis retains only those wavelet modes corresponding to the coherent vortices. It is remapped at each time step in order to follow their motions, in both space and scale. In fact, this numerical scheme combines the advantages of both the Eulerian representation, since it projects the solution onto an orthonormal basis, and the Lagrangian representation, since it follows the coherent vortices by adapting the basis at each time step.

9.3.3   CVS with turbulence model

Up to now no modelling has been done, and equation (9.9) is not closed as long as τ depends on the incoherent unresolved terms. To close it we propose two possibilities to model τ.

- A Boussinesq ansatz as for the LES method:

  we assume that τ is proportional to the negative gradient of the coherent vorticity, *i.e.* $\tau = -\nu_T \nabla \omega_C$ with ν_T a turbulent viscosity coefficient. The turbulent viscosity ν_T can be estimated, either using Smagorinsky's model [24], or taking ν_T proportional to the enstrophy fluxes in wavelet space, such that, where enstrophy flows from large to small scales, ν_T is positive, and, where enstrophy flows from small to large scales (*i.e.* backscatter), ν_T becomes negative. This second method for estimating the turbulent viscosity is in the spirit of Germano's dynamical procedure used for LES [24].

- A Gaussian stochastic forcing term:

  we choose τ to be proportional to the incoherent enstrophy Z_I computed at the previous time step. This modelling is made possible since the time evolution of the incoherent background, characterized by the time scale $t_I = (Z_I)^{-1/2}$, is much slower than the characteristic time scale $t_C = (Z_C)^{-1/2}$ of the coherent motions, because $Z_C \gg Z_I$. This behaviour of the incoherent background had already been noticed, and discussed in comparison to Fourier filtering in [16,53].

The CVS method relies on the assumption that the incoherent velocity remains Gaussian, which is true as long as the nonlinear interactions between the incoherent modes remains weak. This assumption is valid in regions where the density of coherent vortices is sufficient, because the strain they exert on the incoherent background flow then inhibits the development of any nonlinearity there [34]. However, there may be regions, although small, where the density of coherent vortices is not sufficient to control the incoherent nonlinear term. In this case, there are two solutions:

- to locally refine the wavelet basis in these regions in order to deterministically compute the effect of incoherent nonlinear term (no longer neglected), which will lead to the formation of new coherent vortices by instability of the incoherent background flow;

- to directly model the formation of new coherent vortices by adding locally to the wavelet coefficients the amount of coherent enstrophy which has been transferred from the incoherent enstrophy by the nonlinear instability. This procedure is similar to the wavelet forcing we have proposed in [50].

10   Adaptive wavelet computation

10.1   Adaptive wavelet scheme for nonlinear PDE's

This section presents in a general form the adaptive discretization procedure for nonlinear parabolic PDE's. We consider initial value problems and restrict ourself to periodic boundary conditions. The time discretization is done using a classical finite difference scheme of semi-implicit type. One obtains a set of ODE's or PDE's in space which are then solved at each time step with a method of weighted residuals. The particular choice of the trial and test functions defines the different kinds of integration methods.

We consider nonlinear parabolic evolution equation

$$\partial_t u + K u - F(u) = 0 \tag{10.1}$$

with an appropriate initial condition $u(t = 0) = u_0$ and periodic boundary conditions. In (10.1) K denotes a linear differential operator in space (*e.g.* $K = -\nabla^2$) and F a nonlinear function of u. Examples fitting in the above framework are the Navier–Stokes or the reaction-diffusion equations.

10.1.1   Time discretization

Equation (10.1) is discretized in time by a semi-implicit finite difference scheme of second order

$$Lu^{n+1} = f(u^n, u^{n-1}), \tag{10.2}$$

with

$$L = \gamma I + K, \tag{10.3}$$

$$f = \frac{4}{3}\gamma u^n - \frac{1}{3}\gamma u^{n-1} + F(2u^n - u^{n-1}) \tag{10.4}$$

time step Δt, $\gamma = 3/(2\Delta t)$, and I representing the identity. The solution of nonlinear equations is avoided by using an explicit scheme, a modified Adams–Bashforth extrapolation scheme, for the nonlinear function F. However, the restrictive stability condition of a pure explicit scheme is eliminated by discretizing the linear terms $K(u)$ with an implicit scheme of Euler–Backwards type. A suitable first order scheme is employed to start the computation.

The method is presented here considering the equation

$$Lu = f \tag{10.5}$$

where L is an elliptic operator with constant coefficients. Typically L is a one or two-dimensional Helmholtz operator ($L = \gamma I - C\partial_{xx}$ or $L = \gamma I - C(\partial_{xx} + \partial_{yy})$), which arises from the time-discretized Navier–Stokes equation, using a semi-implicit scheme.

The symbol representing L in Fourier space is $\sigma(k)$. In the case of the Helmholtz operator we get $\sigma(k) = \gamma I + 4\pi^2 C k^2$ and $\sigma(k_x, k_y) = \gamma I + 4\pi^2 C(k_x^2 + k_y^2)$ for the one and two-dimensional case, respectively. Observe that $\gamma > 0$ yields $\sigma > 0$, which is another way of expressing that L is an elliptic operator with inhomogenous symbol. Hence, in each time step an elliptic problem has to be solved.

10.1.2   Wavelet decomposition

To simplify our presentation we first describe in some detail the spatial wavelet discretization for the one-dimensional case. The complete algorithms together with a mathematical justification can be found in [30].

For the spatial discretization we use a method of weighted residuals (Petrov–Galerkin scheme). The trial functions are orthogonal wavelets and the test functions are operator adapted wavelets, called vaguelettes [38]. To solve the elliptic equation $Lu^{n+1} = f$ at time step t^{n+1} we develop u^{n+1} into an orthogonal wavelet series, i.e. $u^{n+1} = \sum_{ji} \tilde{u}_{ji}^{n+1} \psi_{ji}$. Requiring that the residuum vanishes with respect to all test functions θ_{kl}, we obtain a linear system for the unknown wavelet coefficients $\tilde{u}_{j,i}^{n+1}$ of the solution u.

$$\sum_{j,i} \tilde{u}_{ji}^{n+1} \langle L\psi_{ji}, \theta_{kl} \rangle = \langle f, \theta_{kl} \rangle. \tag{10.6}$$

The test functions θ are defined in such a way that the stiffness matrix $\langle L\psi_{ji}, \theta_{kl} \rangle$ turns out to be the identity, therefore the solution of equation (10.5) reduces to a simple change of the basis:

$$u^{n+1} = \sum_{kl} \langle f, \theta_{kl} \rangle \psi_{kl}. \tag{10.7}$$

The right hand side f is then expanded into a biorthogonal vaguelette basis

$$f(x) = \sum_{ji} \langle f, \theta_{ji} \rangle \mu_{ji}(x) \tag{10.8}$$

with $\theta_{ji} = L^{\star -1} \psi_{ji}$ and $\mu_{ji} = L\psi_{ji}$ (* denotes the adjoint operator). By construction θ and μ are biorthogonal, $\langle \theta_{ji}, \mu_{kl} \rangle = \delta_{j,k} \delta_{i,l}$. We check that θ and μ have actually similar localization properties in physical and in Fourier space as ψ [26, 30].

To get an adaptive space discretization for the problem $Lu^n = f$ we consider only the significant wavelet coefficients of the solution, $u^n = \sum_{ji} \tilde{u}_{ji}^n \psi_{ji}$, i.e. we retain only the coefficients $\tilde{u}_{j,i}^n$ which have an absolute value larger than a given threshold ϵ. Hence the index set of all active wavelet coefficients is restricted to some subset $(j, i) \in \Lambda_\epsilon^n$, cf. Figure 5. The light-grey entries correspond to those wavelet coefficients which are retained for being larger than ϵ.

To be able to evolve the equation in time we have to account for the evolution of the solution in wavelet coefficient space. Starting at time step t^n (light-grey entries in Fig. 5) with the index set $\Lambda_\epsilon^n = \{(j, i) \big| |\tilde{u}_{ji}^n| > \epsilon\}$, we switch on all neighbours in wavelet coefficient space (dark-grey entires in Fig. 5) to obtain the new index set Λ_ϵ^{n+1} at time step t^{n+1}. This strategy is performed dynamically as it automatically follows the time evolution of the solution in scale and space. The width of the security region (dark-grey entries in Fig. 5) being added in each time step depends directly on the time sampling Δt which of course has to be sufficiently small (CFL condition) as the nonlinear part of the equation is discretized explicitly.

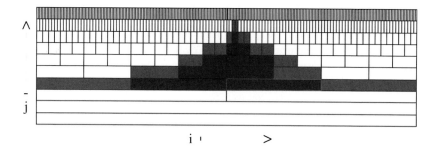

Fig. 5. Scale (vertical axis) and space (horizontal axis) representation of the wavelet coefficients $\tilde{u}_{j,i}$. The light-grey entries correspond to the wavelet coefficients larger than the threshold ϵ. Their neighbors, the dark-grey entries, correspond to the secutity region.

10.1.3   Evaluation of the nonlinear term

For the evaluation of the nonlinear term $f(u^n)$ in (10.5), where the wavelet coefficients of u^n are given, there are two possibilities.

- Evaluation in wavelet coefficient space:

  This technique is appropriate for simple polynomial nonlinearities, but can be extended to general cases using Taylor series expansion. As illustration we consider a quadratic nonlinear term, *i.e.* $f(u) = u^2$. The wavelet/vaguelette coefficients of f can be calculated using the connection coefficients, *i.e.* one has to calculate the bilinear expression, $\sum_{ji} \sum_{kl} \tilde{u}_{ji} T_{jiklmn} \tilde{u}_{kl}$ with the interaction tensor $T_{jiklmn} = \langle \psi_{ji} \psi_{kl}, \theta_{mn} \rangle$. Although many coefficients of T are zero or very small, the size of T leads to a computation which is quite untractable in practice.

- Evaluation in physical space:

  This approach is very similar to the pseudo-spectral evaluation of nonlinear terms used in spectral methods, and therefore this method is called pseudo-wavelet technique. The advantage of this scheme is that more general nonlinear terms, *e.g.* $f(u) = (1 - u)e^{-C/u}$, can be treated more easily. The prerequesites however are that fast adaptive wavelet decomposition and reconstruction algorithms are available. This means that functions can be reconstructed on a locally refined grid from a sparse set of their significant wavelet coefficients and *vice versa*. The algorithms are described and analysed in detail in [30]. The method can be summarized as follows: starting from the significant wavelet coefficients of u, *i.e.* $|\tilde{u}_{ji}| > \epsilon$, one reconstructs

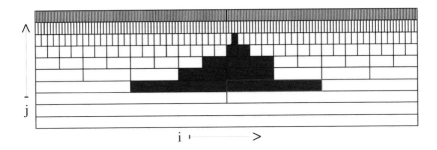

Fig. 6. Scale space representation of the active wavelet coeffcients to be computed (dark entries).

u on a locally refined grid, $u(x_{kl})$. Then one can evaluate $f(u(x_{kl}))$ pointwise and the wavelet coefficients of f can be calculated using the adaptive decomposition to get $\tilde{f}_{ji}$.

Finally, we have to calculate those scalar products of the r.h.s. f with the test functions θ, $\tilde{u}_{ji} = \langle f, \theta_{ji} \rangle$ for the reduced index set $\{(j, i) \in \Lambda_\epsilon^{n+1}\}$, *i.e.* the dark-grey and the light-grey entries in Figure 5, to advance the solution in time.

10.1.4 Substraction strategy

The adaptive vaguelette decomposition is based on the hierarchical representation of the r.h.s. We decompose f into

$$f_J = \sum_i \langle f_J, \theta_{J-1,i} \rangle \mu_{J-1,i} + \sum_i \langle f_J, \theta_{J-2,i} \rangle \mu_{J-2,i} + ... \qquad (10.9)$$

and introduce then hierarchical grids $x_{Jk} = k/2^J$ (see Fig. 7). Starting with the function values on the locally refined grid $f_J(x_{Jk})$ we calculate first the fine scale wavelet coefficients $\tilde{f}_{J-1,i} = \langle f_J, \theta_{J-1,i} \rangle$ (Fig. 6) using an interpolatory quadrature rule on the locally refined grid $\{x_{Jk}\}$ (Fig. 7). Then we coarsen the grid and subtract the fine scale contributions of f, *i.e.* we compute $f_{J-1} = f_J - \sum_i \tilde{f}_{J-1,i} \mu_{J-1,i}$ on the grid $\{x_{J-1,k}\}$ (Fig. 7). Hence we get a coarser scale approximation f_{J-1}. Using $f_{J-1}(x_{J-1,k})$ the wavelet coefficients on the next coarser scale $\tilde{f}_{J-2,i} = \langle f_{J-1}, \theta_{J-2,i} \rangle$ (Fig. 6) can be calculated using the grid $\{x_{J-1,k}\}$ (Fig. 7). The above algorithm is iterated down to the coarsest scale where then a regular grid can be used.

Remark: The above algorithm uses the cardinal function $S_{L;J}(x)$ of the operator adapted approximation space, $V_{L;J} = \text{span}\{\mu_{ji}\}_{j<J}$. The quadrature rule is based on the use of filters $D_{L;m}^J = \langle S_{L;J,m}, \theta_{J-1,0} \rangle$ which are

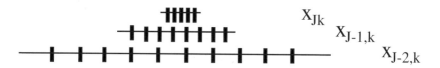

Fig. 7. Corresponding hierarchical grid in physical space.

well localized and hence can be truncated for a given accuracy [30]. This finite filter length leads to an $O(N)$ algorithm where N denotes the number of degrees of freedom, *i.e.* the number of light-grey wavelet coefficients in Figure 6. Let us furthermore mention that the method works for arbitrary index sets, *i.e.* no tree structure of the wavelet coefficients is required.

The algorithms for the vaguelette decomposition, sketched above, and for the adaptive wavelet reconstruction, which is analogous, can be found in [30].

10.1.5 Summary of the algorithm

The essential ingredients of the above algorithm can be summarized as follows:

- the use of orthogonal wavelets and of operator adapted vaguelettes diagonalizes the stiffness matrix and avoids assembling and solving of a linear system;

- the cardinal function allows an easy projection and a decomposition compatible with the operator;

- the localization properties of wavelets lead to a fast decay of the functions, and the associated filters which therefore can be truncated up to a given precision;

- the hierarchical organisation of the basis enables the construction of fast pyramidal algorithms, with $O(N)$ complexity.

10.2 *Adaptive wavelet scheme for the 2D Navier–Stokes equations*

Using a two-dimensional Multi-Resolution Analysis (MRA) obtained through tensor product of two one-dimensional MRA's the above algorithm can be extended to two dimensions. The solution is then developed into a two-dimensional wavelet series $u(x,y) = \sum_j \sum_{i_x,i_y} \sum_{\kappa=1}^{3} \tilde{u}_{j,i_x,i_y}^{\kappa} \psi_{j,i_x,i_y}^{\kappa}(x,y)$.

Due to the fact that the operator L is not separable, *i.e.* $L \neq L_x L_y$, *e.g.* for $L = Id - \partial_{xx} - \partial_{yy}$ the same holds for the operator adapted biorthogonal functions: $\theta(x,y) = L^{\star-1}\psi(x,y) \neq \theta(x)\theta(y)$ and $\mu(x,y) = L\psi(x,y) \neq \mu(x)\mu(y)$ and similarly for the filters S_L and D_L.

Now we extend the previouly described adaptive wavelet-vaguelette algorithm to solve the two-dimensional Navier–Stokes equations written in vorticity-velocity formulation. Using this method the vorticity is decomposed into a lacunary basis of orthogonal wavelets. Adaptivity means that the simulation uses only the minimum number of wavelet modes necessary to represent the vorticity field at any given time and position (within a given precision ϵ). We will now use the reduced set of wavelet basis functions to directly compute the evolution of the flow using the discretization procedure described in the previous section.

We expand the vorticity ω into an orthogonal wavelet series (7.1) and apply a Petrov–Galerkin scheme with test functions

$$\theta^\kappa_{j,i_x,i_y}(x,y) = (\gamma - \nu\nabla^2)^{-1}\psi^\kappa_{j,i_x,i_y}(x,y) \tag{10.10}$$

where $\gamma = 3/(2\Delta t)$ The vaguelettes θ defined in (10.10) can be calculated explicitly in Fourier space and have localization properties similar to those of wavelets [27, 42], *i.e.* in the case of spline wavelets they have exponential decay. The solution then reduces to the calculation of the coefficients

$$\begin{aligned}
\tilde{\omega}^{\kappa,n+1}_{j,i_x,i_y} &= \left\langle \omega^{n+1}, \psi^\kappa_{j,i_x,i_y} \right\rangle \tag{10.11} \\
&= \left\langle \frac{4}{3}\gamma\omega^n - \frac{1}{3}\gamma\omega^{n-1}, \theta^\kappa_{j,i_x,i_y} \right\rangle - \left\langle \vec{v}^* \cdot \nabla\omega^*, \theta^\kappa_{j,i_x,i_y} \right\rangle
\end{aligned}$$

with $\vec{v}^* = 2\vec{v}^n - \vec{v}^{n-1}$ and $\omega^* = 2\omega^n - \omega^{n-1}$, using an adaptive two-dimensional vaguelette decomposition [4], *i.e.* only the coefficients $\tilde{\omega}^{\kappa,n+1}_{j,i_x,i_y}$ larger than the threshold ϵ are calculated. The pyramidal decomposition algorithm is based on a subtraction strategy applied to hierarchical nested grids using the cardinal Lagrange function of the operator-adapted multi-resolution analysis [27]. Subsequently, the vorticity field is reconstructed on a locally refined grid using the adaptive wavelet reconstruction.

The nonlinear term $\vec{v}^\star \cdot \nabla\omega^\star$ is computed by partial collocation [52]. This pseudo-wavelet scheme can be sketched as follows: starting from the wavelet coefficients of ω^n, ω^{n-1} we obtain the values of $\omega^\star$ on a locally refined grid through an inverse wavelet transform. Solving the Poisson equation using a Petrov–Galerkin scheme with test functions $(\nabla^2)^{-1}\psi^\kappa_{j,i_x,i_y}(x,y)$, we get the wavelet coefficients of the stream function $\Psi^\star$. Note that these test functions are rapidly decaying in physical space due to the vanishing moments of the wavelets. Applying an inverse adaptive wavelet transform, the stream function is reconstructed on a locally refined grid. Then the velocity $\vec{v}^\star$, the vorticity gradient $\nabla\omega^\star$ and the nonlinear term $\vec{v}^\star \cdot \nabla\omega^\star$ are calculated using finite differences of 4th order on the adaptive grid. Finally, the right hand side of (10.11) is summed up in physical space using the adaptive grid and then the wavelet coefficients of the vorticity ω^{n+1} are calculated using the adaptive vaguelette decomposition [27].

If the smoothness of ω varies strongly in space and time, it is appropriate to use an adaptive spatial discretization that is re-calculated at each time-step. Adaptivity is also important in order to follow the strong gradients produced by the coherent vortices as the move around. For the wavelet discretization used here the tracking procedure is accomplished by restricting the full index set

$$\Lambda_J = \{(j, i_x, i_y, \kappa)|(j = 0, \ldots, J-1), (i_x = 0, \ldots, 2^j - 1), (i_y = 0, 2^j - 1),$$

$$(\kappa = 1, 2, 3)\} \tag{10.12}$$

to some subset $\Lambda_\epsilon^n \subset \Lambda_J$ which depends on the required tolerance ϵ. The elements designated by Λ_ϵ^n are termed the "lacunary basis" and correspond to the "compressed" wavelet coefficients $|\tilde{\omega}_{j,i_x,i_y}^{\kappa,n}| > \epsilon$. The orthogonality of the basis and the decay of the wavelet coefficients relate this to the L^2-approximation error, which can therefore be evaluated.

The adaptivity algorithm then operates as follows: from the previous time-step one determines those indices with coefficients larger than the threshold ϵ. One prepares a "security zone", to allow for the evolution of the solution in time by adding to each index of this set its neighbouring indices in wavelet space. The solution is then advanced in time by performing a reconstruction of the solution $\omega^\star$ onto the locally refined grid (this minimizes energy loss due to re-gridding). The r.h.s. is then evaluated at these points. Finally, the decomposition onto the operator-adapted basis is applied to determine the new wavelet amplitudes $\tilde{\omega}^{n+1}$. Thus the adaptive algorithm calculates only those wavelet modes that are active (plus a security zone), while the inverse transform ensures that the energy of the neglected modes is re-injected back onto the locally refined grid. Recall that in the inverse transform all wavelet coefficients which have been calculated are evaluated, so that no coefficients are set to zero. The reduction of the index set is only performed afterwards. Finally, let us mention that the complexity of the algorithm is of $O(N_c)$, where N_c denotes the number of wavelet coefficients which are calculated.

Remark: In the case of two-dimensional turbulence the complexity of a DNS increases like $Re = CN$ (with $C = 1$ for Fourier and $C = 3$ to 7 for grid point methods, depending on the order of the scheme employed). The pseudo-spectral methods have a computational cost of $O(N \log_2 N)$, while the adaptive wavelet method has a computational cost of only $O(N_c)$.

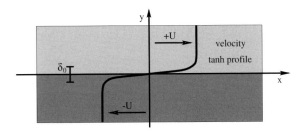

Fig. 8. Initial configuration for the mixing layer.

10.3 Application to a 2D turbulent mixing layer

10.3.1 Adaptive wavelet computation

In [21,55] we studied a temporally developing mixing layer, *cf.* Figure 8. The initial velocity has a hyperbolic-tangent profile $u(y) = U \tanh(2y/\delta_0)$ which implies a vorticity thickness $\delta_0 = 2U/(\mathrm{d}u/\mathrm{d}y)|_{y=0}$. From linear stability analysis the mixing layer is known to be inviscidly unstable. A perturbation leads to the formation of vortices by Kelvin–Helmholtz instability, where the most amplified mode corresponds to a longitudinal wavelength $\lambda = 7\delta_0$. The initial vorticity thickness δ_0 is chosen such that 10 vortices should develop in the numerical domain of size $[0, 2\pi]^2$. To trigger the instability we superimposed a weak white noise in the rotational region. The velocity is $U \approx 0.1035$ and the viscosity is $\nu = 5 \times 10^{-5}$. For the numerical simulation we employ a maximal resolution of 256×256, which corresponds to $L = 8$, and cubic spline wavelets of Battle–Lemarié type. The time step is $\Delta t = 2.5 \times 10^{-3}$. The threshold for the wavelet coefficients was $\epsilon_0 = \epsilon\sqrt{Z} = 10^{-6}$ and 10^{-5}.

In Figure 9 (left) we compare the energy spectrum at $t = 37.5$ of a DNS computation using a classical pseudo-spectral method, with two wavelet computations using different thresholds ($\epsilon_0 = 10^{-6}$ and 10^{-5}).

Figure 9 (left) shows that all scales of the flow are well-resolved for both thresholds. The underlying grid Figure 9 (right) which corresponds to the centers of active wavelets for the computation with $\epsilon_0 = 10^{-6}$ at $t = 37.5$ shows a local refinement in regions of strong gradients.

10.3.2 Comparison between CVS and Fourier pseudo-spectral DNS

In Figure 10 (bottom) we show the evolution of the vorticity field for the adaptive wavelet simulation with threshold $\epsilon_0 = 10^{-6}$ and for the reference pseudo-spectral computation (top). In both simulations, as predicted by the linear theory, 10 vortices are formed, which subsequently undergo

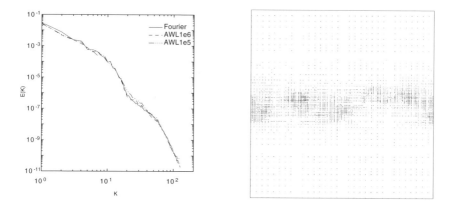

Fig. 9. Left: energy spectra for the pseudo-spectral reference run and for the adaptive wavelet simulations with thresholds $\epsilon_0 = 10^{-6}, 10^{-5}$. Right: adaptive grid reconstructed from the index set of the retained wavelet coefficients. Both at time $t = 37.5$.

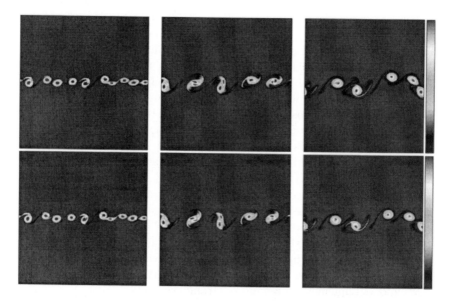

Fig. 10. ω at $t = 12.5$, 25, 37.5. Top: pseudo-spectral method. Bottom: adaptive wavelet method ($\epsilon_0 = 10^{-6}$).

successive mergings. In Figure 11 the active wavelet coefficients (gray entries) are plotted using a logarithmic scale. The coefficients $\tilde{\omega}_\lambda$ are placed at position $(x_1, x_2) = (2^j(1 - \delta_{d,1}) + k_x, 2^j(1 - \delta_{d,2}) + k_y)$ with the origin in

Fig. 11. Active wavelet coefficients at $t = 12.5,\ 25,\ 37.5$ ($\epsilon_0 = 10^{-6}$).

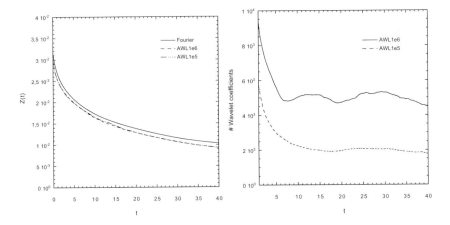

Fig. 12. Left: evolution of enstrophy. Right: evolution of #DOF ($\epsilon_0 = 10^{-6}, 10^{-5}$).

the lower left corner and the y-coordinate oriented upwards, from coarser to finer scales. We observe that the basis dynamically adapts to the flow evolution during the computation with only 8% of the coefficients being used. We observe that in the wavelet simulation the formation and evolution of vortices are well captured, although we find that at later times a slight phase shift appears with respect to the DNS (top). This might be due to the fact that the retained wavelet coefficients contain 94% of the total enstrophy, as observed in Figure 12 (left) which shows the time evolution of the total enstrophy using the different thresholds. The 6% loss of enstrophy comes from the fact that in the wavelet simulations we have not modelled the effect of the discarded modes onto the retained ones. This will be considered in future work, where the enstrophy of the discarded wavelets will be reinjected into the coherent vortices using the wavelet forcing method we have proposed [50].

Finally, we plot the time evolution of the number of degrees of freedom for the two wavelet runs in Figure 12 (right). First, we observe an initial phase, up to $t = 7$ s, where there is a strong reduction in the number of active modes, which corresponds to the formation of the coherent vortices. Then the number of active modes remains almost constant, and represent a significant reduction of the number of modes, with $N_{ad} = 5000$ for $\epsilon_0 = 10^{-6}$ and $N_{ad} = 2000$ for $\epsilon_0 = 10^{-5}$ out of $N = 65536$ initial modes.

Part IV
Conclusion

In these lecture notes we have reviewed wavelet techniques for analyzing and computing turbulent flows. We summarized the continuous and orthogonal wavelet transforms and presented the algorithms for their numerical implementation. Furthermore we presented classical and some recently developed wavelet-based statistical tools for analyzing turbulent flows.

We then described a wavelet-based method designed for computing turbulent flows, called CVS (Coherent Vortex Simulation). The CVS filtering consists of projecting the vorticity field onto an orthogonal wavelet basis and decomposing it into two orthogonal components using a nonlinear thresholding of the wavelet coefficients. The coherent vorticity field is reconstructed from the few wavelet coefficients larger than a given threshold, which depends only on the resolution and on the total enstrophy, while the incoherent vorticity is reconstructed from the many remaining weak wavelet coefficients. The coherent and incoherent velocity fields are then derived from the coherent and incoherent vorticity fields using Biot–Savart's integral equation.

We applied the CVS filtering to a 3D turbulent mixing layer. We have shown that few strong wavelet coefficients represent the coherent vortices, the whole energy spectrum and the whole vorticity PDF. In contrast, many weak wavelet coefficients represent the incoherent background flow, which is structureless, exhibits an energy equipartition spectrum and an exponential distribution for the vorticity PDF. This demonstrates the advantage of the CVS filtering in comparison to the LES filtering, since we have shown that the small scale flow, which is discarded in LES, exhibits many coherent structures, has a much wider PDF of vorticity and does not present an energy equipartition spectrum.

We then presented an adaptive wavelet computation of a 2D mixing layer and shown the dynamical adaption of the grid in physical space which allows to follow the flow evolution with a reduced number of active degrees of freedom. We showed that the adaptive wavelet method produces accurate results with fewer active modes than the classical pseudo-spectral method, since it exploits the vortical structure of turbulent flows, while the adaptive wavelet method does, which leads to greater efficiency at high Reynolds numbers.

In conclusion, although the numerical method we propose is Eulerian, based on a Galerkin scheme, its adaptive character, in both space and scale, allows to track the displacements and deformations of active flow regions, as Lagrangian methods do.

We thankfully acknowledge partial support from the European Program TMR on "Wave-lets in Numerical Simulation" (contract FMRX-CT98–0184), the Pluri-Formation Program of École Normale Supérieure-Paris (contract 15407) and the French–German Program Procope (contract 99090). We thank G. Pellegrino for developing the 3D wavelet code we have used.

References

[1] W.T. Ashurst, A.R. Kerstein, R.M. Kerr and C.H. Gibson, *Phys. Fluids* **30** (1987) 2343.

[2] G.K. Batchelor, *Homogeneous turbulence* (Cambridge University Press, 1953).

[3] G. Beylkin, R. Coifman and V. Rokhlin, *Comm. Pure Appl. Math.* **44** (1991) 141.

[4] H. Bockhorn, J. Fröhlich and K. Schneider, *Combust. Theory Modelling* **3** (1999) 177-198.

[5] C. Canuto, M.Y. Hussaini, A. Quaternioni and T.A. Zang, *Spectral Methods in Fluid Dynamics* (Springer-Verlag, Berlin, 1988).

[6] P. Charton and V. Perrier, *Comp. Appl. Math.* **15** (1996) 139-160.

[7] A.J. Chorin, *Vorticity and turbulence* (Springer, 1994).

[8] I. Daubechies, *Ten Lectures on Wavelets* (SIAM, Philadelphia, 1992).

[9] P.E. Dimotakis and H.J. Catrakis, Turbulence, fractals and mixing, *Mixing, chaos and Turbulence*, edited by H. Chaté, E. Villermaux and J.M. Chomaz, NATO ASI Series, Serie B: *Physics* **373** (Kluwer, 1999) pp. 59-144.

[10] M. Do–Khac, C. Basdevant, V. Perrier and K. Dang–Tran, *Physica D* **76** (1994) 252-277.

[11] D. Donoho, Wavelet shrinkage and Wavelet -Vaguelette Decomposition: A 10–minute tour, in *Progress in Wavelet Analysis and Applications*, Proc. Int. Conf. Wavelets and Applications (Toulouse, 1992) pp. 109-128.

[12] D. Donoho, I. Johnstone, G. Kerkyacharian and D. Picard, *J. Royal Stat. Soc. Ser. B* **57** (1995) 301-369.

[13] D.G. Dritschel and D.W. Waugh, *Phys. Fluids* A **4** (1992) 1737.

[14] M. Farge, *Ann. Rev. of Fluid Mech.* **24** (1992) 395-457.

[15] M. Farge, J.-F. Colonna and M. Holschneider, in *Topological Fluid Mechanics*, edited by H.K. Moffatt and A. Tsinober (Cambridge University Press, 1989) pp. 765-776.

[16] M. Farge, E. Goirand, Y. Meyer, F. Pascal and M.V. Wickerhauser, *Fluid Dynamics Research* **10** (1992) 229-250.

[17] M. Farge and G. Rabreau, *C. R. Acad. Sci. Paris Série II* **307** (1988) 1479.

[18] M. Farge, Y. Guezennec, C.M. Ho and C. Meneveau, Continuous wavelet analysis of coherent structures, *Proceedings of the 1990 Summer Program, Center for Turbulence Research* (NASA-Ames and Stanford University, 1990) pp. 331-348.

[19] M. Farge, K. Schneider and N. Kevlahan, Coherent structure eduction in wavelet–forced two–dimensional turbulent flows, *IUTAM Symposium on Dynamics of slender vortices*, edited by E. Krause and K. Gersten (Kluwer Academic Publishers, 1998) pp. 65-83.

[20] M. Farge, N. Kevlahan, V. Perrier and K. Schneider, Turbulence analysis, modelling and computing using wavelets, *Wavelets in Physics*, edited by J.C. van den Berg (Cambridge University Press, 1999) pp. 117-200.

[21] M. Farge, K. Schneider and N. Kevlahan, *Phys. Fluids* **11** (1999) 2187-2201.

[22] M. Farge and K. Schneider, Coherent Vortex Simulation (CVS), a semi-deterministic turbulence model using wavelets, *Flow, Turb. Combust.* (submitted).

[23] M. Farge, G. Pellegrino and K. Schneider, Coherent vortex extraction in 3D turbulent flows using orthogonal wavelets. Preprint (2000) *Phys. Rev. Lett.* (submitted).

[24] J.H. Ferziger, Large Eddy Simulation, *Simulation and Modeling of Turbulent Flows*, edited by T.B. Gatski, M.Y. Hussaini and J.L. Lumley, *ICASE Series in Computational Science and Engineering* **109** (1992).

[25] U. Frisch, *Turbulence. The legacy of A.N. Kolmogorov* (Cambridge University Press, 1995).

[26] J. Fröhlich and K. Schneider, *Eur. J. Mech. B* **13** (1994) 439-471.

[27] J. Fröhlich and K. Schneider, *C. R. Math. Rep. Acad. Sci. Canada* **17** (1995) 283.

[28] J. Fröhlich and K. Schneider, *Appl. Comput. Harm. Anal.* **3** (1996) 393-397.

[29] J. Fröhlich and K. Schneider, *Physica D* **134** (1999) 337-361.

[30] J. Fröhlich and K. Schneider, *J. Comput. Phys.* **130** (1997) 174-190.

[31] A. Grossmann and J. Morlet, *SIAM J. Math. Anal.* **15** (1984) 723-736.

[32] M. Germano, U. Piomelli, P. Moin and W.H. Cabot, A dynamic subgrid scale model eddy viscosity model, in *Proceedings Summer Workshop* (Center for Turbulence Research, Stanford, 1990).

[33] M. Holschneider and Ph. Tchamitchiam, *Régularité locale de la fonction non-différentiable de Riemann*, in *Les ondelettes en 1989*, Lecture Notes in Mathematics, edited by P.G. Lemarié (Springer, 1989).

[34] N.K.-R. Kevlahan and M. Farge, *J. Fluid Mech.* **346** (1997) 49-76.

[35] A.N. Kolmogorov, *C. R. Acad. Sc. USSR* **30** (1941) 301-305.

[36] R. Kraichnan, *Phys. Fluids* **10** (1967) 1417-1423.

[37] A. Leonard, *Adv. Geophys.* **18 A** (1974) 237.

[38] J. Liandrat and Ph. Tchamitchian, *Resolution of the 1D Regularized Burgers Equation Using a Spatial Wavelet Approximation Algorithm and Numerical Results* (ICASE, 1990).

[39] S. Mallat, *A wavelet tour of signal processing* (Academic Press, 1997).

[40] C. Meneveau, *J. Fluid Mech.* **232** (1991) 469-520.

[41] M.V. Melander, N.J. Zabusky and J.C. McWilliams, *J. Fluid Mech.* **195** (1988) 303.

[42] Y. Meyer, *Ondelettes et Opérateurs I/II* (Hermann, Paris, 1990).

[43] A.S. Monin and A.M. Yaglom, *Statistical Fluid Mechanics: Mechanics of Turbulence* (The M.I.T. Press 1975).

[44] R. Murenzi, *Wavelet transforms associated to the n–dimensional Euclidean Group with dilatations*, Ph.D. Thesis (UCL, Louvain-la-Neuve, 1989).

[45] J. von Neumann, Recent theories of turbulence, in *Collected works (1949–1963)* **6**, edited by A.H. Taub (Pergamon Press, 1963) p. 437.

[46] V. Perrier and C. Basdevant, *Rech. Aérosp.* **3** (1989) 53.

[47] V. Perrier, T. Philipovich and C. Basdevant, *J. Math. Phys.* **36** (1995) 1506-1519.

[48] M. Rogers and R. Moser, *Phys. Fluids* **6** (1994) 903-923.

[49] P.G. Saffman, *Vortex Dynamics* (Cambridge University Press, 1972).

[50] K. Schneider and M. Farge, *C. R. Acad. Sci. Paris Série II* **325** (1997) 263-270.

[51] K. Schneider, N. Kevlahan and M. Farge, *Theoret. Comput. Fluid Dynamics* **9** (1997) 191-206.

[52] K. Schneider, N. Kevlahan and M. Farge, An adaptive wavelet method compared to nonlinearly filtered pseudo-spectral methods for two-dimensional turbulence, *Advances in Turbulence VII*, edited by U. Frisch (Kluwer Academic Publishers, 1998) pp. 147-150.

[53] K. Schneider and M. Farge, Wavelet approach for modelling and computing turbulence. *Lecture Series 1998–05 Advances in turbulence modelling* (von Karman Institute for Fluid Dynamics, Bruxelles, 1998) 132 pages.

[54] K. Schneider, M. Farge and N. Kevlahan, *Intermittency and coherent vortices in fully-developed two-dimensional turbulence*, Preprint (ICT Universität Karlsruhe, 1999).

[55] K. Schneider and M. Farge, *C. R. Acad. Sci. Paris Série II b* (2000) 263-269.

[56] R. Teman, *Navier–Stokes equations and nonlinear functional analysis* (SIAM, Philadelphia, 1983).

[57] A. Vincent and M. Meneguzzi, *J. Fluid Mech.* **225** (1991) 1-20.

[58] A. Vincent and M. Meneguzzi, *J. Fluid Mech.* **258** (1994) 245.

COURSE 10

LAGRANGIAN DESCRIPTION OF TURBULENCE

G. FALKOVICH

Weizmann Institute of Science,
Rehovot 76100, Israel

Contents

LAGRANGIAN DESCRIPTION OF TURBULENCE

G. Falkovich[1], K. Gawedzki[2] and M. Vergassola[3]

Abstract

This lecture course is intended to bring home to a reader two main lessons: power of the Lagrangian approach to fluid turbulence and importance of statistical integrals of motion for systems far from equilibrium. We present the description of turbulence from a Lagrangian viewpoint that is based on the motion of fluid particles. Section 1 describes the statistics of one, two and many-particle configurations. Section 2 describes the statistics of the passive scalar that can be inferred from the particle analysis. In Section 3, we discuss the Navier–Stokes equation from a Lagrangian viewpoint. Conclusion lists the most important open problems. We restrict ourselves to incompressible flows. Lagrangian description of compressible flows (as well as other subjects like magnetic dynamo, Lagrangian numerics etc.) can be found in the extended version [1].

1 Particles in fluid turbulence

In this section, we consider the time-dependent statistics of the set of Lagrangian trajectories defined by the equation $d\mathbf{R}_i/dt = \mathbf{v}(t, \mathbf{R}_i)$ assuming the statistical properties of the velocity field to be known. For a single trajectory, we show that the motion is a simple diffusion at times larger than velocity correlation time in the Lagrangian frame. The separation law of two trajectories depends on the scaling properties of the velocity field. If the velocity is smooth, that is $\mathbf{v}(\mathbf{R}_i) - \mathbf{v}(\mathbf{R}_j) \propto \mathbf{R}_i - \mathbf{R}_j$, then the separation grows exponentially. When the velocity can be considered non-smooth at such distances, that is $\mathbf{v}(\mathbf{R}_i) - \mathbf{v}(\mathbf{R}_j) \propto |\mathbf{R}_i - \mathbf{R}_j|^\alpha$ with $\alpha < 1$, then the separation grows by a power law. In considering a set of more than two trajectories, two new (and related) topics appear: geometry and conservation laws.

This course was partially written in the ITP, UCSB, Santa Barbara.

[1] Weizmann Institute of Science.
[2] CNRS, IHES, Bures-sur-Yvette, France.
[3] Observatoire Nice, France.

1.1   Single-particle diffusion

The Lagrangian trajectory $\mathbf{R}(t)$ of a fluid particle advected by a prescribed incompressible velocity field $\mathbf{v}(\mathbf{r}, t)$ and undergoing molecular diffusion with diffusivity κ is governed by the stochastic equation customarily written for differentials [2]:

$$\mathrm{d}\mathbf{R} = \mathbf{v}(\mathbf{R}, t)\mathrm{d}t + \sqrt{2\kappa}\mathrm{d}\beta(\mathbf{t}). \qquad (1.1)$$

Here, $\beta(\mathbf{t})$ is the d-dimensional standard Brownian motion with zero average and covariance function $\langle \beta^i(t)\beta^j(t') \rangle = \delta^{ij}\min(|t|, |t'|)$. The solution of (1.1) is fixed by prescribing the particle position at $t = 0$.

The simplest instance of (1.1) is the Brownian motion when the advection is absent. The probability density $\mathcal{P}(\mathbf{r}, t)$ of the displacement $\mathbf{R}(t) - \mathbf{R}(0)$ satisfies the heat equation $(\partial_t - \kappa\nabla^2)\mathcal{P}(\mathbf{r}, t) = 0$ whose solution is the Gaussian distribution $\mathcal{P}(\mathbf{r}, t) = (4\pi\kappa t)^{-d/2}\mathrm{e}^{-r^2/(4\kappa t)}$. The other limiting case is a pure advection without noise. Here, the displacement $\mathbf{R}(t) - \mathbf{R}(0) = \int_0^t \mathbf{V}(s)\,\mathrm{d}s$, with $\mathbf{V}(t) = \mathbf{v}(\mathbf{R}(t), t)$ being the Lagrangian velocity. The properties of the displacement depend on the specific trajectory under consideration. We shall always work in the frame of reference with no mean flow: $\langle \mathbf{v} \rangle = 0$. We assume statistical homogeneity of the Eulerian velocities which implies that the stochastic process $\mathbf{V}(t)$ does not depend on the initial position $\mathbf{R}(0)$ of the trajectory. If, additionally, the Eulerian velocities are statistically stationary, then so is $\mathbf{V}(t)$. This follows by averaging the expectations involving $V(t + \tau)$ over the initial position $\mathbf{R}(0)$ (on which they do not depend) and the change of variables $\mathbf{R}(0) \mapsto \mathbf{R}(\tau)$ under the velocity ensemble average. Note that the Jacobian of the change of variables is supposed to be unity which requires incompressibility. For $\kappa = 0$, the mean square displacement satisfies the equation:

$$\frac{\mathrm{d}}{\mathrm{d}t}\langle [\mathbf{R}(t) - \mathbf{R}(0)]^2 \rangle = 2\int_0^t \langle \mathbf{V}(0) \cdot \mathbf{V}(s) \rangle \mathrm{d}s. \qquad (1.2)$$

The behavior of the displacement is crucially dependent on the range of temporal correlations of the Lagrangian velocity. Let us define the correlation time τ of $\mathbf{V}(t)$ by

$$\int_0^\infty \langle \mathbf{V}(0) \cdot \mathbf{V}(s) \rangle \mathrm{d}s = \langle \mathbf{v}^2 \rangle \tau. \qquad (1.3)$$

The value of τ provides a measure of the Lagrangian velocity memory, its divergence being symptomatic of persistent correlations. No general relation between the Eulerian and the Lagrangian correlation times has been established, except for the case of short-correlated velocities. For times $t \ll \tau$, the 2-point function in (1.2) is approximately equal to $\langle \mathbf{V}(0)^2 \rangle = \langle \mathbf{v}^2 \rangle$. The fluid particle transport is then ballistic with $\langle [\mathbf{R}(t) - \mathbf{R}(0)]^2 \rangle \simeq \langle \mathbf{v}^2 \rangle t^2$.

When the correlation time of $\mathbf{V}(t)$ is finite (a generic situation in a turbulent flow) an effective diffusive regime is expected to arise for $t \gg \tau$ with $\langle (\mathbf{R}(t) - \mathbf{R}(0))^2 \rangle \simeq 2\langle \mathbf{v}^2 \rangle \tau t$ [2]. Indeed, the particle displacements over time segments much larger than τ are almost independent. At long times, the displacement $\delta \mathbf{R}(t)$ behaves then as a sum of many independent variables and falls into the class of stationary processes governed by the Central Limit Theorem. In other words, $\delta \mathbf{R}(t)$ for $t \gg \tau$ becomes a Brownian motion in d dimensions, normally distributed with $\left\langle \delta R^i(t) \delta R^j(t) \right\rangle \simeq 2 D_{\mathrm{e}}^{ij} t$, where

$$D_{\mathrm{e}}^{ij} = \tfrac{1}{2} \int\limits_0^\infty \langle V^i(0)\, V^j(s) + V^j(0)\, V^i(s) \rangle \, \mathrm{d}s. \tag{1.4}$$

The same arguments carry over to the case of a non-vanishing molecular diffusivity. The symmetric second order tensor D_{e}^{ij} has the trace equal to $\langle \mathbf{v}^2 \rangle \tau$, *i.e.* to the large time value of the integral in (1.2), while its tensorial properties reflect the rotational symmetries of the advecting velocity field. If the latter is isotropic, the tensor reduces to a diagonal form characterized by a single scalar value D_{e}. The main problem of turbulent diffusion is to obtain the effective diffusivity tensor given the velocity field $\mathbf{v}$ and the value of the diffusivity κ. A huge amount of work has been devoted to it, both from the applied and the mathematical point of view, and exhaustive reviews of the problem are available in the literature [3–6].

The other general issue in turbulent diffusion is the condition on the velocity $\mathbf{v}(\mathbf{r}, t)$ ensuring that the Lagrangian correlation time τ is finite and an effective diffusion regime is taking place for large enough times. A sufficient condition valid for $\kappa \neq 0$ and both static and time-dependent flow, is a finite vector potential variance $\langle \mathbf{A}^2 \rangle$, where the three-dimensional incompressible velocity $\mathbf{v} = \nabla \times \mathbf{A}$ [7–9]. The correlation time is experimentally known to be finite in developed turbulence whereas both subdiffusion (due to particle trapping) and superdiffusion (due to infinite Lagrangian correlation time) are possible in low-Reynolds-number flows.

For the long-time description of the diffusion in finite-correlated flows, it is useful to consider the extreme case of random homogeneous and stationary Eulerian velocities with a very short correlation time. This case may be regarded as describing the sped-up-film view of velocity fields with temporal decay of correlations or, more formally, as the scaling limit $\lim\limits_{\mu \to \infty} \mu^{\frac{1}{2}} \mathbf{v}(\mathbf{r}, \mu t)$.

For a sufficiently rapid decay, such a limit is governed by the Central Limit Theorem and we recover when $\mu \to \infty$ a velocity field which is Gaussian and white noise in time, characterized by the 2-point function

$$\langle v^i(\mathbf{r}, t)\, v^j(\mathbf{r}', t') \rangle = 2\, \delta(t - t') D^{ij}(\mathbf{r} - \mathbf{r}'). \tag{1.5}$$

It is common to call the Gaussian ensemble of velocities with 2-point function (1.5) the Kraichnan ensemble [10]. For the Kraichnan velocities $\mathbf{v}$, the Lagrangian velocity $\mathbf{V}(t)$ has the same white noise temporal statistics as the Eulerian velocity $\mathbf{v}(\mathbf{r}, t)$ for fixed $\mathbf{r}$ and the displacement along a Lagrangian trajectory $\mathbf{R}(t) - \mathbf{R}(0)$ is a Brownian motion for all times. The eddy diffusivity tensor $D_e^{ij} = D^{ij}(\mathbf{0})$, a special case of the relation (1.4). In the presence of molecular diffusion, the overall diffusivity is the sum of the eddy diffusivity $D^{ij}(\mathbf{0})$ and of the molecular one $\kappa\,\delta^{ij}$. We shall see in the sequel that the Kraichnan ensemble of velocities constitutes an important theoretical laboratory of the particle behavior in fluid turbulence.

1.2   Two-particle dispersion in a spatially smooth velocity

The separation $\mathbf{R}_{12} = \mathbf{R}_1 - \mathbf{R}_2$ between two fluid particles with trajectories $\mathbf{R}_i(t) = \mathbf{R}(t; \mathbf{r}_i)$ passing at $t = 0$ through points $\mathbf{r}_i$ satisfies (in the absence of noise) the equation

$$\dot{\mathbf{R}}_{12} \;=\; \mathbf{v}(\mathbf{t}, \mathbf{R_1}) - \mathbf{v}(\mathbf{t}, \mathbf{R_2}). \qquad (1.6)$$

We consider an incompressible flow where particles generally separate. In this section, we start from the smallest distances where the velocity field can be considered spatially smooth due to viscosity. In the next section, we treat the dispersion problem for larger distances (in the inertial interval of turbulence) where the velocity field has a nontrivial scaling and cannot be considered as spatially smooth.

1.2.1   General consideration

If the distance R_{12} is smaller than the viscous scale of turbulence then the velocity field can be considered smooth on such a scale and we may expand: $\mathbf{v}(\mathbf{t}, \mathbf{R_1}) - \mathbf{v}(\mathbf{t}, \mathbf{R_2}) = \sigma(\mathbf{t}, \mathbf{R_1})\mathbf{R_{12}}$ introducing the strain matrix σ which is traceless due to incompressibility. As a function of its spatial argument, σ changes on a scale that is supposed to be much larger than R_{12}. Then, σ can be treated as independent of $\mathbf{R}_{12}$ which thus satisfies locally a linear ordinary differential equation (we omit subscripts replacing $\mathbf{R}_{12}$ by $\mathbf{R}$)

$$\dot{\mathbf{R}}(t) = \sigma(t)\mathbf{R}(t). \qquad (1.7)$$

This equation, with the strain treated as given, may be explicitly solved for arbitrary $\sigma(t)$ only in the 1D case

$$\ln[R(t)/R(0)] = \ln W(t) = \int_0^t \sigma(s)\,\mathrm{d}s \equiv X, \qquad (1.8)$$

expressing $W(t)$ as the exponential of the time-integrated strain. When t is much larger than the correlation time τ of the strain, the variable X is a sum of many independent equally distributed random numbers $X = \sum_1^N y_i$ with $N = t/\tau$. The generating function $\langle e^{zX} \rangle$ of X is thus the N-th power of that for y, denoted $e^{S(z)}$. By inverse Laplace transform we find then the PDF $\mathcal{P}(X)$:

$$\mathcal{P}(X) = \frac{1}{2\pi i} \int e^{-zX + NS(z)} \, dz \; \propto \; e^{-NH(X/N - \langle y \rangle)}, \qquad (1.9)$$

with the integration over the imaginary axis. The second relation holds in the limit $N \to \infty$, where the integral is dominated by the saddle-point z_0 given by $S'(z_0) = X/N$ (provided S has no linear pieces). A few important properties of the PDF $\mathcal{P}(X)$ can be established independently of the distribution $\mathcal{P}(y)$. The function $H = -S(z_0) + z_0 S'(z_0)$ of the variable $X/N - \langle y \rangle$ is called entropy function as it appears also in the thermodynamic limit in statistical physics [11]. The general properties of H (also called rate or Cramér function) are as follows. It is convex and takes its minimum at zero, corresponding to the mean value $\langle X \rangle = N\langle y \rangle = NS'(0)$. The minimal value of H is also vanishing since $S(0) = 0$. The entropy is quadratic around its minimum with $H''(0) = \Delta^{-1}$, where $\Delta = S''(0)$ is the variance of y. The possible non-Gaussianity of the y's leads to a non-quadratic behavior of H for deviations of X/N from the mean of the order of $\Delta/S'''(0)$. Coming back to the interparticle distance $R(t)$ in (1.8), its growth (or decay) rate $\lambda = \langle X \rangle / t$ is called the Lyapunov exponent. The moments $\langle [R(t)]^p \rangle$ behave exponentially as $\exp[E(p)t]$. The convexity of the entropy function leads to the convexity of $E(p)$. This implies, in particular, that even for $\lambda = E'(0) < 0$, high-order moments of R may grow exponentially in time.

In the multidimensional case, the behavior of the vector $\mathbf{R}$ is determined by the product of random matrices rather than just random numbers. Still, the main properties of R statistics (sufficient for most physical applications) can be established for an arbitrary strain. The basic idea is coming back to Lyapunov [12] and it found further development in the Multiplicative Ergodic Theorem of Oseledec [13]. Introduce the evolution matrix W such that $\mathbf{R}(t) = W(t)\mathbf{R}(0)$. The modulus R is expressed *via* the positive symmetric matrix $W^T W$. The main result states that in almost every realization of the strain, the matrix $t^{-1} \ln W^T W$ stabilizes at $t \to \infty$, *i.e.* its eigenvectors tend to d fixed orthonormal eigenvectors $\mathbf{f}_i$. To understand that intuitively, consider some fluid volume, say a sphere, which evolves into an elongated ellipsoid at later times. As time increases, the ellipsoid is more and more elongated and it is less and less likely that the hierarchy of the ellipsoid axes will change. The limiting eigenvalues

$$\lambda_i = \lim_{t \to \infty} t^{-1} \ln |W \mathbf{f}_i| \qquad (1.10)$$

are called Lyapunov exponents. The major property of the Lyapunov exponents is that they are realization-independent (ergodicity of the strain is assumed). We arrange the exponents in non-increasing order.

The relation (1.10) tells that two fluid particles separated initially by $\mathbf{R}(0)$ pointing into the direction $\mathbf{f}_i$ will separate (or converge) asymptotically as $\exp(\lambda_i t)$. The incompressibility constraints $\det(W) = 1$ and $\sum \lambda_i = 0$ imply that a positive Lyapunov exponent will exist whenever at least one of the exponents is nonzero. Consider indeed

$$E(p) = \lim_{t \to \infty} t^{-1} \ln \langle [R(t)/R(0)]^p \rangle, \qquad (1.11)$$

whose derivative at the origin gives the largest Lyapunov exponent λ_1. Because of incompressibility and isotropy, $E(-d) = 0$ [14, 15] that is $\langle R^{-d} \rangle$ is a statistical integral of motion at $t \to \infty$. Apart from $p = 0, -d$, the convex function $E(p)$ cannot have other zeroes if it does not vanish identically. It follows that $dE/dp(0)$, and thus λ_1, is positive. The simplest way to appreciate intuitively the existence of a positive Lyapunov exponent is to consider the saddle-point 2D flow $v_x = \lambda x, v_y = -\lambda y$ with the axes randomly rotating after time interval T. A vector initially at the angle ϕ with the x-axis will be stretched after time T if $\cos \phi \geq [1 + \exp(2\lambda T)]^{-1/2}$, *i.e.* the measure of the stretching directions is larger than $1/2$ [15].

A major consequence of the existence of a positive Lyapunov exponent for any random incompressible flow is the exponential growth of the interparticle distance $R(t)$. In a smooth flow, it is also possible to analyze the statistics of the set of vectors $\mathbf{R}(t)$ and to establish a multidimensional analog of (1.9) for the general case of a nondegenerate Lyapunov exponent spectrum. The idea is to reduce the d-dimensional problem to a set of d scalar problems excluding the angular degrees of freedom. We describe this procedure following [16]. Consider the matrix $I(t) = W(t)W^T(t)$, representing the tensor of inertia of a fluid element like the above mentioned ellipsoid. The matrix is obtained by averaging $R^i(t)R^j(t)/\ell^2 d$ over the initial vectors of length ℓ and $I(0) = 1$. In contrast to $W^T W$ that stabilizes, the matrix I rotates in every realization. To account for that rotation, we represent the matrix as $O^T \Lambda O$ with the orthogonal O composed of the eigenvectors of I and the diagonal Λ having the eigenvalues $e^{2\rho_1}, \ldots e^{2\rho_d}$ arranged in non-increasing order. The evolution equation $\partial_t I = \sigma I + I \sigma^T$ takes then the form

$$\partial_t \rho_i = \tilde{\sigma}_{ii}, \quad \tilde{\sigma} = O \sigma O^T, \qquad (1.12)$$

$$\partial_t O = \Omega O, \quad \Omega_{ij} = \frac{e^{2\rho_i} \tilde{\sigma}_{ji} + e^{2\rho_j} \tilde{\sigma}_{ij}}{e^{2\rho_i} - e^{2\rho_j}}, \qquad (1.13)$$

with no summation over repeated indices. We assume isotropy so that at large times the $SO(d)$ rotation matrix O is distributed uniformly over the sphere. Our task is to describe the statistics of the stretching and the contraction, governed by the eigenvalues ρ_i. We see from (1.12, 1.13) that the evolution of the eigenvalues is generally entangled to that of the angular degrees of freedom. As time increases, the eigenvalues will however become widely separated ($\rho_1 \gg \ldots \gg \rho_d$) for the majority of the realizations and $\Omega_{ij} \to \tilde{\sigma}_{ji}$ for $i < j$ (the upper triangular part of the matrix follows from antisymmetry). The dynamics of the angular degrees of freedom becomes then independent of the eigenvalues and the set of equations (1.12) reduces to a scalar form. The solution $\rho_i = \int_0^t \tilde{\sigma}_{ii}(s)\, \mathrm{d}s$ lends to the application of the large deviation theory, giving the asymptotic PDF:

$$
\begin{aligned}
\mathcal{P}(\rho_1,\ldots,\rho_d;t) \quad &\propto \quad \exp\left[-t\,H(\rho_1/t - \lambda_1,\ldots,\rho_{d-1}/t - \lambda_{d-1})\right] \\
&\times \quad \theta(\rho_1 - \rho_2)\ldots\theta(\rho_{d-1} - \rho_d)\,\delta(\rho_1 + \ldots + \rho_d)\,. \quad (1.14)
\end{aligned}
$$

The Lyapunov exponents λ_i are related to the strain statistics as $\lambda_i = \langle \tilde{\sigma}_{ii}\rangle$. The expression (1.14) is not valid near the boundaries $\rho_i = \rho_{i+1}$ in a region of order unity. At $t \gg (\lambda_i - \lambda_{i+1})^{-1}$ we can disregard this region as it is much smaller than $\lambda_i t$.

The entropy function H depends on the details of the statistics of σ and has the same general properties as above: it is non-negative, convex and it vanishes at zero. Near the minimum, $H(\mathbf{x}) \approx (C^{-1})_{ij}x_i x_j/2$ with the coefficients of the quadratic form given by the integrals of the connected correlation functions of $\tilde{\sigma}$ defined in (1.12):

$$
C_{ij} = \int \langle\langle \tilde{\sigma}_{ii}(t), \tilde{\sigma}_{jj}(t')\rangle\rangle \; \mathrm{d}t'\,, \quad i,j = 1,\ldots,d-1\,. \quad (1.15)
$$

In the δ-correlated case, H is everywhere quadratic – see below. For a generic initial vector $\mathbf{r}$, the long-time asymptotics of $\ln(R/r)$ coincides with that of ρ_1 whose PDF also takes the large-deviation form (1.14) at large times. The quadratic expansion of the entropy near its minimum corresponds to the lognormal distribution for the distance between two particles

$$
\mathcal{P}(r;R;t) \propto \exp\left\{-\left[\ln(R/r) - \lambda_1 t\right]^2/(2t\Delta)\right\}\,, \quad (1.16)
$$

with $r = R(0)$ and $\Delta = C_{11}$.

Molecular diffusion is incorporated into the above picture by replacing the differential equation (1.7) by its noisy version[4]

$$
\mathrm{d}\mathbf{R}(t) = \sigma(t)\mathbf{R}(t)\, \mathrm{d}t + 2\sqrt{\kappa}\, \mathrm{d}\beta(t)\,. \quad (1.17)
$$

---

[4]The noise in the separation vector is a sum of the independent noises of two particles, hence the change in the noise coefficient with respect to (1.1).

This is an inhomogeneous linear stochastic equation whose solution is easy to express *via* the matrix $W(t)$. The tensor of inertia of a fluid element $I^{ij}(t) = R^i(t)R^j(t)/\ell^2 d$ is now averaged both over the initial vectors of length ℓ and the noise, thus obtaining [16]:

$$I(t) = W(t)W(t)^T + \frac{4\kappa}{\ell^2 d} \int_0^t W(t)\,[W(s)^T W(s)]^{-1} W(t)^T\,\mathrm{d}s. \quad (1.18)$$

The matrix $I(t)$ evolves according to $\partial_t I = \sigma I + I\sigma + 4\kappa/\ell^2 d$ and the elimination of the angular degrees of freedom proceeds as previously. An additional diffusive term $2\kappa \exp(-2\rho_i)$ appears in (1.12) and its asymptotic solution becomes

$$\rho_i(t) = \int_0^t \tilde{\sigma}_{ii}(s)\,\mathrm{d}s + \frac{1}{2} \ln \left\{ 1 + \frac{4\kappa}{\ell^2 d} \int_0^t \exp\left[-2\int_0^s \tilde{\sigma}_{ii}(s')\,\mathrm{d}s' \right]\mathrm{d}s \right\}. \quad (1.19)$$

The last term in (1.19) is essential for the directions corresponding to negative λ_i. The molecular noise will indeed start to affect the motion of the marked fluid volume when the respective dimension gets sufficiently small. If ℓ is the initial size, the required condition $\rho_i < -\rho_i^* = -\ln(\ell^2|\lambda_i|/\kappa)$ is typically met for times $t \simeq \rho_i^*/|\lambda_i|$. For longer times, the Brownian motion does not allow the respective ρ_i to decrease much below $-\rho_i^*$, while the negative λ_i prevents it from increasing. As a result, the corresponding ρ_i becomes a stationary random process with a mean of the order $-\rho_i^*$. The relaxation times to the stationary distribution are determined by $\tilde{\sigma}$, which is diffusion independent, and they are thus much smaller than t. On the other hand, the components ρ_j corresponding to non-negative Lyapunov exponents are the integrals over the whole evolution time t. Their values at time t are thus not sensitive to the latest period of evolution lasting of the order of the relaxation times for the contracting ρ_i. Fixing the values of ρ_j at times $t \gg \rho_i^*/|\lambda_i|$ will not affect the distribution of the contracting ρ_i and the whole PDF is thus factorized [16–18]. For example, there are two positive and one negative Lyapunov exponents in 3D developed Navier–Stokes turbulence [19]. For times $t \gg \rho_3^*/\lambda_3$ we have then

$$\mathcal{P} \propto \exp\left[-t\,H\left(\rho_1/t - \lambda_1, \rho_2/t - \lambda_2\right)\right] \mathcal{P}_{\mathrm{st}}(\rho_3), \quad (1.20)$$

with the same function H as in (1.14) since ρ_3 is independent of ρ_1 and ρ_2. The account of the molecular noise violates the condition $\sum \rho_i = 0$ as fluid elements at scales smaller than $\sqrt{\kappa/|\lambda_3|}$ cannot be distinguished. To avoid misunderstanding, note that (1.20) does not mean that the fluid is getting compressible: the simple statement is that if one tries to follow any marked volume, the molecular diffusion makes this volume growing.

Note that we have implicitly assumed ℓ to be smaller than the viscous length $\eta = \sqrt{\nu/|\lambda_3|}$ but larger than the diffusion scale $\sqrt{\kappa/|\lambda_3|}$. Even though ν and κ are both due to molecular motion, their ratio widely varies depending on the type of material. The theory of this section is applicable for the materials having the Schmidt number ν/κ large.

The universal forms (1.14) and (1.20) for the two-particle dispersion are basically everything we need for physical applications. The reader interested in the description of passive scalar may go directly to the Section 1.2. We show there that the senior Lyapunov exponent determines the small-scale statistics of a passively advected scalar in a smooth incompressible flow. For other problems, the whole spectrum of exponents and even the form of the entropy functions are relevant.

1.2.2 Solvable cases

Generally, the Lyapunov spectrum and the entropy function cannot be derived from a given statistics of σ except for few limiting cases. The case of a short-correlated strain allows for a complete solution. As far as finite-correlated strain is concerned, one can express analytically λ_1 and Δ *via* the correlators of σ only in two dimensions for a long-correlated strain and at large space dimensionality.

i) **Short-correlated strain**. Consider the case where the strain $\sigma(t)$ is a stationary white in time Gaussian process with zero mean and the 2-point function

$$\langle \sigma^{ij}(t)\sigma^{k\ell}(t')\rangle = 2\,\delta(t-t')C^{ijk\ell}. \tag{1.21}$$

This case may be considered as the long-time scaling limit $\lim_{\mu\to\infty} \mu^{\frac{1}{2}}\sigma(\mu t)$ of a general strain along a Lagrangian trajectory, provided that the temporal correlations of the latter decay fast enough. It may be also viewed as describing the strain in the Kraichnan ensemble of velocities decorrelated in time and smooth in space: $C^{ijk\ell} = -\nabla_j\nabla_\ell D^{ik}(0)$, see (1.5). We assume $D^{ij}(\mathbf{r})$ to be smooth in $\mathbf{r}$ (or at least twice differentiable), the property that is assured by a fast falloff of its Fourier transform $\hat{D}^{ij}(\mathbf{k})$. The incompressibility, isotropy and parity invariance impose the form $D^{ij}(\mathbf{r}) = D_0\delta^{ij} - \frac{1}{2}d^{ij}(\mathbf{r})$, where

$$d^{ij} = D_1[(d+1)\delta^{ij}r^2 - 2r^ir^j] + o(r^2) \tag{1.22}$$

or, for the 2-point function of σ,

$$C^{ijk\ell} = D_1[(d+1)\delta^{ik}\delta^{j\ell} - \delta^{ij}\delta^{k\ell} - \delta^{i\ell}\delta^{jk}], \tag{1.23}$$

where the constant D_1 has the dimensionality $time^{-1}$.

The solution of the stochastic differential equation (1.7) is given by the matrix $W(t)$ involving stochastic integrals over time. The matrices

$W(t)$ describe a diffusion process on the group $SL(d)$ of real matrices with determinant one. Its generator is a second order differential operator $M = D_1[d\,H^2 - (d+1)J^2]$ acting on $SL(d)$ [20]. Here H^2 is the quadratic Casimir operator of $SL(d)$ and J^2 is the quadratic Casimir of its $SO(d)$ subgroup. In other words, the PDF $\mathcal{P}(w,t)$ of $W(t)$ satisfies the evolution equation $(\partial_t - M)\mathcal{P}(w,t) = 0$.

If we are interested in the statistics of stretching-contraction variables only, then $W(t)$ may be projected to the diagonal matrix Λ with positive non-increasing entries $e^{2\rho_1}, \ldots, e^{2\rho_d}$ by the decomposition $W = O\Lambda^{\frac{1}{2}}O$, where O and O' are orthogonal matrices in $SO(d)$. Note that $W^T W = O^T\Lambda O$ and $W W^T = O'\Lambda O'^T$. One has the stochastic Langevin equation

$$\partial_t \rho_i = D_1 d \sum_{j \neq i} \coth\left(\rho_i - \rho_j\right) + \eta_i. \tag{1.24}$$

Here η is a white noise with the 2-point function $\langle \eta_i(t)\eta_j(t')\rangle = 2D_1(\delta_{ij}\,d - 1)\delta(t - t')$. At long times, where one expects the separation between the ρ_i's to be large, we may approximate $\coth(\rho_i - \rho_j)$ by ± 1. The equation reduces then to $\partial_t \rho_i = D_1 d(d - 2i + 1) + \eta_i$ leading to the long time form (1.14) of the probability distribution function of ρ_i with the exactly quadratic entropy function

$$H(\mathbf{x}) = \frac{1}{4D_1 d} \sum_i [x_i - D_1 d(d - 2i + 1)]^2. \tag{1.25}$$

The Lyapunov exponents are $\lambda_i = D_1 d(d - 2i + 1)$ and the dispersion $\Delta = D_1(d - 1)$.

ii) **2d slow strain**. In 2d, one can turn the vector equation (1.7) into the scalar equation of the second order. In a particular case of a slow strain satisfying $\dot{\sigma} \ll \sigma^2$, we differentiate the equation $\dot{\mathbf{R}} = -\hat{\sigma}\mathbf{R}$ with respect to time and neglect $\dot{\sigma}$ in comparison with σ^2. And here a little miracle happens: because of incompressibility, the matrix $\hat{\sigma}$ is traceless so $\hat{\sigma}^2$ is proportional to the unit matrix in 2D. We thus come to the scalar equation for $\Psi = R_x + iR_y$ instead of the matrix one:

$$\partial_t^2 \Psi = (\sigma_{11}^2 + \sigma_{12}\sigma_{21})\Psi. \tag{1.26}$$

One can consider (1.26) as a Schrödinger equation for a particle in a random potential $U = \sigma_{11}^2 + \sigma_{12}\sigma_{21}$; time plays the role of coordinate. Our problem is thus equivalent to the localization problem in the quasiclassical limit. As long as U is random, the modulus $|\Psi| = R$ grows unlimited with t as $\exp(\lambda_1 t)$. This exponential growth is described by the same exponent as the exponentially decaying tails of a localized quantum Ψ-function.

The problem can be solved using semi-classical methods. The flow is partitioned in elliptic ($\Omega > S$) and hyperbolic ($S > \Omega$) regions [21], corresponding to classical allowed ($U < 0$) and forbidden ($U > 0$) regions. The wave function Ψ is given by two oscillating exponentials or one decreasing and one increasing, respectively. Furthermore, the typical length of the regions is the correlation time τ, assumed much larger than the inverse of the rms strain and vorticity S_{rms}^{-1} and Ω_{rms}^{-1}. It follows that the increasing exponentials in the forbidden regions are large and dominate the growth of $R(t)$. With exponential accuracy we have:

$$\lambda(t) = \ln\left(\frac{R(t)}{R(0)}\right) = \frac{1}{t}\,\text{Re}\int_0^t \sqrt{U(s)}\,ds\,, \qquad (1.27)$$

where the real part restricts the integration to the hyperbolic regions. The parameters λ_1 and Δ in the lognormal expression (1.16) are immediately read from (1.27):

$$\lambda_1 = \left\langle \text{Re}\sqrt{U} \right\rangle, \qquad \Delta = \int \left\langle\left\langle \text{Re}\sqrt{U(0)},\, \text{Re}\sqrt{U(t')} \right\rangle\right\rangle dt'. \qquad (1.28)$$

Note that the vorticity gave no contribution in the δ-correlated case. For a finite correlation time, it suppresses the stretching by rotating the fluid element with respect to the axes of expansion. Indeed, taking a real part in (1.28) filters out the elliptic regions. The Lyapunov exponent is given by a single-time average in a slow case, while in the δ-correlated case it was expressed by the time-integral of a correlation function. It follows that λ_1 does not depend on the correlation time τ (for S_{rms} and Ω_{rms} given) and it can be estimated as S_{rms} for $\Omega_{\text{rms}} < S_{\text{rms}}$. The estimate of the variance is $\Delta \sim \langle S^2 \rangle \tau$. As the vorticity increases, the rotation takes over, the stretching is suppressed and $\bar{\lambda}$ reduces. The correlation time τ_s of the stretching rate is the minimum between $1/\Omega_{\text{rms}}$ and τ [22]. For $\Omega_{\text{rms}}\tau \gg 1$ we are back to a δ-correlated case and $\lambda_1 \sim \langle S^2 \rangle / \Omega_{\text{rms}}$. Those estimates can be made systematic for Gaussian strain [22].

iii) **Large space dimensionality**. The key remark for this case is that scalar products like $R^i(t_1)R^i(t_2)$ are sums of a large number of random terms. Such sums do not fluctuate and obey closed equations in the large-d limit. Those equations can be effectively studied for an arbitrary strain statistics. This approach developed in [23] is inspired by large N methods in quantum field theory [24] and statistical mechanics [25]. Here one can relate strain to the behavior of the interparticle distance and find explicitly λ_1 and Δ that enters (1.16). We consider a Gaussian strain with a general different-time correlation function $\langle \sigma_{\alpha\mu}(t)\sigma_{\beta\nu}(0)\rangle = D\delta_{\alpha\beta}\delta_{\mu\nu}$ $g(t/\tau)/d\tau + o(d^{-2})$, which satisfies only the normalization condition

$\int g(\zeta)\mathrm{d}\zeta = 1$. We introduce dimensionless $\zeta = t/\tau$. At large d, the correlation function $F(\zeta_1, \zeta_2) = \langle R_\alpha(t_1)R_\alpha(t_2)\rangle$ satisfies the equation

$$\frac{\partial^2 F}{\partial \zeta_1 \partial \zeta_2} = \tau^2 \langle \sigma_{\alpha\beta}(t_1)\sigma_{\alpha\gamma}(t_2)R_\beta(t_1)R_\gamma(t_2)\rangle = \beta g(\zeta_1 - \zeta_2)F, \qquad (1.29)$$

with the initial condition $\partial_\zeta F(\zeta, 0) = 0$. The limit of large d is needed for the factorization of the average leading to the second equality in (1.29). Here $\beta = D\tau$ is the dimensionless parameter measuring whether the strain is long- or short-correlated. Since (1.29) is linear and the coefficient at the right-hand side explicitly depends only on the difference $\zeta_1 - \zeta_2$ a solution can be written as a sum of harmonics $F_\lambda(\zeta_1, \zeta_2) = \exp[\lambda\tau(\zeta_1 + \zeta_2)/2]\Psi(\zeta_1 - \zeta_2)$. Since $F(\zeta_1, \zeta_2) = F(\zeta_2, \zeta_1)$ then $\Psi(t)$ is an even function which satisfies the Schrödinger equation with the strain correlation function as a potential

$$\left[(\lambda\tau)^2/4 - \partial_\zeta^2\right]\Psi(\zeta) = \beta g(\zeta)\Psi(\zeta), \qquad (1.30)$$

At large time, only the largest exponent λ_1 contributes, which corresponds to the ground state in the potential $g(\zeta)$. The relation between the stretching rate λ_1 (related to the energy of the ground state) and the parameter β (which determines the depth of the well) can be readily established. For a shallow well $\lambda_1\tau \propto \beta$, while for a deep well the lowest level $\lambda_1\tau \propto \sqrt{\beta}$. Comparing the behavior of $F(\zeta, \zeta) = d^{-1}\langle R^2\rangle$ with (1.16) we conclude that $\lambda_1 \simeq D$ for the fast-fluctuating strain and $\lambda_1 \simeq \sqrt{D/\tau}$ in the slow case. Considering $\zeta_1 - \zeta_2 \gg 1$ we can neglect the right-hand side in (1.30) and see that the same stretching rate λ determines both the growth of $\langle R^2(\zeta)\rangle$ and the decay of different-time correlation at $\zeta_1 + \zeta_2$ fixed. That also shows that the correlation function is getting independent of the larger time when time difference exceeds τ.

For the short-correlated strain, one can put $g(\zeta) = \delta(\zeta)$ and the solution of (1.30) is amazingly simple: $F(\zeta_1, \zeta_2) = R^2(0)\exp[\beta\min(\zeta_1, \zeta_2)]$. The Lyapunov exponent $\lambda_1 = D$, in agreement with the result $\lambda_1 = D_1 d^2 + \mathcal{O}(d)$ obtained for the Kraichnan ensemble. For the slow case, the stretching rate is independent of τ at a given value of D/τ (determining the simultaneous correlation function of the strain). The analysis of (1.30) with a deep potential also gives the correlation time τ_s of the stretching rate, which does not generally coincide with the strain correlation time τ [23].

1.3   Two-particle dispersion in a nonsmooth incompressible flow

We now assume the Reynolds numbers sufficiently high and study the separation between two trajectories in the inertial interval of scales $\eta \ll r \ll L$, where L denotes the integral scale at which the flow is induced.

Let us describe first the usual phenomenology of two-particle dispersion. In the inertial interval, the velocity differences exhibit an approximate scaling that is the power law behavior of the structure functions $\langle(\delta\mathbf{v})^n\rangle \propto r^{\zeta_n}$. Low-order exponents are close to the Kolmogorov prediction $\zeta_n = \alpha n$ with $\alpha = 1/3$. The linear dependence of ζ on n would signal the scaling

$$|\delta\mathbf{v}(\mathbf{r}, t)| \propto r^\alpha \qquad (1.31)$$

for $\eta \ll r \ll L$ with a sharp value of α. The nonlinear dependence of ζ_n indicates a whole spectrum of exponents α, depending on the realization of the velocities, the moment of time and the positions of points (the phenomenon called multiscaling). The 2D and 3D inverse and direct energy cascades provide concrete examples of the two possible situations. Rewriting equation (1.6) for the fluid particle separation as $\dot{\mathbf{R}} = \delta\mathbf{v}(\mathbf{R}, \mathbf{t})$, we infer that $dR^2/dt = 2\mathbf{R} \cdot \delta\mathbf{v}(\mathbf{R}, t) \propto R^{1+\alpha}$ if (1.31) holds with a fixed value of α. For $\alpha < 1$, this is solved (ignoring the time and point dependence in the proportionality constant) by

$$R(t)^{1-\alpha} - R(0)^{1-\alpha} \propto t \qquad (1.32)$$

implying that $R(t) \propto t^{1/(1-\alpha)}$ for large t with the dependence of the initial separation quickly wiped out.

Of course, for the random process $\mathbf{R}(t)$, relation (1.32) is of the mean field type and should pertain to the large time behavior of the averages:

$$\langle R(t)^p \rangle \propto t^{p/(1-\alpha)} \qquad (1.33)$$

for $p > 0$ implying their superdiffusive growth, faster than the diffusive one $\propto t^{p/2}$. The power-law scaling (1.33) may be amplified to the scaling behavior of the PDF of the interparticle distance:

$$\mathcal{P}(R, t) = \lambda \mathcal{P}(\lambda R, \lambda^{1-\alpha} t) . \qquad (1.34)$$

Possible deviations from a linear behavior in the order ζ of the exponents in (1.33) should be interpreted as a signal of multiscaling of the Lagrangian velocity $\Delta\mathbf{v}(\mathbf{R}(t), t) \equiv \Delta\mathbf{V}(t)$. The power-law growth (1.33) for $p = 2$ and $\alpha = 1/3$, *i.e.* $\langle R(t)^2 \propto t^3$, is the celebrated Richardson dispersion law stating that

$$\frac{d}{dt}\langle R(t)^2 \rangle \propto \langle R(t)^2 \rangle^{2/3}. \qquad (1.35)$$

The Richardson law was the first quantitative phenomenological prediction in developed turbulence. It seems to be confirmed by experimental data [26, 27] and by the numerical simulations [28, 29]. The more general property of self-similarity (1.34) (with $\alpha = 1/3$) has been observed in the inverse cascade of two-dimensional turbulence [27]. It is likely that (1.35) is exact within the inverse cascade of 2d turbulence while it may be only approximately correct in 3d. More about that in Section 3 below.

It is important to remark that, even assuming the validity of the Richardson law, it is impossible to establish general properties of the PDF $\mathcal{P}(R; t)$ such as those in Section 1.2.1 for the single particle PDF. The physical reason becomes clear looking at the Lagrangian velocity difference correlation time

$$\tau_t = \int_0^t \langle \delta \mathbf{V}(t) \cdot \delta \mathbf{V}(s) \rangle \, \mathrm{d}s / \langle (\delta \mathbf{V})^2 \rangle. \tag{1.36}$$

The numerator coincides with $\mathrm{d}\langle R^2 \rangle / \mathrm{d}t$ and is thus proportional to $\langle R^2 \rangle^{2/3}$, while the denominator $\propto \langle R^2 \rangle^{1/3}$. It follows that τ_t grows as $\langle R^2 \rangle^{1/3} \propto t$, i.e. the random process $\delta \mathbf{V}(t)$ has a correlation time comparable with its whole span. The absence of decorrelation explains why the Central Limit Theorem and the large deviation theory cannot be applied. There is in fact no a priori reason to expect $\mathcal{P}(R; t)$ to be Gaussian with respect to a power of R either, although we shall see that this happens to be the case in the Kraichnan ensemble.

It is instructive to contrast the exponential growth (1.11) of the distance between the trajectories within the viscous range with the power-law growth (1.33) in the inertial range. In the viscous regime, the closer two trajectories are initially the more time is needed to effectively separate them. As a result, the infinitesimally close trajectories never separate and trajectories in a fixed realization of the velocity field are continuously labeled by the initial conditions. They depend, however, in a sensitive way on the latter due to the exponential magnification of small deviation of the initial point. This sensitive dependence is usually considered as the defining feature of the chaotic behavior. On the other hand, in the inertial range the trajectories separate in a finite time independent of their initial distance $R(0)$, provided that the latter is also in the inertial range. The speed of this separation may depend on the detailed structure of the turbulent velocities but the very fact of the explosive separation is a consequence of the scaling behavior (1.31) with fractional exponents. For very high Reynolds numbers, the viscous scale η is negligibly small (a fraction of a millimeter in the turbulent atmosphere) and setting it to zero (or equivalently, setting the Reynolds number to infinity) is an appropriate abstraction if we want to concentrate on the behavior of the fluid trajectories in the inertial range. In such a limit, however, the power law separation extends down to infinitesimal distances between the trajectories: the infinitesimally close trajectories still separate in a finite time. This points to a marked difference in the behavior of trajectories in comparison to that in the chaotic regime: developed turbulence and chaos are clearly different phenomena. This explosive separation of trajectories results in a breakdown of the deterministic Lagrangian flow in the limit $Re \to \infty$, a rather dramatic effect [30–32]. Indeed, in this limit the trajectories cannot be labeled by the initial conditions. The sheer existence of the Lagrangian trajectories $\mathbf{R}(t; \mathbf{r})$ depending, as functions of random velocities, continuously on the initial position $\mathbf{r}$ would imply that $\lim_{\mathbf{r}_1 \to \mathbf{r}_2} \langle |\mathbf{R}(t; \mathbf{r}_1) - \mathbf{R}(t; \mathbf{r}_2)|^p \rangle = 0$ and contradict the persistence of a power law separation of the Richardson type for infinitesimally close trajectories. The breakdown of the deterministic Lagrangian flow at $Re \to \infty$ does not contradict the theorem about solutions of the ordinary differential equation $\dot{\mathbf{R}} = \mathbf{v}(\mathbf{R}, t)$. The theorem requires that $\mathbf{v}(\mathbf{r}, t)$ be Lipschitz in $\mathbf{r}$, *i.e.* that $|\delta \mathbf{v}(\mathbf{r}, t)| \leq \mathcal{O}(r)$, for a solution to be unique. At $Re = \infty$, however, as first noticed by Onsager [33], the velocities are only Hölder continuous: $|\delta \mathbf{v}(\mathbf{r}, t)| \simeq \mathcal{O}(r^\alpha)$ with the exponent $\alpha < 1$ ($\alpha \simeq 1/3$ in Kolmogorov's phenomenology). As is shown by the classical example of equation $\dot{x} = |x|^\alpha$ with two solutions $x = [(1 - \alpha)t]^{\frac{1}{1-\alpha}}$ and $x = 0$ starting at zero, one should expect multiple Lagrangian trajectories starting or ending at the same point for velocity fields with $\alpha < 1$. Does then the Lagrangian description of the fluid breaks down completely at $Re = \infty$?

Even though the deterministic Lagrangian description breaks down, the statistical description of trajectories is still possible. As we have seen above, certain probabilistic questions concerning the flow, like the ones about the averaged powers of the distance between initially close trajectories, should still have well defined answers in this limit. We expect that for typical velocity realization at $Re = \infty$, one can maintain a probabilistic description of Lagrangian trajectories and make sense of such objects as the PDF $P(\mathbf{r}, s; \mathbf{R}, t | \mathbf{v})$ of the position $\mathbf{R}$ of the trajectory at the time t, given its position $\mathbf{r}$ at the time s. For a regular velocity with deterministic trajectories, one has

$$P(\mathbf{r}, s; \mathbf{R}, t | \mathbf{v}) = \delta(\mathbf{R} - \mathbf{R}(t; \mathbf{r}, s)), \tag{1.37}$$

where $\mathbf{R}(t; \mathbf{r}, s)$ denotes the unique Lagrangian trajectory passing at time s through $\mathbf{r}$. In the presence of small molecular diffusion, equation (1.1) for the Lagrangian trajectories perturbed by the white noise has always a solution as a Markov process in each velocity realization, whether the latter is Lipschitz or Hölder continuous [34]. The resulting Markov process is characterized by the transition probabilities $P_\kappa(\mathbf{r}, s; \mathbf{R}, t | \mathbf{v})$ satisfying the advection-diffusion equation

$$\left[\partial_t - \nabla_{\mathbf{R}} \cdot \mathbf{v}(\mathbf{R}, t) \pm \kappa \nabla_{\mathbf{R}}^2 \right] P_\kappa(\mathbf{r}, s; \mathbf{R}, t | \mathbf{v}) = 0, \tag{1.38}$$

where the sign in front of the Laplacian is the one of $(s - t)$. The mathematical difference between the cases of smooth and rough velocities is that in the latter case the transition probabilities are weak solutions of (1.38) rather than strong ones. What happens if we turn off molecular diffusion? If the velocity $\mathbf{v}(\mathbf{r}, t)$ is Lipschitz in $\mathbf{r}$ then the Markov process describing the noisy trajectories concentrates when $\kappa \to 0$ on the deterministic Lagrangian trajectories and the transition probabilities $P_\kappa(\mathbf{r}, s; \mathbf{R}, t|\mathbf{v})$ converge to (1.37). It has been conjectured in [32] that for a generic $Re = \infty$ turbulent velocity field, the Markov process describing noisy trajectories still tends to a limit when $\kappa \to 0$, but that the limit stays diffuse. Such convergence would be signaled by the convergence of the transition probability to the limiting one $P(\mathbf{r}, s; \mathbf{R}, t|\mathbf{v})$ which is a weak solution of the pure advection equation,

$$[\partial_t - \nabla_{\mathbf{R}} \cdot \mathbf{v}(\mathbf{R}, t)] \, P(\mathbf{r}, s; \mathbf{R}, t|\mathbf{v}) = 0, \qquad (1.39)$$

that is a solution not concentrated at a single trajectory $\mathbf{R}(t; \mathbf{r}, s)$. We will say then that the limiting Markov process[5] defines a stochastic Lagrangian flow. This way the roughness of turbulent velocities resulting in the explosive separation of the Lagrangian trajectories would assure the persistence of stochasticity of the noisy trajectories in a fixed generic realization of the velocity field even in the limit $\kappa \to 0$. To avoid misunderstanding, let us stress again that, according to this claim, in the limit of large Reynolds numbers the Lagrangian trajectories behave stochastically already in a fixed velocity field and for negligible molecular diffusivity and not only due to a random noise or to random fluctuations of the velocities. This intrinsic stochasticity of fluid particles seems to constitute an important aspect of developed turbulence, an inescapable consequence of the Richardson dispersion law or of the Kolmogorov-like scaling of velocity differences in the limit $Re \to \infty$ and a natural mechanism assuring the persistence of dissipation in the inviscid limit: $\lim_{\nu \to 0} \nu \langle |\nabla \mathbf{v}|^2 \rangle \neq 0$.

The general conjecture about the existence of stochastic Lagrangian flows is known to be true for the Kraichnan Gaussian ensemble (1.5) of velocities decorrelated in time. Setting, for instance, the Fourier transform of the spatial part $D^{ij}(\mathbf{r} - \mathbf{r}')$ of the 2-point function

$$\hat{D}^{ij}(\mathbf{k}) \propto \left(\delta^{ij} - \frac{k^i k^j}{k^2} \right) \frac{e^{-(\eta \mathbf{k})^2}}{(\mathbf{k}^2 + L^{-2})^{(d+\xi)/2}} \qquad (1.40)$$

with ξ between 0 and 2, one obtains $D^{ij}(\mathbf{r}) = D_0 \delta^{ij} - (1/2) d^{ij}(\mathbf{r})$ with $D_0 = \mathcal{O}(L^\xi)$ and $d^{ij}(\mathbf{r})$ scaling as r^ξ for $\eta \ll r \ll L$, as r^2 for $r \ll \eta$ and approaching $2D_0 \delta^{ij}$ for $r \gg L$. Clearly, η plays here the role of the viscous scale and L of the integral one with the inertial range $\eta \ll r \ll L$. As we discussed in Section 1.1, D_0 gives the eddy diffusivity pertaining to

---

[5]More exactly, the collection of such processes starting from different initial points.

the average behavior of a single fluid particle. Notice that D_0 is dominated by the integral scale indicating that the effective diffusion of a single fluid particle is driven by the velocity fluctuations at the largest scales present. On the other hand, $d_{ij}(\mathbf{r})$ describes the statistics of the velocity differences: $\langle \delta v^i(\mathbf{r},t)\delta v^j(\mathbf{r},t')\rangle = 2\delta(t-t')d^{ij}(\mathbf{r})$. It picks up contributes of all scales. In particular, it has a purely scaling limit when $\eta \to 0$ and $L \to \infty$

$$\lim_{\substack{\eta\to 0 \\ L\to\infty}} d^{ij}(\mathbf{r}) = D_1[(d-1+\xi)\delta^{ij}r^\xi - \xi r^i r^j r^{\xi-2}] . \qquad (1.41)$$

The normalization constant D_1 has the dimensionality of $length^{2-\xi}time^{-1}$. For $\eta > 0$, the typical velocities are smooth in space (after smearing in time) with the behavior (1.31) visible only for $r \gg \eta$. For $\eta = 0$ and $0 < \xi < 2$, however, the Kraichnan ensemble is supported on the velocities that are Hölder continuous in space with a fixed exponent α arbitrarily close to $\xi/2$. It mimics this way the main property of the infinite Reynolds number turbulent velocities characterized by fractional Hölder exponents. The rough (distributional) behavior of Kraichnan velocities in time, although not very physical, is not expected to modify essentially the qualitative picture of the trajectory behavior (it is the spatial regularity, not the temporal one, of a vector field that is crucial for the uniqueness if its trajectories).

Averaging the product of a Gaussian variable v and an arbitrary functional $f\{v\}$ is done by virtue of the well-known formula of Gaussian integration by parts: $\langle v(x)f(x)\rangle = \int dy \langle v(x)v(y)\rangle\langle \delta f/\delta v(y)\rangle$. That means that, in the Kraichnan ensemble, one can directly average the equation (1.38) over Gaussian velocity: $[\partial_s - (D_0 + \kappa)\nabla_{\mathbf{r}}^2]\mathcal{P}(\mathbf{r},s) = 0$. The above calculation agrees with the result discussed at the end of Section 1.1 about the all-time diffusive behavior of a single fluid particle in the Kraichnan ensemble characterized by the enhancement of the molecular diffusivity κ by the eddy diffusivity D_0.

In order to study the two-particle dispersion, one should examine the joint PDF of the simultaneous values of the coordinates of two fluid particles averaged over the velocity ensemble:

$$\mathcal{P}_{2,\kappa}(\mathbf{r}_1,\mathbf{r}_2,s;\mathbf{R}_1,\mathbf{R}_2,t) = \left\langle P_\kappa(\mathbf{r}_1,s;\mathbf{R}_1,t|\mathbf{v})P_\kappa(\mathbf{r}_2,s;\mathbf{R}_2,t|\mathbf{v})\right\rangle. \quad (1.42)$$

For the Kraichnan ensemble, it satisfies the equation

$$(\partial_t - \mathcal{M}_2)\mathcal{P}_{2,\kappa}(\mathbf{r}_1,\mathbf{r}_2,s;\mathbf{R}_1,\mathbf{R}_2,t) = \delta(t-s)\delta(\mathbf{R}_1-\mathbf{r}_1)\delta(\mathbf{R}-\mathbf{r}_2)$$

with an explicit elliptic second-order differential operator

$$\mathcal{M}_2 = -\sum_{n,n'=1}^{2} D^{ij}(\mathbf{r}_n - \mathbf{r}_{n'})\nabla_{r_n^i}\nabla_{r_{n'}^j} - \kappa\sum_{n=1}^{2}\nabla_{\mathbf{r}_n}^2, \qquad (1.43)$$

a result which goes back to the original work of Kraichnan [10]. If we are interested only in the separation $\mathbf{R} = \mathbf{R}_1 - \mathbf{R}_2$ of two fluid particles at time t, given their separation $\mathbf{r}$ at time s, then the relevant PDF $\mathcal{P}_{2,\kappa}(\mathbf{r}, s; \mathbf{R}, t)$ is obtained by averaging over the simultaneous translations of the final (or initial) positions of the particle and is governed by the operator $\mathcal{M}_2$ restricted to the translationally invariant sector. The latter is equal to $-[d^{ij}(\mathbf{r}) + 2\kappa\delta^{ij}]\nabla_{r^i}\nabla_{r^j}$. Note that the eddy diffusivity D_0, dominated by the integral scale, drops out in the action on translation-invariant functions. The above result shows that the relative motion of two fluid particles in the Kraichnan ensemble of velocities is an effective diffusion with a distance-dependent diffusivity tensor scaling like r^ξ in the inertial range. This is a precise realization of the scenario for the turbulent diffusion put up by Richardson as far back as 1926 [35]. Similarly, the PDF $\mathcal{P}_{2,\kappa}(r, s; R, t)$ of the distance R between two particles satisfies the equation

$$(\partial_t - M_2)\mathcal{P}_{2,\kappa}(r, s; R, t) = \delta(t - s)\delta(r - R), \tag{1.44}$$

where M_2 is the restriction of $\mathcal{M}_2$ to the homogeneous and isotropic sector. In the scaling regime, i.e. for $\kappa = 0$, $\eta = 0$ and $L = \infty$, it has the form

$$M_2 = -D_1(d - 1)r^{1-d}\partial_r r^{d-1+\xi}\partial_r \tag{1.45}$$

and (1.44) can be readily solved [10, 36]. At $r \ll R$, the PDF has particularly simple form

$$\lim_{r \to 0} \mathcal{P}_{2,\kappa}(r, s; R, t) \propto \frac{R^{d-1}}{|t - s|^{d/(2-\xi)}}\exp\left[-\text{const.}\frac{R^{2-\xi}}{|t - s|}\right]. \tag{1.46}$$

That confirms the diffusive character of the limiting process describing the Lagrangian trajectories in fixed non-Lipschitz velocities: the endpoints of the process stay at finite distance when the initial points converge. If we set $\eta = 0$ but maintain finite integral scale L, then the behavior (1.46) is modified for $R \gg L$ and crosses over to the simple diffusion with the diffusivity $2D_0$: at distances much larger than the integral scale two fluid particles undergo independent Brownian walks driven by the velocity fluctuations on scale L.

Note how the PDF (1.46) changes from Gaussian to lognormal when ξ changes from zero to two. The PDF has the scaling form (1.34) for $\alpha = \xi - 1$ and implies the power law growth (1.33) of the averaged powers of the distance between trajectories. The Richardson dispersion $\langle R^2(t) \rangle \propto t^3$ is reproduced for $\xi = 4/3$ rather than for $\xi = 2/3$ when the spatial Hölder exponent of the typical Kraichnan ensemble velocities takes the Kolmogorov value 1/3. The reason is that the velocity temporal decorrelation cannot be ignored and we should replace the time t in the right hand side of (1.32) by the Brownian motion $\beta(t)$. That replacement indeed reproduces for $\alpha = \xi/2$ the large-time PDF (1.46) up to a geometric power-law prefactor.

1.4 Multiparticle configurations and breakdown of scale-invariance

This section is a highlight of the course. We describe here the time-dependent statistics of multi-particle configurations with the emphasis on conservation laws of turbulent diffusion. As we have seen in the previous sections, the two-particle statistics is characterized by a simple behavior of the single separation vector. In nonsmooth velocities, the length of the vector grows by a power law, while the initial separation is forgotten. In contrast, the many-particle evolution in nonsmooth velocities exhibits non-trivial statistical conservation laws that involve geometry and are proportional to the positive powers of the distances. The distance growth is balanced by the decrease of the shape fluctuations in those integrals. The existence of multiparticle conservation laws indicates the presence of a long-time memory and is a reflection of the coupling among the particles due to the simple fact that they all are in the same velocity field. The conserved quantities may be easily built for the limiting cases. Already for a smooth velocity, very close particles become aligned at sufficiently long times with the distances R_{nm} between them growing exponentially so that the ratios of distances do not change. In the opposite case of a very irregular velocity, the fluid particles undergo a Brownian motion. The distances between the Brownian particles grow according to $\langle R_{nm}^2(t) \rangle = R_{nm}^2(0) + Dt$. The statistical integrals of motion are $\langle R_{nm}^2 - R_{pr}^2 \rangle$, $\langle 2(d+2)R_{nm}^2 R_{pr}^2 - d(R_{nm}^4 + R_{pr}^4) \rangle$, and an infinity of similarly built polynomials (zero modes of Laplacian) where all powers of t cancel out. Another trivial case is the infinite-dimensional flow where the distances between particles do not fluctuate. The two-particle law $R_{nm}(t)^{1-\alpha} - R_{nm}(0)^{1-\alpha} \propto t$, implies then that the expectation of any function of $R_{nm}^{1-\alpha} - R_{pr}^{1-\alpha}$ does not change with time. Away from the degenerate limiting cases, the conserved quantities continue to exist yet they cannot be generally constructed so easily and they depend substantially on the number of particles. We thus see that the very existence of conserved quantities is natural. What is nontrivial in a general case is their precise form and their scaling. The intricate statistical conservation laws of multiparticle dynamics were first discovered for the Kraichnan velocities [37, 38]. That came as a surprise since the Kraichnan velocity ensemble is Gaussian and time-decorrelated, with no structure built in except for the spatial scaling in the inertial range. The discovery has led to a new qualitative and quantitative understanding of intermittency. Even more importantly, it has pointed to the aspects of the multiparticle evolution that seem both present and relevant in generic turbulent flows [39].

1.4.1   Absolute and relative evolution of particles

Consider the joint PDF of the simultaneous positions $\underline{\mathbf{R}} = (\mathbf{R}_1, \dots, \mathbf{R}_N)$ of N fluid trajectories

$$\mathcal{P}_{N,\kappa}(s, \underline{\mathbf{r}}; \underline{\mathbf{R}}, t) = \left\langle \prod_{n=1}^{N} P_\kappa(\mathbf{r}_n, s; \mathbf{R}_n, t | \mathbf{v}) \right\rangle \qquad (1.47)$$

with the average over the velocity ensemble. Such PDF, that we shall call multiparticle Green function, accounts for the overall statistics of the many-particle systems.

As for many-body problems in other branches of physics (*e.g.* in kinetic theory or in quantum mechanics), the multi-particle dynamics may bring about new aspects due to the cooperative behavior of particles. In turbulence, such behavior is mediated by the velocity fluctuations correlated at large scales. If the velocities are statistically homogeneous, it is convenient to separate the absolute motion of particles from the relative one, as in the other many-body problems with spatial homogeneity. For single particle, there is nothing but the absolute motion, which is diffusive on time scales longer that the Lagrangian correlation time (Sect. 1.1). For N particles, we may define the absolute motion as the one of the mean position $\overline{\mathbf{R}} = \sum \mathbf{R}_n / N$, that motion is also expected to be diffusive for times long enough for the particles to separate beyond the correlation length of velocities. Indeed, for such time scales the particles may be considered as moving independently. The diffusivity of the absolute motion will then be N times smaller than that of a single particle. The statistics of the relative motion of N particles is described by the joint PDF (1.47) averaged over rigid translations $\underline{\rho} = (\rho, \dots, \rho)$:

$$\mathcal{P}_{N,\kappa}^{\mathrm{rel}}(s, \underline{\mathbf{r}}; \underline{\mathbf{R}}, t) = \int \mathcal{P}_{N,\kappa}(s, \underline{\mathbf{r}}; \underline{\mathbf{R}} + \underline{\rho}, t) \mathrm{d}\rho. \qquad (1.48)$$

The PDF $\mathcal{P}_{N,\kappa}^{\mathrm{rel}}$ describes the distribution of the separations $\mathbf{R}_{nm} = \mathbf{R}_n - \mathbf{R}_m$ or the relative positions $\underline{\mathbf{R}}^{\mathrm{rel}} = (\mathbf{R}_1 - \overline{\mathbf{R}}, \dots, \mathbf{R}_N - \overline{\mathbf{R}})$. Similarly as for two particles, we expect that when $\kappa \to 0$ the PDF tends to (possibly distributional) limits. We shall denote those limits by dropping the subscript κ. The PDF $\mathcal{P}_N$ are again expected to show a different short-distance behavior for smooth and nonsmooth velocities.

If all the distances between the particles are much less than the viscous length, one may consider velocity smooth and approximate the velocity field differences by linear expressions:

$$\mathcal{P}_N^{\rm rel}(\mathbf{r}, 0; \underline{\mathbf{R}}, t) = \int \left\langle \prod_{n=1}^{N} \delta(\mathbf{R}_n + \rho - W(t)\mathbf{r}_n) \right\rangle \mathrm{d}\rho . \tag{1.49}$$

Clearly, the above PDF depend only on the statistics of the evolution matrix $W(t)$ that has been discussed in Section 1.2. Under the evolution governed by $W(t)$, all distances between points grow exponentially for large times while their ratios R_{nm}/R_{kl} tend to a constant. For whatever initial positions, asymptotically in time, the points tend to be situated on the line. This behavior and its dramatic consequences for passive scalar statistics are further discussed in Section 2.2.1.2.

It is natural to ask if in the nonsmooth case one may also build integrals of motion that compensate the increase in the distance between the particles by the decrease in fluctuations of the interparticle configuration geometry. This might indeed be done explicitly for the Kraichnan ensemble of velocities.

1.4.2 Multiparticle motion in Kraichnan velocities

The great simplification of the Kraichnan model consists in the Markov character of the effective N-trajectory processes which is due to the time decorrelation of the velocities. In other words, the PDF $\mathcal{P}_{N,\kappa}(s, \mathbf{r}; \underline{\mathbf{R}}, t)$ and the relative version (1.48) give, for fixed N, the transition probabilities of Markov processes $\underline{\mathbf{R}}(t)$ and $\underline{\mathbf{R}}^{\rm rel}(t)$. The process $\underline{\mathbf{R}}(t)$ is characterized by its generator, a $2^{\rm nd}$ order differential operator $\mathcal{M}_N$ whose explicit form may be deduced by a straightforward generalization of the arguments that we have employed for two particles, see (1.43):

$$(\partial_{\rm t} - \mathcal{M}_N)\mathcal{P}_{N,\kappa}(\mathbf{r}, s; \underline{\mathbf{R}}, t) = \delta(t - s)\delta(\underline{\mathbf{R}} - \mathbf{r}), \tag{1.50}$$

$$\mathcal{M}_N = -\sum_{n,m=1}^{N} D^{ij}(\mathbf{r}_{nm})\nabla_{r_n^i}\nabla_{r_m^j} - \kappa \sum_{n=1}^{N} \nabla_{\mathbf{r}_n}^2, \tag{1.51}$$

where, $D^{ij}(\mathbf{r})$ is the spatial part of the velocity 2-point function. For the relative process $\underline{\mathbf{R}}^{\rm rel}(t)$, $\mathcal{M}_N$ is replaced by its translation-invariant version

$$M_N = \sum_{n<m} \left(d^{ij}(\mathbf{r}_{nm}) + 2\kappa\delta^{ij} \right) \nabla_{r_n^i}\nabla_{r_m^j}, \tag{1.52}$$

where $D^{ij}(\mathbf{r}) = D_0\delta^{ij} - \frac{1}{2}d^{ij}(\mathbf{r})$. Note the multibody structure of the generators $\mathcal{M}_N$ and M_N.

As we have seen previously, the Kraichnan ensemble may be used to model both smooth and Hölder continuous velocities. In the first case, one keeps the "viscous" cutoff η in the 2-point function of velocities (1.40) with the result that $d^{ij}(\mathbf{r}) = \mathcal{O}(r^2)$ for $r \ll \eta$. In particular, retaining only the terms of the second order in $\mathbf{r}$ is equivalent to the approximation (1.49) with $W(t)$ becoming a diffusion process on the group $SL(d)$ of unimodular matrices with an explicitly known generator, as discussed in Section 1.2.2 (i). In this case, the right hand side of (1.49) may be found rather effectively with the help of the harmonic analysis on $SL(d)$ [20, 30].

From the form (1.52) of the generator of the process $\mathbf{R}^{\mathrm{rel}}(t)$ we infer that, in the Kraichnan model, N fluid particles undergo an effective diffusion with the diffusivity depending on the interparticle distances. In the inertial interval of distances $\eta \ll r \ll L$, where $d^{ij}(\mathbf{r}) \propto r^\xi$, and for small molecular diffusivity κ, the effective diffusivity grows as the power ξ of the distance. It should be intuitively clear that, in comparison to the standard diffusion with constant diffusivity, the particles will tend to spend longer time together when they are close and to separate faster when they become distant. Of course, both tendencies may coexist and dominate the motion of different clusters of fluid particles. It remains to find a more analytic and quantitative way to capture this behavior. The effective short-distance attraction slowing down the separation of close particles is a robust phenomenon that should be present also in time-correlated and non-Gaussian velocity fields. We believe that it is responsible for intermittency in the transport of scalars by high Reynolds number flows, as shall be discussed in Section 2.

As for a single particle, the absolute motion of the particles is dominated by the velocity fluctuations on scales of order L. In contrast, the relative motion within the inertial range is approximately independent of the cutoffs η and L and it is then convenient to take directly the scaling limits $\eta = 0$, $L = \infty$ and $\kappa = 0$. In these limits, M_N scales as $length^{\xi-2}$ implying that $time$ should scale as $length^{2-\xi}$ and that

$$\mathcal{P}_N^{\mathrm{rel}}(\underline{\mathbf{r}}, 0; \underline{\mathbf{R}}, t) = \lambda^{(N-1)d}\mathcal{P}_N^{\mathrm{rel}}(\lambda\underline{\mathbf{r}}, 0; \lambda\underline{\mathbf{R}}, \lambda^{2-\xi}t). \quad (1.53)$$

The relative motion of N fluid particles may be tested by studying the time evolution of the Lagrangian averages

$$\langle f(\underline{\mathbf{R}}(t)) \rangle = \int f(\underline{\mathbf{R}}) \, \mathcal{P}_N^{\mathrm{rel}}(\underline{\mathbf{r}}, 0; \underline{\mathbf{R}}, t) \, d\underline{\mathbf{R}}' \quad (1.54)$$

of translation-invariant functions f of the simultaneous positions of the particles (we assume that $\underline{\mathbf{R}}(0) = \underline{\mathbf{r}}$ and set $\underline{\mathbf{R}}' \equiv (\mathbf{R}_1, \dots \mathbf{R}_{N-1})$). Think about the evolution of N fluid particles as of that of a discrete cloud of marked points in the physical space. There are two elements in the relative evolution of the cloud: the growth of its size and the change of its shape. We shall define the overall size of the cloud as $R = [(2N)^{-1}\sum \mathbf{R}_{nm}^2]^{1/2}$ and its "shape" as $\underline{\widehat{\mathbf{R}}} = \underline{\mathbf{R}}^{\mathrm{rel}}/R$. For example, 3 particles form a triangle in the space, with labeled vertices, and the notion of shape that we are using includes the orientation of the triangle. The growth of the size of the cloud may be studied by looking at the evolution of the Lagrangian average of $f(\mathbf{R}) = R^p$ with $p > 0$. More generally, let f be a scaling function of scaling dimension p, *i.e.* such that $f(\lambda \mathbf{R}) = \lambda^p f(\mathbf{R})$. The change of the variables $\mathbf{R} \mapsto t^{\frac{1}{2-\xi}} \mathbf{R}$ and the use of (1.53) and of the scaling of f allow to trade the Lagrangian time t PDF in (1.54) for $t^{\frac{p}{2-\xi}} \mathcal{P}_N^{\text{rel}}(t^{-\frac{1}{2-\xi}} \mathbf{r}, 0; \mathbf{R}, 1)$ (in the arbitrary time units). As for two points, the limit of $\mathcal{P}_N$ when the initial points approach each other is non-singular for nonsmooth velocities and we infer that

$$\langle f(\mathbf{R}(t)) \rangle = t^{\frac{p}{2-\xi}} \int f(\mathbf{R}) \, \mathcal{P}_N^{\text{rel}}(\mathbf{0}, 0; \mathbf{R}, 1) \, \mathrm{d}\mathbf{R}' + o\left(t^{\frac{p}{2-\xi}}\right). \qquad (1.55)$$

In particular, we obtain the N-particle generalization of the Richardson-type behavior (1.33): $\langle R^p(t) \rangle = \mathcal{O}(t^{\frac{p}{2-\xi}})$. Hence in the Kraichnan model the size of the cloud of Lagrangian points grows superdiffusively $\propto t^{\frac{1}{2-\xi}}$. What about its shape?

1.4.3   Zero modes and slow modes

We should be able to test the evolution of the shape of the cloud by comparing the Lagrangian averages of different scaling functions f. But (1.55) shows that they all become proportional to $t^{\frac{p}{2-\xi}}$ at large times, in agreement with the dimensional analysis. All but those for which the prefactor of $t^{\frac{p}{2-\xi}}$ in (1.55) vanishes. Such scaling functions, whose evolution violates the dimensional prediction, may then be better suited for testing the evolution of the shape of the cloud. Do such functions exist? Suppose that f is a scaling function of dimension $p \geq 0$ annihilated by M_N, *i.e.* that $M_N f = 0$. The Lagrangian averages of such scaling zero modes, instead of growing like $t^{\frac{p}{2-\xi}}$, do not change in time: $\langle f(\mathbf{R}(t)) \rangle = f(\mathbf{r})$. Indeed, the time derivative of $\langle f(t) \rangle$ vanishes since it brings down M_N acting on f on the right hand side of (1.54). Thus the zero modes of M_N are conserved in mean by the Lagrangian evolution. The importance of such conserved modes for the transport properties of short correlated velocities has been recognized independently in [20, 37, 38]. The above mechanism can be easily generalized. Suppose that f_k is of scaling dimension $p + (2-\xi)k$ and that it is a zero mode of the $(k+1)^{\text{th}}$ power of M_N (but not of a lower one). Then the Lagrangian average $\langle f_k(\mathbf{R}(t)) \rangle$ is a polynomial of degree k in time since its kth time derivative vanishes. If $p > 0$, its time growth is slower than $\mathcal{O}(t^{\frac{p}{2-\xi}+k})$ predicted by the dimensional analysis so that the integral coefficient in (1.55) must vanish. Following [30] we shall call such scaling functions slow modes. The slow modes may be organized into "towers" each having a zero mode at the bottom. One descends down the tower by applying the operator M_N which lowers the scaling dimension by $2 - \xi$ (note that $(M_N)^k f_k$ is a zero mode of scaling dimension p). The zero modes and the slow modes are natural candidates for probes of the shape evolution of the Lagrangian cloud. There is an important general feature of the zero modes and of the slow modes due to the multibody structure of operators M_N: the zero modes of M_{N-1} are also zero modes of M_N and the same for slow modes. Of course, only those modes that depend non-trivially on all (relative) positions of N points[6] may give a novel information on the N-particle evolution which cannot be read from the evolution of a smaller number of particles. We shall call such zero modes and slow modes irreducible. The examples of such modes are given in the next subsection for limiting cases.

To get convinced that zero and slow modes do exist, let us first consider the limiting case $\xi \to 0$ of very rough velocity fields. In this limit, the operator M_N becomes proportional to $\underline{\nabla}^2$, the (Nd)-dimensional Laplacian restricted to the translation-invariant sector. The relative motion of particles becomes pure diffusion. With R denoting the size-of-the-cloud variable,

$$\underline{\nabla}^2 = R^{-d_N+1} \partial_R \, R^{d_N-1} \partial_R + R^{-2} \widehat{\underline{\nabla}}^2 , \qquad (1.56)$$

where $d_N \equiv (N-1)d$ and $\widehat{\underline{\nabla}}^2$ is the angular Laplacian on the $(d_N - 1)$-dimensional unit sphere of shapes $\widehat{\mathbf{R}}$. The spectrum of the latter may be analyzed using the properties of the rotation group. Its eigenfunctions ϕ_ℓ have eigenvalues $-\ell(\ell+d_N-2)$ where $\ell = 0, 1, \ldots$ is the angular momentum. The averages of the angular eigenfunctions decay as follows: $\langle \phi_\ell(\widehat{\mathbf{R}}) \rangle \propto t^{-\ell/2}$. To compensate for the decay, we introduce the functions $f_{\ell,0} = R^\ell \phi_\ell(\widehat{\mathbf{R}})$ which are zero modes of the Laplacian with the scaling dimension ℓ – the contributions coming from the radial and the angular parts in (1.56) indeed cancel out. The averages $\langle f_{\ell,0} \rangle$ are thus conserved. The polynomials $f_{\ell,k} = R^{2k} f_{\ell,0}$ form the corresponding (infinite) tower of slow modes. All the scale-invariant zero and slow modes of the Laplacian are of that form.

General multiparticle zero modes of M_n are of the same nature: when the size increases, the average of a generic function of the shape tends to a constant with the difference decreasing as a combination of negative powers of t (or R), with the zero modes giving the modes of relaxation [40].

It was argued in [30] that zero and slow modes determine the asymptotics of the multiparticle PDF's describing their factorization when the initial

---

[6] *I.e.* are not combinations of functions depending on positions of smaller number of points.

points become close:

$$\lim_{\lambda \to 0} \mathcal{P}_N^{\text{rel}}(\lambda \underline{\mathbf{r}}, 0; \underline{\mathbf{R}}, t) = \sum_\beta \sum_{k=0}^\infty \lambda^{p_\beta + (2-\xi)k} \, f_{\beta,k}(\underline{\mathbf{r}}) \, g_{\beta,k}(\underline{\mathbf{R}}, t). \qquad (1.57)$$

Here β is a label of the zero mode of scaling dimension p_β and $f_{\beta,k}$ for different k form the corresponding tower of the slow modes. The slow modes $f_{\beta,k}$ and the counterpart functions $g_{\beta,k}$ may be normalized so that

$$f_{\beta,k-1} = M_N f_{\beta,k} \qquad \text{and} \qquad g_{\beta,k+1} = -\partial_t g_{\beta,k} = M_N g_{\beta,k}. \qquad (1.58)$$

The leading term in the expansion (1.57) comes from the constant zero mode $f_{0,0} = 1$ which corresponds to $g_{0,0}(\underline{\mathbf{R}}, t) = \mathcal{P}_N^{\text{rel}}(\mathbf{0}, 0; \underline{\mathbf{R}}, t)$. The asymptotic expansion is easy to establish for $\xi = 0$ and for general ξ for $N = 2$. For general ξ and N, it has been obtained under some plausible but unverified regularity assumptions. Note that, due to (1.53), the expansion (1.57) describes also the asymptotics of the multiparticle PDF's when the final points get far apart and the times become large. The use of (1.57) allows to extract the complete long-time asymptotics of the Lagrangian averages:

$$\left\langle f\big(\underline{\mathbf{R}}(t)\big) \right\rangle = \sum_\beta \sum_{k=0}^\infty t^{\frac{p-p_\beta}{2-\xi} - k} \, f_{\beta,k}(\underline{\mathbf{r}}) \int f(\underline{\mathbf{R}}) \, g_{\beta,k}(\underline{\mathbf{R}}, 1) \, \mathrm{d}\underline{\mathbf{R}}' \qquad (1.59)$$

which is a detailed refinement of (1.55), the latter corresponds to $f_{0,0}$ term. Note that the pure polynomial in time behavior of the Lagrangian averages of slow modes implies partial orthogonality relations between the $f_{\beta,k}$ and the $g_{\beta,k}$ modes.

It is instructive to compare the shape-*versus*-size stochastic evolution of the Lagrangian cloud to the imaginary-time evolution of the quantum-mechanical many-particle systems governed by the Hamiltonians $H_N = \sum_n \frac{p_n^2}{2m} + \sum_{n<m} V(\mathbf{r}_{nm})$. The (Hermitian) imaginary-time evolution operators $\mathrm{e}^{-t H_N}$ decompose in the translation-invariant sector as

$$\sum_a \mathrm{e}^{-t E_{N,a}} |\psi_{N,a}\rangle\langle\psi_{N,a}|$$

with the ground state energy $E_{N,0}$ and the sum replaced by an integral for the continuous spectrum contributions. An attractive potential between the particles may lead to the creation of bound states at the bottom of the spectrum of H_N. What this means is that breaking the system into subsystems of N_i particles by removing the potential coupling between them one raises the ground state energy: $E_{N,0} < \sum_i E_{N_i,0}$. A very similar phenomenon occurs in the stochastic shape evolution in the Kraichnan model.

For simplicity, let us only consider the case of even number of particles and of the isotropic sector. Let $p_{N,0}$ be the lowest value of the scaling dimension of the irreducible (*i.e.* dependent on positions of all N particles) zero mode invariant under d-dimensional translations, rotations and reflections. For $N = 2$, where there are no invariant irreducible zero modes, we shall set $p_{2,0} = 2 - \xi$. Suppose now that we break the system into subsystems of N_i particles (with even N_i) by removing in M_N the derivative terms $d(\mathbf{r}_{nm})\nabla_{\mathbf{r}_n}\nabla_{\mathbf{r}_m}$ coupling the subsystems, see (1.52). The recalculation of the smallest dimension of the invariant irreducible zero modes gives now the value $\sum_i p_{N_i,0}$. Indeed, if $N_i \geq 4$, the lowest irreducible zero mode for the broken system is the product of such modes for the subsystems. If some of the subsystems contain only 2-particles, the irreducible zero mode for the broken system is obtained by combining the irreducible zero modes and the slow modes. In particular, its part dependent on all variables is equal to the product of the irreducible zero modes for the bigger subsystems and of the first slow mode $\propto r^{2-\xi}$ for the 2-particle ones. The crucial observation, confirmed by perturbative and numerical analysis discussed below, is that the breaking of the system raises the minimal dimension of the irreducible zero modes: $p_{N,0} < \sum_i p_{N_i,0}$. In particular, $p_{N,0} < \frac{N}{2}(2 - \xi)$. One even expects $p_{N,0}$ to be a concave function of (even) N. Interesting that for $N \gg d$, the dependence of $p_{N,0}$ on N saturates that is adding extra particles does not change the energy at all [36]. By analogy with the multibody quantum mechanics we may say that the irreducible zero modes are bound states of the shape evolution of the Lagrangian cloud. The effect is at the root of the anomalous scaling of the structure functions of the passive scalar advected by nonsmooth Kraichnan velocities, as we shall see in Section 2.3. It is a cooperative phenomenon exhibiting a short-distance attraction of close Lagrangian trajectories diffusing with the diffusivity proportional to a power of the distance, superposed on the overall repulsion of the trajectories. Similar bound states of the shape evolution persist in general turbulent flows and are responsible for intermittency in the scalar advection [39].

1.4.4   Perturbative schemes

Kraichnan model contains two parameters $d \in [2, \infty)$ and $\xi \in [0, 2]$ so it is natural to ask if the problem is simplified at the limiting values of the parameters and if one can then use perturbation theory to get the zero modes near the limits. No significant simplifications has been recognized for $d = 2$ at arbitrary ξ. Three other limits do allow for a consistent treatment since the particle interaction is weak and anomalous scaling disappears there. The perturbation theory is regular around the limits $\xi \to 0$ and $d \to \infty$ while it is singular at $\xi \to 2$.

Let us start by considering the case $\xi = 0$, when the velocities become very rough and the operator M_N is proportional to the Laplacian $\underline{\nabla}^2$. As was mentioned, the zero modes are polynomials of scaling dimension ℓ whose restrictions to the unit sphere $R = 1$ give eigenfunctions of angular momentum ℓ of the spherical Laplacian $\widehat{\nabla}^2$. The polynomials invariant under d-dimensional translations, rotations and reflections can be reexpressed as polynomials in $\mathbf{R}^2_{nm}$. Let us assume that N is even. The irreducible $O(d)$-invariant zero mode with the lowest scaling dimension has then the form

$$f_{N,0}(\underline{\mathbf{R}}) = \mathbf{R}^2_{12}\mathbf{R}^2_{34}\dots\mathbf{R}^2_{(N-1)N} + [\dots] \qquad (1.60)$$

where $[\dots]$ denotes a combination of terms that depend on positions of $(N-1)$ or less particles. *E.g.* for four points, the zero mode is $\mathbf{R}^2_{12}\mathbf{R}^2_{34} - \frac{d}{2(d+2)}(\mathbf{R}^4_{12} + \mathbf{R}^4_{34})$, the example already mentioned before. The terms $[\dots]$ are not uniquely determined since we may add to them degree N zero modes for smaller number of points. Besides, the functions differing from $f_{N,0}$ by a permutation of points are also zero modes so that we may symmetrize the above expressions and look only at the permutation invariant modes. Clearly, the scaling dimension $p_{N,0} = N$. The linear in N growth of the dimension signals the absence of the extra attractive effect between the particles diffusing with a constant diffusivity (no particle binding in the shape evolution). As we shall see in Section 2.3, this leads to the disappearance of intermittency in the advected scalar which becomes a Gaussian field in the limit $\xi \to 0$. For small but positive ξ, the scaling dimension of the irreducible 4-point zero mode $f_{4,0}$ was first calculated to the linear order in ξ by Gawędzki and Kupiainen using a version of the degenerate Rayleigh-Schrödinger perturbation theory [38]. Parallelly, a similar calculation in the linear order in $1/d$ was performed by Chertkov *et al.* [37]. Those two papers present the first ever analytic calculations of the anomalous exponents in turbulence. A generalization for larger N has been achieved in [41,42]. Let us sketch here the main lines of those calculations.

To the linear order in ξ, $M_N \propto \underline{\nabla}^2 - \xi V$, where V is a second order differential operator with logarithmic terms involving $\log(r_{nm})$. The scaling relation $[R\partial_R, M_N] = (\xi - 2)M_N$ implies that

$$[R\partial_R, V] = -\underline{\nabla}^2 - 2V. \qquad (1.61)$$

Let us expand the zero mode $f_{N,0}$ as $f_0 + \xi f_1$ and its scaling dimension $p_{N,0}$ as $N + \xi p_1$. The zeroth order term f_0 must be given by the symmetrization of (1.60). Of course, one of the problems is that the $\xi = 0$ zero modes of order N are degenerate, with the degeneration hidden in the $[\dots]$ terms. As usually in such problems, the perturbation may lift this degeneration fixing the term f_0 for each zero mode. Inserting the above decompositions to the equations defining the scaling zero modes: $M_N f_{N,0} = 0$ and $R \partial_R f_{N,0} = p_{N,0} f_{N,0}$, one obtains the relations

$$\underline{\nabla}^2 f_1 \; = \; V f_0, \qquad (R \partial_R - N) f_1 \; = \; p_1 f_0. \qquad (1.62)$$

Given an arbitrary $\xi = 0$ zero mode f_0, one shows that the first equation has a solution of the form $f_1 = h + \sum_{n < m} h_{nm} \log(r_{nm})$ with $O(d)$-invariant, degree N polynomials h and h_{nm}. Of course, h is determined up to a zero mode of $\underline{\nabla}^2$. Note that the function $(R \partial_R - N) f_1 = \sum_{n < m} h_{nm}$ is also a zero mode of the Laplacian. Indeed,

$$\begin{aligned}
\underline{\nabla}^2 (R \partial_R - N) f_1 & = & (R \partial_R - N + 2) \underline{\nabla}^2 f_1 \; = \; (R \partial_R - N + 2) V f_0 \\
& = & ([R \partial_R, V] + 2V) f_0 + V (R \partial_R - N) f_0 \\
& = & 0, \qquad\qquad\qquad\qquad\qquad\qquad\qquad (1.63)
\end{aligned}$$

where the last equality follows by virtue of (1.61) and of the scaling of f_0. One obtains this way a linear map Γ on the space of the zero modes of $\underline{\nabla}^2$ of degree N with $\Gamma f_0 = (R \partial_R - N) f_1$. The second equation of (1.62) states that we have to choose f_0 which is an eigenstate of the map Γ. Besides, f_0 should not belong to the subspace of codimension 1 of the zero modes that do not depend on all points (*i.e.* are $\xi = 0$ zero modes corresponding to $N - 1$ particles). It is easy to see that such subspace is preserved by the map Γ. As the result, the eigenvalue p_1 is equal to the ratio of the coefficients of $\mathbf{R}_{12}^2 \mathbf{R}_{34}^2 \ldots \mathbf{R}_{(N-1)N}^2$ in Γf_0 and in f_0. The latter is easy to extract (see [42] for the details) and yields the result $p_1 = -\frac{N(N+d)}{2(d+2)}$ or

$$p_{N,0} \; = \; \frac{N}{2}(2 - \xi) \; - \; \frac{N(N-2)}{2(d+2)} \xi \; + \; \mathcal{O}(\xi^2) \qquad (1.64)$$

giving the leading correction $\propto \xi$ to the scaling dimension of the lowest irreducible zero mode. Note that, to that order, the scaling dimension $p_{N,0}$ is, indeed, a concave function of N.

For large dimensionality d, it is convenient to use the variables $x_{nm} = R_{nm}^{2-\xi}$ as the independent coordinates[7]. Rewriting the operator M_N in terms of these variables makes their d-dependence of the operator explicit. Up to the linear oder in $\frac{1}{d}$, $M_N \propto \mathcal{L} - \frac{1}{d} U$, where

$$\mathcal{L} = \sum_{n < m} \left(\frac{d-1}{d} \nabla_{x_{nm}} + \frac{2-\xi}{d} x_{nm} \nabla^2_{x_{nm}} \right)$$

---

[7]Their values are restricted only by the triangle inequalities between the interparticle distances.

and U is a second order d-independent differential operator mixing derivatives over different x_{nm}. When $d \to \infty$, $\mathcal{L}$ reduces to the first order operator $\mathcal{L}' \equiv \sum_{n<m} \nabla_{x_{nm}}$ signaling that the particle evolution becomes deterministic with all $R_{nm}^{2-\xi}$ growing linearly in time. The irreducible zero modes of $\mathcal{L}$ of the lowest dimension are given by the expression similar to (1.60):

$$f_{N,0}(\mathbf{R}) = x_{12}x_{34}\ldots x_{(N-1)N} + [\ldots], \qquad (1.65)$$

of scaling dimension $\frac{N}{2}(2-\xi)$ and the permutations thereof. *E.g.* for $N = 4$, one may take $f_{4,0} = x_{12}x_{34} - \frac{d-1}{2(2-\xi)}(x_{12}^2 + x_{34}^2)$. As in the ξ-expansion, in order to take into account the perturbation U, one has to solve the equations

$$\mathcal{L}f_1 = Uf_0, \qquad \left(\sum_{n<m} x_{nm}\partial_{x_{nm}} - \frac{N}{2}\right)f_1 = \frac{p_1}{2-\xi}f_0. \qquad (1.66)$$

One checks again that $\Gamma f_0 \equiv \left(\sum_{n<m} x_{nm}\partial_{x_{nm}} - \frac{N}{2}\right)f_1$ is annihilated by $\mathcal{L}$. In order to calculate p_1, it remains to find the coefficient of $x_{12}\ldots x_{(N-1)N}$ in Γf_0. In its dependence on x_{nm}'s, the function Uf_0 scales with power $(\frac{N}{2}-1)$. One finds f_1 by applying the inverse of the operator $\mathcal{L}$ to Uf_0. Γf_0 is then obtained by gathering the coefficients of the logarithmic terms in f_1, see [37] for the details. Alternatively, one may redecompose: $\mathcal{L}-\frac{1}{d}U = \mathcal{L}'-\frac{1}{d}U'$. The leading zero modes of $\mathcal{L}'$ have also the form (1.65). Although the second order operator U' constitutes a singular perturbation of the first order one $\mathcal{L}'$, the zero mode $\Gamma'f_0$ may be obtained directly as the coefficient of the logarithmically divergent term in $\int_0^\infty \left(e^{-t\mathcal{L}'}U'f_0\right)(x_{nm})\mathrm{d}t = \int_0^\infty U'f_0(x_{nm}-t)\mathrm{d}t$, see [41]. In both approaches, the final result is

$$p_{N,0} = \frac{N}{2}(2-\xi) - \frac{N(N-2)}{2d}\xi + \mathcal{O}\left(\frac{1}{d^2}\right) \qquad (1.67)$$

which is consistent with the small ξ calculation (1.64). The non-isotropic zero modes, as well as those for odd N, may be studied similarly [37,43–45].

The effective expansion parameter in the large d approach turns out to be $\frac{N}{d(2-\xi)}$ so that neither the small ξ nor the large d expansions are applicable to the region of ξ close to 2, *i.e.* to almost smooth velocity fields. Both the limit $\xi = 2$ and the perturbation theory around it are singular. The main reason is the fact that advection by a smooth velocity field preserves straight lines. Admitting a little roughness of the velocity does not change significantly the statistics of particles' motion almost everywhere except near-collinear geometry. That calls for a boundary-layer approach when one separately studies a zero mode at near-collinear geometry and then matches it with a (regular) perturbation expansion for a general geometry [46]. For almost smooth velocities, the very close particles separate very slowly and the collective behavior of particles is masked by this effect which leads to an accumulation of zero and slow modes with very close scaling dimensions. At $\xi = 2$, a continuous spectrum and infinite degeneration of the zero modes appear which is another source of difficulties in the perturbative treatment of for ξ close to 2. Not surprisingly, the dimensions of the zero modes are series in non-integer powers of $2 - \xi$ [47, 48].

To conclude, one is able to build statistical conserved quantities by compensating the growth of interparticle distances by the decrease of the shape fluctuations of the particle configurations. The size of the cloud increases with time while the average of a generic functions of shape relaxes to a constant as a combination of negative powers (of R or t). The scaling exponents of the zero modes depend in a nontrivial way of the number N of the particles which is the manifestation of particle interaction.

2   Passive fields in fluid turbulence

The qualification "passive" means that we disregard the back reaction of the advected fields on the advecting velocity. The first section of this chapter is devoted to the statistical initial value problem: how an initially created distribution of a passive scalar evolves in a statistically steady turbulent environment? The rest of the chapter treats steady cascades under the action of a permanent pumping.

2.1   Unforced evolution of passive fields

For simplicity, we restrict ourselves by smooth velocity (non-smooth case can be solved as well). We consider the tracer field that satisfies

$$\partial_t \theta + (\mathbf{v}\nabla)\theta = \kappa\nabla^2\theta. \tag{2.1}$$

Let us start from the simplest problem: consider a small spherical spot of the tracer θ released in a spatially smooth incompressible 3d velocity field. Physically, we imply the Schmidt number $Sc = \nu/\kappa$ to be large that is the viscous scale of the flow η is much larger than the diffusion scale of the scalar defined as $r_d = \sqrt{-\kappa/\lambda_3}$. The initial size of the spot L is assumed to satisfy $\eta \gg L \gg r_d$. The spot is stretched and contracted by the velocity field. As we have shown in Section 1.2, during the time less that $-\lambda_3^{-1}\ln(L/r_d)$, diffusion is unimportant and θ inside the spot does not change. At larger time, the dimensions of the spot with negative Lyapunov exponents are frozen at

$r_{\rm d}$, while the rest keep growing exponentially, resulting in an exponential growth of the total volume. That leads to an exponential decay of scalar moments averaged over velocity statistics: $\langle[\theta(t)]^\alpha\rangle \propto \exp(-\gamma_\alpha t)$. The decay rates γ_α can be expressed *via* the PDF (1.20) of stretching variables ρ_i. Since θ decays as the inverse volume which increases as $\exp(\rho_1 + \rho_2 + \rho_3)$ then

$$\langle[\theta(t)]^\alpha\rangle \propto \int \mathrm{d}\rho_1\mathrm{d}\rho_2 \exp\left[-tH(\rho_1/t - \lambda_1, \rho_2/t - \lambda_2) - \alpha(\rho_1 + \rho_2)\right]. \quad (2.2)$$

At large t, the integral is determined by the saddle point. At small α, the saddle-point lies within the parabolic domain of H so γ_α increases with α quadratically. At large α, the main contribution is due to the realization with smallest possible spot which has the volume L^3 so γ_α is independent of α [16, 17, 49].

Let us consider now an initial random distribution of $\theta(0, \mathbf{r})$ statistically homogeneous in space. We pass to the reference frame which moves with the Lagrangian point $\mathbf{R}(t|T, \mathbf{r}_0)$ coming to $\mathbf{r}_0$ at T. Such $\theta(t, \mathbf{r}) = \tilde\theta(t, \mathbf{r} - \mathbf{R}(t|T, \mathbf{r}_0))$ satisfies

$$\partial_t\tilde\theta + \tilde\sigma_{\alpha\beta}r_\beta\nabla_\alpha\tilde\theta = \kappa\nabla^2\tilde\theta. \quad (2.3)$$

Since the correlation functions of θ and $\tilde\theta$ coincide at the moment of observation we omit the tilde sign in what follows. One may treat diffusion in two equivalent ways: either by introducing Brownian motion or by making Fourier transform in (2.3). Here we choose the second way defining the time-dependent wavevector $\mathbf{k}(t') = W^T(t, t')\mathbf{k}(t)$ and solving (2.3) as follows

$$\theta(t, \mathbf{k}) = \theta_0\left(W^T(t)\mathbf{k}\right)\exp\left[-Q_{\mu\nu}k_\mu k_\nu\right], \quad (2.4)$$

$$Q(t) = \kappa\int_0^t \mathrm{d}t'\, W(t)W^{-1}(t')\left[W(t)W^{-1}(t')\right]^T. \quad (2.5)$$

The moments of $\theta(t, 0) = \int \mathrm{d}\mathbf{k}(2\pi)^{-3}\theta(t, \mathbf{k})$ are to be averaged both over velocity statistics and over the initial statistics of the scalar. As the long-time limit is independent of the statistics of $\theta(0, \mathbf{r})$ [16], we take it Gaussian with $\langle\theta(0, \mathbf{r})\theta(0, 0)\rangle = \chi(r) = \chi_0\exp[-r^2/(8L^2)]$. Then the moments of θ are as follows

$$\langle[\theta(t)]^\alpha\rangle \propto \int \mathrm{d}\rho_1\mathrm{d}\rho_2 \exp\left[-tH - \alpha(\rho_1 + \rho_2)/2\right]. \quad (2.6)$$

Notice that in (2.6) the scalar amplitude is proportional to the square root of the volume factor as distinct from (2.2). This difference can be intuitively understood by imagining initially different blobs of size L with uncorrelated values of θ. At time t those blobs overlap. The mutual cancellations of θ

from different blobs leads to the law of large numbers with initial statistics forgotten and the rms value of θ being proportional to the square root of the number of blobs. The number of blobs is inversely proportional to the volume $\exp(\rho_1 + \rho_2 + \rho_3)$. Similarly to (2.2), the same qualitative conclusions about the decay rates $\gamma_\alpha = \lim_{t\to\infty} t^{-1} \ln\{\langle[\theta(t)]^\alpha\rangle\}$ can be drawn from (2.6). In particular, for the Kraichnan model $\gamma_\alpha \propto \alpha(1 - \alpha/8)$ for $\alpha < 4$ and $\gamma_\alpha = \text{const.}$ for $\alpha > 4$ [16].

Note that in both cases (single spot and random homogeneous distribution) γ_α is not a linear function of α. That means that the scalar decay is not self-similar in a smooth velocity. On the contrary, one may show that in a non-smooth velocity field (1.40), $\mathcal{P}(\theta, t) = \mathcal{P}(\theta^{4-2\xi}t^{\mathrm{d}})$ if the statistical Corrsin invariant $\int\langle\theta(0, t)\theta(bfr, t)\rangle\,\mathrm{d}\mathbf{r}$ is nonzero (see [1] for the details).

2.2   Cascades of a passive scalar

This section describes forced turbulence of the passive tracer which is statistically stationary in time and homogeneous in space. To the advection-diffusion equation

$$\partial_t\theta + (\mathbf{v} \cdot \nabla)\theta = \kappa\nabla^2\theta + \varphi \tag{2.7}$$

we added the pumping φ, characterized by the variance $\langle\varphi(t, \mathbf{r})\varphi(0, 0)\rangle = \Phi(r)\delta(t)$ with $\Phi(r)$ constant at $r < L$ and decaying fast at $r > L$. The below consideration is valid for a finite-correlated pumping too, as long as the pumping correlation time in Lagrangian frame is much smaller than the time of stretching from a given scale to the pumping correlation scale L. Note that in most physical situations the sources do not move with the fluid so that the Lagrangian correlation time of the pumping is either it's Eulerian correlation time or L/V, depending on which one is smaller, here V is the typical fluid velocity.

Advection by an incompressible flow preserves θ so that pumping has to be balanced by diffusive dissipation in a steady state. Stretching and contraction by an inhomogeneous velocity provides for a cascade of a scalar from the pumping scale L to the diffusion scale r_{d} (where diffusion is comparable to advection). Assuming $L \gg r_{12} \gg r_{\mathrm{d}}$, Yaglom derived the flux relation of θ^2

$$\langle(\mathbf{v}_1 \cdot \nabla_1 + \mathbf{v}_2 \cdot \nabla_1)\theta_1\theta_2\rangle = \Phi(r_{12}) \approx \Phi(0). \tag{2.8}$$

The relation (2.8) states that the mean flux of θ^2 stays constant within the convective interval and expresses analytically the downscale scalar cascade. For velocity fields scaling as $\delta v \propto r^\alpha$, dimensional arguments suggest that $\delta\theta \propto r^{(1-\alpha)/2}$ [50, 51]. This relation gives a proper qualitative understanding that the degrees of roughness of the scalar and the velocity are complementary, yet it suggests a wrong scaling for the scalar structure functions of order higher than two (see Sect. 2.3 below).

To go beyond dimensional estimates, we apply Lagrangian description. The scalar field along the Lagrangian trajectories $\mathbf{R}(t)$ changes as

$$\frac{\mathrm{d}}{\mathrm{d}t}\theta(\mathbf{R}(t),t) = \varphi(\mathbf{R}(t),t). \tag{2.9}$$

The N-th order scalar correlation function $\langle\theta(\mathbf{r}_1,t)\ldots\theta(\mathbf{r}_N,t)\rangle$ is therefore given by

$$\int_0^t \ldots \int_0^t \left\langle \varphi(\mathbf{R}_1(s_1),s_1)\ldots\varphi(\mathbf{R}_N(s_N),s_N)\right\rangle \mathrm{d}s_1\ldots\mathrm{d}s_N, \tag{2.10}$$

with the Lagrangian trajectories satisfying the final conditions $\mathbf{R}_i(t) = \mathbf{r}_i$. For the sake of simplicity we have written down the expression for the case where the scalar field was absent at $t = 0$. Averaging over the Gaussian pumping we get for the second moment:

$$\langle\theta(\mathbf{r}_1,t)\theta(\mathbf{r}_2,t)\rangle = \left\langle \int_0^t \Phi\left(\mathbf{R}_{12}(s)\right)\mathrm{d}s \right\rangle. \tag{2.11}$$

Higher-order correlations are obtained similarly by using the Wick rule to average over the Gaussian forcing statistics and the remaining average is made over the ensemble of Lagrangian trajectories. The function Φ restricts integration to the time intervals where $R_{12} < L$. Simply speaking, the pair correlation function of the tracer is proportional to the average time two particles spent in the past within the correlation scale of the pumping [52]. Correlation functions probe the statistics of the times spent by fluid particles at distances $\mathbf{R}_{ij}$ smaller than L. Because of explosive separation of the trajectories in nonsmooth flows, the time to go from one distance to another is not sensitive to the value of the smaller distance. Therefore, the correlation functions at small scales, $r_{ij} \ll L$, are close to the single-point contributions, corresponding to initially coinciding particles. To pick up a dependence on the positions $\mathbf{r}$, one has to study the structure functions $S_n(r_{12}) = \langle(\theta_1 - \theta_2)^n\rangle$ which are determined by the time differences between different initial configurations. Conversely, the correlation functions at scales larger than L are strongly dependent on the positions, as it will be shown below. For the scale-invariant velocity statistics with $\delta v \propto r^\alpha$ and the Lagrangian correlation time $\tau \propto r/\delta v \propto r^{1-1/\alpha}$, one has $S_2 \propto r_{12}^{1-\alpha}$ according to (1.32).

2.2.1 Passive scalar in a spatially smooth velocity

In this section, all the scales are supposed to be much smaller than the viscous scale of turbulence so that the velocity field can be assumed spatially smooth and we may use the Lagrangian description developed in Section 1.2.

We pass to the reference frame which moves with the Lagrangian point $\mathbf{R}(t|\mathbf{T}, \mathbf{r}_0)$ coming to $\mathbf{r}_0$ at T. Similar to (2.3), we get $\partial_t \tilde{\theta} + \tilde{\sigma}_{\alpha\beta} r_\beta \nabla_\alpha \tilde{\theta} = \varphi + \kappa \nabla^2 \tilde{\theta}$ so that (2.11) takes the form

$$\langle \theta(T, 0)\theta(T, \mathbf{r})\rangle = \int_0^T dt \langle \Phi(R(t|T, \mathbf{r}))\rangle \tag{2.12}$$

$$= \int_0^T dt \int d\rho_1 \ldots d\rho_d \Phi\left(re^{\rho_d(t)}\right) \mathcal{P}(t, \rho_1, \ldots, \rho_d).$$

The behavior of the interparticle distance crucially depends on the sign of λ_d. For $\lambda_d < 0$ the limit $T \to \infty$ leads to a well-defined steady state with $\langle \theta(t, 0)\theta(t, \mathbf{r})\rangle \approx -\lambda_d^{-1}\Phi(0)\ln(L/r)$ at $r < L$. Note in passing that in a compressible flow, one may have $\lambda_d > 0$, the pair correlation function then has a part which is independent of r and grows proportional to T: when Lagrangian particles cluster rather than separate, tracer fluctuations grow at larger and larger scales – phenomenon that can be loosely called an inverse cascade of a passive tracer [53, 54].

We restrict ourselves by the incompressible case when $\lambda_d < 0$ so that particles do separate and the steady state exists. We first treat the interval of scales between the diffusion scale r_d and the pumping scale L, which is called convective interval. Deep inside the convective interval when $r \ll L$, the statistics of passive scalar approaches Gaussian. Indeed, the n-point correlation function is as follows

$$\langle \theta(T, \mathbf{r}_1) \ldots \theta(T, \mathbf{r}_{2n})\rangle = \int_0^T dt_1 \ldots dt_n$$
$$\times \langle \Phi(R(t_1|T, \mathbf{r}_{12})) \ldots \Phi(R(t_n|T, \mathbf{r}_{2n-1,2n}))\rangle + \ldots, \tag{2.13}$$

where we average over $\mathcal{P}(\rho)$ and perform summation over all sets of the pairs of the points $\mathbf{r}_i$. The reducible part in

$$\left\langle \Phi\left[r_{12}e^{\tilde{\rho}_d(t_1)}\right] \ldots \Phi\left[r_{2n-1,2n}e^{\tilde{\rho}_d(t_n)}\right]\right\rangle$$

prevails for n less that the ratio between the transfer time $|\lambda_d|^{-1}\ln(L/r)$ and the correlation time τ_s of the stretching rate fluctuations. The reason is that the irreducible contributions have less large logarithmic factors than the reducible ones [22]. Therefore, for $n \ll n_{cr} \simeq (\lambda_d \tau_s)^{-1}\ln(L/r)$, the statistics of the passive tracer is Gaussian. Since $L \gg r$, then $n_{cr} \gg 1$. The single-point statistics is Gaussian up to $n_{cr} \simeq (\lambda_d \tau_s)^{-1}\ln(L/r_d)$. Larger n correspond to the exponential tails of tracer's pdf. The physics behind this is transparent and most likely valid also for a non-smooth velocity (even though the consistent derivation is absent in the non-smooth case). Indeed, large values of the scalar can be achieved only if during a large time the pumping works uninterrupted by advection (which eventually brings diffusion into play). When the time in question is much larger than the typical stretching time from r_d to L then the stretching events can be considered as a Poisson process and the probability that no stretching occurs during time t is $\exp(-ct)$. Integrating that with a pumping-produced distribution we get: $\mathcal{P}(\theta) \propto \int dt \exp(-ct - \theta^2/2\Phi t) \propto \exp(-\theta\sqrt{2c/\Phi})$. The detailed derivation for a smooth case can be found in [16, 17, 22, 30].

We consider now the scales $r > L$ [55]. From a general physical viewpoint, it is of interest to understand the properties of turbulence at scales larger than the pumping scale. If only direct cascade exists, one may expect equilibrium equipartition at large scales with the effective temperature determined by small-scale turbulence [56, 57]. The peculiarity of our problem is that we consider scalar fluctuations at the scales that are larger than the scale of excitation yet smaller than the correlation scale of the velocity field, which provides for mixing of the scalar. Although one finds simultaneous correlation functions of different orders, it is yet unclear if such a statistics can be described by any thermodynamics-like variational principle.

As has been told, the correlation functions of the scalar are proportional to the time spent by respective particles within the pumping scale. The scalar statistics at scales larger than L is thus related to the probabilities of initially distant particles to come close. Such statistics is rather peculiar at spatially smooth random flow: it demonstrates strong intermittency and non-Gaussianity at large scales. Another unexpected feature of the scalar statistics in this limit is a total breakdown of scale invariance: not only the scaling exponents are anomalous (*i.e.* don't grow linearly with the order of correlation function) yet even any given correlation function is not generally scale invariant (that is the scaling exponents depend on the angles between the vectors connecting the points).

A non-zero two-point correlation function (2.12) at $r > L$ appears only when two fluid particles manage to come there that were in the past within the pumping correlation length. We thus have to estimate the probability for the vector $\mathbf{R}(t)$ that was once within the pumping correlation length L to come exactly to the prescribed point $\mathbf{r}$ which is far away. Since the volume is conserved, then all the particles from the pumping volume L^d will evolve in such a way to be stretched in a narrow strip with the length r. Assuming ergodicity [which requires that the stretching time $\lambda_1^{-1} \ln(r/L)$ is much larger than the strain correlation time] we conclude that two points separated by r belong to a "piece" of scalar originated from within L with the probability given by the volume fraction $(L/r)^d$. The two-point correlation function thus decreases with the distance as follows: $F_2 \propto r^{-d}$.

Since an advection by a smooth velocity preserves straight lines then the same answer is true for the correlation function of arbitrary order if all the points lie on a line. In this case, the history of stretching is the same for all the distances. Looking backward in time we may say that when the largest distance between points was within L then all other distances were as well. Therefore, the n-point correlation function for collinear geometry is determined by the largest distance: $F_n \propto r^{-d}$. This is true also when different pairs of points lie on parallel lines. Note that the exponent is n-independent which corresponds to a strong intermittency and an extreme anomalous scaling. The fact that for collinear geometry $F_{2n} \gg F_2^n$ is due to strong correlation of the points along the line.

When we consider a non-collinear geometry, the opposite takes place, namely the stretching of different non-parallel vectors is generally anti-correlated because of incompressibility and volume conservation. Indeed, an advection by a smooth velocity field preserves a number of invariants. A d-volume $\epsilon_{\alpha_1\alpha_2...\alpha_d}\rho_1^{\alpha_1}\ldots\rho_d^{\alpha_d}$ is conserved for any d Lagrangian trajectories $\rho_i(t)$. In particular, for $d = 2$ there are area conservation laws $\epsilon_{\alpha\beta}\rho_1^\alpha\rho_2^\beta$ for any two vectors relating three points. Let us now consider a two-dimensional flow where the anti-correlation due to area conservation can be easily understood and the scaling for non-collinear geometry can be readily appreciated.

Consider the contribution from $\int dt_1 dt_2 \langle \Phi[R_{12}(t_1)]\Phi[R_{34}(t_2)]\rangle$ into the fourth-order correlation function. Since the area $|\mathbf{R}_{12} \times \mathbf{R}_{34}|$ is conserved, the answer is crucially dependent on the relation between $|\mathbf{r}_{34} \times \mathbf{r}_{12}|$ and L^2. When $|\mathbf{r}_{34} \times \mathbf{r}_{12}| \ll L^2$ we have a collinear answer $F_4 \propto r^{-2}$. Let us now consider the case of non-collinear geometry and find the probability of an event that during evolution R_{12} became of the order L, and then, at some other moment of time, R_{34} reached L (only such events will contribute into F_4). There is a reducible part in pumping, which makes F_4 nonzero (decaying as power of r_{ij}) even when $|\mathbf{r}_{34} \times \mathbf{r}_{12}| \gg L^2$. The probability that R_{12} came to L is L^2/r_{12}^2. Due to area conservation, there is an anti-correlation between R_{12} and R_{34}: if $R_{12} \sim L$, than $R_{34} \sim r_{12}r_{34}/L$. So probability for R_{34} to come back to L is $L^2/(r_{12}r_{34}/L)^2 = L^4/r_{12}^2 r_{34}^2$. Therefore, the total probability can be estimated as L^6/r^6, which is much smaller than the naive Gaussian estimation L^4/r^4 while the collinear answer L^2/r^2 is much larger than Gaussian.

That consideration can be readily generalized for arbitrary number of non-collinear pairs, $F_{2n} \propto (L/r)^{4n-2}$, and for arbitrary geometry [55]. Unfortunately, not much can be argued qualitatively about the scaling at $d > 2$. The crucial point for our considerations in $d = 2$ was the conservation of the area. In other terms, it is related to the fact that there is a single Lyapunov exponent at two dimensions. When $d > 2$ we have only the conservation of the d-dimensional volumes and hence more freedom in the dynamics. Nevertheless, the anti-correlation between different Lagrangian trajectories exists yet the analysis has been done only for Kraichnan model [55].

2.3 Passive scalar in a spatially nonsmooth velocity

In this section we shall analyze the steady cascade of a scalar in the inertial interval of scales where the velocities become effectively nonsmooth. As discussed in Section 1.3, in nonsmooth velocities the explosive separation of trajectories blows up small neighborhoods to finite sizes in a finite time. As a result, the single-point values of the advected quantities depend on the whole realization of the field in the past. This phenomenon plays an essential role in maintaining the dissipation of the scalar when the diffusivity tends to zero but it makes the statistics of the advected fields more difficult to analyze. Such analysis is still possible, however, for the Kraichnan model (1.5, 1.41). Making a straightforward Gaussian averaging of (2.7) over the statistics of pumping and velocity one gets the following equation for the n-point simultaneous correlation function of the scalar $F_n(t, \mathbf{r}_1 \dots \mathbf{r}_n) = \langle \theta(t, \mathbf{r}_1) \dots \theta(t, \mathbf{r}_n) \rangle$ [17]

$$\partial_t F_n + M_n F_n = \sum_{k,l} \Phi(r_{kl}) F_{n-2}. \tag{2.14}$$

Here the operator M_n is given by (1.52) and, of course (2.14) can be derived in a Lagrangian way by using the propagator (1.50) [30, 58]. The great simplification of scalar description in the Kraichnan model is due to the fact that the set of (2.14) for different n presents a recursive problem since the rhs is expressed in terms of lower-order correlation functions. There is no closure problem and any correlation function satisfies closed equation after the lower-order functions are found. We consider steady state and drop the time derivative.

One starts from the pair correlation function that depends on a single variable and satisfies an ordinary differential equation $M_2 F_2(r) = \Phi(r)$ [10] which is the Yaglom flux relation (2.8) for the Kraichnan model. This equation with two boundary conditions (zero at infinity and finiteness at zero) can be explicitly integrated

$$r^{1-d} \partial_r \left[(d-1) D_1 r^{d-1+\xi} + 2\kappa r^{d-1} \right] \partial_r F_2(r) = \Phi(r), \tag{2.15}$$

$$F_2(r) = \int_r^\infty \frac{x^{1-d} \mathrm{d}x}{x^\xi + r_\mathrm{d}^\xi} \int_0^x \Phi(y) y^{d-1} \mathrm{d}y \ ,$$

where we introduced the diffusion scale $r_\mathrm{d}^\xi = 2\kappa/D_1(d-1)$. Let us remind that we consider the pumping correlated on the scale L assumed to be much larger than r_d. There are thus three intervals of the distinct behavior. At $(D_0/D_1)^{1/\xi} \gg r \gg L$ the pair correlation function is given by the zero mode of $M_2(\kappa = 0)$: $F_2(r) = r^{2-\xi-d} \bar{\Phi}/d(d-1)(d+\xi-2)D_1$ which may be thought of as Rayleigh–Jeans equipartition $\langle \theta_k \theta_{k'} \rangle = \delta(k+k') \bar{\Phi}/\omega_k$ with the temperature $\bar{\Phi} = \int \Phi(x) x^{d-1} \mathrm{d}x$ and $\omega_k = k^{2-\xi} d(d-1)(d+\xi-2) D_1$ being an inverse stretching rate. At the convective interval, $L \gg r \gg r_\mathrm{d}$, F_2 is equal to a constant (another zero mode of M_2) plus the inhomogeneous part given by the rhs of (2.15): $S_2(r) = \langle [\theta(\mathbf{r}) - \theta(0)]^2 \rangle = 2F_2(0) - 2F_2(r) = r^{2-\xi} \Phi(0)/d(d-1)(d+\xi-2)(2-\xi) D_1$. Note that in the convective interval the degrees of roughness of the scalar and velocity are indeed complementary, a smooth velocity corresponds to a roughest scalar and *vice versa*. And finally, $S_2(r) \approx r^2 \Phi(0)/4\kappa d$ at the diffusive interval. Note though that $S_2(r)$ is not analytic at zero since it expansion contains noninteger powers $r^{2n+\xi}$, $n = 1, 2, \ldots$ This is an artifact of extending velocity nonsmoothness to the smallest scales that is setting the viscous scale to zero (*i.e.* Schmidt/Prandtl number to infinity).

Consider now high-order correlation functions in the convective interval. Solving recursively the stationary version of (2.14) one finds that F_n generally contains powers from $r^{2-\xi}$ to $r^{n(2-\xi)}$ plus a constant and other zero modes of M_n [37,38,41,42]. Note that one cannot satisfy the boundary conditions at large scales without the zero modes. In the structure function, $S_n(r) = \langle [\theta(\mathbf{r}) - \theta(0)]^n \rangle$, all the terms cancel except for the irreducible zero mode. We thus conclude that $S_n(r) = A_n r^{\zeta_n}$ with $\zeta_n = p_{n,0}$. Note that only S_2 is universal (that is determined by the flux only), all the other A_n depend on the pumping statistics [1]. As we have seen in Section 1.4, the anomalous exponents $\Delta_{2n} = n\zeta_2 - \zeta_{2n} = n(2-\xi) - \zeta_{2n}$ are positive for any $d < \infty$ and $\xi \neq 0, 2$. That means an anomalous scaling and small-scale intermittency of the scalar field: the ratio S_{2n}/S_2^n grows as r decreases. In the perturbative domain, $n\xi/d(2-\xi) \ll 1$, the scaling exponents are given by (1.64). At $n \gg d(2-\xi)/\xi$, the dependence $\zeta(n)$ saturates which means that the sharp fronts of the scalar determine high moments. The saturation value has been calculated for large d: $\zeta_n \to d(2-\xi)^2/8\xi$ [36].

It is instructive to discuss the limits $\xi = 0, 2$ and $d = \infty$ from the viewpoint of the scalar statistics. Since the scalar field at any point is the superposition of fields brought from d directions then it follows from a central limit theorem that scalar's statistics approaches Gaussian when space dimensionality d increases. In the case $\xi = 0$, an irregular velocity field acts like Brownian motion so that turbulent diffusion is much like linear diffusion: scalar statistics is Gaussian provided the input is Gaussian. What is general in both limits $d = \infty$ and $\xi = 0$ is that the degree of Gaussianity (say, flatness S_4/S_2^2) is independent of the ratio r/L. Quite contrary, we have seen in Section 2.2.1 that $\ln(L/r)$ is the parameter of Gaussianity in the Batchelor limit so that statistics is getting Gaussian at small scales whatever the input statistics. At $\xi = 2$ the mechanism of Gaussianity is temporal rather than spatial: since the stretching is exponential in a smooth velocity field then the cascade time grows logarithmically as the scale decreases.

That leads to the essential difference: at small yet nonzero ξ/d, the degree of non-Gaussianity increases downscales while at small $(2-\xi)$ the degree of non-Gaussianity first decreases downscales until $\ln(L/r) \simeq 1/(2-\xi)$, and then starts to increase, the first region grows with ξ approaching 2. Already that simple reasoning shows that the perturbation theory is singular at the limit $\xi = 2$, which formally is manifested by the many-point correlation functions having singularity (smeared by molecular diffusion only) at the collinear geometry [59].

Note that the dependence $\Delta_n(\xi)$ has to be nonmonotonic since $\Delta_n(0) = \Delta_n(2) = 0$. There is a transparent physics behind the nonmonotonic dependence $\Delta(\xi)$ because the influence of velocity nonsmoothness (measured by ξ) on scalar intermittency is twofold: if one considers scalar fluctuation of some scale then velocity harmonics with comparable scales produce intermittency while small-scale harmonics act like diffusivity and smooth it out. At $\xi_* < \xi < 2$ the first mechanism is stronger while at $0 < \xi < \xi_*$ the second one takes over. Still, our understanding is only qualitative here, we don't know how the maximum position ξ_* depends on n and d.

The anomalous exponents determine also the moments of the dissipation field $\epsilon = \kappa|\nabla\theta|^2$. By a straightforward analysis of (2.14) one can show that $\langle\epsilon^n\rangle = c_n\langle\epsilon\rangle^n(L/r_{\mathrm{d}})^{\Delta_{2n}}$ [37,41]. Here the mean dissipation $\langle\epsilon\rangle = \Phi(0)$ while the dimensionless constants c_n are determined by the fluctuations of dissipation scale, most likely they are of the form n^{qn} with yet unknown q. In the perturbative domain, $n \ll d(2-\xi)/\xi$, the main factor is $(L/r_{\mathrm{d}})^{\Delta_{2n}}$ and the dissipation PDF is close to lognormal since Δ_{2n} is a quadratic function of n [41], the form of the distant PDF tails are unknown.

The scalar correlation functions decay by power laws at scales larger than that of the pumping. The pair correlation function is $\propto r^{2-\xi-d}$. Note that $\mathcal{M}_2 r^{2-\xi-d} \propto \delta(r)$. The analysis of higher-order correlation functions is simplified in a non-smooth case since straight lines are not preserved and no strong angular dependencies of the type encountered in the smooth case are thus expected. To determine the scaling behavior of the correlation functions, it is therefore enough to focus on a specific geometry. Consider for instance the equation $M_4 C_4 = \sum\chi(r_{ij})C_2(r_{kl})$ for the fourth order correlation function. A convenient geometry to analyze is that with one distance among the points, say r_{12}, much smaller than the other distances which are of order R. At the dominant order in r_{12}/R, the solution of the equation is $C_4 \propto C_2(r_{12})C_2(R) \sim (r_{12} R)^{2-\xi-d}$. Similar arguments apply to arbitrary orders. We conclude that the scalar statistics at $r \gg \ell$ is scale-invariant, *i.e.* $C_{2n}(\lambda\mathbf{r}) = \lambda^{n(2-\xi-d)}C_{2n}(\mathbf{r})$ as $\lambda \to \infty$. Note that the statistics is generally non-Gaussian when the distances between the points are comparable. As ξ increases from zero to two, the deviations from the

Gaussianity starts from zero and reach their maximum for the smooth case described in Section 2.2.1.

3   Navier–Stokes equation from a Lagrangian viewpoint

All the previous sections were written under the assumption that the velocity statistics (whatever it is) is given. In this Chapter, we shall describe what one can learn about the statistics of the velocity field itself by considering it in the Lagrangian frame. As we learnt from the study of the passive fields, treating the dissipation does not really present much difficulty as it is a linear mechanism. The main problem is in proper understanding advection. For incompressible turbulence, the problem is even more complicated due to the spatial nonlocality of the pressure term. The Euler equation for the Lagrangian trajectory $\mathbf{R}(t; \mathbf{r})$ may indeed be written as

$$\partial_t^2 \mathbf{R} = \mathbf{f}(\mathbf{R}, t) - \nabla P, \qquad (3.1)$$

where $\mathbf{f}$ is the external force and the pressure field is determined by the incompressibility condition $\nabla^2 P = -\nabla \cdot [\mathbf{v} \cdot \nabla \mathbf{v}]$ with $\mathbf{v} = \partial_t \mathbf{R}$ and the spatial derivatives taken with respect to $\mathbf{R}$. The Laplace operator in the previous relation brings in the nonlocality via its kernel decaying as a power. We thus have a system of infinitely many particles interacting strongly and nonlocally. In such a situation, any attempt at an analytic descriptions looks unavoidably dependent on possible small parameters and simplifications in the limiting cases. The only parameter of the incompressible Euler equation is the space dimensionality, varying between two and infinity. The two-dimensional case indeed presents important simplifications since the vorticity is a scalar Lagrangian invariant of the inviscid dynamics. It is very tempting to exploit the opposite limit of the infinite-dimensional Euler equation by describing the interaction between the fluid particles in some mean-field approximation. Nobody yet managed to do so.

3.1   Enstrophy cascade in two dimensions

Besides the energy, the Euler equation in an even-dimensional space has an infinite set of vorticity integrals of motion. One may indeed show that the determinant of the matrix $\omega_{ij} = \partial_j v^i - \partial_i v^j$ is the nonnegative density of an integral of motion, *i.e.* that $\int F(\det \omega_{ij})d\mathbf{x}$ is conserved for any function F. In the presence of an external pumping ϕ injecting energy and vorticity, it is clear that both quantities may flow throughout the scales. If both cascades are present, they cannot go in the same direction: the different dependence of the energy and the vorticity on the scale prevents their fluxes to be both constant in the same interval of scales. Since one cannot provide a turbulent cascade by a potential flow (completely determined by boundaries in 2d) then energy cannot flow to small scales where a finite energy dissipation would mean an infinite vorticity dissipation at the limit $\nu \to 0$. The natural conclusion is that, given a single pumping at some intermediate scale, the energy and the vorticity cascades flow toward the large and the small scales, respectively [60–62].

In this section, we consider the direct vorticity cascade in 2D. The basic knowledge of the Lagrangian dynamics presented in the Sections 1.2.1 and 2.2 is essentially everything one needs for understanding the direct cascade. The vorticity in 2D is a scalar and the analogy between the cascades of the vorticity and the passive scalar was noticed by Batchelor and Kraichnan already in the sixties. The vorticity is not passive though and such analogies may be very misleading, as is the case for the analogy between the vorticity and the magnetic field in 3D.

The basic flux relation for the enstrophy cascade is analogous to (2.8):

$$\langle (\mathbf{v}_1 \cdot \nabla_1 + \mathbf{v}_2 \cdot \nabla_2)\omega_1\omega_2 \rangle = \langle \varphi_1\omega_2 + \varphi_2\omega_1 \rangle = P_2. \tag{3.2}$$

The subscripts indicate the spatial points $\mathbf{r}_1$ and $\mathbf{r}_2$ and the pumping is assumed to be Gaussian with $\langle \varphi(\mathbf{r}, t)\phi(\mathbf{0}, 0) \rangle = \delta(t)\tilde{\Phi}(r/L)$ decaying rapidly for $r > L$. The constant $P_2 \equiv \tilde{\Phi}(0)$, of dimensionality $time^{-3}$, is the input rate of the enstrophy ω^2. Equation (3.2) states that the ensrophy flux is constant in the inertial range, that is for r_{12} much smaller than L and much larger than the viscous scale. A simple power counting suggests that the velocity differences and the vorticity scale as the first and the zeroth power of r_{12}, respectively. That fits the idea of a scalar cascade in a spatially smooth velocity: scalar correlation functions are indeed logarithmic in that case, as was discussed in Section 2.2.

Even though one can imagine hypothetical solutions with the power-law vorticity correlators [63–65], they are structurally unstable [52]. Indeed, let us imagine for a moment that the pumping at L produces the spectrum $\langle (\omega_1 - \omega_2)^p \rangle \propto r_{12}^{\zeta_p}$ at $r_{12} \ll L$. Regularity of the Euler equation in 2D then requires $\zeta_p > 0$ [66]. In the spirit of the stability theory of Kolmogorov spectra [67], let us add infinitesimal pumping at some ℓ in the inertial interval with the only condition that small yet nonzero flux of enstrophy is produced. Small perturbation $\delta\omega$ obeys the equation $\partial\delta\omega/\partial t + (v\nabla)\delta\omega + (\delta v\nabla)\omega = \nu\nabla^2\delta\omega$. Here δv is the velocity perturbation related to $\delta\omega$. The ratio of the third term to the second one could be estimated as $(\ell/L)^2$ since the vorticity at the main spectrum is concentrated at L. Since $\zeta_2 > 0$ then the third term could be neglected so that $\delta\omega$ behaves as a passive scalar convected by the main turbulence *i.e.* the Batchelor regime from the Section 2.2 takes place. The correlation functions of the scalar are logarithmic in this case for any velocity statistics. The relative share of the perturbation in any vorticity correlation function thus grows downscales. That means that any hypothetical power-law spectrum is structurally unstable with respect to pumping variation. Only the logarithmic regime is neutrally stable so it represents the universal small-scale asymptotics of the steady forced turbulence. Experiments and numerics are compatible with that conclusion [68, 69, 71].

The physics of the enstrophy cascade is thus basically the same as that for a passive scalar: a fluid blob embedded into a larger-scale velocity shear is extended along the direction of positive strain and compressed along its negative eigendirection; such stretching provides for the vorticity flux toward the small scales, with the rate of transfer proportional to the strain. The vorticity rotates the blob decelerating the stretching due to the rotation of the axes of positive and negative strain. One can show that the vorticity correlators are indeed solely determined by the influence of larger scales (that give exponential separation of the fluid particles) rather than smaller scales (that would lead to a diffusive growth as the square root of time). The subtle differences from the passive scalar case come from the active nature of the vorticity. Consider for example the relation (2.11) expressing the fact that the passive scalar correlations are essentially the time spent by the particles at distances smaller than L. The passive nature of the scalar makes Lagrangian trajectories independent of scalar pumping which is crucial in deriving (2.11) by two independent averages. For an active scalar, the two averages are coupled since the forcing affects the velocity and thus the Lagrangian trajectories. In particular, the statistics of the forcing along the Lagrangian trajectories $\phi(\mathbf{R}(t))$ involves nonvanishing multipoint correlations at different times. Falkovich and Lebedev argued that, as far as the dominant logarithmic scaling of the correlations is concerned, the active nature of the vorticity simply amounts to the following: the field can be treated as a passive scalar, but the strain and the vorticity acting on it must be renormalized with the scale [52]. Their arguments are based on the analysis of the infinite system of equations for the variational derivatives of the vorticity correlations with respect to the pumping and the relations between the strain and the vorticity correlations. The law of renormalization is then established as follows, along the line suggested earlier by Kraichnan [60, 61]. From (1.7), one has the dimensional relation that *times* behave as $\omega^{-1} \ln(L/r)$. Furthermore, by using the relation (2.11) for the vorticity correlation, one has $\langle \omega \, \omega \rangle \propto P_2 \times time$. Combining the two previous relations, the scaling $\omega \sim [P_2 \ln(L/r)]^{1/3}$ follows. The consequences are that the distance between two fluid particles satisfies: $\ln(R/r) \sim P_2^{1/2} t^{3/2}$, and that the pair correlation function $\langle \omega_1 \omega_2 \rangle \sim [P_2 \ln(L/r_{12})]^{2/3}$.

What about the fluxes of the higher powers of the vorticity, that are also inviscid integrals of motion? One might think that the $(2n)^{\text{th}}$ order correlations involve the corresponding injection rate P_n of ω^{2n}, independent of P_2

for a non-Gaussian forcing. As for a passive scalar in the Batchelor regime, this is not the case since higher-order fluxes are not constant in the inertial range. Correlations of order higher than the second are indeed controlled by the effective pumping coming from lower-order correlation functions. The flux relation for ω^4 is obtained similarly to (3.2) from the equation for the time derivative of $\langle\omega_1^2\omega_2^2\rangle$. Its forcing term is $\langle\varphi_1\omega_1\omega_2^2 + \varphi_2\omega_2\omega_1^2\rangle$. Besides the constant irreducible part P_4, it must necessarily contain reducible contributions of the form $\langle\varphi_1\omega_2\rangle\langle\omega_1\omega_2\rangle = P_2\langle\omega_1\omega_2\rangle$ (in the inertial range). Since the pair correlation function grows logarithmically as r_{12} decreases, the reducible part will dominate at small enough scales. It is worth emphasizing that the effective forcing for the fourth-order correlation is thus acting throughout the inertial range and not just at the large scales as for the second order, the phenomenon called "distributed pumping" [52]. The same happens for all higher orders ω^{2n} whose effective forcing behaves as $P_2^n[\ln(L/r_{12})]^{2(n-1)/3}$. The fluxes of ω^{2n} are also non-constant in the inertial range and determined uniquely by the enstrophy production rate P_2.

3.2   On the energy cascades in incompressible turbulence

The phenomenology of the energy cascade suggests that the energy flux $\bar\epsilon$ is a major quantity characterizing the velocity statistics. It is interesting to understand the difference between the direct and the inverse energy cascades from the Lagrangian perspective. The mean Lagrangian time derivative of the squared velocity difference is minus twice the flux:

$$\left\langle\frac{\mathrm{d}(\Delta\mathbf{v})^2}{\mathrm{d}t}\right\rangle = 2\left\langle 2\mathbf{f}\cdot\mathbf{v} - \mathbf{f}_1\cdot\mathbf{v}_2 - \mathbf{f}_2\cdot\mathbf{v}_1 + \nu(2\mathbf{v}\cdot\nabla^2\mathbf{v} - \mathbf{v}_1\cdot\nabla^2\mathbf{v}_2 - \mathbf{v}_2\cdot\nabla^2\mathbf{v}_1)\right\rangle.$$

In the 2D inverse energy cascade, there is no energy dissipative anomaly and the right hand side is determined by the injection term $4\langle\mathbf{f}\cdot\mathbf{v}\rangle$. The energy flux is thus negative (*i.e.* directed upscales) and the mean Lagrangian derivative is positive. On the contrary, in the 3D direct cascade the injection terms cancel and the right hand side becomes equal to $-4\nu\langle(\partial_i v_j)^2\rangle$ so that the mean Lagrangian derivative is negative while the flux is positive (directed downscales). This is natural as a small-scale stirring causes effects opposite to those of a small-scale viscous dissipation. The negative sign of the mean Lagrangian time derivative in 3D does not contradict the fact that any couple of Lagrangian trajectories eventually separates and their velocity difference increases. It tells however that the squared velocity difference between two trajectories generally behaves in a nonmonotonic way: the transverse contraction of a fluid element makes initially the velocities approach each other, while eventually the stretching along the trajectories takes over. Such behavior was observed in numerical simulations by Pumir, Chertkov and Shraiman (private communication).

If one assumes (after Kolmogorov) that $\bar{\epsilon}$ is the only pumping-related quantity that determines the statistics then the separation between the particles $\mathbf{R}_{12} = \mathbf{R}(t; \mathbf{r}_1) - \mathbf{R}(t; \mathbf{r}_2)$ has to obey the already mentioned Richardson law: $\langle R_{12}^2 \rangle \propto \bar{\epsilon} t^3$. The equation for the separation immediately follows from the Euler equation (3.1): $\partial_t^2 \mathbf{R}_{12} = \mathbf{f}_{12} - \nabla P_{12}$. The corresponding forcing term $\mathbf{f}_{12} \equiv \mathbf{f}(\mathbf{R}(t; \mathbf{r}_1)) - \mathbf{f}(\mathbf{R}(t; \mathbf{r}_2))$ has completely different properties for an inverse energy cascade in 2D and for a direct energy cascade in 3D. For the former, R_{12} in the inertial range is much larger than the forcing correlation length. The forcing can therefore be considered short-correlated both in time and in space. Was the pressure term absent, one would get the separation growth: $\langle R_{12}^2 \rangle / \bar{\epsilon} t^3 = 4/3$. The experimental data give a smaller numerical factor $\simeq 0.5$ [70], which is quite natural since the incompressibility constraints the motion. What is however important to note is that already the forcing term prescribes the law $\langle R_{12}^2 \rangle \propto t^3$, consistent with the scaling of the energy cascade. Conversely, for the direct cascade the forcing is concentrated at the large scales and $f_{12} \propto R_{12}$. The forcing term is thus negligible and even the scaling behavior comes entirely from the advective terms (the viscous term should be accounted as well). Another amazing aspect of the 2D inverse energy cascade can be inferred if one considers it from the viewpoint of the vorticity. The latter is transferred toward the small scales and there is no flux of it at the large scales where the inverse energy cascade is taking place. By analogy with the passive scalar behavior at the large scales discussed in Section 2.2.1, one may expect the behavior $\langle \omega_1 \omega_2 \rangle \propto r_{12}^{1-\alpha-d}$, where α is the scaling exponent of the velocity. The self-consistency of the argument dictated by the relation $\omega = \nabla \times \mathbf{v}$ requires $1 - \alpha - d = 2\alpha - 2$ which indeed gives the Kolmogorov scaling $\alpha = 1/3$ for $d = 2$. The experiments [72] as well as the numerical simulations [73] indicate that the inverse energy cascade has a normal Kolmogorov scaling for the correlation functions of all orders. No consistent theory is available yet, but the previous arguments based on the enstrophy equipartition might give an interesting clue. To avoid misunderstanding, note that in considering the inverse cascades one ought to have some large-scale dissipation (like bottom and wall friction in the experiments with a fluid layer) to avoid the growth of condensate modes on the scale of the container.

Qualitatively, it is likely that the scale-invariance of an inverse cascade is physically associated to the growth of the typical times with the scale. As the cascade proceeds, the fluctuations have indeed time to get smoothed out and not multiplicatively transferred as in the direct cascades, where the typical times decrease in the direction of the cascade.

An interesting Lagrangian insight into the difference between direct and inverse energy cascades is provided by the different signs of the third-order structure function. From the generalization of the energy flux law to the

d-dimensional case, we have: $\langle (\Delta v)^3 \rangle = -12\bar{\epsilon}r/d(d+2)$, where $\bar{\epsilon}$ is positive for the direct cascade and negative for the inverse one. Since the average velocity difference vanishes, negative $\langle (\Delta v)^3 \rangle$ means that small longitudinal velocity differences are predominantly positive, while large ones are negative. In other words, in 3D if the longitudinal velocities of two particles differ strongly, then they are likely to attract each other; if the velocities are close, then the particles preferentially repel each other. The opposite behavior takes place in 2D, where the third-order moment of the longitudinal velocity difference is positive. Another Lagrangian meaning of the 4/5-law in 3D can be appreciated by extrapolating it down to the viscous interval. Here, $\Delta v^i \approx \sigma_{ij} r^j$ and the positivity of the flux is likely to be related to the fact that the negative Lyapunov exponent is the largest one (in absolute value) in 3D incompressible turbulence.

The observed scaling is anomalous in the 3d energy cascade: $\langle (\Delta v)^p \rangle \propto r^{\sigma(p)}$ with $\sigma(p) \neq p/3$. The primary target here is to understand the nature of the statistical integrals of motion responsible for it. Note that $\sigma(3) = 1$ and the experiments demonstrate that $\sigma(p) \to ap$ as $p \to 0$ with a exceeding 1/3 beyond the measurement error (see [74] and the references therein). The convexity of $\sigma(p)$ then means that $\sigma(2) > 2/3$ that is already the pair correlation function is determined by some nontrivial conservation law. Although $\langle \mathbf{v}_1 \mathbf{v}_2 \rangle$ involves two Lagrangian particles, each of them is now carrying its own vector $\mathbf{v}$. In other words, we are somehow dealing with a four-particle problem, where the vectors bring the geometrical degrees of freedom needed for the appearance of nontrivial zero modes.

4   Conclusion

We hope that the reader have absorbed by now the two main lessons: the power of the Lagrangian approach to fluid turbulence and the importance of statistical integrals of motion for systems far from equilibrium.

As it was shown in the Sections 1 and 2, the Lagrangian approach allows the analytical description of most important aspects of the statistics of particles and fields for velocity fields either spatially smooth or temporally decorrelated (or both). In a spatially smooth flow, the Lagrangian chaos with the ensuing exponentially separating trajectories is generally present. The respective statistics of passive scalar and vector fields is related to the statistics of large deviations of the stretching and contraction rates in a way that is well understood. The theory finds a natural physical domain of application in the viscous range of scales. The most important open problem here seems to be the understanding of the back reaction of the advected field on the velocity. That would include an account of the buoyancy force in inhomogeneously heated fluids, the saturation of the small-scale magnetic dynamo and the polymer drag reduction. In nonsmooth velocities, pertaining to the inertial interval of developed turbulence, the main Lagrangian phenomenon is the intrinsic stochasticity of the fluid particle trajectories that accounts for the dissipation at short distances. These phenomena are fully captured in the Kraichnan ensemble of nonsmooth time-decorrelated velocities. It is an open problem to exhibit them for more realistic nonsmooth velocities and to relate them to hydrodynamical evolution equations obeyed by the latter. The spontaneous stochasticity of Lagrangian trajectories enhances the interaction between fluid particles leading to intricate multi-particle stochastic conservation laws. There are open problems already in the framework of the Kraichnan model. First, there is the issue of whether one can build an operator product expansion, classifying the zero modes and revealing their possible underlying algebraic structure, both at large and small scales. The second class of problems is related to a consistent description of high-order moments of scalar, vector and tensor fields, especially in the situations where their amplitudes are growing, in a further attempt to describe feedback effects.

Our inability to derive the Lagrangian statistics directly from the Navier–Stokes equations of motion for the fluid particles is related to the coupling among the particles that is both strong and nonlocal, due to pressure effects. We would like to stress that most strongly coupled systems, even if local, are not analytically solvable and not all the measurable quantities may be derived from first principles. In fluid turbulence, what seems more important than the numerical value of the scaling exponents is to reach a basic understanding of the underlying physical mechanisms, such as that achieved in the passive scalar through the conservation laws of the Lagrangian statistics. We consider the notion of statistical conservation law to be of central importance for fluid turbulence and general enough to apply to other systems in non-equilibrium statistical physics. As we explained throughout the course, the conservation laws are related to the geometry either of the particle configurations (for the scalar fields) or the particle-plus-field configurations (for the vector fields, where already the two-point correlation function may have an anomalous scaling). It is a major open problem to identify the appropriate configurations for the active and the nonlocal cases. We see there the potential direction of progress: coupling analytical, experimental and numerical studies to investigate the geometrical statistics of fluid turbulence with the primary aim to identify the underlying conservation laws.

References

[1] G. Falkovich, K. Gawedzki and M. Vergassola, *Rev. Mod. Phys.* (2001).
[2] G.I. Taylor, *Proc. London Math. Soc.* **20** (1921) 196.

[3] A. Bensoussan, J.-L. Lions and G. Papanicolaou, *Asymptotic Analysis for Periodic Structures* (North-Holland, Amsterdam, 1978).

[4] S.B. Pope, *Ann. Rev. Fluid Mech.* **26** (1994) 23-63.

[5] A. Fannjang and G.C. Papanicolaou, *Prob. Theory Rel. Fields* **105** (1996) 279.

[6] A.J. Majda and P.R. Kramer, *Phys. Rep.* **314** (1999) 237.

[7] R.H. Kraichnan, *The Padé Approximants in Theoretical Physics* (Acad. Press, New York, 1970).

[8] M. Avellaneda and A. Majda, *Phys. Rev. Lett.* **62** (1989) 753.

[9] M. Avellaneda and M. Vergassola, *Phys. Rev. E* **52** (1995) 3249.

[10] R.H. Kraichnan, *Phys. Fluids* **11** (1968) 945-963.

[11] R.S. Ellis, *Entropy, large deviations, and statistical mechanics* (Springer-Verlag, Berlin, 1995).

[12] M.A. Liapounoff, *Ann. Fac. Sci. Toulouse* **9** (1907).

[13] V.I. Oseledets, *Trans. Moscow Math. Soc.* **19** (1968) 197-231.

[14] H. Furstenberg, *Trans. Am. Math. Soc.* **108** (1963) 377-428.

[15] Ya. Zeldovich, A. Ruzmaikin, S. Molchanov and V. Sokolov, *J. Fluid Mech.* **144** (1984) 1-11.

[16] E. Balkovsky and A. Fouxon, *Phys. Rev. E* **60** (1999) 4164-4174.

[17] B. Shraiman and E. Siggia, *Phys. Rev. E* **49** (1994) 2912.

[18] M. Chertkov, G. Falkovich and I. Kolokolov, *Phys. Rev. Lett.* **79** (1998) 2121.

[19] S. Girimaji and S. Pope, *J. Fluid Mech.* **220** (1990) 427.

[20] B. Shraiman and E. Siggia, *C.R. Acad. Sci.* **321** (1995) 279-284.

[21] J. Weiss, *Physica D* **48** (1991) 273.

[22] M. Chertkov, G. Falkovich, I. Kolokolov and V. Lebedev, *Phys. Rev. E* **51** (1995) 5609.

[23] G. Falkovich, V. Kazakov and V. Lebedev, *Physica A* **249** (1998) 36.

[24] G. 't Hooft, *Nucl. Physics B* **72** (1974) 461.

[25] H.E. Stanley, *Phys. Rev.* **176** (1968) 718.

[26] A. Monin and A. Yaglom, *Statistical Fluid Mechanics* (MIT Press, 1979).

[27] M.-C. Jullien, J. Paret and P. Tabeling, *Phys. Rev. Lett.* **82** (1999) 2872-2875.

[28] N. Zovari and A. Babiano, *Physica D* **76** (1994) 318.

[29] F.W. Elliott Jr. and A.J. Majda, *Phys. Fluids* **8** (1996) 1052-1060.

[30] D. Bernard, K. Gawędzki and A. Kupiainen, *J. Stat. Phys.* **90** (1998) 519-569.

[31] U. Frisch, A. Mazzino and M. Vergassola, *Phys. Rev. Lett.* **80** (1998) 5532-5535.

[32] K. Gawędzki, *Advances in Turbulence VII*, edited by U. Frisch (Kluwer, Dordrecht, 1998) 493-502; preprint chao-dyn/9907024.

[33] L. Onsager, *Nuovo Cim. Suppl.* **6** (1948) 279.

[34] D.W. Stroock and S.R.S. Varadhan, *Multidimensional Diffusion Processes* (Springer, Berlin 1979).

[35] L.F. Richardson, *Proc. R. Soc. Lond. A* **110** (1926) 709-737.

[36] E. Balkovsky and V. Lebedev, *Phys. Rev. E* **58** (1998) 5776.

[37] M. Chertkov, G. Falkovich, I. Kolokolov and V. Lebedev, *Phys. Rev. E* **52** (1995) 4924-4941.

[38] K. Gawędzki and A. Kupiainen, *Phys. Rev. Lett.* **75** (1995) 3834-3837.

[39] A. Celani and M. Vergassola, *Phys. Rev. Lett.* (2000) (to appear).

[40] O. Gat and R. Zeitak, *Phys. Rev. E* **57** (1998) 5511-5519.

554   New Trends in Turbulence

[41] M. Chertkov and G. Falkovich, *Phys. Rev Lett.* **76** (1996) 2706-2709.
[42] D. Bernard, K. Gawȩdzki and A. Kupiainen, *Phys. Rev. E.* **54** (1996) 2564-2572.
[43] A. Pumir, *Europhys. Lett.* **34** (1996) 25-29; **37** (1997) 529-534.
[44] E. Balkovsky and D. Gutman, *Phys. Rev. E* **54** (1996) 4435-4438.
[45] I. Arad, V.S. L'vov and I. Procaccia, *Phys. Rev. E* **81** (1999) 6753.
[46] B.I. Shraiman and E.D. Siggia, *Phys. Rev. Lett.* **77** (1996) 2463-2466.
[47] A. Pumir, B.I. Shraiman and E.D. Siggia, *Phys. Rev. E.* **55** (1997) R1263-R1266.
[48] E. Balkovsky, G. Falkovich and V. Lebedev, *Phys. Rev. E* **55** (1997) R4881-R488.
[49] D.T. Son, *Phys. Rev. E* **59** (1999) 3811-3815.
[50] A.M. Obukhov, *Izv. Akad. Nauk SSSR, Geogr. Geofiz.* **13** (1949) 58-69.
[51] S. Corrsin, *J. Appl. Phys.* **22** (1951) 469-473.
[52] G. Falkovich and V. Lebedev, *Phys. Rev. E* **50** (1994) 3883-3899.
[53] M. Chertkov, I. Kolokolov and M. Vergassola, *Phys. Rev. Lett.* **80** (1998) 512-515.
[54] K. Gawȩdzki and M. Vergassola, *Physica D* **138** (2000) 63-90.
[55] E. Balkovsky, G. Falkovich, V. Lebedev and M. Lysiansky, *Phys. Fluids.* **11** (1999) 2269.
[56] A. Forster, D. Nelson and M. Stephen, *Phys. Rev. A* **16** (1977) 732.
[57] E. Balkovsky, G. Falkovich, V. Lebedev and I. Shapiro, *Phys. Rev. E* **52** (1995) 4537.
[58] B. Shraiman and E. Siggia, *Nature* **405** (2000) 639-646.
[59] E. Balkovsky, M. Chertkov, I. Kolokolov and V. Lebedev, *JETP Lett.* **61** (1995) 1049.
[60] R.H. Kraichnan, *Phys. Fluids* **10** (1967) 1417.
[61] R.H. Kraichnan, *J. Fluid Mech.* **47** (1971) 525; **67** (1975) 155.
[62] G.K. Batchelor, *Phys. Fluids Suppl. II* **12** (1969) 233-239.
[63] P.G. Saffman, *Stud. Appl. Math.* **50** (1971) 277.
[64] H.K. Moffatt, *Advances in turbulence*, edited by G. Comte–Bellot and J. Mathieu (Springer-Verlag, Berlin, 1986) p. 284.
[65] A. Polyakov, *Nucl. Phys. B* **396** (1993) 367.
[66] G. Eyink, *Phys. Rev. Lett.* **74** (1995) 3800-3803.
[67] V. Zakharov, V. Lvov and G. Falkovich, *Kolmogorov Spectra of Turbulence* (Springer-Verlag, Berlin, 1992).
[68] J.C. Bowman, B.A. Shadwick and P.J. Morrison, *Phys. Rev. Lett.* **83** (1999) 5491.
[69] J. Paret, M.-C. Jullien and P. Tabeling, *Phys. Rev. Lett.* **83** (1999) 3418.
[70] M.-C. Jullien, J. Paret and P. Tabeling, *Phys. Rev. Lett.* **82** (1999) 2872-2875.
[71] M.-C. Jullien, P. Castiglione and P. Tabeling, *Phys. Rev. Lett.* **85** (2000) 3636.
[72] J. Paret and P. Tabeling, *Phys. Rev. Lett.* **79** (1997) 4162-4165; *Phys. Fluids* **10** (1998) 3126-3136.
[73] G. Boffetta, A. Celani and M. Vergassola, *Phys. Rev. E* **61** (2000) R29-32.
[74] K.R. Sreenivasan and R.A. Antonia, *Ann. Rev. Fluid Mech.* **29** (1997) 435.

Achevé d'imprimer par Corlet, Imprimeur, S.A. - 14110 Condé-sur-Noireau (France)
N° d'Imprimeur : 55775 - Dépôt légal : décembre 2001 - *Imprimé en U.E.*

RETURN TO: ENGINEERING LIBRARY
110 Bechtel Engineering Center 642-3366

LOAN PERIOD	1	2	3
4		5	6

ALL BOOKS MAY BE RECALLED AFTER 7 DAYS.
Please return books early if they are not being used.

DUE AS STAMPED BELOW.

AUG 2 1 2003		
SEP 2 5 2004		
MAY 2 7 2009		

FORM NO. DD 11
5M 5-00

UNIVERSITY OF CALIFORNIA, BERKELEY
Berkeley, California 94720–6000